COMPUTER APPLICATIONS IN CONSTRUCTION

Boyd C. Paulson, Jr.
Stanford University

McGRAW-HILL, INC.

New York   St. Louis   San Francisco   Auckland   Bogotá   Caracas
Lisbon   London   Madrid   Mexico City   Milan   Montreal
New Delhi   San Juan   Singapore   Sydney   Tokyo   Toronto

This book was set in Times Roman by Publication Services, Inc.
The editors were B. J. Clark and John M. Morriss;
the production supervisor was Louise Karam.
The cover was designed by Jane Paulson and Rafael Hernandez.
Project supervision was done by Publication Services, Inc.
R. R. Donnelley & Sons Company was printer and binder.

COMPUTER APPLICATIONS IN CONSTRUCTION

This book is printed on recycled, acid-free paper containing
10% postconsumer waste.

1 2 3 4 5 6 7 8 9 0   DOC   DOC   9 0 9 8 7 6 5 4

ISBN 0-07-048967-X

Library of Congress Cataloging-in-Publication Data

Paulson, Boyd C.
    Computer applications in construction / Boyd C. Paulson, Jr.
        p.   cm. — (McGraw-Hill series in construction engineering and
    project management)
    Includes bibliographical references and index.
    ISBN 0-07-048967-X
    1. Engineering — Management — Data processing.   2. Contractors'
operations — Data processing.   3. Construction industry — Management-
Data processing.   I. Title.   II. Series.
TA190.P37   1995
004' .024624 — dc20   94-2793

COMPUTER APPLICATIONS IN CONSTRUCTION

McGraw-Hill Series in Construction Engineering and Project Management

Consulting Editor
Raymond E. Levitt, Stanford University

Barrie and Paulson: *Professional Construction Management: Including CM, Design-Construct, and General Contracting*
Bockrath: *Contracts and the Legal Environment for Engineers and Architects*
Callahan, Quackenbush, and Rowings: *Construction Project Scheduling*
Hinze: *Construction Contracts*
Jervis and Levin: *Construction Law: Principles and Practice*
Koerner: *Construction and Geotechnical Methods in Foundation Engineering*
Oberlender: *Project Managment for Engineering and Construction*
Oglesby, Parker, and Howell: *Productivity Improvement in Construction*
Paulson: *Computer Applications in Construction*
Peurifoy and Ledbetter: *Construction Planning, Equipment, and Methods*
Peurifoy and Oberlender: *Estimating Construction Costs*
Schuette and Liska: *Building Construction Estimating*
Shuttleworth: *Mechanical and Electrical Systems for Construction*
Stevens: *Techniques for Construction Network Scheduling*

Also Available from McGraw-Hill

Schaum's Outline Series in Civil Engineering

Most outlines include basic theory, definitions, and hundreds of example problems solved in step-by-step detail, and supplementary problems with answers.

Related titles on the current list include:

Descriptive Geometry
Dynamic Structural Analysis
Engineering Economics
Engineering Mechanics
Fluid Dynamics
Fluid Mechanics & Hydraulics
Introductory Surveying
Mathematical Handbook of Formulas & Tables
Reinforced Concrete Design
Statics & Mechanics of Materials
Strength of Materials
Structural Steel Design, (LRFD Method)

Schaum's Solved Problems Books

Each title in this series is a complete and expert source of solved problems with solutions worked out in step-by-step detail.

Related titles on the current list include:

3000 Solved Problems in Calculus
2500 Solved Problems in Differential Equations
2500 Solved Problems in Fluid Mechanics & Hydraulics
3000 Solved Problems in Linear Algebra
2000 Solved Problems in Numerical Analysis
700 Solved Problems in Vector Mechanics for Engineers: Dynamics
800 Solved Problems in Vector Mechanics for Engineers: Statics

Available at most college bookstores, or for a complete list of titles and prices, write to: Schaum Division
McGraw-Hill, Inc.
1221 Avenue of the Americas
New York, NY 10020

ABOUT THE AUTHOR

Boyd C. Paulson, Jr., holds the endowed Charles H. Leavell Professorship of Civil Engineering in Stanford Unversity's Graduate Program in Construction Engineering and Management. He served on the civil engineering faculty at the University of Illinois in 1972 and 1973. He was also a Visiting Professor at the University of Tokyo in 1978, the Technical University of Munich in 1983, and the University of Strathclyde in Glasgow, Scotland, in 1990–91. He earned his B.S. (1967), M.S. (1969), and Ph.D. (1971) in civil engineering from Stanford University. He is the author or coauthor of two books and over 90 papers.

Paulson's research and teaching interests are primarily in computer applications in construction, including automated data acquisition, operations simulation, and automated process control. He has had numerous research projects in these and other areas sponsored by the National Science Foundation, the U.S. Department of Transportation, and others.

Paulson's professional activities include past chairman of ASCE's Committee on Professional Construction Management and the ASCE Task Committee on Computer Applications in Construction, past vice chairman of ASCE's Construction Research Council, and past chairman of the ASCE Construction Division Executive Committee. He was twice elected secretary of the Project Management Institute. He is a member of Tau Beta Pi, Sigma Xi, ACM, ASCE, ASEE, and the IEEE Computer Society.

His honors and awards include ASCE's 1980 *Walter L. Huber Civil Engineering Research Prize*, West Germany's *Alexander von Humbolt Foundation Research Fellowship* in 1983, ASCE's 1984 *Construction Management Award*, selection in 1984 as a *Distinguished Scholar* by the U.S. National Academy of Sciences Committee on Scholarly Communication with the People's Republic of China, the 1986 *Henry M. Shaw Lecturer* at the North Carolina State University, the Project Management Institute's 1986 *Distinguished Contributions Award*, 1990–91 faculty research and teaching scholarships from *The Fulbright Foundation* and *The British Council*, the 1992 *Kudroff Memorial Lecturer* at Pennsylvania State University, and ASCE's 1993 *Peurifoy Construction Research Award*.

To Jane, Jeff, and Laura

CONTENTS

Part III Application Planning, Development, and Management

Part V   Application Packages

Part VI Future Trends

LIST OF FIGURES

LIST OF TABLES

PREFACE

Computer knowledge has become increasingly essential to the success of today's construction engineers and managers and will become even more so in the future. The question for those going into this field is how best to acquire an understanding of this subject that will be most useful in construction practice.

THE NEED FOR THIS BOOK

In colleges and universities, even three decades after the importance of the subject began gaining recognition in professional construction engineering and management programs, the most common way of including computer studies in the curriculum still seems to be a class in computer programming. When the subject is taught well, students can acquire a general knowledge of computer science or data processing and learn how computer methods can assist in structured problem solving. Too often, however, the students just learn the rules and syntax for a programming language they may never use again and apply it only to the rote solution of abstract problems. It is no wonder that for many people their introduction to computers was an experience they would just as soon forget, and that later in their careers they become wary when anyone talks about applying computers to solve problems related to their job environment.

I have taught computer applications in construction for over 20 years and have grappled with the questions of what to teach and how to teach it to construction professionals. Like many, I began by teaching computer languages and focused on construction mainly via examples and assignments to program solutions to simple construction-oriented problems. Especially in the days of card-fed mainframe computers, the results were hardly worth the effort, and I fear that I too left many students disillusioned by the ordeal of late nights at the computer center debugging their way through the minutia of programming language syntax. Over time things improved, first with interactive minicomputers, whose online editing capabilities and quick-response program compilers at least took some of the drudgery and delays out of the homework, and they improved even more as microcomputers made the tools accessible in our own teaching environment. Things got much better still when programmable spreadsheets and databases came along in the early 1980s, because such tools made the actual programming simple enough that students could focus more directly on developing professional-looking and useful construction applications. But as a teacher I still did not feel as satisfied as the students seemed to be with the course.

The problems that remained were how to select the most useful body of computer knowledge for *construction*, not computer science, and to find a text to support that goal. The problem of selection is that there is now far more to be learned about rapidly evolving construction applications than can fit into a college course or a book, so one must strike a balance between fundamental concepts and useful applications. The problem of a suitable text has dogged me for all of the years I have taught my course. Every year or two I would select a new text, mostly from those intended for introductory software engineering or data processing courses, but too little of their content seemed relevant to construction from the students' and practitioners' viewpoint. There have been a few attempts at books oriented toward computer applications in construction, but those available seemed to date from the earlier era when teaching a language—or more recently a few application packages—with a few construction-flavored examples, was viewed as sufficient.

This book is a result of my years of struggling with the basic questions of how to teach computer concepts and applications in a way that is most useful to students and practitioners in construction. It follows a formula that has achieved considerable success in my own teaching in recent years, and I have wanted for some time to write it down in a way that might help others. A recent year's sabbatical in Scotland gave me the opportunity and time to make a good start toward achieving this goal.

STRUCTURE AND CONTENT

A quick glance at the table of contents shows the method I have chosen. There is a logical progression in the overall structure, reflected as Parts I to VI, but within the major applications-oriented sections (Parts IV and V)—about 60 percent of the book—chapters are modular and self-contained. One can pick and choose as desired.

Inevitably a book has compromises. In selecting and ordering the material, I have tried to make my compromises in favor of entry-level students or practitioners who would like to start with little or no prior background in computers—no programming skill or computer science course is needed as a prerequisite for this book—and reach a level where they have a sound grasp of how computers are being used in construction today and know better how to go about acquiring or developing useful applications for the future.

Part I begins with an overview of the subject. Chapter 1 shows why computer technology has been such an important force in changing the way the construction industry does business and why it is strategically important for companies in the future. For overall perspective, Chapter 2 offers a quick tour through a wide range of construction applications, many of which will be examined in more detail in the eight chapters that make up Parts IV and V.

Part II provides a background introduction to computer technology. The focus is on practical knowledge and basic terminology needed by construction professionals in making acquisition and development decisions that involve computers. Chapter 3 focuses on hardware topics, including the central processor, peripheral storage, input and output devices, automated data collection and control systems, and computer communications. Chapter 4 concentrates on system software and application development software, plus information storage in files and databases. These chapters sufficiently define the basic concepts and terminology for the level needed in later chapters. Readers having prior computer experience or course work might wish just to skim through them.

Part III is analogous to the planning and design stages of a construction project—stages that should take place before the equipment, materials, workers, and methods of construction are employed to construct the project itself. Chapters 5, 6, and 7 deal with the analysis and design of applications and with methods and concepts that can more successfully guide development and implementation. Only the basic and more practical aspects are covered here, but there is enough information to improve the success of planning and acquisition decisions for computer applications.

Part IV, in Chapters 8 to 11, deals with the development tools—the materials, equipment, and methods, if you like—that are commonly used by people to create construction applications today. They include programming languages, spreadsheet and database packages, and new methods based on artificial intelligence and knowledge-based systems. Although these four chapters are not a substitute for the specialized textbooks that are available on each tool, the basic concepts are introduced and illustrated with practical examples and problems that show a variety of ways in which these tools can effectively help create useful construction applications.

Whereas Part IV deals with general development tools that can produce a wide range of construction applications, Chapters 12 to 15 in Part V introduce application packages more specifically focused on certain types of construction problems: estimating, planning and scheduling, accounting and cost control, and simulation. Each of these chapters includes at least two packages to demonstrate different aspects of the application's spectrum.

The final chapter of the book, in Part VI, explores trends in an area that is literally creating its own future! Computers have already had a major impact on the ways that companies, projects, and people work, but they are increasingly going to be changing the structure of the construction industry itself. Already, projects are being designed as electronic three-dimensional models of reality, with the ability to simulate very realistically the types of production problems that might be anticipated—and avoided—in the field. With the advent of expert systems, automated machines, and even robots, we are fairly close to computers doing parts of the field work themselves. Increased computer-based integration of design, construction, and facility management will continue to change the relationships among design consultants, contractors, and owners alike. Chapter 16 briefly examines technologies behind these trends so that the reader will have basic concepts and terminology needed to follow the changes that will be taking place.

HOW TO USE THIS BOOK

Taken as whole, this book provides the reader with a sound background on what is going on with computers in construction, and reading for a deeper understanding will enable construction engineers and managers to make better decisions in acquiring or developing useful construction applications. However, nothing beats hands-on experience, so in addition to studying this book it is highly recommended that the reader try some of the applications and solve some of the problems on a computer. Of course, whereas the book itself assumes no prerequisite courses, use of a computer, development tool, or application package will require access to the relevant manuals or help from a colleague who can quickly introduce a few fundamentals. Do not be daunted by the thick bulk of typical software manuals, however, since one can become productive with much if not most software of the type used in this book with only a brief tutorial (whether online, by following a manual, or with a teacher) and can thus acquire enough knowledge to do the types of exercises given in this book.

Particularly in the chapters that make up Parts IV and V, the development tools and several of the application packages that have been selected for illustration are those that are often used in colleges and universities and in construction companies. Most are also fairly economical to buy, and even some of the commercially more expensive ones are made available to colleges and universities at substantial discounts or even as grants. All of the commercial programs used for illustration in Parts IV and V of this book are available for readily accessible and easily used microcomputers. Furthermore, I have made a deliberate effort where practical to select a tool or application in each chapter for Apple Macintosh computers, which are commonly used in colleges and universities for their ease of use in teaching, and for IBM PCs and compatibles, which are also used extensively in academia and dominate the construction business world; PCs further account for most of the commercial application packages in construction.

This book, combined with exercises on microcomputers, is designed for a one-quarter course of four credit hours or a one-semester course of three credit hours. It would be most suitable for upper division undergraduates or master's degree students who already have a working knowledge of some of the construction application areas (estimating, planning and scheduling, etc.). Where time is more constrained, the material in the book is modularly packaged so that a teacher can easily skip chapters out of Parts IV and V as desired and can also focus within them on the examples most pertinent to the type of computer and software being used to supplement the reading. Similarly, if the curriculum already includes a basic computer programming or data processing course, readers might skim or skip Part II. However, do not skip Parts I and III, since they are fundamental. Part VI is the most speculative and could be included as time permits or be left to independent reading.

For a practitioner to get the most out of this book, he or she should ideally already have some familiarity with the type of microcomputers found in offices and job sites, including the use of one or two packages such as spreadsheet, database, or scheduling programs. Even though you may not have the same software as that used for illustration in this book, the types of examples and exercises given here could be done using almost any similar software. A few hours spent using the office computer at lunch or after work would considerably deepen your understanding and provide a real personal satisfaction in studying this subject. The design and implementation of applications on computers provide rewarding opportunities for personal creativity, and once implemented the results of your own ideas can also take much of the tedium out of routine construction engineering and administrative chores.

Within this text, computer terms that may not yet be familiar to the general reader are presented in boldface italic type on their first usage. Besides their definition in the text, these terms are also defined in the glossary of computer terms in Appendix B.

ACKNOWLEDGMENTS

I began writing this book on a cold November night in Bearsden, Scotland, not far from Loch Lomond. I was there, at home with my family, through the good fortune of my 1990–91 sabbatical at the University of Strathclyde in Glasgow. That year for thought, research, and writing was granted to me by the Fulbright Commission, the National Science Foundation, and The British Council. I am deeply grateful to them all for that opportunity.

My hosts at Strathclyde were Professor George Fleming and Professor Iain MacLeod, who extended their kind and gracious Scottish hospitality during our stay and made that year a wonderful cultural experience for our family. In that environment I was able to get about halfway through the first draft of the manuscript. More importantly, I had freedom from day-to-day academic pressures—freedom that enabled me to think through the whole concept of the book that you now have in hand.

After returning to Stanford, it took me two and a half more years of weekends, vacations, and seemingly endless evenings to being this book to completion. It takes an enormously kind, supportive, and loving family to make that much time possible, and to them I am most grateful of all. I cannot restore to them all that lost time, but they can take some satisfaction if you, the reader, find this book useful. It is their book, too.

Several people gave generously of their time to help me with this book. Ray Levitt was my sounding board for Chapter 11, and his research is reflected in the safety example. Mohan Manavazhi prepared two of the figures in Chapter 8. Hossam El-Bibany wrote the Prolog program in Chapter 11. John Fletcher and Erv

Marriott provided the Sitework Engineering screen shots used in Chapter 12, and Marty Rosen provided the screen images for Figure 12-2. Curtis Peltz helped with the Timberline example in Chapter 12. My wife, Jane, a CPA and controller in construction, helped me with accounting in Chapter 14. Others reviewed parts of the book at various stages of production and gave constructive feedback and suggestions. I wish to express appreciation to the following reviewers for their many helpful comments and suggestions: Martin Fischer, Stanford University; Dan Halpin, Purdue University; Chris Hendrickson, Carnegie–Mellon University; Craig Howard, Stanford University; R. Raymond Issa, University of Florida; Dean Kashiwagi, Arizona State University; Victor Sanvido, Pennsylvania State University; James Stevens, University of Kentucky; and Jorge Venegas, Purdue University.

It has been a pleasure working with the folks who handled the management and production of this book—B.J. Clark, John Morriss, Louise Karam, and Meredith Hart at McGraw-Hill, and Doris Cadd, Peter Feely, Jan Fisher, and Marguerite Torrey at Publication Services. To some people such an effort might be just a job, but these professionals showed a genuine personal interest and excelled through long months of copyediting, galleys, page proofs, and final production.

Finally, I salute my students from the past two decades of CE 243 at Stanford. Too often, as we struggled together through rapid changes in computer technology—including some things that did not work very well—they ended up spending too many "all nighters" with computers. To them I confess that the course deserved its reputation for excessive demands on student time. Next time it is definitely going to be easier than before!

Boyd C. Paulson, Jr.

PART

I

OVERVIEW
AND
BACKGROUND

"Toto, I've a feeling we're not in Kansas anymore!"

The Wizard of Oz*

I n acquiring this book—even if you did so at the request of an employer or a professor—you have taken a step in recognizing that computers are becoming one of the most pervasive technical and economic influences in the construction industry. The way in which our projects are designed, managed, and built will never again be the same. Part I of this book, which consists of two short chapters, explains why.

Chapter 1 puts into perspective the overwhelming economic force of computer technology relative to others, including construction technology. The sharp contrast between the rates of technological progress in computers and in construction could work against our industry in the competitive marketplace. But computer technology is also a force that can be harnessed to improve the competitiveness

---

*MGM, 1939. (Based on *The Wonderful Wizard of Oz,* by L. Frank Baum, 1900.)

of construction. Those construction professionals who recognize and acquire the relevant knowledge about computers and their applications will be better equipped to be the leaders in using computers for strategic advantage. Chapter 1 briefly reviews the kinds of knowledge such professionals should seek, and it then provides an overview of what the succeeding chapters will contribute toward that quest.

Chapter 2 offers a quick tour through a wide variety of productive ways in which computers are being applied in construction today. Broad categories include management and administration, construction engineering, and automated process control and data acquisition. From computer-aided design through planning and scheduling to the actual operation of construction machinery, computers are already at work in the everyday world of construction.

Leading-edge construction professionals have already mastered sophisticated computer-based application tools, and their knowledge is growing continually so that they can keep up with the rapid pace of developments in this field. This book is designed to give others a head start in catching up, and it should broaden the perspective of those who are already deeply into some of the existing applications.

CHAPTER
1

THE IMPACT
OF COMPUTERS

Top management must ensure that an appropriate information infrastructure is built. It's already starting to fall into place in many of our better-run corporations because information is increasingly regarded as a strategic resource—not merely the meat of routine data processing. The how and when of extraction are the basis of a workable infrastructure.

James Martin*

Who could have guessed 15 or 20 years ago that the typical construction estimator or job-site engineer today would have a desktop computer as powerful as a large general contractor's central computer around 1970? Who might have dreamed that a software tool called a spreadsheet would become as common as a pickup truck for handling day-to-day project chores? Would anyone have imagined that, in a typical employment interview for a new construction engineer, the interviewer would probe for what the prospect knows about computer applications for construction or that the astute interviewee might be equally interested to learn whether the company has the latest computer productivity tools?

After decades of slow evolution, the world of construction has changed rapidly in recent years. Even so, the impact of computer technology has only begun to be felt. Thus far we have seen gains in efficiency mainly at the level of individual tasks and departments (see Chapter 2), with modest though perceptible effects on projects and on companies as a whole. In the decades ahead, the structure of the whole industry will change; the time-honored relationships between owners, consultants, contractors, subcontractors, and suppliers will yield to technological and economic forces that are already building. The leading construction professionals of the future will be as familiar with computers and their effective

---

*We're Managing a Revolution, 1985.

applications as with telephones, and those who best use their computers will be valued as the most productive and important in the industry.

THE FORCES FOR CHANGE

With an economic force generated by a technology moving as rapidly as that of the computer, change is inevitable. Some experts have said that the historical impact of the microprocessor itself, which is but one product of the computer revolution, will be greater than the influence of movable type or the steam engine. For perspective, consider the economic implications of the different technologies illustrated in Fig. 1-1. The ordinate of the graph corresponds to a normative measure of the cost per unit of value that a given industry produces, and the abscissa shows the change in cost over time.

Relative to a constant dollar, the cost per unit of construction—be it a mile of highway, a hundred square meters of house, or a megawatt of power plant—has generally gone *up* since the mid-twentieth century, in spite of more mechanization, prefabrication, and so on. Certainly, in some absolute sense, things have probably improved; but relative to the economy as a whole, which has improved even more rapidly, construction has become more expensive.

Things have been somewhat better for automobiles—today's cars are better engineered and more fuel efficient, yet are still within the purchasing power of most workers in industrialized countries (unlike new homes!)—but the gains have still been modest on a relative scale. Progress has been greater in commercial aircraft. Around 1950 air travel was mostly for wealthier people, and by today's standards the aircraft were slow and expensive to run. Today's efficient commercial jets fly on less than half the fuel of earlier models, and international as well as domestic air travel has become accessible to people of average means. But compared to computers, even these gains have been modest. It has been said that if the economics and productivity of aircraft had evolved as rapidly as those

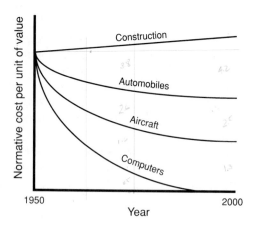

FIGURE 1-1
Economic progress over time in different industries.

of computers, one could travel from San Francisco to New York for just a few cents and arrive within seconds of taking off. What if we extended the analogy to construction? You should then be able to buy a single-family house for under a dollar, and have it constructed while reaching for your wallet. This is the meaning of that lower line in Fig. 1-1, which essentially disappears into the abscissa when normative cost is plotted on a scale that accommodates construction, automobiles, aircraft, or almost any other product.

An economic force like that is going to have an impact, and indeed it has. The world of banking and international finance differs radically today from that of a few decades ago, as do telecommunications, manufacturing, and defense. Partly as a result of other industries' more rapid advances in keeping their products economically attractive, construction's share of the gross domestic product in most of the technologically advanced countries has declined. If a middle-class family, for instance, sees little hope of affording a house, they may give up saving and instead buy a new car, an island vacation, or perhaps even a personal computer. An industrialist might think it better to invest in new machinery than a new factory building to improve a company's productive capital. And how many cities have planned new roads, libraries, and other improvements, but found they could not afford them in the face of demands for social and recreational programs; new police, fire and utility vehicles; office equipment; and other needs?

It is possible to use advanced technology to boost an industry that might otherwise remain stagnant. For example, computer-aided design, factory automation, and microcomputer-controlled engines have helped keep down the cost of cars while enabling manufacturers to cope economically with government fuel-efficiency and pollution regulations that pessimists two decades ago had forecast would break the industry. Today's complex and efficient telecommunications systems have grown to become essentially manifestations of large computer networks; even the desktop telephone often contains its own computer, and it can reach its counterparts in any corner of the world without human intervention, transmitting data and fax pictures as well as bringing people together.

The full economic force of computer technology can similarly be made to work for the construction industry, and the potential benefits for doing this are great. Construction is oriented toward custom products, with unique designs adapted to specific sites and client requirements. Unfortunately, in other industries some of our products would be viewed as prototypes, with considerable need for improvement. Computer-aided design can readily support the creation of unique projects of high technical and aesthetic quality. Computer-based simulations can catch mistakes, facilitate changes, and improve production methods, and do so while working with realistic electronic prototypes before they are rendered in steel and concrete. Construction is a fragmented industry, with numerous design firms, contractors, subcontractors, and suppliers involved in almost any project—even a house. Computers can provide unifying modeling, management, and communications systems to bring the unique talents of these parties together in a more productive and integrated manner.

In these and many other ways, the leading consulting and construction firms are increasingly recognizing computers as a strategic technology, and it is very probable that these firms will be the ones who will ensure the industry's success in the next century. In the major design and construction companies, computer applications more advanced than those typically found in any university civil engineering or construction program already exist. Those entering and working in this industry will have to master those technologies if they are to succeed as professionals and go on to provide the leadership that will be needed to keep this industry competitive in the economy.

THE KNOWLEDGE TO SUCCEED

It is unnecessary for a construction engineer or a construction manager to have the knowledge of a computer scientist in order to use computers effectively to make their work not only more productive but also more enjoyable. One does not need to be a technological prophet to be a valuable leader in employing computers as a strategic force to make a company more competitive. But one must know the fundamentals of the technology, have an interest in staying abreast of its advances, be aware of how it is being most usefully directed at present, and be willing to use some imagination to foresee where it might best be applied in the future. It is also useful, and personally rewarding, to know how to use some of the tools oneself.

Finding information about computer technology is not very difficult. Walk into any good bookstore, and you will find at least a rack and probably a wall of shelves full of books about spreadsheet and database programs, personal computers of all kinds, business aspects of computers, social aspects of computers, artificial intelligence, the electronic innards of the machines, and the most esoteric aspects of their software. The problem is finding where to start, for indeed one can soon become bewildered with the barrage of information available and become discouraged about starting at all.

For people interested in using computers in construction, this book is intended as a starting place. It provides an introduction to computer hardware, software, and communications technology, at least at the component level, including a basic idea of what the pieces do and how they work together. Necessarily, this involves learning some of the terminology of computers and even some of the industry's jargon, in order to learn from other sources of knowledge.

This text reflects the belief that the following kinds of knowledge are important for construction engineers and managers:

- The ability to analyze systematically a potential area of need, such as a function in a department, to determine whether a computer application might help, and to assess the costs and benefits of alternative proposals.
- The ability to proceed from that analysis to a clear specification for what the proposed solution should accomplish.
- At least a general knowledge of how computer professionals go about evaluating hardware and designing the software and systems that will implement a

given application, or of how they select commercial "package" applications to perform the desired function; the ability to interact with these computer professionals intelligently to assess how well the design or recommendation satisfies the need that was specified.

- The ability to evaluate and compare alternative solutions—whether choices among various vendors' computer hardware or different features of similar software packages—and to consider the long-term ramifications for things like maintenance, growth, and compatibility when making seemingly short-term choices.

- Knowledge of how to manage the people, resources, time, and money required in the development process for creating a given application program or for adapting a package application to the needs specified in the design. This is analogous to managing the field construction process that brings an architect's drawings into reality, and most construction managers are well equipped for this already.

- Understanding of the organizational and psychological issues that must be faced when implementing changes in an individual's or a group's work methods, or even when changing standard procedures for a whole company.

- Knowledge of alternative approaches to implementing a major change and of the advantages, disadvantages, and appropriateness of each approach in the context of a given type of function.

- Familiarity with the common tools of today's business computers, such as databases, spreadsheets, and so on. It is not necessary to become a "power user" to gain the required knowledge, but some hands-on experience is almost essential to appreciate the concepts and capabilities found in these programs.

- An awareness of systems and applications available in the marketplace to perform specific functions, such as scheduling, estimating, or cost control; some understanding of how they might apply in a construction company; and the ability to evaluate their capabilities relative to the magnitude and complexity of a particular company's needs and ways of working.

- Some idea of emergent technologies and trends and some imagination to foresee how an individual, group, or company might plan ahead and exploit the trends to advantage.

In reviewing the preceding list, notice that most of the entries reflect general management principles that are not necessarily specific to computers, or they reflect a type of learning that applies to any area of technology. When planning and managing a development project, many construction managers are well experienced already, and the need is more to adapt this knowledge to a new field. Certainly this involves learning some new concepts and terminology, but we can still build upon what we know. Computer technology moves ahead much more rapidly than anything we are used to, and engineers and managers have to be fast on their feet to make the best decisions. This book is a starting place, but to be successful one must continue devoting time to this subject.

OVERVIEW OF THIS BOOK

The remainder of this book provides information in six main parts:

1. A general overview and background on the range of construction applications
2. Explanation of the underlying computer technologies that make construction applications possible
3. Guidelines for the successful analysis, development or acquisition, and implementation of construction applications
4. An introduction to the software development tools commonly used to create construction applications
5. Examples of major types of commercially developed application packages that perform common functions required by construction professionals
6. A look at advanced concepts and technologies that are already beginning to define some trends for future applications in the industry

The remainder of Part I, together with Part II, provides a foundation upon which later chapters can build. Chapter 2 offers a quick tour through a broad range of construction tasks that already employ computer applications. Some of these will be examined in more detail either as examples created by the tools described in Part IV or as package applications in Part V, but others are included for an overall perspective for the later parts of the book. In Part II, Chapters 3 and 4 introduce concepts and terms needed to understand computer hardware, software, and communications technologies.

Part III addresses many of the important management and development issues listed in the previous section. Chapter 5 provides guidelines for identifying functions that might be performed or enhanced with the aid of computers, for assessing their technical and economic feasibility, and for defining the scope of and preparing specifications for what is to be expected. Chapter 6 explains general criteria and methods that computer professionals employ to design computer applications and shows how construction management techniques apply in managing the development of computer applications. Chapter 7 looks at the alternative of buying package applications and systems and shows how to evaluate systematically alternatives that might be available. It then offers guidelines for implementing and managing applications in an organization.

Part IV describes software development tools used both by computer professionals and by construction people to create new applications specifically designed to implement the designs for specified needs. These are mostly very flexible tools that can help create a wide range of applications. Each chapter includes examples not only to illustrate the methods and capabilities of the tools but also to show some simplified but useful types of construction applications. Chapter 8 discusses procedural languages, such as BASIC, C, FORTRAN, and Pascal. Although these are mainly used by computer professionals or engineers with sound knowledge of how they work, it is important for all of us to be knowledgeable of what these languages are and what they do and of the implications of using them to develop a construction application. Chapter 9 introduces spreadsheets, which are perhaps the most common computer tools applied by engineers and managers to solve complex problems and create useful applications. Chapter 10 explains file management and database software that users can employ to build simple as well as complex information systems of many types. Finally, Chapter 11 shows how some of the tools that have become available from research in the *artificial intelligence* branch of computer science, such as declarative languages and knowledge-based system (expert system) shells, can handle the qualitative judgmental types of situations that are so important in construction but that previously have not lent themselves to computer applications.

Part V (Chapters 12 to 15) examines four important areas where readily available computer application packages perform useful work in construction: estimating, project planning and scheduling, accounting and cost control, and operations simulation. Most of these applications are ones on which major contractors used to spend large sums of money in-house to develop using computer programming tools of the type described in Chapter 8, yet many of the low-cost packages are so good that these same contractors have scrapped the results of their own years of development to implement the packaged solutions instead. There is enough demand in most areas—and not just in construction—that several alternatives are available, and strong competition among developers keeps making the software better and more capable. The problem is not so much one of finding a suitable package, but rather choosing the one that best serves the needs of a particular company or project. Packages suitable for specialist engineers working on a large power plant or high-rise building may not be the ones most suited for the general engineer or supervisor working on a smaller project. The packages used for illustration in Part V give an idea of the range of possibilities, but diligent application of the methods from Part III is required to select the ones that fit a particular set of needs.

Part VI examines trends in technology that either are already underway in some of the more advanced design and construction companies or are likely to have an impact in the near future. Chapter 16 explores areas that are now carrying over into the construction phase, such as advanced computer-aided design, modeling, and simulation systems. This chapter further indulges in some speculation about a few areas that are presently at the research stage in computer science that may have important implications for construction. Chapter 16 also addresses automation and robotics, since robots are basically the electromechanical manifestation of computer programming for planning, sensing, and control. Initial prototypes have already received considerable publicity, but some major conceptual issues need to be dealt with before they can have a big impact in the field. Finally, Chapter 16 addresses what is probably the most important trend of all: the computer-based integration of the many parties and processes involved in the planning, design, construction, and management of constructed facilities. This is the area of computer technology that is most likely to change the basic relationships among the people and organizations involved, and it is the one that is most promising for maintaining and improving the industry's long-term health.

Each chapter offers a set of short review questions and a few suggested readings at the end. Appendix A provides answers to odd-numbered review questions. Appendix B is a glossary containing some of the more important and commonly used computer terms found in this book. When first defined in the text, such terms are printed in boldface italic type.

Although this book is intended as a starting place for construction professionals wishing to learn about computer applications in their field, it also provides a useful body of knowledge upon which to build. From here you can go in many directions. Some construction professionals who have started at this level have gone on to become managers of the whole computer operation for large general contractors. Others have started their own successful construction software development and marketing companies or consulting companies. Several have been given the responsibility for getting their companies started on implementing computer-based applications that have significantly improved their companies' competitive positions. Most, however, have simply been able to do a better job in their normal construction engineering and management tasks, using computers to cut through some of the administrative drudgery, solve problems more quickly, provide easier access to important information, submit more accurate and reliable bids, communicate more clearly and effectively, and make better decisions. Knowledge of computers, to quote Apple's advertisement, can indeed provide "the power to be your best."

SUMMARY

The economic importance of computer applications in construction has been emphasized in this chapter. The related technology is moving so fast that the number and variety of applications in construction will have increased markedly even by the time this book is published. Certainly, change will be rapid and continuing. Today's managers and supervisors must be alert to these changes and to new applications that will improve their effectiveness and efficiency.

A promising trend is that, through microcomputers and user-oriented software, the computer tools are getting directly into the hands of the project managers and engineers who need them most. At the same time, however, such people need to become more knowledgeable about the capabilities and limitations of computers and their related software for construction.

This chapter has provided an introduction to the knowledge that is important for construction engineers and managers. The most important type builds upon the management and professional skills already essential to this field, such as sound planning, evaluation of alternatives, decision making, and management of people, resources, money, and time. Beyond this, one needs some specific knowledge of computer technology, the tools used for developing applications, and the types of application software already available in the marketplace. Those providing leadership and long-term direction in their organizations must also be aware of future trends in computer technology and plan ahead to employ that technology for competitive advantage.

REVIEW QUESTIONS

1. Consider the economic implications of the cost trends reflected in Fig.1-1. Briefly describe what the advancement of computer technology has meant for people involved in each of the following.
   (*a*) Preparing architectural or engineering drawings
   (*b*) Construction estimating
   (*c*) Cost engineering
   (*d*) Planning and scheduling
2. A pessimistic view of the construction industry is that, as more "advanced" industries like computers and electronics take an increasing share of the gross domestic product, the market share of construction will continue to decline. Briefly describe a more positive scenario wherein technologies from computers and electronics actually can enhance the competitiveness of the construction industry.
3. Indicate how advances in construction might also benefit the computer and electronics industries.
4. Briefly describe how a good working knowledge of computer technologies and their applications can enhance the career development of a construction engineering and management professional.

SUGGESTIONS FOR FURTHER READING

The following books offer selected readings on the history (Goldstine, Kidder, Slater), impacts (Forester, Meindl), and future (Leebaert, Sproul and Kiesler) of computer science and technology.

Forester, Tom. *High-Tech Society*. Cambridge, Mass.: MIT Press, 1987. A future-oriented look at the role of advanced technology, especially computers, in contemporary and future society.

Goldstine, Herman H. *From Pascal to Von Neumann*. Princeton, N. J.: Princeton University Press, 1972. A history of the early days of computers by a man who was a pioneering collaborator with John Von Neumann on the ENIAC project at the University of Pennsylvania (1943–45).

Kidder, Tracy. *Soul of a New Machine*. New York: Avon, 1981. An acclaimed documentary about the people involved, and about the intense development pressures that they experienced, in bringing a powerful new computer to a highly competitive market.

Leebaert, Derek, ed. *Technology 2001*. Cambridge, Mass.: MIT Press, 1991. A variety of informed perspectives on where technology, including computers, will be taking us in the near future.

Meindl, James D., ed. *Brief Lessons in High Technology,* Stanford, Calif.: Stanford Alumni Association, 1991. Seven essays by leading researchers in a range of advanced technologies, including the developer of the RISC chips that are powering some of today's most advanced computers.

Slater, Robert. *Portraits in Silicon*. Cambridge, Mass.: MIT Press, 1987. Concise and well-written biographies of 31 pioneers and current leaders in computer science and technology.

Sproul, Lee, and Sara Kiesler. *Connections*. Cambridge, Mass.: MIT Press, 1991. Places special emphasis on the role of networking and telecommunications in changing organizations and the way we live and work within them.

CHAPTER
2

OVERVIEW OF CONSTRUCTION APPLICATIONS

Our rockets can find Halley's comet and fly to Venus with amazing accuracy, but side by side with these scientific and technical triumphs is an obvious lack of efficiency in using scientific achievements for economic needs.

Mikhail Gorbachev*

Computers can assist us in almost all aspects of construction engineering and management, for example, estimating, scheduling, operations simulation, safety, structural analysis, and even direct field applications like automated data collection and robotics. This chapter will examine a wide range of possibilities in three main categories. Most applications now used by construction professionals fall under *construction management and administration,* which is the first and largest category in this chapter. Second, depending on the type of contractor, computers have been employed extensively in *construction engineering.* Third, although many of the possibilities are still in the future, there are already practical applications in *automated data acquisition and process control,* and the potential for applications in this category is great.

This chapter provides a broad perspective on computer application possibilities in construction, but no one application will be discussed in detail. However, several of them will appear again in the development examples in Part IV, in the major application packages in Part V, and in the future trends of Part VI. The categories used in this chapter and the types of applications within them should provide a good framework for understanding the more detailed sections of this book.

Perestroika, 1988.

MANAGEMENT AND ADMINISTRATION

It is difficult to fit some applications neatly into any finite category. For example, construction estimating blends engineering analysis and management thinking, but here we present it under the construction engineering section. Some of the applications covered in this management section also have engineering components but are considered predominantly managerial or administrative. Those described in this category include the following.

- Accounting and payroll
- Cost engineering
- Company and project finance
- Project planning and scheduling
- Materials management
- Equipment management
- Human resources management
- Office administration
- Education and training

Accounting and Payroll

In larger construction companies, computers often appeared first in accounting departments. Therefore, it was not surprising that accounting and finance people retained control not only of the money-related functions of their companies but also of the computer itself. This centralized control of the computer evolved in most firms even as information systems proliferated throughout the organization; but accounting and payroll systems generally have remained among the most important in the business. In many cases they directly integrate with other functions, such as cost control and materials procurement.

At a minimum, the accounting system maintains the general ledger accounts, permits development of financial reports such as profit and loss statements and balance sheets, and enables a company to pay its taxes. The most common extension to this is an accounts payable system, whereby the company can receive, check, and pay invoices from materials and equipment suppliers and subcontractors, of which there are many even in a smaller firm. Accounts receivable systems are also common in larger companies; they receive progress and expenditure reports from job sites and process this information into invoices to project owners, with due allowance for retention and other contract provisions.

The most complex part of a contractor's accounting system is often payroll. Owing to the large number of specialty trades and the numerous geographical jurisdictions that are encountered even in the operations of smaller and medium-size firms, the variations in wage rates, fringe benefits, taxes, work schedules, and other factors quickly become enormous and require frequent updating. Various types of insurance payments are also linked to the payroll system. The frequent hirings and terminations that are common with craftworkers in construction require systems to be flexible and easy to use. There are also strict time constraints for paying workers, insurance companies, tax authorities, and others, and numerous possible deductions are needed for health and vacation plans, retirement programs, and such. Compounding matters, some firms extend the payroll system to handle part of the personnel records.

Cost Engineering

Over the years, job-costing and cost-engineering systems have evolved from reports that might be printed as part of the weekly payroll to more effectively designed systems that provide timely information and put the focus on tasks that need prompt attention to keep a project on budget. As a minimum, a cost-control system enables a contractor (1) to break a project down into categories that make sense from the point of view of the crews and supervisors doing the work and (2) to accurately reflect progress with respect to the planned budget. Common approaches include *work packaging* or a *work breakdown structure* that is compatible with the project schedule and even the materials management system. With schedule and procurement information thus linked, cash flow projections become possible, which help the financial managers of a company to maximize the effective use of capital.

Cost-control systems differ from routine accounting and payroll systems in that they place special emphasis on timeliness and are deliberately oriented toward the needs of operations and project management. Although accurate but week-old information is still useful for accountants for financial reporting and tax functions, it is of little value to people managing operations planned and in progress, where timeliness and future projections are far more important. This disparity in needs between project management and accounting has been a source of contention and in part has led to cost-control systems being managed separately, typically under a project engineer.

More detailed information about accounting and project cost control is given later in this book, especially in Chapter 14.

Company and Project Finance

Financial modeling and forecasting in construction have become increasingly sophisticated over the years, and companies have developed and acquired powerful computer software in this area. At a high level, developer-builders can compare alternative sites, building types, design and construction schedules, and possible types of tenants or buyers in a local real estate market and try to project expenditure and income streams to develop discounted cash flows. Once models have been built, say with a computer spreadsheet program, one can explore alternative scenarios and contingencies, such as ranges of interest rates for borrowing, the impact of delays in getting permits, difficulties in leasing space, and the like, to judge the risk versus the potential return of various alternatives.

General contractors use financial models at the bid or proposal stage to evaluate the impact of contractual clauses, such as retention and timing of progress payments, and to examine the risks associated with changes in material and labor costs, weather problems, and other contingencies. On the basis of this information and the extent of competition, they can make more informed decisions about the amount of markup to put on the bid to make the project an attractive investment of company resources. Such financial modeling can continue while the project is underway to help in decision making when contingencies arise, and to manage better the financial resources that are available.

The most common tools used to develop such models are microcomputer-based spreadsheet programs that put power, flexibility, and ease of use in the hands of the decision makers themselves. Specialty programs are also available, especially in areas like real estate investment analysis.

Project Planning and Scheduling

Critical path scheduling, one of the earliest computer applications in construction, took decades to gain wide acceptance. Since this application, like cost control, should be directly in the hands of management, its slow acceptance may be a result of the fact that earlier computers were not easily used by non-data-processing people. Powerful scheduling tools are now available on personal computers, so the trend toward wider acceptance and application of quantitative scheduling methods has accelerated.

Scheduling programs on computers typically start with basic *CPM* or *PERT* computations. As their level of sophistication increases, the programs include resource loading and aggregation, resource allocation and leveling, network-based cost control, and graphic output on interactive computer screens, printers, and plotters.

Since examples of scheduling applications and packages are given in Chapter 13, further details are deferred to that chapter.

Materials Management

To a layperson, obtaining materials of the right type and quality in time to keep field construction moving in a timely manner might seem straightforward, but on a project of any size the process is complex indeed. More importantly, effective materials management is essential to the technical and financial success of any project. Fortunately, this task lends itself well to organization via a computer.

A typical sequence of steps for materials procurement is as follows:

- Recognize need and issue requisition
- Prepare or adapt specifications
- Advertise for bids or solicit price quotations
- Receive and evaluate bids or quotations

- Issue purchase order or subcontract
- Prepare shop drawings or samples
- Approve shop drawings or samples; done by contractor and architect/engineer
- Fabricate
- Ship
- Deliver and unload
- Inspect and accept or reject
- Store
- Use item on the project

Simpler procurement systems are just extensions of the accounting department's accounts payable program. However, larger organizations have separate programs for procurement scheduling and expediting to ensure that steps such as the preceding are not overlooked and that problems and delays come to management's attention before they become acute. A typical materials management system sets up a schedule for the bulk and individual items needed on the project and maintains planned and actual dates to guide administrators and engineers through each of the foregoing steps to be sure materials will be available when needed in the field. Should any step fall behind, the computer system will print exception reports or otherwise notify managers, who can evaluate the effect of the delay on construction activities and the project as a whole and then either reschedule activities or expedite the materials. Typically, this software takes the form of a materials-management database. Materials and procurement systems can also include simple or sophisticated inventory control systems for job materials, tools, and supplies. When linked via a database, the procurement applications can interface to the quality assurance application for testing and documentation, to the accounts payable system for vendor payments, and to the scheduling system to assess the impact of procurement delays on overall project status.

A simplified example of a materials management system is given as a problem in Chapter 10 to illustrate the development of an application using relational database software.

Equipment Management

Particularly for a heavy construction or highway contractor, the successful management of the firm's fleet of equipment is crucial to the success of the business as a whole. Computers can be applied effectively in many ways, including maintaining ownership and utilization status records, computing ownership and operating costs, dispatching, maintenance, and economic policy studies. Integrating these functions lends itself well to a relational database program of the type described in Chapter 10.

Maintaining records for each machine starts with the basic asset accounting data of acquisition dates and costs, records of capital enhancements such as an engine rebuild, depreciation schedules, and disposal information. Records of the current status and expected future availability dates should also be maintained, including location, assignment, transfers, and the like.

Accurate records of ownership and operating costs are important for internal transfer pricing, project billings to owners in cost-reimbursable projects, accurate pricing of work in estimating and bidding, and decisions about the disposal of machines at the end of their economic life. Records can be kept for each machine and also be aggregated within categories of machines.

Dispatching is the function of deciding which pieces of equipment are to be assigned to which projects and tasks within projects, scheduling how long they will work, and planning where they will go next. It requires having accurate knowledge of the present and projected status of each machine and of the type and schedule of the work going on in the field. It also entails setting priorities on incoming requests from field supervisors for the equipment they think they will need. Dispatchers not only allocate the equipment owned by their company but also know the sources, status, quality, and costs of equipment from rental firms. Thus, they can meet field needs when company resources are inadequate.

With the major share of some contractors' net worth tied up in equipment, effective maintenance is essential to getting the most out of that asset. Even more important is minimizing the impact of poor performance, breakdowns, and accidents and the consequent disruption of work that results from poor maintenance. Computer-based maintenance systems keep track of the manufacturer's specified schedules for preventive maintenance and service, monitor the engine hours for each machine in operation, and ensure that the specified maintenance and service are scheduled, preferably at times that minimize interruption to productive work. More sophisticated systems identify weaknesses and idiosyncrasies in various machines and can enable contractors to anticipate problems or even help identify where design modifications might overcome them.

Several computer programs have been written to perform sophisticated engineering economic analyses regarding the optimum economic life of construction equipment. People in this business discovered long ago that the life that minimizes overall costs or maximizes profits occurs well before a machine reaches the end of its mechanical life, but there are so many parameters and assumptions involved that the calculations may not be feasible in some economic models without a computer. Other types of engineering economic studies involve decisions as to whether to lease, buy, or rent equipment, and which machines to acquire when there are choices.

Human Resources Management

Beyond payroll, numerous benefits accrue from using computers to improve human resources management. Construction by nature involves considerable turnover of field craft personnel and even technical and supervisory personnel. This business also involves many specialty skills, and in some cases the better workers can greatly outproduce the poorer ones and with much better quality. Particularly in the open-shop sector, which does not have the advantage of union-based training and dispatching of workers, contractors will maintain personnel records that have 10 or more times the number of people that they employ at any one time. Such a database is invaluable when it comes time to staff a new project or hire workers for a new phase of an existing one.

In addition to skills inventories, employment history, and contact information, human resources management systems can also incorporate the various benefit programs (insurance, retirement, vacation, investment plans, etc.) that companies often make available. Another use, admittedly more controversial, is to maintain records of personal information, such as substance abuse problems, personality difficulties, accident history, or other factors that can affect employment and safety. Although computer technology makes these things possible, it does not answer the accompanying policy and ethical questions.

Office Administration

Personal microcomputers can almost immediately pay for themselves in various kinds of construction office tasks. For example, numerous lists must be maintained on projects for drawing logs, tool inventories, safety equipment, and so on. Any number of microcomputer spreadsheet or file systems can handle such applications, as we will see in Chapters 9 and 10. *Word-processing* software on small or large computers can assist with letters, transmittals, reports, and so forth. In effect, most of the administrative drudgery that is typically assigned to junior engineers and office clerks can be greatly mitigated through the intelligent application of microcomputers on the job site.

A particularly important area is safety. There are numerous administrative procedures related to safety, ranging from award incentive systems for safe workers and supervisors to accident reporting when there are problems. Computers can have databases ranging from those that just keep track of who has what safety equipment to those that enable a site engineer to find what rules, regulations, and hazards to watch out for in designing a forthcoming field operation. Advanced expert system software has also been applied to this field. Chapter 11 provides an example.

Computers can maintain records for all transactions for contract administration. In an increasingly litigious industry, it is essential to keep track of drawing revisions, claims from subcontractors, claims to owners, filing deadlines, transmittals, changes to the scope of work, decisions or requests that affect schedule or cost, and so on. Computer databases can help organize these records, making sure that actions required by the contract are taken when needed. Pessimistically, these records help build a case should disputes arise; optimistically, the availability of clear, well-supported factual records can help resolve differences before they evolve into disputes and claims.

Education and Training

Construction is a business of many skills, and these can be improved continuously through training at all levels, from field crafts to president. Computers are severely underutilized for training in construction; but recently, powerful user-guided multimedia educational systems with lots of graphical interaction, and even sound and video pictures to enhance the presentations, have made this type of training interesting and attractive to a wide range of users. These systems can teach subjects ranging from equipment operation to boardroom strategy—and keep users interested. Furthermore, relative to the cost of hiring instructors and bringing people together in conference rooms, computer-based instruction can be economical and effective, and it can take place at workstations right in the field.

Related to this are simulations for production and management decision making that can serve as a basis for team-building exercises, for checking the impact of real decisions before they are implemented in fact, and so on. Although such tools have by no means become common in construction, in seminars and other settings they have been realistic enough to capture the enthusiastic attention of sophisticated engineering and management audiences. In a typical case, the simulation might define a project as a logical network or define a company and markets, allow participants to augment resources, schedule overtime or double shifts, postpone activities, change personnel, and the like. The simulation may bring in probabilistic factors related to weather, the economy, labor, management, and such. The objective may be to maximize profit on a job, increase a company's market share in a region, and so on.

CONSTRUCTION ENGINEERING

Construction engineering applications are those that require knowledge of the engineering analysis or design, or a more technical understanding of the work that goes on in the field. Applications include the following:

- Estimating
- Productivity improvement
- Operations simulation
- Quality assurance
- Surveying
- Computer-aided engineering analysis
- Computer-aided design and drafting

Estimating

Construction estimating is a broad subject, and computerized estimating thus means different things to different people. In the following list are some functions where computers significantly reduce the time needed to prepare an estimate, enhance the quality of the underlying analysis, improve the accuracy of computations, and help estimators cope with the stressful time pressure that is typical of this part of the construction business:

- Maintenance and preparation of master checklists
- Quantity takeoff
- Productivity and cost analyses
- Compilation and organization of summary and backup computations
- Compilation and analysis of subcontractor bids
- Distribution of overhead and indirect costs
- Modeling for sensitivity analysis of alternatives and contingencies in determining markup
- Preparation and delivery of the bid or proposal

Experienced estimators can tell "war stories" about the times when, under the pressure of deadlines, they forgot to include a significant portion of the work. Some contractors thus maintain master checklists that include all of the many items that may go into work of a particular type, such as school buildings. With a computer these checklists can be kept in a database, spreadsheet, or estimating package that can be adapted directly when the quantity takeoff and cost estimate get underway. A well-written software program can improve the productivity and accuracy of estimating by partially automating a quantity takeoff. Estimators using automated takeoff report that they can work at least four times as fast as before, and do so more accurately. In developing crews and evaluating the productivity of labor and equipment, computers can first provide data from files of past projects and then assist with specialized engineering programs for calculations such as cableway cycles, earthmoving fleet simulation, formwork quantities, and so on. To support the costing phase, computers can maintain current cost information. With remote access or information supplied on optical data disks, it is now becoming possible for computers to tie into various databases maintained by commercial estimating data sources.

With the calculations accurately and reliably under control, management can also indulge in sensitivity analysis to determine which aspects of the bid contain the greatest risk, and thus it can more intelligently determine the best markup. Through computer programs that provide a rough cash flow the markup can also be based upon a desired rate of return rather than simply a percentage of gross profit. As the estimate nears completion, a computer can be particularly valuable for developing summary cost sheets and in preparing bid sheets. These are especially important in the final hectic hours of competitive bidding, when newly arrived subcontractor bids and materials quotations can make frequent recalculations of the estimate necessary. Modems and fax machines even make it possible to transmit the bid or proposal to the owner electronically.

An example of a spreadsheet in Chapter 9 illustrates how several different types of calculation forms can be linked for automatic updating of summary sheets when input data and calculations change. Chapter 12 describes applications that help us to perform many more of the preceding functions.

Productivity Improvement

Deliberate and systematic efforts to improve productivity on construction projects are becoming increasingly common. Computers can assist with the compilation and analysis of questionnaires distributed to workers and supervisors, with the design and simulation of operations before they are implemented, with analysis of the results of time studies and videotapes of work in progress, and so on. This is a new area, but with the greater availability of computers on job sites, it is expected to become increasingly common in the future.

Operations Simulation

A powerful computer tool that has too seldom been exploited in construction is simulation. Historically, this sophisticated application required sound technical knowledge in order to set up models that correctly represented construction operations and to interpret and apply the results. Recently, however, graphically oriented microcomputer *simulation programs* have made simulation modeling as easy to use as spreadsheet programs, making them accessible to most construction professionals. The interpretation of results is also made clear through dynamic graphical displays of the models, which can look much like the real construction operations being simulated. With such software, we can represent the workers, machines, and materials of a typical construction operation, compute the cycle times of each step, determine the overall production output, and introduce variations and uncertainties to simulate what happens in the real world. Observing the models while running the simulation quickly helps the user identify production bottlenecks, reduce underutilized resources, reorder operation steps to a more productive sequence, adjust travel distances, and make many other changes to improve the output and reduce the cost of the production system. All of this can be done quickly and economically, either to design and implement a better operation in the field or to diagnose and correct problems in an existing operation. Chapter 15 gives two examples of this type of simulation software.

Even more impressive, large engineering contractors have developed software that builds upon three-dimensional computer-aided design (CAD) representations of the project and includes three-dimensional electronic pictures of construction equipment (trucks, cranes, etc.) displayed to the same scale as the structure's design. The equipment can be "operated" as if it were real, and thus engineers can find the best installation sequence for a particular system, check close tolerances for lifting large pieces into place, and even anticipate potential field problems in time to suggest constructibility changes to the designers before they have gone too far to change the design. Chapter 16 examines this type of software.

Quality Assurance

Quality assurance can begin with online retrieval of specifications, codes, and standards. Quality assurance systems also assist in documenting procedures and testing requirements and in reporting test results and completion of administrative steps to various interested agencies and parties. Some of the most advanced applications involve not only administrative procedures but also direct production control. For example, modern automated concrete batch plants enable the operator to call up any of several predefined mixes; the computer then operates the plant until the correct mix discharges into a waiting concrete truck; batch information is printed out and copies are given to the truck driver to take to the point of delivery for an inspector's confirmation and approval before the concrete goes into the pour. Copies are also attached to samples made at the pour site, and the loop closes following testing, when sample results are logged and sent back to the quality assurance department. Similar applications are gaining acceptance in welding, asphalt paving, and the like.

Surveying

Construction surveying has benefited enormously from computer applications. Microprocessors integrated into electronic theodolites and distance-measuring devices automatically record the data produced. Upon returning to the office, surveyors can transfer the recorded information to sophisticated computation and analysis software running on powerful microcomputers. Not only can simple traverses be computed easily but also whole subdivision layouts can be developed, and staking diagrams can be drawn automatically on high-quality plotters to guide the earthmoving process almost the next day!

The work of the field layout surveyor and grade-checker has also been changing rapidly. Enhancements to surveying technology have led to some of the first practical applications of automated construction machines. Spinning laser beams guide field tasks from setting footing forms at the beginning of the job to installing hung ceilings at the finish stage. In some cases just a few reference hubs can control site grading, while laser transmitters send to optical sensors mounted on the blades of scrapers, bulldozers, graders, and compactors all the information they need to maintain close tolerances for a high-quality product. Global positioning devices reading signals from satellites are increasingly finding roles in construction. With automatic control and fewer delays for surveying checks, excavation and grading can also proceed far more quickly than in the past.

Computer-Aided Engineering Analysis

One normally thinks of computer applications for engineering analysis as primarily the domain of engineers working on the design of a project, such as structural analysis, electrical power distribution, and pipe layout. Many of these same programs have applications in the field, however, such as designing temporary construction facilities, designing concrete formwork, or checking stresses during a critical erection sequence. Other analytical programs apply specifically to the construction stage. Examples of both types of applications include the following:

- Structural design and dynamic analysis (e.g., temporary falsework)
- Concrete form design
- Pipe network layout and flow calculations (e.g., for air or water supplies)
- Soil mechanics and foundation engineering (e.g., for trench safety, temporary bracing of excavations, etc.)
- Cofferdams and diversion tunnels (e.g., to compute how high or how big, predict likely water flows, and assess the risks and consequences of failure)
- Crane and cableway cycles (e.g., for erecting a building or constructing a dam)
- Drill and blast computations (e.g., size and spacing of holes, type and amount of explosives to produce a desired range of sizes of material from rock in a quarry)
- Design of screening and crushing plants (e.g., what size of plant and sizes of screens yield the most material for a given application)

Computer-Aided Design and Drafting

As with engineering analysis, one normally thinks of the popular computer-aided design and drafting (**CAD**) software as being the domain of architects and engineering consultants. However, there are several reasons why it is also important for construction professionals to be familiar with this software. First, many engineering and architectural drawings need to be produced even at the construction stage. Concrete formwork is one common example, and other needs include drawings for temporary buildings and structures and the overall layout of a construction site. Second, shop drawings need to be produced for submittal to the design consultants, building departments, and owners for approvals. Some contractors have gone further and redesigned parts of structures for greater economy and constructibility, and computer support for this work has been helpful in obtaining the approval of other professionals for the related changes.

Third, and even more important, there is a growing tendency for design consultants to submit CAD-based project documents to contractors in electronic form, and contractors are then expected to know how to use the software that goes with them. As time goes by, more and more owners expect contractors to maintain and update these electronic designs to reflect changes, include the added detail of shop drawings, and accurately show the "as-built" status when the project is completed. Some owners want these electronic designs to be compatible with the graphical and information systems they use for facility management, and the contractor may be caught in the middle translating between different CAD system standards. This situation is but one facet of the trend toward increasing design-construction integration that computers are making possible. Side benefits to contractors will be better interfacing to computer-based estimating systems and project management software such as materials management and quality assurance. Although there are tremendous benefits to be realized from such trends for all parties, they place much higher technical demands on the construction contractors. Knowledge of CAD software is thus becoming important for construction professionals.

Still to come is the closer integration of design and engineering databases and the field construction processes themselves. Automated construction machines and even robots are now evolving. If their development follows the pattern of factory automation (i.e., CAD/*CAM*, or the integration of computer-aided design and computer-aided manufacturing), machines will obtain their instructions more directly from computer-based designs and engineering information and query this information themselves to make plans for performing their tasks. Chapter 16 will comment further on this subject.

AUTOMATED DATA ACQUISITION AND PROCESS CONTROL

As construction equipment and field operations move toward advanced methods of data collection and automated control of equipment and even production processes, there will be more and more need for construction professionals to understand the technologies and computer applications behind automated data acquisition and automated process control and robotics.

Automated Data Acquisition

Automated data acquisition in construction means to use sensors and instruments to observe or measure things going on in the field, to use computers to monitor these instruments, and to store and report the information that is relayed to them. On a large scale, difficult foundation excavations and tunnels have been instrumented to monitor and record changes in soil pressures and stresses in bracing members as excavation proceeds. In the immediate case, alarms sound if shoring members move or stresses build up beyond specified tolerances so that remedial action can prevent the situation from becoming more serious. For the longer term, the recorded information can help design engineers improve their analytical modeling and calculation procedures. Smaller-scale applications of instrumentation are found in cranes, where loads, boom lengths and angles, stresses in outriggers, and such are continuously monitored by an onboard computer and reported to the operator via a display in the cab so that the crane can operate within safe limits and avoid catastrophic accidents. Chapter 3 has more information on this subject.

Automated Process Control and Robotics

Given the difficult and complex environment of construction, it is remarkable that robots are already performing routine tasks on some job sites. The first construction robots were derived either by adding sensors and computer-based controls to construction equipment (e.g., to control the cutting edges or screeds on various types of earthmoving and paving equipment), by adapting comparatively rigid factory-type robots to construction (e.g., for spraying fireproofing material or painting), or by developing hybrids of the two (e.g., robot arms mounted on tunnel machines). Although the sophistication of their mechanisms and sensors has often been quite high, these robots have had only the most rudimentary forms of programmed "intelligence." Some machines that have been called robots are really just teleoperated devices without any programmed automation at all.

Most of the construction robots developed to date are standalone devices designed to perform narrowly defined tasks without the need to communicate or cooperate with other machines. The concept of a construction "crew" does not apply yet to construction robots. However, coordinated teams of robots quite commonly perform sequential operations on factory assembly lines, and communication mechanisms link them together. It was only a matter of time before integrated automation technology started moving into construction, and it has already happened in Japan, with automated high-rise building construction and tunneling systems. Chapter 16 further explores this subject and provides a general overview of progress in this important field.

SUMMARY

This chapter has provided an overview of construction applications in three main categories: construction management and administration, construction engineering, and automated data acquisition and process control. The range of possible applications in construction management and administration includes accounting and payroll, cost control, finance, planning and scheduling, materials management, equipment management, human resources management, office administration, and training. No one contractor, large or small, has the best in all of these areas, but sophisticated applications exist in both large and small firms. If intelligently used, all applications can improve the efficiency and effectiveness of project management and reduce the drudgery that has become an increasing burden in administration.

Applications in construction engineering include estimating, productivity improvement, operations simulation, quality assurance, surveying, engineering analysis, and design and drafting. Some of these applications, such as the quantity takeoff stage of estimating, and field surveying and layout, have already experienced multiples of improvement in productivity and quality, and others, such as productivity improvement and operations simulation, have similar potential in store. Software for engineering analysis definitely has its place in construction, as does computer-aided design. The integration of engineering, design, and facility management that is being fostered by computers places a great demand for technical growth on construction professionals.

Although a bit more in the future for commonplace applications, there are already projects that have put extensive automated field data collection systems in place, and automated construction machines have multiplied production by factors of two to 10 with relatively small investments in machine automation. Although not yet very practical or widespread, robots have found their way onto construction sites, and they may have the greatest potential of all for changing work in the field.

This overview should provide a broad context in later chapters for a more cohesive appreciation for computer applications in construction as a whole. Later chapters provide much more information and some examples for most of these application areas as well as technical and management guidance for their successful implementation.

REVIEW QUESTIONS

1. For the nine categories of applications described in the *Management and Administration* section, prepare a categorical list showing which ones are most likely to be used by (*a*) the corporate or home office, (*b*) project management, and (*c*) field operations people. Briefly annotate your list to indicate the type of user who fits into the organizational level where you placed each application.

2. For the seven types of applications described in the *Construction Engineering* section, prepare a categorized list as described in Question 1. Briefly annotate your list in a similar manner.

3. Most of the applications described in the *Automated Data Acquisition and Process Control* section are more likely to be used in the future than at present. For each of the two main categories mentioned, state what construction professionals can do now to prepare themselves and their companies the better to take advantage of these technologies as they become more practical and economical, and even to accelerate their rate of development and use in the industry.

SUGGESTIONS FOR FURTHER READING

Adrian, James J. *Microcomputers in the Construction Industry.* Reston, Va.: Reston Publishing Company, Inc., 1985. This 318-page book was one of the first to examine microcomputer applications in construction. It briefly explains computer technology then quickly covers a variety of construction applications. All of its examples are written in the BASIC language, and it includes the source code.

Barton, Paul, ed. *Information Systems in Construction Management: Principles and Applications.* London, England: Batsford Academic and Educational, 1985. Written from a British perspective, this 228-page book consists of 14 papers by various authors on management topics and applications.

Construction Software Directory. *Constructor.* Washington, D.C.: Associated General Contractors, published each December. The AGC publishes one of the more comprehensive annual reviews and tabulations of commercially available construction software packages.

Tidwell, Mike C. *Microcomputer Applications in Field Construction Projects.* New York: McGraw-Hill, 1992. The first five chapters of this 339-page book review computer hardware and MS-DOS system software at a somewhat descriptive level and include several photographs of equipment. The next four chapters briefly introduce construction applications using spreadsheets and word processors, followed by scheduling (*Primavera*) and administration (*Expedition*) application packages. Two final chapters discuss computer management and maintenance. Appendix A provides spreadsheet listings for examples used in the book.

PART
II

COMPUTER
TECHNOLOGY

So long as we represent technology as an instrument, we remain held fast in the will to master it. We press on past the essence of technology.

Martin Heidegger*

In computer technologies, the last few decades have been like a time warp. Technical terms that once shielded the computer elite from mere mortals have become household words. Whimsically named companies founded by young prodigies with capital derived from hocked mementos of the counterculture have grown by megabucks to become producers of mass consumer goods. Little machines more powerful than prestigious blue-chip business processors of two decades ago are mistaken for—indeed, often used as—toys. However, the toy that plays spacewar and adventure games in millions of homes is also doing practical work for the nation's largest companies and for small ones as well. For most companies, the computer has become indispensable. But how much does the construction professional need to know about this rapidly advancing technology to put it to work?

---

*The Question Concerning Technology, 1955.

In almost any field, one can get more out of a technology by knowing something about it. For example, although one can certainly use a car for transportation without understanding much about auto mechanics, knowing a bit about engines, transmissions, and brakes helps one drive more safely and efficiently. In purchasing a new car, the informed person can also make more intelligent decisions about which features and options are useful and which are merely expensive fads or gimmicks. If things go wrong, the knowledgeable amateur can either fix it at home or at least be less likely to be exploited at the auto repair shop. Similarly with computers and software, technical knowledge helps a person make better decisions in obtaining the systems that will best fit anticipated applications without wasting money on overspecified components or unnecessary options, and more effectively use those systems once acquired.

Part II provides a brief overview of computer hardware and software technology. It is intended primarily for readers who have not had prior introductory course work in computer science, programming, or data processing, or who lack equivalent working experience. More knowledgeable readers may wish to skip or skim this section, but even some of those may find these two chapters useful as a review and as a means to see how we relate the technology to construction needs.

Explaining these technologies in detail is beyond our scope, but outlining the subjects that should be studied to become knowledgeable in this area is worthwhile. Four major categories presented are (1) computer hardware, (2) computer communications, (3) the software or programs that make it possible to use the computer in various applications, and (4) information organization and storage in files and databases. The first two will be covered in Chapter 3, and the second two in Chapter 4.

CHAPTER
3

COMPUTER HARDWARE

In 1950, it was generally agreed by executives of firms producing data-processing equipment that eight or ten of the big electronic brains would satisfy the entire demand for such devices! This monumental blunder in market forecasting today boggles the mind.

Donald H. Sanders*

Hardware is the part of computer systems most familiar to the general public—it is tangible and is certainly the easiest to show in photographs and movies. We will examine hardware in six main categories: (1) the central processing unit (CPU) and its high-speed electronic memory; (2) peripheral devices such as disk and tape drives for mass storage of information; (3) input devices such as keyboards and mice; (4) output devices like cathode-ray-tube displays, printers and plotters; (5) *real-time* data collection and control techniques for field instrumentation and automated machinery; and (6) computer communications. Figure 3-1 is a schematic diagram of a microcomputer that contains some of these components. Concepts and devices within each of these main categories will be discussed in the following sections.

PROCESSOR AND MEMORY

This section describes the main components used for data manipulations within a **computer**, that is, the central processor, memory, and some devices that enhance their performance. Before describing these parts, however, we first need to introduce some terminology related to how computers store and process data.

---

*Computers and Management, 1970.

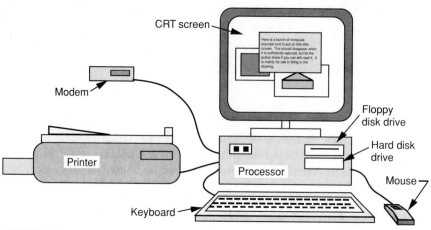

FIGURE 3-1
Hardware components of a computer system.

Terminology

The smallest unit of data storage or manipulation in a computer is called a *bit*, which can be thought of as a switch with only two positions, "on" or "off." Numerically, "on" might correspond to 1 and "off" to 0, and logically, "on" might represent True and "off" False, thus providing minimal arithmetic and logical ability.

The amount of memory in most computers is measured in a unit called a *byte*, which is a group of eight bits that, in the binary number system, can represent numbers from 00000000 to 11111111, equivalent to decimal numbers 0 to 255. Eight bits can also represent codes for up to 256 distinct symbols; thus used, a byte is approximately equal to one alphabetic, numeric, or symbolic character of storage. In most computers, the 256 possible codes represent upper and lower case letters, numeric digits 0 to 9, punctuation marks and some special computer control characters in what is called *ASCII* code (American Standard Code for Information Interchange).

For the next level of data organization beyond bytes, one sometimes hears the term *word*, which can be a multiple of 2 to 8 bytes (16 to 64 bits) or more, thus holding a set of characters or a much larger or more precise number than a byte. The size of the word in a particular computer design is commonly associated with the size of the main storage registers in the processor and the width of the path for moving data between memory and the processor (some of these new terms will be defined later). Word size also makes a difference in the degree of arithmetic precision to which the computer can operate directly in hardware (although software can be written to allow any degree of precision desired, even on eight-bit computers, but at a substantial trade-off in speed) and the range of memory locations that can be accessed directly (thus also increasing processing speed). In Fig. 3-2 the hierarchy of this low-level data organization is shown.

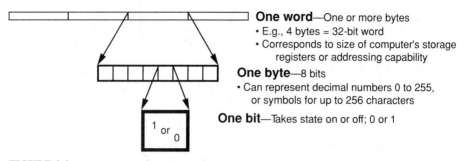

One word—One or more bytes
- E.g., 4 bytes = 32-bit word
- Corresponds to size of computer's storage registers or addressing capability

One byte—8 bits
- Can represent decimal numbers 0 to 255, or symbols for up to 256 characters

One bit—Takes state on or off; 0 or 1

FIGURE 3-2
Low-level organization of computer data.

Another useful type of terminology has to do with the way computer people write very large and very small numbers. Large numbers (used for quoting computer speeds and memory capacities, for example) use the prefixes *kilo* (for thousands, e.g., 800 kilobytes—or K—of floppy disk storage, or a 50-K file size), *mega* (for millions, e.g., a 20-megahertz—or MHz—processor speed), and *giga* (for billions, e.g., four gigabytes of optical disk storage). Very small numbers use the prefixes *milli* (for thousandths, e.g., 30 milliseconds to transfer a data file), *micro* (for millionths, e.g., 10 microseconds for disk access time), and *nano* (for billionths, e.g., a memory cycle time of 80 nanoseconds). Again, the terms used in these examples will be introduced shortly.

Large numbers used in computers may not be precisely what they seem. At the fundamental electronic level, computers employ the binary number system—or base 2—where the only two digits are 0 and 1. In converting to our familiar decimal system, we use powers of 2 to represent the positions in a number. Thus the binary number 11010011 converted to decimal would be

$$2^7 + 2^6 + 0 + 2^4 + 0 + 0 + 2^1 + 2^0$$

or

$$128 + 64 + 0 + 16 + 0 + 0 + 2 + 1$$

which sums to 211 in base 10. Now, 2^{10} is 1024, which is approximately 1000, or 1 kilo or K of something, and 2^{20} is 1,048,576, which is about a million, or a mega-something. Since things like memory and disk storage are organized in base-2 multiples to get the most out of each device's capacity, it has become convenient to use these "kilo" and "mega" decimal-number approximations to state their capacities in terms that people (who do not relate well to numbers like 11000111010101111100110101) can understand more easily. However, when we take multiples of these kilos and megas, the approximation errors accumulate. Thus "64 megabytes" of memory actually is 67,108,864 bytes, closer to 67 million than 64. You need not worry about this distinction, but it is nice to know that you actually have more of something than the kilos and megas would lead you to believe.

A final custom that we will mention is that computer professionals often use *octal* and *hexadecimal* (hex) numbers, which correspond to base 8 and base 16, respectively. Although this may seem a bit odd, when one is programming these are quite useful, and indeed even essential if programming in a computer's native language. Base 8 exactly corresponds to binary digits taken three at a time (e.g., 0 to 7 corresponds to 000 to 111), whereas base 16 corresponds to binary digits taken in groups of 4 (e.g., hexadecimal numbers 0 to F correspond to binary 0000 to 1111). Number F, you ask? Well, because Arabic digits only go from 0 to 9, it is customary to keep going after 9 with the first six capital letters of the alphabet (A = 10, B = 11, ... F = 15). Most computer users do not need to know about base 8 and base 16, but sometimes it is good to have some idea about what is going on when computer programmers start talking about "hex addresses" or "octal memory dumps."

Processor

The *central processor* performs a computer's arithmetic and logical functions. Relationships among some of its components are shown in Fig. 3-3. The *arithmetic logic unit* (ALU) contains circuitry to manipulate and perform computations on the data in the *registers* and *accumulators*, which temporarily hold instructions, data, and memory addresses. The *control unit* manages the ALU and its interactions with data in memory. The combination of the ALU, control unit, and registers is the **central processing unit**, or **CPU**, of a computer. Think of the *data path* as a set of parallel wires that acts as a multilane highway between the memory (described below) and the registers. The *address path* is the set of wires that select a specific memory location or interface device for input to (a *read*) or output from (a *write*) the CPU. Sometimes the data path and address path functions are **multiplexed** onto the same physical set of wires, with precise timing from the control unit to keep the functions apart. The *address register* holds the address of the memory location or device that is presently being accessed on the data path. The *control path* is a set of wires that transmits signals to regulate the traffic on the data "highway." Where data coming into the CPU represent instructions, the *instruction decoder* and *controller sequencer* translate and signal the appropriate actions to the processor. The *program counter* maintains the current location in memory of the next instruction to be executed. The *condition codes* are set or cleared for operations that do or do not result in a *carry* (C), *overflow* (V), attempted *divide by zero* (Z), and *negative result* (N). Other common codes are *device interrupt* (I) and *half-carries* (H). In the computer is a very high-speed *crystal clock* that provides the beat to synchronize the various components, and its precise frequency of oscillation is called the *clock speed*. The design and speed of the CPU, combined with the width and speed limits on the data path and address path, help determine the speed and power of the computer as a system, which should not be confused with the clock speed itself.

An important distinction in hardware design relates to the number of fundamental arithmetic and logical instructions that the electronic circuitry of a CPU

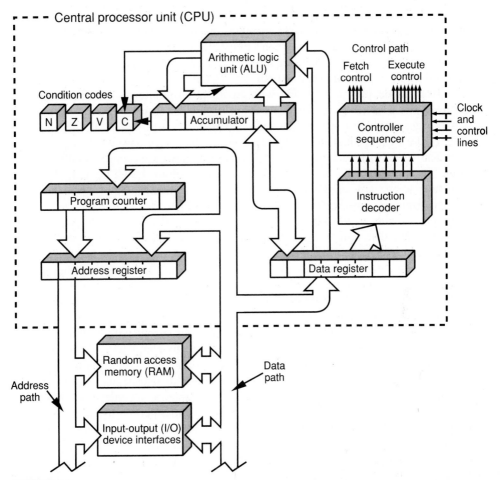

FIGURE 3-3
Organization of the central processor unit and memory. (Adapted from *Microprocessors*, Heath Company, 1977, Figure 5-1.)

can perform. At one end is a design philosophy that says to keep the number of instructions small, say 40 to 80 or so, but make each very fast (typically one machine clock cycle), leaving more complex functions to combinations of these instructions implemented in software. Such hardware is called ***reduced instruction set computers*** (or ***RISC*** machines). They are easy to program and economical to build, but the extra instruction steps also tend to take up more space in memory and storage devices. At the other extreme is ***complex instruction set computers*** (***CISC***), which may have 400 or more instructions implemented with over a million devices (transistors, etc.) on their ***integrated circuits*** (usually called ***chips***). Although some of the more complex instructions might take two or more clock cycles to execute, they do it all in hardware, which is inherently faster than equivalent software. These processors are expensive to design, build, and make error-free, and system programmers who fully appreciate all that can be done with their instruction sets are rare, but they can reduce the need for some support devices in computer hardware and they can support very high-speed applications. Modern hybrid designs exploit the best of both approaches.

Even laypeople have to deal with technical terminology when reading advertisements, brochures, or magazine reviews or when going to buy a computer. Ads or salespeople may claim that such and such machine has a "30-megahertz, 32-bit, RISC CPU," and the prospective buyer may not know what to make of that. The speed in megahertz (i.e., millions of cycles per second) usually refers to the rate at which the basic logic cycles take place in the CPU, and "32-bit" may refer to the size of the storage registers or the width of the data path. As a rough approximation, a 66-megahertz version will work twice as fast as a 33-megahertz version of the same CPU design, and a model with a data path of 16 bits may move data half as fast as one with 32. However, many other parameters in CPU design have an influence on actual processor throughput, such as the number of fundamental operations (add, subtract, multiply, divide, store, jump, etc.) designed into its hardware. A RISC *microprocessor* may need to run at a faster clock speed than a CISC *microprocessor* to get equivalent throughput, depending on the type of application software.

One also hears abbreviated model numbers associated with particular processor designs, such as "486 machines," referring to computers whose central microprocessor is an 80486 chip made by Intel, or an "040," referring to a 68040 chip made by Motorola. For the uninitiated, listening to salespeople or colleagues at work carrying on conversations where the vocabulary includes lots of these numbers can be quite confusing, especially when most people using (or misusing) these terms do not understand their meaning either. The underlying concepts are not difficult, but the jargon can be a barrier.

As commercial products, central processors are generally classified into three main ranges, primarily on the basis of size, cost, and capability. The largest computers are called *mainframes*, and they still exist in the central computing facilities of major universities, government agencies, and large companies. Very high-speed mainframes, usually optimized for large-scale scientific computing, are called *supercomputers*. Large mainframes and supercomputers typically fill one or more cabinets, and their processors consist of several discrete components on multiple circuit boards. Specially designed rooms with air conditioning and raised floors usually house mainframes and their storage devices, and they are tended by staffs of computer professionals. A typical installation appears in Fig. 3-4.

The next category down in size is *minicomputers*, and these range from small but powerful single-user *workstations* on up to superminis, whose capabilities can exceed those of some mainframes. Until recently the processors of minicomputers consisted of several logic chips on one or more circuit boards, but now they differ from microcomputers (discussed in the next paragraph) mainly in the speed and capacity of their components. Their applications range from powerful CAD workstations to servers connected to several terminals for multi-user

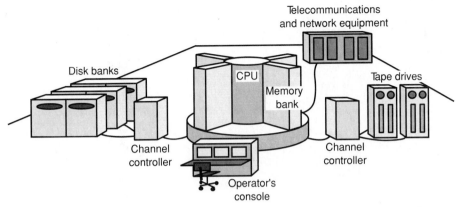

FIGURE 3-4
Mainframe computer.

office systems to controllers for automated factory production lines. Most are managed without specialist staff support, but typically they require a more advanced level of computer knowledge from at least one of the users. Figure 3-5 is a sketch of a multi-user office configuration.

The smallest processors are found in *microcomputers*. The main distinctions of this category are that the processor normally is implemented on a single small silicon microprocessor chip and the whole machine is often configured in an enclosure about the size of a typewriter. Microcomputers range from dedicated processors used as controllers for home appliances and automobile engines on up to machines whose capabilities exceed those of the minicomputers of but a few years ago. Figure 3-6 is a sketch of a typical microcomputer system.

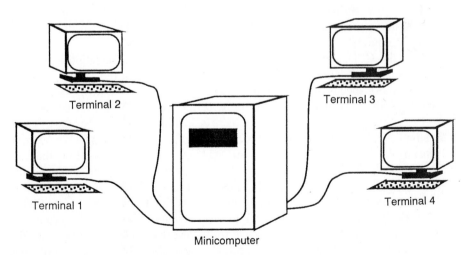

FIGURE 3-5
Multi-user office minicomputer installation.

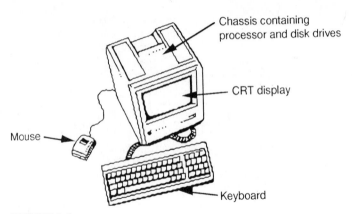

FIGURE 3-6
Basic microcomputer system.

Given the rapid changes in technology, some distinctions between these categories of machines are disappearing. Today's more advanced microcomputers are more powerful than all but the largest mainframes of two decades ago.

Memory

In most computers, primary ***memory*** consists of sets of semiconductor chips, each typically capable of storing a million or more bits, and organized such that groups of chips store a series of bytes or words in a manner logically accessible to the processor. Two general categories are ***random-access memory (RAM)***, where data can be both stored in and retrieved from memory, and ***read-only memory (ROM)***, where data are permanently encoded in the chip hardware and can be retrieved by the processor but cannot be written or replaced. ROM generally stores parts of the computer ***system software*** rather than applications and data, although application programs are sometimes sold as credit-card-size ROM packages for use in small portable computers.

Like the CPU, memory also has different types of organization and speed characteristics, and these should be matched to the design of the CPU. Instead of quoting memory speeds in terms of millions of cycles per second (i.e., megahertz), memory speeds are usually given in terms of the time it takes for one memory access cycle (e.g., in nanoseconds, which is in billionths of a second). Thus, very simplistically, a 20-megahertz processor might be matched to 50-nanosecond memory (i.e., $1/20,000,000 = 50 \times 10^{-9}$). If the memory is too slow, then the CPU may have to insert *wait states*—that is, skip one or more cycles—until memory responds. The other key chip parameters are capacity and organization. In 1994, memory chips used in microcomputers typically had from 1,000,000 to 16,000,000 bits of storage each (or 1 to 16 megabits) and were organized as a linear one-dimensional array. Alternative forms of organization for a one-megabit chip might be $4 \times 250,000$ or $8 \times 125,000$ arrays. In the latter case, only one chip would be needed for 8-bit memory on a small computer, such as that found embedded in consumer audio or automotive control products.

Performance Enhancements

Of course, one might enhance an application's performance by installing a faster CPU, more memory, or even a second computer. However, engineers often find better ways to enhance something by adding capabilities that may not be present in the basic design. There are several such ways to improve the performance of a computer, and some of these are either built-in or available as options on today's computers. Common examples include floating-point coprocessor units (*FPUs*), cache memory buffers, and dedicated input/output (I/O) processors.

Floating-point coprocessors are auxiliary chips that implement extensions to the logical and arithmetic instructions designed into the main CPU. Floating-point coprocessors perform additional operations in hardware (which is fast) rather than software (which would require the equivalent of several instructions when performed on the base CPU). The most common extensions are those for working with fractional (i.e., floating point) numbers rather than just integers, so such chips are called floating-point coprocessors ("co-" because they work in conjunction with the main CPU). Typically they implement floating-point addition, multiplication, and division directly, and possibly some logarithmic and trigonometric functions as well.

Cache memory buffers are an economic trade-off between costly high-speed memory and less expensive lower-speed memory. The basic idea is to use a smaller amount (e.g., 64 kilobytes) of high-speed memory as a buffer between the CPU and the main memory (which might have 16 megabytes of RAM and be too expensive to implement in high-speed memory chips). The computer's programmed logic is designed to look ahead and predict which bank of the main memory may be needed next (pretty easy if a program is just working its way sequentially through a contiguous table of numbers, but harder in other cases), and then copy that bank of memory into the high-speed memory, usually while the main CPU is still working on a problem. The main CPU accesses the high-speed cache memory, which is matched to the CPU speed, but has to slow down by inserting wait states if it needs to go directly to main memory. The term *hit rate* is sometimes used to refer to the fraction of times that the computer finds what it needs in cache memory, and this hit rate can usually be increased (but with diminishing returns) by adding more cache, an expensive proposition.

Another function for cache memory is to serve as a buffer between main memory and data stored on a peripheral device such as a disk drive, but we will look at this in the next section.

Dedicated I/O processors are specialized computers that handle specific input/output functions, such as moving data between memory and a disk, or waiting for somebody to type something on a keyboard. Delegating these tasks to specialized processors optimized specifically for the purpose can free the main processor to continue its primary program work with fewer interruptions and thus enhance the overall throughput of the computer. Why pay a powerful $200 CPU chip to wait around for a human to type when a $2 chip can do the job well enough? This type of design keeps the costs down and improves performance, especially in computers designed to do several things simultaneously, such as print a report from your word processor (with the help of a dedicated printer output processor) while you are working with a drawing on the screen using a CAD program. A common type of optional dedicated I/O processor is a fast graphics processor to handle the computations for dynamic and colorful screen displays.

Floating-point coprocessors and cache memories are often available as options for computers, whereas dedicated I/O processors are more likely to be part of the basic design. Whether or not they are needed, however, depends on the application. Word processing benefits little from floating-point chips, but spreadsheets and graphics programs can speed up by a factor of 10 or more this way. As technology advances, more and more of these capabilities are being pulled into the main chips of advanced CISC CPUs such as Intel's Pentium and Motorola's 68040, reducing system costs and enhancing performance.

PERIPHERAL MEMORY AND STORAGE

Computers used for data processing, word processing, computer-aided design and other applications involving large amounts of information normally have most of their data kept in *files* on *peripheral* storage devices. The information is only brought into the main memory of the computer when it is actually being created, searched, modified, moved, printed, displayed, or when it is otherwise currently active. Peripheral storage is used mainly because it is much cheaper per unit of storage, the information is preserved when the computer is turned off (whereas normal RAM memory "blanks out" at that stage), and some storage media can be conveniently moved from computer to computer. The most important peripherals are the *disk* and *tape* devices that store information on magnetic media. Optical disk storage is also becoming common. These types of devices and their controllers are described in this section.

Magnetic Tape

Tape drives for computers are similar in concept to the audio or video tape decks found in many homes. Tape drives range in size from small cassettes that quite literally are the same as those used for audio storage on up to large dedicated units that store information on $\frac{1}{2}$-inch-wide magnetic tape at densities of 65,000 characters per inch or more. Figure 3-7 is a sketch of a tape drive and a detail of a length of tape showing how data are organized into parallel tracks about one-byte (i.e., eight tracks) wide. Since the length of a tape may be several hundred feet, over one billion characters of information can be stored in a relatively compact package. This is equivalent to about two million pages like this one, or thousands of books. As with audio tape drives, access to the information is sequential, so tape drives normally are used for storing large quantities of information in a continuous stream. However, it can take several minutes to reach any particular item of data, which is like geologic time to a computer. Thus tape normally is not used where *direct access* to information is needed in a flexible or unpredictable order (such as looking up airline reservation records for customers calling on the

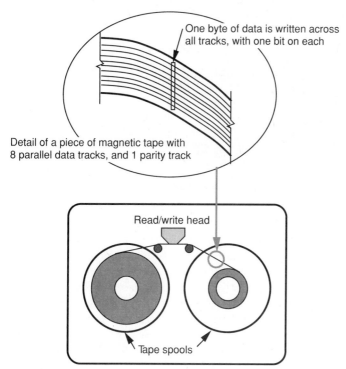

One byte of data is written across
all tracks, with one bit on each

Detail of a piece of magnetic tape with
8 parallel data tracks, and 1 parity track

Read/write head

Tape spools

FIGURE 3-7
Magnetic tape for data storage.

telephone). Rather, tape is most often used to make duplicate (i.e., backup) copies
of information kept on disk or other media so that it can be retained for security
in case the main files are lost or damaged, or so that information can be easily
transported to another computer.

Magnetic Disk Storage

Disk drives have a storage medium that looks somewhat like a phonograph record
or a compact disc in an audio hi-fi system. However, the method of encoding and
retrieving information is usually magnetic, like that for tape drives, and rather
than the information being along spiral grooves, the tracks or circles storing the
information on disks are concentric. Furthermore, unlike either compact discs or
phonograph records, the drive can write information to the disk as well as retrieve
information already stored there.

A disk drive has one or more circular platters spaced one above another as
shown in Fig. 3-8, and one or more *read/write heads* on an arm that reaches in
from the side to position a head over the track containing the data to be read or
written as the disk spins beneath the head. Because the read/write heads can move
from point to point relatively directly, somewhat like the tone arm on a record

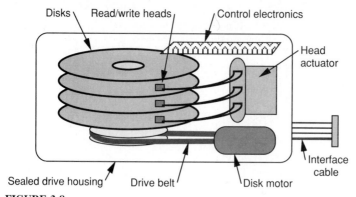

FIGURE 3-8
Internal configuration of a disk drive for data storage.

player, disk drives are especially appropriate where information is to be accessed in a nonsequential or, as it is sometimes loosely called, "random" manner. Storage in the form of disk technology ranges from small $3\frac{1}{2}$-inch diameter *floppy diskettes* with capacities of 800,000 to 2,000,000 characters or so, through removable *cartridge disks* with capacities ranging from 20 million to 500 million characters or more, on up to fixed **hard disks** (or **Winchester disks**) with capacities that can be over one billion characters. The cost of comparable drives for tape or disk media is similar, but regarding the storage media itself (disk cartridges or magnetic tape), it is usually cheaper to store on tape than disks, based upon cost per character.

Because disk drives involve mechanical motion as well as electronics, there is usually some delay (typically 10 to 50 milliseconds) until the head is positioned over the part of the disk surface where information is to be transferred. Specifications you may see related to disk drive performance may state the *seek time* (time needed to position the heads over a certain track), *rotational delay* (time elapsed while the beginning of the data on the track rotates into position under the head), and *transfer rate* (the electronic rate at which data then move). Once the head is mechanically positioned, transfer rates can exceed one million characters per second.

Optical Disk Storage

By the late 1980s, *optical disk* storage was also becoming common for computers, just as compact audio discs were taking over the hi-fi market. The drive mechanisms look not unlike those of magnetic disks and come in both fixed and removable forms. However, the data access is usually via a laser beam rather than a magnetic head, so positioning to a particular location on the disk can potentially be much quicker. Initially, optical disks were only able to read data previously encoded on the disk by a special recording device (as is the case with CDs), but recent advances in technology have also made user-writable disks available. Initially encoding could only be done once and it could not be erased. Optical drives of this type were called "write once/read many" or *WORM drives*. Full read/write capabilities have since become available. Optical storage volumes can be very high—over one billion characters on a 5-inch disk—but costs have been higher than and access speeds have lagged those of magnetic media.

Peripheral Controllers

Disk or tape drives are normally connected to the computer's processor via an interface device called a **controller**. Typically, the part of the control logic that deals with mechanical aspects of the disk are in the drive itself, whereas the parts that make it compatible with a particular type of computer are on either the computer's main circuit board or an optional board that plugs into the computer. There are numerous standards for such control logic, including one commonly called **SCSI** (pronounced "scuzzy," but meaning **Small Computer System Interface**) which allows several peripheral devices of similar or different types (disks, tapes, printers, etc.) to be strung together from one plug on the computer. To manage each peripheral device there is a **device driver**, which, although it sounds like a physical piece of hardware, is actually a small computer program that runs under the system software of the computer.

The controller also may contain circuitry to manage a *cache memory input/output (I/O) buffer*. This behaves somewhat like the high-speed RAM cache described earlier. The main purpose in this case is to overcome the mechanical delays in the disk drive mechanism (seek and rotation times). We accept as a given that information is transmitted rapidly once the disk head is correctly positioned, at which time the idea is to transmit large blocks of information, while anticipating future instruction and data needs to the extent possible to minimize the number of disk accesses. For example, four kilobytes of output might be accumulated in the cache buffer then be transmitted to the disk in one burst rather than a character or line at a time. Physically, the cache memory may be either dedicated RAM included within the controller circuitry or a portion of the computer's primary memory reserved for this purpose. Either way, applications involving frequent disk input and output will run appreciably faster than when such circuitry is absent.

Other important categories of peripherals are **input and output (I/O)** devices. Broadly speaking, disk and tape drives fall into this category, but they are classified separately because information stored in them is not directly readable by human beings. On the other hand, human beings normally do help prepare data and interact with programs via input devices such as a keyboard or a mouse. Similarly, output devices such as cathode-ray-tube (CRT) screens, printers, and plotters exist mainly for converting computer data into a form humans can see. The next two sections explain some of basic concepts and applications of the more commonly used I/O devices.

INPUT DEVICES

There are numerous devices for interacting with a computer and for entering information into it. Apart from the aforementioned mice and keyboards, these include touch-sensitive overlays for CRT screens of the type found in tourist information displays in airports, light pens that can point to objects on a screen, scanners that can enter whole pages of pictures or text, *joysticks* and buttons of the type found on videogame machines, special matrix pads and analog dials for controlling some graphics machines, electronic *digitizer* panels and overlays used as tablets or sketch pads for drawing or handwriting input, bar code readers, and even speech recognition devices. In earlier times, there were also punched card readers and paper tape readers, but these have mostly disappeared from modern computing. Most input devices are rather specialized in their application, so only the two most common ones will be explained here—the keyboard and the mouse.

Keyboards

Keyboards for computers are somewhat like those of typewriters, enabling a person to select and press keys in an order that makes sense for a program running on the computer—sequences to make words or numbers for data entry—or perhaps just single keys to choose items from a menu displayed on the screen.

In countries that use the Roman alphabet, most computer keyboards are laid out in the QWERTY format (named after the first six keys in the upper left alphabetic row of such keyboards) that is familiar to touch typists, but this standard was created a century ago to keep the mechanical keys of typewriters from getting tangled up, not to make typing easier for humans. Alternatives sometimes encountered are a straight alphabetic layout or a layout called Dvorak, after its inventor, who designed it to be more efficient for people whose minds are not yet locked into the QWERTY standard. Minor variations of QWERTY layouts correspond to the different characters, accent marks, and symbols of various languages and are designed to suit the conventions of individual countries. Keyboards exist for Arabic, Greek, Russian, Hebrew, and other alphabets, and even the tens of thousands of characters that make up Chinese and Japanese Kanji pictorial writing are now encoded via keyboards with about the same number of keys as those found in languages with smaller alphabets.

Compared with typewriters, computer keyboards have a few differences that should be noted. First, a typewriter has shift keys that, when held down or locked down while typing, make the keyboard type upper-case letters and the symbols that appear above the lower numbers or punctuation marks on other keys. Computer keyboards also have shift and lock keys (often a separate lock for upper-case letters alone), but they have other keys as well that similarly alter the normal function of various keys on the keyboard. Typically labeled as CONTROL (or CTRL), ALT, OPTION, COMMAND, or even with some special symbols, they may enable users to access another character or symbol set (such as Greek letters for mathematics or science writing), issue ***commands*** to the computer (e.g., CTRL with S may stop a stream of text being transmitted to the screen), or perform functions equivalent to commands or menu choices that are specific to a particular application program (e.g., advance one word). Besides these keys for altering the standard behavior of other keys on the keyboard, most computer keyboards have programmed function keys (typically labeled F1, F2, ... in a row along the top or clustered at one side of the keyboard). These can be programmed in application software or by users to make a single keystroke do things that might otherwise take several keystrokes to type on the main keyboard.

It is important to be aware of what happens when you type on a computer keyboard. As illustrated in Fig. 3-9, when you press a key on a computer keyboard, an electrical contact between a certain set of wires arranged in a matrix encodes a unique electrical signal that is sent to a storage buffer in the computer (which may be a small processor in the keyboard itself that relays a signal to the main computer). What gets sent to the computer may be an ASCII code for the specific character shown on the key, or about eight bits of data. The character does not necessarily appear on paper or on a screen. As far as the computer is concerned, the character code is just data in memory. It only gets displayed if the logic of the computer connected to the keyboard is designed or programmed to take the additional step of displaying the character. Probably millions of hours have been spent by human users trying to resolve confusion resulting from mismatched keyboard and computer settings, mostly because these users have a mechanical typewriter paradigm in mind and really do not understand what is going on with the deceptive appearance of the computer's keyboard.

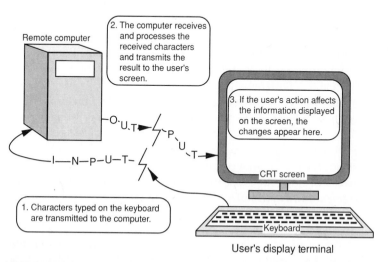

FIGURE 3-9

Character transmission from terminal to remote computer, with echo back.

The Mouse

A poignantly amusing misunderstanding about a computer mouse appeared during a backward time warp in the movie *Star Trek IV*. While the crew was stranded in San Francisco, the chief engineering genius of starship *Enterprise*, Mr. Scott, wanted to use a "primitive" Apple Macintosh computer to perform some desperately needed calculations. Accustomed to a far more advanced breed of computers, he walked up to the machine, put the mouse to his lips and said, "Computer! Computer!"—and of course nothing happened! Although voice recognition is nearer at hand than the Star Trek era, the mouse is not yet used like a microphone.

The computer *mouse*, invented in the 1960s by Douglas Engelbart of the Stanford Research Institute, is aptly named for looking somewhat like its rodent namesake. It is small and cute, it has a wire coming out that looks like a tail, and the multibuttoned ones even seem to have eyes or ears or a nose. But there the resemblance ends. The computer version generates electronic signals when it is moved along a surface, and these signals encode motion that computer software can interpret to move a *cursor* (a small graphic image such as a cross, square, or an arrow) proportionately on the screen. One or more buttons on the mouse can send other signals that enable the software to control actions such as "attaching" the cursor to an object so it can be moved, making a choice from a menu, noting where to insert text, and so on.

Encoding signals to deduce motion is done in various ways, usually involving a combination of optics and electronics. Figure 3-10 is an illustration of the type of mouse included with Apple Macintosh computers. A palm-size plastic shell houses a ball that protrudes from the bottom of the shell, letting it roll along the surface of a table. As it rolls, it drives two shafts set at a 90° angle to each other inside the housing. The shafts turn disks that have either small holes or notches around their perimeters. For each disk, a light beam is projected on the radius where the notches or holes pass, and a photocell receives light on the other side of the disk. As the holes and intervening solid material on the disk roll by, the number and pace of interruptions to the light beam are observed by the photocells (one each for the vertical and horizontal axes) and encoded for transmission down the wire to the computer. The computer program can interpret the direction, speed, and displacement of these motions and move the screen cursor in a way that soon appears intuitively natural to the user—a very nice piece of human engineering.

Another type of mouse is more strictly optical—no mechanical tracking ball on the bottom. Instead, it must be moved on a special reflective surface that has a closely spaced matrix of horizontal and vertical mirrored ridges. As the mouse moves, the light beamed downward reflects back from the surface into photocells in a way that can be encoded, transmitted, and interpreted by the computer to deduce direction, speed, and displacement.

Although our discussion of input devices ends here, it is worth noting that several of the other devices mentioned at the beginning of this section work on similar principles. Depressing buttons, twisting dials, or moving levers produces electronic signals that the computer interprets in some meaningful way. Some are

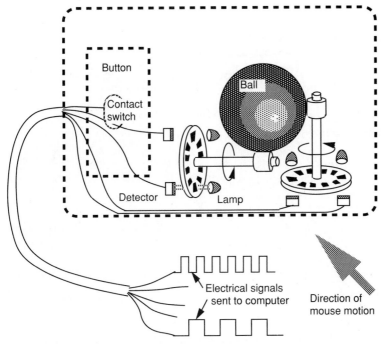

FIGURE 3-10

Workings of a computer mouse (top view). (Adapted in part from Figure 10-19, *Technical Introduction to the Macintosh Family,* ©1987 by Apple Computer, Inc. Reprinted by permission of Addison-Wesley Publishing Company, Inc.)

fairly simple, such as the analog potentiometers that are found in many videogame joysticks; others are enormously complex, such as the combination of an optical scanner and the type of character and pattern recognition software that can actually convert the images from paper into electronic versions of real text and figures in the computer. All, however, can be reduced to a set of fundamental principles, most of which are fairly simple. Understanding what is going on can help you get more out of using computers.

OUTPUT DEVICES

Numerous kinds of devices enable computers to present information in a form intelligible to humans. These range from the common display screens and audio speakers to several kinds of printers and graphic plotters. A few will be described in this section.

Display Screens

Several technologies can display information on a screen (or *monitor*) connected to or built in to a computer. The primary one uses a *cathode-ray tube (CRT)*, which looks like a television screen. These are found in office desktop computers or

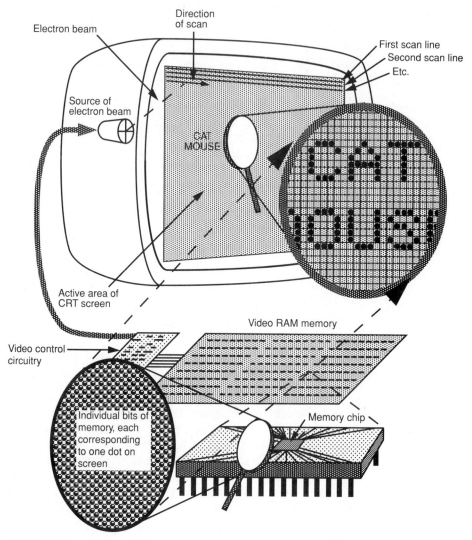

FIGURE 3-11
Raster-scan screen display matrix mapped from memory.

personal computers kept at home, since they are not easy to carry around. Other types are *flat-panel **displays*** made from **liquid crystals** (**LCDs**), gas plasmas, or similar thin-plate technologies. These are mostly used in portable computers and in certain applications where CRT displays would be too vulnerable to breakage. What most CRTs and almost all flat panel displays have in common is that they divide the screen up into tiny dots (called **pixels**) arranged in a two-dimensional matrix such as that shown conceptually in Fig. 3-11. The *resolution* of the screen can be expressed in terms of its display area dimensions in pixels, such as

640h × 480v (horizontal × vertical), its linear pixel density in dots per inch (e.g., 72 dpi), and *pitch* between pixel center points (e.g., 0.26 mm).

The screen matrix maps to a defined area of memory located in computer hardware, and as computer programs modify locations in memory, the output display electronics cause corresponding changes to appear in the dots on the screen. If the display is strictly two-color (usually called *monochrome*, ignoring the background color), where each dot can only be on or off (i.e., white or black, or perhaps green or black, or even orange or black), then only one bit of memory is needed to map each screen dot. A screen 640 dots wide by 480 dots high could thus be mapped into $640 \times 480 = 307,200$ bits, which divided by 8 is 38,400 bytes, much less than the equivalent of one typical memory chip. However, as will be explained later, much more information can be mapped onto each dot, such as color or intensity, and thus 8, 16, 24, or even 32 bits of memory may be necessary to store the information for each dot on the screen.

CRT screens themselves fall into several categories. Early computer CRTs were a natural outgrowth of the technology of the oscilloscope and were called *vector graphic* CRT displays. They used analog voltages under the control of mathematical programs in the computer to move the electron beam directly from one coordinate to another on the screen. The main alternative to vector graphics in CRT displays is the *raster scan*, which uses technology similar in many ways to regular television. The electron beam (or three beams in color systems) moves in a constant pattern, starting in the upper left of the screen, sweeping across the first row of dots, then coming back to the beginning of the next row down, sweeping across again, and continuing this way for the several hundred or thousand rows that are displayed on the screen. This pattern is shown in Fig. 3-11. When the beam reaches the bottom, it blanks out, returns to the upper left, and starts again. As it passes over a simple black-and-white screen, the beam turns on for an instant if a dot is to be illuminated, stays off if the dot is to remain off, and does so according to the status of the corresponding bits in the computer's display memory. This process repeats, from top to bottom, for hundreds of thousands or millions of individual dots, from 60 to 80 times per second, which is the *refresh rate* of the CRT. (Indeed, if the refresh rate is much below this, the human eye takes notice, at first subconsciously, leading to eyestrain and headaches.) Since the computer program is concurrently modifying the corresponding bits in memory and does so almost independently of the CRT display controller, which just keeps mapping to the screen whatever it finds in memory each time it passes through a screen refresh cycle, the complexity of the picture is not limited by the refresh rate as it is in vector graphic displays. Complexity is limited only by the number of dots available on the screen.

If it is a **gray-scale** monochrome display, the *intensity* of the beam can be varied for each dot to look like shades of gray. With two bits of memory per dot, allowing binary mappings of 00, 01, 10, and 11, there can be four shades, from black to white. Four bits allow 16 shades, 8 bits allow 256, and so on. Finer shading controls can make an image look almost like a black-and-white photograph if the matrix density of screen dots is fine enough. In the case of color, three convergent beams or sources of electrons are scanning the screen, corresponding to red, green, and blue (thus these are sometimes called *RGB displays*). These colors can be mixed in various intensities to create almost any other color. However, the shadings of these colors are limited by the number of bits allocated to support each dot in memory (4 bits allow 16 colors, 8 bits allow 256, etc.). The high-quality displays have 24 or 32 bits per dot, allowing millions of colors, and again the results can look strikingly like a photograph. But the cost of larger and higher-quality displays goes up rather quickly. In the world of the IBM *PC* and its compatibles, increasing resolutions and color depths have led to a succession of de facto video display standards called CGA, EGA, VGA, SVGA, XGA, and the like.

Another difference in technology in screen displays is whether they map character text patterns or whether they allow full bit-mapped graphics. Earlier *character-based displays* found in microcomputers such as the old Apple II and IBM PC used electronics that generated fixed dot patterns to represent text characters. These displays were economic, simple to program, and fast. The preferred alternative today is full *bit-mapped* graphics. In this case, every dot in the screen is directly accessible by computer software. Characters and pictures alike are all produced in software. The flexibility thus provided has made possible the window-type displays found on modern microcomputers and workstations and has produced whole subindustries of text font designers, desktop publishers, computer artists, and so on. The driving economic progress discussed in Chapter 1 has made the costs competitive with the old text display systems, technology has brought the hardware components into the mass market, and programmers' skills have grown to meet the software challenge. The advantages are such that almost all computer-display technology has moved to full bit-mapped graphics; and the cost, technology, and complexity questions have moved on to how we can do ever more realistic color.

In this discussion of computer displays, old-timers may have noticed that I omitted *hard-copy terminals* where the output came back from the computer and was typed on paper as one worked. I spent long hours on those machines and felt security in knowing I had a printed copy of my work in case the computer's electrons ceased to dance, but it has been a long time since I saw anybody actually using one. The type of interactive bit-mapped graphics that we take for granted today existed only in the imagination when we still worked on such antiquated devices.

Printers

Printers for computers range from small dot-matrix impact and ink-jet printers used with microcomputers through high-quality laser printers and even color printers for offices and professionals on up to large-volume printers used in business data processing. Most use dots to form characters and graphic patterns, much like the dots that form the images on a raster-scan CRT display. Differences are mostly in the technology used to move the paper and put the ink on the page.

The early ***impact printers*** had solid metal character images that were moved into position before being hammered against a ribbon that transferred the character image to paper. For office use these types of impact printers have largely been replaced by other technologies because they were noisy and were limited to the character types and images that could be molded onto their bands, daisy wheels, and golf balls. The main surviving advantage of impact technology is for printing multipart carbon forms, but office procedures are changing so that even these forms are becoming obsolete as more and more documents are transmitted electronically.

Some ***dot-matrix printers*** also use the impact of small hammers to drive tiny pins against an inked ribbon to form patterns on the paper. A sketch showing the basic technology appears in Fig. 3-12. Early printers of this type typically had nine pins in a vertical row on a print head that moved across the paper. As it moved, the appropriate pins were hammered forward to form successive vertical rows of dots that successively built up images of characters much like those on a CRT screen. The output of these early models indeed looked very "dotty" and usually was not acceptable for business letters. Dot-matrix technology has evolved, using more pins and small offset motions to produce very good-quality output. They can produce graphics as easily as text, and they can indeed bang out multiple carbon copies of forms. But most are still very noisy.

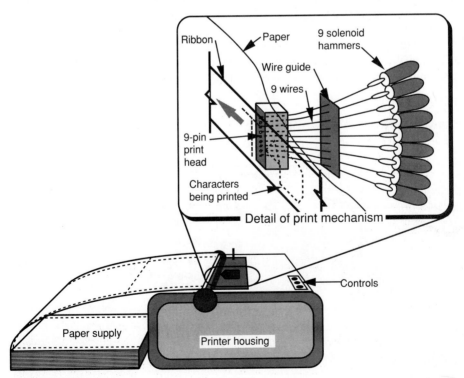

FIGURE 3-12
Dot-matrix printer technology.

Ink-jet printers are also based on the matrix principle and have become a good alternative to impact printers. Instead of a vertical row of pins, they have a vertical row of tiny holes through which minute quantities of ink squirt out directly onto the paper. A simple but quite marvelous bit of cheap thermoelectronic technology that is built right into the disposable ink cartridges performs this process. The ink dries by the time the paper is removed from the printer. Lacking the heavier mechanics and power demands of impact printers, some of these ink-jet models have shrunk to battery-powered units weighing about one kilogram that are a convenient companion to portable computers. The main disadvantage relative to impact matrix printers is that they cannot print multiple carbon forms, but in most cases this is overridden by the advantage that they are very quiet. Both impact and ink-jet matrix printers share the disadvantage of being somewhat slow (although there is a huge range of speeds within these technologies), since they print character-row by character-row rather than a whole page at a time. Their print speed also varies inversely with the quality of the print image selected.

A growing alternative to the preceding technologies is *laser printers*, which have become the preferred output device for most office uses. Their technology is somewhat like that of copiers, where a whole page image is sent to sensitize a drum that then picks up ink (or "toner") in the areas sensitized, transfers the ink image to paper, and fixes it by heat. See Fig. 3-13 for a sketch of a laser printer. The underlying graphic technology is still a dot matrix, but usually the densities (e.g., 90,000 to 360,000 dots per square inch are common) are much higher than for more mechanical types of printers. Initially they were very expensive both to buy and to operate, but keen competition has now made them one of the most cost-effective alternatives. They are an excellent complement to the publication-quality text and graphics work that can now be done on high-quality microcomputers and

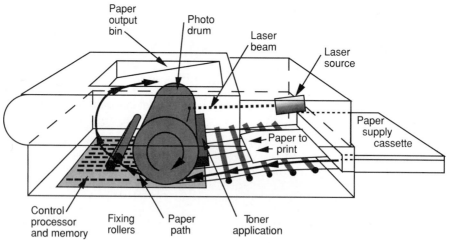

FIGURE 3-13
Laser printer.

workstations. However, they still do not print on carbon forms (or they make a horrible mess if you try).

Many other interesting printer technologies are now in use or developing, especially those that can produce high-quality color pages and even 35-mm slides that do justice to the photograph-quality color images now appearing on the best color CRT screens. But we can only cover so much in an introductory chapter, so we will leave these topics for you to study elsewhere.

Plotters

Computer *plotters* are of limited interest in most computer applications, but because this is a book about computer applications in construction, and drawings are still the lingua franca of this business, they are important here. Basically, there are two main underlying technologies, and they correspond somewhat to the differences between the aforementioned vector-graphic CRT display and the raster-scan variety.

Analog pen plotters correspond to vector graphics, and they are really fun to watch when they are drawing. They literally pick up pens (from a "stable" containing pens of one or more colors and/or line widths), and mechanical carriers move the pens around and lift them up and put them down to make a drawing, much as you would yourself. However, they are much faster, more accurate (they typically work to 0.0001 inch), and neater than the most superhuman draftsperson, and they don't take long coffee breaks. Since they literally *draw* the lettering as well as the lines and symbols, they can be rather slow (compared to the type of plotter described next) when doing complex drawings. But the ink work of a high-quality analog pen plotter can be as nice as it comes in architectural or engineering design work. Once a drawing has been stored in the computer, it is also very easy to make corrections and economically print another original rather than to keep erasing and patching an old sheet of vellum.

Analog pen plotters come in two main shapes: flatbed and drum type. *Flatbed plotters* even look somewhat like a high-technology drafting table. The user fastens down an E-size sheet (or whatever size is desired or fits), and the carriage mechanism, which is like a gantry crane with a transverse carrier where the pen is attached, moves the pen in both the X and Y directions. Figure 3-14 is a sketch of such a plotter.

Drum plotters (see Fig. 3-15) stand vertically and load the paper on a roller; the back-and-forth motion of the roller provides the X dimension used in the drawing. The pen mounts on a carrier that moves along a track parallel to the axis of the roller, and this provides motion in the Y direction. The coordinated motion of the roller and the pen carrier permits lines and curved figures to be drawn in any direction. This type of analog pen plotter is usually less expensive than a flatbed for equivalent quality output and has the further advantage of not limiting the lengthwise dimension of paper that can be mounted on a roll (or as cut sheets if still desired). They have thus become the most widely used type of pen plotter in architectural and engineering offices.

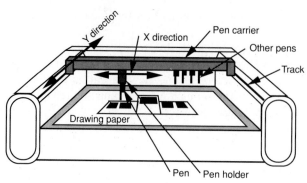

FIGURE 3-14
Flatbed plotter.

The main alternatives to pen-based analog plotters are plotters based on dot-matrix images. Early ones used special photo- or heat-sensitive papers mounted in rolls. As the paper passed over a bar containing a row of closely spaced thermal or electrostatic contact points, dots would be transferred a row at a time, building an image as the paper rolled on through the printer. More modern versions do a similar sort of thing but have benefited from xerographic copier technology that has brought costs down and improved quality to copier standards, which is quite acceptable. In size and appearance, xerographic-like plotters are similar to

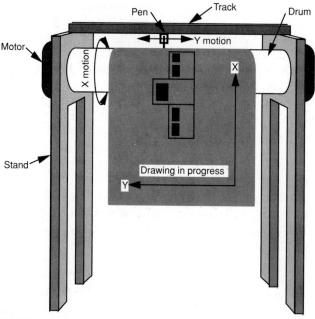

FIGURE 3-15
Drum plotter.

a drum-type plotter, but this is a one-pass operation where an image stored in the computer's memory transfers directly as the paper rolls through the plotter. For most drawings, especially complex ones with lots of lettering, they are much faster than pen plotters, and speed is a major advantage. They do not normally draw with colors, however, which makes them less attractive for some applications such as drawing critical path method (CPM) charts and perspective renderings. Color is coming, but it will be expensive for many years. With a pen plotter, changing color is just a matter of picking up a different pen.

AUTOMATED DATA COLLECTION AND PROCESS CONTROL

As yet relatively few construction field applications require knowledge of how to interface computers to *sensors*, instruments, and control devices of the type found in advanced factories, aircraft computers, and robots. But some applications do exist already that point to the tremendous economic potential of applying these technologies in the field. The more construction professionals can learn about this subject, the better they will be able to guide the development and successful implementation of the supporting technologies. As a side benefit, several of these technologies are used in computer devices themselves, so this section will provide an even better understanding of disk drives, plotters, and other such computer peripherals.

Before looking at either automated data collection or control, we need first to understand some concepts and terms that describe how computers can be connected (or *interfaced*, to use the appropriate computer jargon) to physical objects on a construction job site. Examples might be a strain gauge or load cell attached to a steel brace to warn if it is overstressed, or a set of electrically controlled valves that regulate the hydraulic control systems for an automated blade on a motor grader.

Matching Machines from Different Worlds

One key problem of interfacing physical construction applications to computers is that there is usually an enormous mismatch in the forces and energies involved. Thus we must take some measures to *amplify* and *reduce* voltages, currents, and so on. Another factor is that many real-world objects are mechanical in nature, whereas computers are primarily electronic. Also, the operating principle of computers is *digital* (that is, there are discrete jumps from one state to another), whereas many real-world phenomena are *analog* (that is, continuously changing) in nature. Finally, there are problems of shielding (or hardening) computer technology to protect it from the rigors of the construction workplace.

Consider first the mismatch in energy levels. Although computers are reliable in normal office applications, they are still delicate machines. Inside them one finds currents in milliamps, and voltages typically are under 5 volts. A whole microcomputer, ignoring its peripherals, typically consumes less power than a

100-watt light bulb. Even connecting standard household AC current to one of the input/output peripheral plugs on the back will do serious damage and quite likely destroy the machine. In the field, of course, we commonly work with loads measured in tons, voltages in the hundreds, power measured in kilowatts or horsepower, and currents of many amps—enough to fry instantly or crush any computer that dares to get near them. On the other hand, there are also energy levels too small to be measured directly by a computer. The output of the aforementioned strain gauge, for example, may be just a few millivolts, even though this may correspond to loads of several hundred tons. One aspect of computer interfacing is thus to amplify small signals and reduce large ones before they reach the computer.

The problem of converting nonelectrical phenomena (mechanical forces, motions and displacements, hydraulic and pneumatic pressures, temperature, etc.) to the electrical currents and voltages used by a computer, or vice versa, is solved with *transducers*. Examples dealing with mechanical phenomena include strain gauges (for changes in dimension), load cells (for force or pressure), and inclinometers (for changes in angles). Transducers typically take the mechanical measure as input and produce a small voltage or current as output. We measure fluid phenomena with pressure gauges, flow meters, and other instruments. Temperature, of course, can be measured with thermometers. Equivalent thermal transducers typically use a material that, when its temperature changes, alters its electrical resistance or conductivity. Measures of time are also important, but computers already have clocks. Other phenomena can also be measured, such as moisture content (e.g., of aggregates feeding into an automated concrete mixer), noise emissions, and the like.

Going the other way—converting computer output voltages or currents to forces—tends to be done more with relays, switches, and electromagnetic devices that in turn control engines, motors, heat sources, jacks, pumps, valves, hopper gates, and other familiar equipment that we already use in the field. Even here, however, transducers often will be used to measure the results of such control actions, and these measures are fed back to the computer to be compared with what was desired so that corrections can be made if necessary—an engineering approach called *feedback control*.

Interfacing Computers to the Physical World

Analog-to-digital (often written *A to D*, or *A/D*) conversion is needed to input continuous real-world phenomena into the digital world of computer electronics. *Digital-to-analog* (*D to A* or *D/A*) applies when the computer needs to send output signals to control something with a range of valid states in the physical world. The difference between analog and digital is perhaps best illustrated by example. The "old-fashioned" kinds of clocks and watches are approximately analog in the way their hands continuously and almost imperceptibly move around a dial. Perhaps a better analog example is change in temperature as measured by a mercury thermometer. It moves very continuously. Fluid flow is also a continuous phenomenon. If you are old enough to remember it, the slide rule that once was the primary calculation tool for engineers was an analog device. In contrast, on a digital watch or clock, you can clearly see the discrete second-by-second advances as the digital circuit inside ticks away. Electronic calculators are also digital. Even some things that have been around for a long time are digital, such as a standard light switch.

Converting from D to A or A to D is done by using a range of digital numbers to approximate an analog phenomenon. The number of bits of logic used to represent the range determines the precision of the conversion, so one hears terms like "8-bit converters," "12-bit resolution," and such. One bit, as discussed in the beginning of this chapter, basically allows only an on-off phenomenon, which is not analog. Anything beyond that, however, could be considered an analog approximation. Two bits give an approximation consisting of 4 discrete steps, 4 bits allow 16 steps, and 8 bits, with 256 discrete steps, are near enough to continuous for many practical applications. For illustration, in Fig. 3-16 is shown a 4-bit analog approximation for the simple linear function X = Y. There are many A/D and D/A devices available in the 8-bit, 12-bit, and 16-bit range, the last allowing 65,536 discrete steps, or a tolerance of under 0.002 percent. This is far more accuracy than is needed for almost anything in construction, and circuit boards implementing 8-, 12-, and 16-bit conversions are available for almost any common type of microcomputer.

Of course in many practical applications conversions of this type are not even necessary, such as turning on and off motors and lights and operating gates

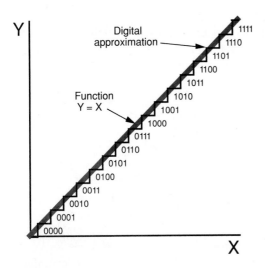

FIGURE 3-16
Four-bit analog approximation of a linear function.

and valves that are designed to be all the way open or all the way closed. Computer interfaces to such phenomena are called *digital-to-digital (D to D*, or *D/D)*. On the other hand, the jobs of running a precise milling machine, controlling the cutting blade on a motor grader, or winching the cable on a hoist or crane lend themselves to D/A control, and measuring loads, speeds, and such require A/D input conversions.

Three factors described thus far—energy amplification, electrical to mechanical conversion or temperature transduction, and analog conversions—are illustrated in Fig. 3-17. The two boxes on the left—computer hardware and software, respectively—are components similar to those of a standard computer found in office applications. Next is the interface circuit, which is an analog-to-digital, a digital-to-analog, or a straight digital-to-digital circuit. Circuit boards are commonly designed to implement several *channels* of input or output on one circuit board to monitor or control several devices at once. For example, a board might combine 16 A/D input channels with 4 D/A output channels, and 12 D/D channels, each programmed selectively to work in either direction—32 channels in all. Buy one of these for your home computer and you can run your lights, keep track of the temperature in several rooms, water your lawn, perk your coffee, and start and stop your microwave oven, all while sitting at your computer. Some people really do this.

The fourth box in Fig. 3-17 represents the device where some form of amplification takes place. The inputs and outputs of the interface circuit board typically operate in the range of −5 v to +5 v, or −12 v to +12 v, with a resolution corresponding to the number of bits. An output power amplifier may be needed to step these up to the 240-volt range to run a variable-speed motor. Going the other way, an input amplifier may be needed to take small voltages from a sensor—say 0 to 50 millivolts—and multiply them by 100 so that incremental changes do not "fall between the cracks" of an eight-bit A/D input converter. (That is, −12 v to +12 v, say, is a range of 24 volts. Dividing by 256 steps possible in an 8-bit converter still leaves almost 100 millivolts per step, double the whole range of the sensor.) Similarly, a 5-volt digital output to turn on a light switch

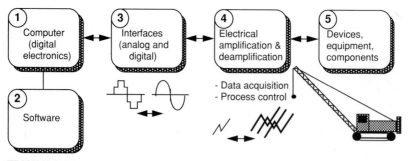

FIGURE 3-17
Computer interfacing to the physical world.

may still need to feed into a device called a *relay*, where basically a small switch flips to trigger an electromagnet that in turn throws the big switch. Amplifiers need not even be entirely electrical. Pneumatic switches, fluid couplers, levers, gears, and other machines can all come into play. Again, the purpose is to get the ranges of phenomena in the physical world matched to the more limited range of computer electronics.

For completeness, the fifth box in Fig. 3-17 is the physical object that is being monitored or operated upon. It could be the household gadgetry mentioned previously, or components of construction machines, or perhaps construction robots. It could even be the machinery that runs computer disk and tape drives, printers and plotters, and the like—so now you know even more about how they work. Applications of these technologies will be discussed briefly in Chapter 16.

As a further illustration of interfacing, Fig. 3-18 is a cutaway view of a multichannel interface board installed in a microcomputer.

Making Real-Time Computers Rugged Enough to Survive in the Field

The other major concern in bringing computer equipment into the field is how to make it rugged enough to survive. Obvious hazards include getting clonked by an errant steel beam or run over by an earthmover, but dust, temperature extremes, vibration, and moisture can also take their toll. For example, a manufacturer of a computer-based control system that is mounted on motor graders, scrapers, and bulldozers for blade control on such machines has gone through several refinements in field-hardening designs. From the start the manufacturer used steel enclosures with military-specification cable connections and vibration-damped mountings and sealed the circuit board itself in about a $\frac{1}{2}$-inch layer of

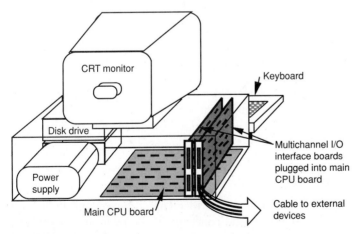

FIGURE 3-18
Multichannel interface boards installed in a microcomputer.

silicone gel. Even then, through capillary action, moisture from early-morning condensation worked its way up the wire leads that necessarily came out of the package, and—accelerated by the electric voltages present on the circuit board— contact points corroded enough to disable the computer. If the computer had been monitoring the overturning moment on a crane boom being lowered (computer-monitored cranes are an increasingly common safety application in construction), the results could have been catastrophic.

Fortunately, many users have solved similar problems. The military embeds computers in tanks, jet fighters, submarines, and warships, and they are designed to withstand even the rigors of combat. NASA sends all kinds of computers into space, and even though occasional malfunctions in these enormously complex systems make the headlines, the overall reliability record is truly astonishing. Probes of deep space send back signals from the outer reaches of our solar system even after a decade or more of frigid travel. In industrial plants and factories, computers have been hardened to work near blast furnaces, oil refineries, and production lines.

The toughest computers actually have specially selected chips and circuit elements. They are assembled using extra-rugged circuit boards and connections and, where possible, mechanical devices like disk drives are replaced by solid-state electronic devices. There are volumes of specifications for these types of machines, and some companies specialize in making them. The costs, needless to say, can be much higher than conventional computer equipment, but these costs are far outweighed by the consequences of failure when computers are working in the physical world where safety is an ever-present concern.

COMPUTER COMMUNICATIONS

Although it involves a mixture of hardware and software, telecommunications is described here because the focus will be on communications hardware. This subject is particularly important to construction companies since their production centers, or projects, tend to be decentralized from the principal place of business. In its most common form, telecommunications technology interfaces between computer systems and the common carriers such as the telephone company to enable information to be transmitted from one computer system or device to another. But computer communications also include local area *networks* (LANs) for offices and job sites, and even the links from computers to printers. This section will begin with some fundamentals, then describe both telecommunications and the technologies of local area networks.

Communications Fundamentals

Before getting into some of the mechanisms of computer communications, we again need to learn a few more terms and concepts. These relate to the physical wiring for data transmission, the means of organizing and carrying the data, and some standards.

One basic difference in transmission is whether data movement is *parallel* or *serial*. For example, parallel connections commonly hook printers to computers. In these the effect is of eight data wires plus a few control signal wires going from the computer to the printer. Thus the eight bits of data that make a byte or character of information can move down the lines in parallel, side by side, and arrive simultaneously. This transmission is simple and reliable for short distances, but the wiring gets expensive for longer distances. Thus for most telecommunications it is more common to use serial data transmission. In this case, the eight bits or so of each character first go into a data buffer, and the bits are sent in a sequence — that is, serially — down a single line. At the receiving end they accumulate in another buffer, and when the character is reassembled it is sent on to the receiving computer. The difference is illustrated in Fig. 3-19.

In some cases, it is possible for several signals to share the same physical line simultaneously. They are separated by a technique called *multiplexing*. For example, if three signals exist as audio wave forms, say in the range of 300 to 3000 Hz, one could use a multiplexer to add to them electronically three distinct carrier frequencies, say 30,000, 34,000, and 38,000 Hz, respectively, as shown in Fig. 3-20. Since they are separate, they do not get muddled together on the wire. The signals are separated at the other end, the carriers are subtracted off, and the three signals thus end up at their respective destinations.

The term *bandwidth* tells what range of frequencies a particular transmission medium (wire, microwave, satellite, etc.) can handle. The wider the bandwidth, the more information it can carry. For example, a 3000 Hz channel will carry a voice over a telephone. (Admittedly, it does not sound as good as a person next to you, whose voice may be able to go from 200 Hz to 15,000 Hz, but it saves money to clip off the extreme highs and lows — a range of 3000 is a good compromise between economy and fidelity for telephone purposes.) On the other hand, a single satellite TV channel might take five megahertz, and into this bandwidth you could multiplex a lot of voice channels.

In digital data transmission, the on/off, 1/0 data can be represented either by timed digital pulses of current or voltage or by modulating between two different frequencies in an analog wave form. The rate at which this pulse turns on or off, or the wave switches between the two frequency levels, is sometimes called the *baud rate*, a term left over from the French Baudot code of early telegraphy days. As an approximation, 1 baud means about 1 bit per second, so a 2400-baud transmitter moves data at about 2400 bits per second or, dividing by 8 bits per byte or character, at about 300 characters per second. There may also be extra check bits called *parity* bits for checking that each character is transmitted correctly. Numerous design details have been added to data communications technology to increase speed and reliability, but this paragraph gives the basic idea.

Parity checking can apply in computer memory organization as well as in communications, and is worth describing in more detail. Basically, it means adding an extra bit to a byte or word as a check to improve the chances that the data bits have remained what they were supposed to be originally. For example, if data are stored in eight-bit bytes, a ninth bit might be added to make the sum of all

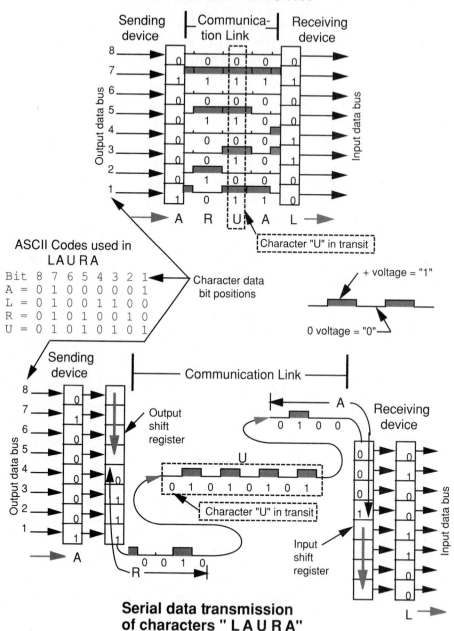

Parallel data transmission of characters "LAURA"

ASCII Codes used in LAURA

Bit	8	7	6	5	4	3	2	1
A =	0	1	0	0	0	0	0	1
L =	0	1	0	0	1	1	0	0
R =	0	1	0	1	0	0	1	0
U =	0	1	0	1	0	1	0	1

Serial data transmission of characters "LAURA"

FIGURE 3-19
Parallel and serial data transmission.

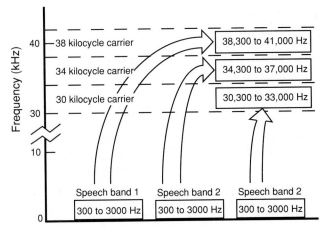

FIGURE 3-20
Multiplexing three bands into modulated higher-frequency carriers.
(James Martin, *Telecommunications and the Computer,* ©1969, p.
192. Adapted by permission of Prentice-Hall, Inc. Englewood
Cliffs, N.J.)

nine bits come out to be an odd number. If the eight data bits were 11010011, the bits of which sum to five, then a zero would be placed in the ninth bit so that the sum is still odd. If the data bits were 11011110, the ninth bit would be made one so that the odd sum would be seven. The convention just described is called *odd parity.* There is also a convention for *even parity* if the sum of the bits should be even. Both even and odd are commonly used (typically they can be chosen by software or by a switch on devices like modems), so one must be sure that the two ends of a data exchange are consistent for a particular conversation. The benefit of the parity bit comes when the data are received and the parity sum is checked at the other end. If any single bit has erroneously changed, say as a result of electromagnetic interference somewhere down the line, the sum will be the opposite of what it was supposed to be (odd changed to even, or vice versa). The receiving device could then tell the sending device to retransmit that character or perhaps just notify the user that a data error has occurred. Of course, if two data bits flip erroneously in the same character transmission (the probability of this happening is much smaller), there would be a compensating error that the single parity bit would not detect. In situations demanding extremely high data reliability there are more sophisticated schemes using two or more parity bits that can detect and even correct more complex types of errors.

Another pair of common communication terms is *half-duplex* and *full-duplex.* In full-duplex data transmission, you can think of there being two wires (or two signals multiplexed on the same wire) that allow simultaneous two-way data transmission. Thus you could be typing the next command to a computer while characters are still coming back for display on your screen. This transmission is analogous to a two-lane highway. Half-duplex, on the other hand, is more like a section of one-lane road where a signal system has been set up to allow traffic to go in only one direction at a time, as happens when a construction crew is repaving one side of a two-way highway. Again, hardware or software switches need to be set consistently so that devices and/or programs agree on the same convention. If you get it wrong, you typically get either double characters on the screen for each one typed (you may have set your communications program for half-duplex whereas the computer at the other end thinks you want characters echoed back in full-duplex) or none at all (the opposite condition), which can be frustrating if you don't know what is going on.

The terms *synchronous* and *asynchronous*, when applied to data communications, describe whether or not both ends are synchronized in such a way that a series of characters is sent through with a regular and predictable timing. Since almost anything involving human beings is irregular and unpredictable, asynchronous communications are used, for example, for applications that involve people typing at a keyboard, or even machine-to-machine communications where there may be interruptions to a regular flow of data. In asynchronous transmission, there is character-by-character checking, each being acknowledged before the next is sent. On the other hand, in machine-timed synchronous transmission, there is a well-controlled system where each end starts processing simultaneously and batches of characters can be sent through more rapidly with much less signal-control flagging going on in between.

Finally, to coordinate something as complex as data communications and enable various devices of different types from different manufacturers to be plugged together requires standards. For representing character data itself, there needs to be agreement on something like the aforementioned eight-bit ASCII standard or others (such as *EBCDIC*—the Extended Binary Coded Decimal Interchange Code that has been used on IBM's mainframes). A mismatch of standards causes chaos. A common hardware and signaling standard is EIA-RS232 (EIA is Electronic Industries Association). Originally it defined a 25-pin plug configuration with two rows of pins or sockets, one wire for data output, one for input, plus various lines for control signals. However, since only four to nine of these lines are commonly used for today's serial devices like modems, printers and terminals, 9-pin, 15-pin, and even round-plug variations of this standard have become common. This change means you can get into adapter problems if there is a mismatch. It is outside the scope of this chapter to go beyond these two basic standards, but you need to be aware that such standards exist. The proliferation of even minor variations on these standards has caused countless hours to be lost in setting up what should be routine data communication links.

Communication over Telephone Lines

The most frequently encountered device for telecommunications is a ***modem*** (short for "modulator-demodulator"), which converts the computer's digital electronic signal into the type of analog voice-grade signal that can be transmitted over a telephone line. In Fig. 3-21 we see how a digital computer signal can be converted to analog voltages that carry telephone audio signals. Basically, the bilevel voltages of digital data—the 1s and 0s—are converted to audio frequencies that fit in the telephone band of 300 Hz to 3000 Hz. For example, the 0s might

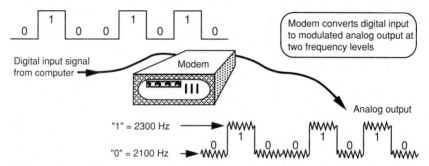

FIGURE 3-21
Converting digital data for an analog telephone carrier.

be encoded at 2100 Hz and the 1s at 2300 Hz. The resultant carrier looks like a fuzzy square wave.

The series of devices used in a typical two-way telephone "conversation" between computers is shown in Fig. 3-22. After the analog data are sent out, they are received at the other end of the line by another modem, which translates the analog signal back into the digital signal the receiving device wants to hear. Common modems have transmission speeds ranging from 300 characters per second (cps) to about 1200 cps over voice-grade lines, with higher rates achievable on dedicated lines.

Typical features on modems include the ability to *originate* a connection (i.e., "dial" a phone and recognize whether the receiving end sounds like a busy signal, a confused human being, or an understanding and compatible modem), *answer* incoming calls from other modems, and switch to the appropriate communication speeds and parameters.

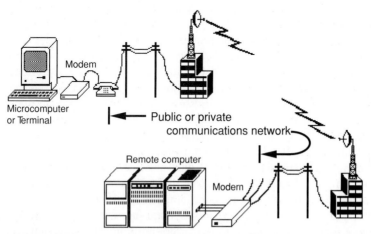

FIGURE 3-22
Computer-to-computer telecommunications.

A recent advance in computer telecommunications technology is the *fax modem*. As the name implies, it has the capability of sending digitized pictures over telephone lines, but it has a number of advantages over standalone fax machines that usually produce images of much degraded quality at the other end of the line. Fax modems can transmit directly from an electronic image of text and drawings stored in a computer, and the image can be captured and reconstituted in its original fidelity by a receiving computer on the other end. If the image is then printed on a laser printer at the receiving end, the quality can be just as good as if it were printed directly from the originating computer. As a side benefit, most fax modems can double as regular character-based modems as well.

With digital communications networks being established by telephone companies and by special-purpose public and private carriers, the speed and reliability of transmission are increasing rapidly, and costs per unit of capacity are decreasing. Ironically, digital data that have been converted into audio signals to be fed into a home telephone circuit can be converted right back to digital data by the telephone company for long-distance transmission, and then back into and out of audio form in the local exchange and modem at the other end. It is becoming possible to bypass this audio modem conversion altogether, but it will take awhile for that change to become universal.

Networks

This section will focus primarily upon *local area networks* (*LANs*) of the type found in office and institutional environments. Before beginning, however, it is worth noting that there are also *wide area networks* spanning national and even international boundaries. For example, in the United States there is a network called Internet that connects most universities and also numerous research centers in industry and government. There are also private networks such as MCI and BITNET, and these can be interlinked into Internet and to each other. Networks in other countries can also tie in, such as JANET (Joint Academic Network) in the UK, so researchers in many fields communicate around the world as readily as they do with colleagues next door. Costs are generally negligible to very low and have seldom been an obstacle to national and international communications. These networks are built in part on dedicated ground or satellite lines provided by commercial carriers such as telephone companies and in part on private lines. The academic networks usually depend on major university computing centers to provide host facilities for software to handle *electronic mail*, for user data storage, and for transmission of data files from one link to the next.

There are numerous standards for local area networks, ranging from proprietary ones like Apple's Appletalk (provisions for which are built into every Macintosh computer) to standards in the public domain like Ethernet (which was developed originally by Xerox). They typically operate just within a specific site (hence *local area*), such as a company headquarters building or a university, or even departments within these. Some construction job sites also have local area networks. Within a standard like Ethernet, there still can be numerous alternative *communications protocols* (i.e., the hardware and software rules and conventions for connecting devices and transmitting data), and thus incompatibilities can arise, but we will not go into that level of detail here. Various types of networks can be interlinked, including Appletalk and some Ethernet protocols, and larger institutions like universities typically have a mixture.

There are several alternative *topologies* (i.e., connection geometries) by which the various cables and devices in a network (i.e., computers, terminals, printers, modems, etc.) can be hooked together. The devices on the network are usually called *nodes*. In Fig. 3-23 are the most common network topologies and some nodes within them. The *linear* configuration is basically a set of devices hooked in series, where data pass from one to another. In *point-to-point* networks, each device has a direct connection to one or more of the other devices with which it needs to communicate. This geometry is impractical for more than a few nodes. The *star* configuration is the most hierarchical, whereby a central computer (which has the role of *server*) has direct links to every other computer, terminal, or other device in the system. All data transmitted from one device to another go via the server. In a *ring* topology the devices are connected in a closed

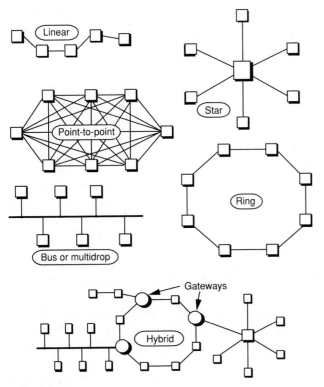

FIGURE 3-23
Examples of network topologies.

loop. Such networks can be quite egalitarian, with no one computer taking the lead role, or one of the machines can be designated as a server. The *bus* topology (also called *multi-drop*) has a trunk line, and feeder lines tap off this to various devices in the network. Again, there may or may not be a server. Finally, there are *hybrid networks*, which are combinations of those just described. If these involve networks from different system standards, then *gateways* are provided to perform protocol translations as data move from one network subsystem to another. The gateways themselves usually contain dedicated computer processors.

Numerous data handling, signaling, and regulatory functions must be performed to keep complex data networks from becoming chaotic. At a simplistic level, one might think that a computer on the network could just start sending out a stream of data and hope that some other computer would recognize them and pick them up. But how should the intended recipient be identified from among the many devices on the network? What if some other computer should start transmitting a stream of data at the same time—will all the bits get mixed up? Should confirmation be sent back to the sending computer when data have been received? Or is there a way to tell the sending computer if the transmission failed? Issues and questions of this type have led to the design of the numerous communications protocols.

Several terms needed to discuss these protocols include packet switching, polling, and collision detection. *Packet switching* indicates that a data file to be transmitted is divided into finite pieces (say 512 bytes); these are given some form of identity and destination address (analogous to putting a letter in an envelope), and sent on their way. If there is a *collision* (i.e., some other device puts a packet on the network at the same time), *collision detection* logic or software takes note of this and, when the line is clear (or possibly after stepping in and imposing some order), tells the sender to try again. However, if the data are divided into smaller packets, the probability of any one of them colliding with another transmission, in the few microseconds it might take them to get through, is smaller than if the originator tried to send a whole large file (perhaps 50,000 bytes), as it might reasonably try to do if it had a dedicated point-to-point line. Thus, even though there is a little more overhead related to dividing the file into packets, the overall *throughput* of such networks can be higher than that with less sophisticated protocols. Indeed, a network with a *communications bandwidth* of two megahertz might actually be able to move more data under heavy loads (i.e., lots of devices competing to send large amounts of data) than one rated at, say, 10 megahertz. The latter, with a less sophisticated protocol, might even break down into chaos if too many devices try to use it at the same time.

In *network polling*, the sending computer or a supervisory computer first checks with the intended recipient to see if it is ready to receive data. Depending on the answer, the transmission can either proceed if ready or wait for the next iteration of polling if not. If the recipient is ready, the sender or supervisor might also first notify all other devices on the network to stay off the line for the time it takes the sending computer to transmit its data to the recipient.

Further protocol issues deal with *verification*, of which the aforementioned parity checking could be a simple kind. Verification techniques can range in sophistication on up to duplicate transmissions (or transmissions back to the sender), which then can be compared so erroneous blocks can be retransmitted as necessary.

These are just a few examples of some network design considerations. Real systems are much more complex and detailed, but exploring further is beyond our scope. Suffice it to say that if an organization is planning to install a network of more than a few nodes, it is well worth hiring some expert and reputable consulting advice. The long-term results will likely be much more satisfactory.

SUMMARY

This chapter has provided a brief overview of the computer hardware technology that is most applicable for a construction professional. Although it is not essential to know all about the "bits and bytes" that have been introduced herein, the more knowledgeable you become, the more effectively you can put the technology to work. Indeed, you are encouraged to explore further in books such as those listed at the end of this chapter.

Just as laypeople would have a hard time conversing in depth with construction professionals about constructing a building or plant without having some of our working vocabulary—including the acronyms and slang—so it is that we need to know some of the terminology used with computers. This chapter began at the most fundamental bit level with the binary on/off, 1/0, true/false nature of computer processing, and worked its way up through bytes and words and on to some of the arithmetic number systems used to represent computer processing.

A look at the components of computers began with the basic design characteristics of the central processor and mentioned some of the many forms they can take, ranging from single-chip microcomputers to supercomputers. Computer memory technologies include RAM, ROM, and variations on these; and different design parameters for memory devices must be matched to the characteristics of the computer processors and the applications for which they are used. Performance enhancements that can boost a computer's capabilities include coprocessors, cache memory buffers, and dedicated I/O processors.

The main peripheral devices used for information storage outside of the central processor include magnetic tapes and magnetic disks. Optical disks are also playing a role, and their use may expand in the future. Peripheral controllers—whether implemented as circuit boards or chips—serve as the interface between a computer and its peripheral storage devices.

Many different types of input devices are used with computers. Two of the most common ones, discussed in this chapter, include the keyboard and the mouse. Keyboards are the source of some confusion to users who too closely compare them to mechanical typewriters. New ways of thinking are helpful when using computers. The mouse, by comparison, seems almost intuitive to most users.

Common output devices used in construction include display screens and printers. There are many different types in each of these general categories. Graphical plotters are also particularly important for communication in construction.

Although applications in the field are not yet nearly so common as in the office, computers and related technologies are increasingly being used for automated data acquisition from, and process control of, construction machines and physical operations in the field. Applying computers in data acquisition and control effectively requires some understanding of transducers (or sensors) and of interfacing analog and discrete phenomena in the field to the digital electronics of computers. Also, one must be able to make the systems rugged enough to survive the rigors of construction.

Computer communications are particularly important in the highly decentralized and mobile world of construction. Fundamental concepts include parallel and serial data transfer, multiplexing several signals into one carrier, and the bandwidth that should be used in different types of applications. Technical terms include baud rate, parity, half- and full-duplex, synchronous and asynchronous transmission, and standards such as ASCII character codes and EIA hardware configurations. Telephone-based communications utilize devices such as modems and a wide variety of private and public network services. Within organizations, local area networks are available using a variety of communication protocols and network topologies. Beyond systems linking a few users, this subject is as complex as computers themselves, and expert consultation is often advisable to establish successful communication systems for construction companies.

REVIEW QUESTIONS

The questions below address concepts and terminology related to computer hardware and data communications.

1. The following two columns list types and characteristics of memory devices for storing binary data. In the blank by each device, write the letter corresponding to the single characteristic that best describes the device. Write no more than one letter per device, and do not use the same letter for more than one device.

Device	Characteristic
___ROM	A—high-speed, primary read/write memory
___WORM optical laser disk	B—removable direct-access storage media
___magnetic tape	C—lowest cost/unit of storage for read-write media
___Winchester disk	D—removable high-density, read-only storage media
___floppy disk	E—instant availability of stored programs
___RAM	F—large-capacity, direct-access secondary storage

2. A computer has 16-bit address and data paths and 128 K bytes of physical memory. Which of the following is the maximum physical memory location that can be addressed with a single CPU instruction cycle?

   (a) byte number 16,385 (from 0) (16K)
   (b) byte number 65,535 (from 0) (64K)
   (c) byte number 131,071 (from 0) (128K)

3. Which of the following statements best describes the *instruction set* of a particular type of microcomputer?
   (*a*) It is a set of machine language program instructions that cause a particular program to perform its intended function.
   (*b*) It is the electronic hardware logic implementation that limits and defines the rudimentary set of operations that the microcomputer can perform.
   (*c*) It is the program steps stored on a disk that tell a computer what to do.

4. True or false: Cache memory serves as a higher-speed buffer area to store parts of a program or data files that are likely to be in immediate use while the other parts are held in main memory or on disk.

5. You are given the following alternatives for computer hardware performance enhancement:

   - Floating-point coprocessor
   - High-speed cache memory buffer
   - ROM-based system software
   - Additional physical memory modules

   Which of these would most improve performance for each of the following situations?
   (*a*) Working with large programs and files that must be subdivided to fit into memory
   (*b*) Relatively small, CPU-intensive, single-user scientific and engineering program execution with lots of decimal arithmetic
   (*c*) An onsite microcomputer controlling a concrete batch plant, where frequent momentary power interruptions and surges force delays to reload the RAM-based system software from floppy disk

6. Let us compare plotters with graphics CRT technologies.
   (*a*) With what kind of a CRT does an analog pen plotter compare most closely?
   (*b*) To what kind of a CRT is a dot-matrix printer most similar?

7. True or false: In a project office environment where there are five microcomputers in different rooms being used for various functions, and no central server, the "bus/multidrop" network topology would be better than a "star" topology for exchanging files and electronic mail.

PROBLEMS AND ESSAY QUESTIONS

1. Assume you are attracted by the design features of a particular critical path scheduling program that runs on several different kinds of microcomputers. In order to maximize execution speed, one design trade-off of this program is that all network data remain in the memory from the time the model is loaded from disk until the user manually gives a save command. Thus network models are limited by the size of memory in the computer.
   Assume you now wish to do a technical evaluation of alternative microcomputers to decide which would be the most suitable for the application. The second and third columns below have specifications corresponding to the hardware design parameters listed in the first column.

Design Parameter	System A	System B	Ratio
Word size (on data path)	16-bit	32-bit	–
System clock (CPU cycles/sec)	12 mHz	24 mHz	–
CPU instruction set	200	300	–
Memory size	1,000,000 bytes	4,000,000 bytes	–
Cache memory for disk I/O	No	16,000 bytes	–
Floating-point coprocessor	No	Yes	–
Memory chip technology	1 megabit RAM	4 megabit RAM	–
Hard disk capacity	80 MB	200 MB	–
Hard disk I/O rate	1,200 KB/sec	2,400 KB/sec	–

(a) Complete the last column of the table by computing the ratio showing the relative performance impact, if any, that each of the given parameters will have on the execution speed of network computations for a large CPM network. Where the ratio cannot be quantified from the given data, write one of the words "High" (over 2.0), "Low" (1+ to 2.0), or "None" for no effect.

(b) Assuming that the operating system software takes 600,000 bytes of memory, the CPM program takes 200,000 bytes, and the data for each activity takes 200 bytes, compute the maximum size of the CPM network that could be handled by each of the two computers.

(c) Which design parameter most strongly affects the time required to start up the CPM program from the disk operating system and load a large network model? If it would fit in memory, how long should it take to load the data for a 500-activity network using System A?

2. Sketch and explain the basic concepts of organization of information and the operation of a magnetic disk storage unit. Include in your explanation definitions of the terms "track" and "seek time."

3. In terms of the technology involved, what are the key differences between computers applied for the direct monitoring and/or control of physical machines or processes in the field and microcomputers used for engineering and administrative purposes in an office? Include in your answer a brief description of the additional elements that a field process monitoring and control system requires and briefly state their function.

SUGGESTIONS FOR FURTHER READING

The following books are suitable for technically oriented readers whose primary field is other than computer science or computer engineering. Most have been designed for introductory courses (Blissmer, Byrd and Pettus, Stallings, Trainor and Krasnewich), but Martin and Chapman are particularly well-known for focusing on those in practice. Blissmer and Trainor and Krasnewich have the most elementary treatment and touch on the broadest range of computer-related topics, Byrd and Pettus focus on computer hardware in more depth but are still fairly concise, and Stallings provides the most depth. Martin has written dozens of computer-oriented books, but *Telecommunications and the Computer* remains one of his best. The book on networks by Martin and Chapman also provides a good, readable overview of that topic.

Blissmer, Robert H. *Introducing Computers: Concepts, Systems, and Applications*. New York: John Wiley, 1993.

Byrd, Joseph S., and Robert O. Pettus. *Microcomputer Systems: Architecture and Programming*. Englewood Cliffs, N. J.: Prentice Hall, 1993.

Martin, James. *Telecommunications and the Computer*, 3d ed. Englewood Cliffs, N. J.: Prentice Hall, 1990.

Martin, James, and Kathleen K. Chapman. *Local Area Networks: Architecture and Implementation*. Englewood Cliffs, N. J.: Prentice Hall, 1989.

Stallings, William. *Computer Organization and Architecture: Principles of Structure and Function*, 3d ed. New York: Macmillan, 1993.

Trainor, Timothy N., and Diane Krasnewich. *Computers!*, 3d ed. Santa Cruz, Calif.: Mitchell Publishing, 1992.

CHAPTER

4

COMPUTER SOFTWARE AND FILE SYSTEMS

We are thinking beings, and we cannot exclude intellect from participating in any of our functions.

William James*

A computer without software is like a skyscraper without tenants—a marvelous piece of technology, but not very useful. It is software that brings a computer to life and defines the functions it can perform. Software, in turn, works with data, which are usually stored in files or databases.

This chapter introduces fundamentals of computer software technology and describes basic methods of information organization and storage. Four major categories to be examined are (1) system software, or programs that control the computer itself and that supervise other programs that perform useful applications; (2) software tools used to develop other programs; (3) data files; and (4) database technologies.

COMPUTER SYSTEM SOFTWARE

Computer *software* refers to programmed instructions written by people and translated into a form intelligible to the electronic logic of computer hardware. Types of software programs include those that manage the computer and its peripheral devices, those that create other programs, and those that are designed for practical applications. The first category is described in this section. Development tools

---

Varieties of Religious Experience, 1902.

will be introduced in general terms in the next section and in more detail in Part IV of this book. Applications software—programs specifically designed for tasks like accounting, estimating, or scheduling—is deferred to Part V.

System software consists of a *program* or group of programs called the *operating system* and *utility programs* for creating, maintaining, deleting, backing up, and moving program and data files and for other miscellaneous system tasks. Program development tools such as text editors, program compilers and linkers, debugging tools, and other aids to programmers are sometimes included under the category of system software, but they will be discussed separately in the next section of this chapter.

Although construction managers need not understand system software in detail, they should appreciate that the success or failure of a computer installation is, if anything, even more dependent upon its system software than upon its hardware. Construction professionals lacking expertise must therefore be certain that experts are involved in the selection of computer hardware and software that best suit the needs of the organization.

Operating Systems

Computer system design is really an integration of a computer's hardware and system software. In many cases, they are designed and optimized simultaneously, with trade-offs in hardware being accommodated in software, and vice versa. In other cases, the hardware is designed to take advantage of existing system software, particularly for widely used standards such as Microsoft's *Disk Operating System (MS-DOS)* or the UNIX system developed originally by Bell Laboratories. Certain types of proprietary system software, such as that used on IBM's mainframes or Apple's Macintosh microcomputers, actually have defined the essence of such computers and have given them their competitive advantages.

The operating system supervises the utilization of the computer's resources. It is responsible for loading programs from storage devices such as disks into memory at the time when the programs are to be executed and in general for handling the traffic between peripherals and the processor. The operating system also divides and allocates the computer's resources among various competing programs and users and protects the users from mistakes that others might make. The operating system also may handle the accounting for the utilization and billing of system resources as well as the security to restrict access to authorized users. Operating systems exist in varying degrees of complexity and functionality; it is important to understand at least a few of the key differences among these systems when selecting computers for specific types of applications.

The simplest type of system control software in common use today is the *single-user disk-based operating system*. Microcomputer users first encountered systems of this type in the late 1970s in programs such as Digital Research's CP/M (no relation to critical path scheduling) for microcomputers based on the S-100 hardware standard, Apple's DOS for the Apple II, and Tandy's TRS/DOS for the Radio Shack Model 80. IBM's **PC-DOS** version of Microsoft's MS-DOS for its popular IBM PC microcomputer in effect made this operating system a de facto standard for microcomputers that "cloned" IBM's design. Although new to most users, these operating systems followed a path blazed some 10 to 15 years earlier by minicomputers, and by mainframes even before that.

The typical single-user operating system contains software conveniently stored on disk. The microcomputer also usually has a small amount of program code stored in ROM that automatically looks for a disk containing the operating system and loads the central parts into memory when the computer is switched on (or *powered up*). This is sometimes called **bootstrapping**, or **booting**, the computer. Other parts, such as code for less frequently used commands, may be loaded from disk only when called for. Early versions had a command-based **user interface** in which the user had to learn to type a couple of dozen or more commands to copy or delete files, initialize new disks, start application programs, check on the amount of memory available, and so forth. Over the years, the number of these commands increased, so developers created graphical **icon** and **menu-based interfaces** (often called **GUIs**, for **graphical user interfaces**), such as those found on the Apple Macintosh or Microsoft Windows system programs, in order to make it easier for most people to use such computers without memorizing numerous cryptic commands. Nevertheless, the underlying functionality is similar; such programs still have the ability to deal with only one user at a time, and predominantly with only one program application at a time.

Moving up to more functionality, *foreground/background operating systems* provide the ability to run a second program in the background during gaps in the use of a primary program in the foreground. For example, one might use a word processor or spreadsheet program in the foreground while a background program supervises printing of a **document** created earlier. People who used to have to wait (or plan to take a meal break) after asking the computer to print a long document have greatly appreciated this added capability, and most of the earlier single-user, single-task operating systems still in use have been extended at least this far.

The next step is a *multitasking operating system* capable of running several programs at once (or apparently at once). Examples are IBM's OS/2, later versions of Microsoft Windows, and Apple's Macintosh system software since version 7. For instance, your spreadsheet program might handle lengthy calculations and your database program might search for a requested set of records while you type a letter with a word processor. While all of this is going on, the computer system software also is working to maintain the fancy **window** displays on the screen, handle disk input and output, supervise printing, and so on. Note that all of these programs are not exactly executing simultaneously since there is still only one main CPU at the heart of such computers, but the operating system does such a good job of allocating hardware resources to each of the programs that it seems to the user as if they are indeed running simultaneously.

The operating systems described thus far are based on the assumption that there is only one human user at the computer, usually with only one keyboard and one display screen. But what happens when several people want to use a single computer at the same time? Here we get into *multi-user **time-sharing** operating systems*. The aforementioned UNIX system is one of the best-known examples; it is offered on a variety of hardware, including microcomputers that one normally associates with single-user operating systems. Other multi-user time-sharing systems function on large mainframes and in proprietary minicomputer systems. A typical multi-user system continuously monitors the activities of all the users and programs working on the computer and, with a complex set of rules and priorities, allocates computer resources in a manner that best tries to serve them all. For example, if you are sitting and thinking about what to write next in a letter you are composing, the operating system just checks on you a few times a second in case you have decided to press a key, but otherwise it moves on to allocate resources to somebody else. Such systems also usually give higher priority to online users directly interacting with the computer than to long-running programs performing computations in the background; this prioritization gives a more responsive feel to the computer. Humans generally interact slowly enough that background programs still proceed with little delay.

Both single-user and time-sharing versions of the preceding types of operating systems sometimes have the capability of ***virtual memory management***. This means that they can manage part of the large disk capacity as if it were an extension of the main memory of the computer and divide up running programs so that only parts of them need be in memory at one time. They *swap* portions of programs (divided into units commonly called *pages*) between disk and RAM memory as needed and thus create the illusion that the main memory is much larger than it really is. The key advantage is an economic one, for disk is a much cheaper storage medium than RAM, but there is a trade-off in performance. When too many programs compete for too little memory, performance deteriorates into a condition called *thrashing*, where more time is consumed swapping pieces of programs in and out of memory than in performing program computations. The cure, if economically and technically feasible, is to install more main memory.

Real-time operating systems can be either single- or multitask in nature. They focus on applications where precise real-world timing is critical, such as running factories, managing laboratory experiments, monitoring hospital patients, and controlling robots. Real-time systems give very high priority to recognizing inputs from interfaces with activities in the real world and to issuing control instructions with millisecond precision. Normal single-user and time-sharing systems are not intended for such applications, so it is important to get the right systems software for applications of this type. Examples include Digital Equipment Corporation's RT-11 (single-user) and RSX-11/M (multi-user, multitask) operating systems.

Finally, we are entering an era when computers with *parallel processors* are emerging from research laboratories, where they have existed for years. In this case there are multiple CPUs, sometimes hundreds or even thousands; and a *parallel operating system* coordinates them and divides and parcels out tasks to be performed in a manner suitable to such hardware. Appropriate applications include large database searches, numerical analysis using finite elements, graphical display processors (e.g., animated CAD), and many others. However, programming for such systems is complex, and they are still at a fairly early stage of development compared with single-processor systems.

Device Drivers

Chapter 3 mentioned the controller circuits that interface between a computer and its peripheral devices such as disk and tape drives, analog-to-digital boards, printers and plotters, and display screens. The specialized pieces of system software that manage the controllers and devices are called **device drivers**. Although they integrate closely with the workings of the operating system, we mention them separately because they usually are added to the operating system only when the corresponding peripheral device is installed.

For purposes of the construction professional, it is mainly important to be aware of device drivers as a necessary and well-proven ingredient when acquiring a new piece of hardware. There have been many disappointed people who, perhaps seeing a dazzling new gadget at a trade show, have bought it, connected it to their computer, and found that it does absolutely nothing. Upon calling the hardware manufacturer, they are told something like, "Didn't you read in the manual that you will have to write your own driver program if you are using your type of computer? If you can't do it yourself, perhaps you could look for a system programmer in the Yellow Pages. Bye!" The main lesson here is that, before you buy any accessory hardware for your computer, you must be sure that it has suitable support software to go with it, that it can be installed easily, and that it works well with your operating system. Even on the same hardware, there may be incompatibilities with different operating systems, or even different versions of the same system, so beware!

Utilities

Utility programs perform routine functions needed in the normal operation of a computer and can be extended to provide useful new functionality and convenience. A basic set of utilities comes with an operating system, and others can be acquired from manufacturers, user groups, or independent software developers.

The basic set of utilities typically performs *disk file management* functions (e.g., copy, delete, rename, and compare files; examine directories of files; create new partitions and directories to subdivide the storage of information on a disk). A convenient file **backup** utility is essential to preserve information kept on disks in case data should inadvertently be deleted or overwritten or if the device should fail. Another important function is *formatting* new disk media (i.e., placing electromagnetic *tracks* on the disk in closely spaced concentric circles, dividing the tracks into *sectors*, and providing **directories** for the information to be stored therein—all in a precise manner compatible with the operating system and device hardware). Other utility commands set the system time and date, alter screen display parameters (number of lines, colors, etc.), assign output devices for printing, and modify usage of memory (e.g., to set up a disk I/O cache).

In addition to single-command functions, most operating systems provide software for writing simple programs consisting of an often-performed series of commands or actions and allow the user to assign a single new command name or menu item to invoke the series when desired, even automatically when the computer starts up. These are sometimes called *batch commands*; a familiar example on Microsoft DOS-based personal computer operating system is a file named AUTOEXEC.BAT, which executes the commands it contains automatically whenever the computer is turned on or reinitialized.

Although the aforementioned functions are part and parcel of the operating system itself, some utilities are supplied with operating systems as separate programs that run like other application software the user might acquire. For example, most operating systems include a simple text *editor* (e.g., EDIT in Microsoft DOS or TeachText on a Macintosh) that provides a minimal ability to create and modify text files (such as a batch command file) or perhaps to read last-minute documentation supplements or corrections supplied only on disk after the computer manuals have been printed.

Other utilities commonly included are of interest mainly to programmers. Examples include error-checking software to *debug* faulty programs while they are being developed, and *linkers* to combine one or more user-written programs with manufacturer-supplied subprograms or others to form a single machine-executable program.

Increasingly, computers include utility software to support communications among computers. Some come with basic network software that enables several machines to share a laser printer and exchange files, and even *electronic mail* software (programs that enable users to address and send messages to each other in a manner analogous to letters through the post) is often included or available as a low-cost option. Electronic mail removes the requirement for both ends of the conversation to be connected at the same time, which is a major advantage given that the average business telephone call takes approximately three iterations of "telephone tag" to complete. Other operating systems include *terminal emulation* software that enables a person to telephone computers of other users or gain access to information utility firms to get stock quotes, publication references, airline schedules, estimating cost data, and the like and *download* (i.e., copy from the remote computer to yours) information of interest. This whole field is developing rapidly, and construction managers who wish to remain competitive must stay up with the technology.

For any widely used brand of computer, third-party software developers soon make available on the aftermarket innumerable utility programs. Some further enhance the operating system (e.g., substitute faster disk-access routines or make more screen display colors available). A popular type of utility is *diagnostic software* that helps the user to troubleshoot possible hardware or software errors in the computer and possibly save a trip to a repair shop, or just to learn more about what makes the machine tick. Others offer some useful new function. (For example, *screen savers*, after a user-selected number of minutes since the last keystroke or mouse click, substitute a predominantly dark and dynamic graphic display in place of the bright static display inadvertently left on the screen. The bright static display could otherwise cause screen phosphors to deteriorate, or even etch a pattern such as a menu or spreadsheet, if left on too long.) Some utilities add a touch of whimsy, such as screen savers that look like a pyrotechnic display or colorful tropical fish swimming in a bowl, and some are just plain corny, such as the *sound generators* that some users install to make weird noises at various stages of normal—or perhaps otherwise tedious—computer usage. In a free and creative market, even needs that nobody dreamed they had will be satisfied by some enterprising developer. If a utility provides real value, or just a good laugh, you will probably hear about it over the computer user grapevine.

APPLICATION DEVELOPMENT SOFTWARE

Until the past decade or so, developers created most application software with tools called procedural **programming** languages or occasionally with native machine-oriented assembly languages designed to implement a specific computer's hardware instructions. These remain prevalent for professionally written application software, but the rapid evolution of microcomputers has also spawned the growth of user-oriented tools such as programmable **spreadsheets** and **databases** that can be applied more easily to develop custom standalone applications as well as to solve the problems for which they were initially created. More recently, software evolving from research in the computer science field called *artificial intelligence* has entered the commercial marketplace in the form of *expert-system shells*, *knowledge-based systems*, declarative languages, and others; these have opened whole new areas of application development.

Most of the development tools just mentioned are explored in more detail in Part IV (Chapters 8 to 11). The main focus there is on describing specific, widely used products and showing how they can help create practical applications. As background, this section gives a conceptual overview of these categories of tools (what they are and how they work) and indicates their relative advantages and limitations.

A Hierarchy of Development Tools

One can define a hierarchy of software development tools that, at the upper end, is closest to machine hardware in speed, flexibility, and compatibility and at the lower is intelligible and easy to use for human beings. In Table 4-1 some of these differences are summarized. The indices shown are rough; they mainly indicate the nature and magnitude of the differences. Specific instances may differ considerably from these at any level.

In Table 4-1, the *index of cost and difficulty* refers to the level of effort that it takes to implement an equivalent application using each of the different approaches. The *index of efficiency* shows that there is a degradation in the execution speed and memory requirements of the resulting program. *Application flexibility*

TABLE 4-1
Hierarchy of development tools

Type of tool	Index of Cost and difficulty	Index of efficiency	Application flexibility
Machine language	1000	1	Maximum
Assembly language	100	1	
Compiled procedural language	10	2	
Interpreted language	3	~ 10–100	
Generic-problem software	1	~ 20–200	Minimum

reflects the variety of applications that may be handled using the given approach (e.g., anything the computer can do using assembly language, but limited to the type of problem suited to a generic package such as a spreadsheet or database package program). The category types are described in the following subsections.

MACHINE LANGUAGE. At the lowest level of the hierarchy we have what is called *machine language*—the arithmetic and logical manifestation of the instructions implemented in the actual electronic circuitry of the computer. To a human, machine code typically appears as binary numbers. The native *instructions* of the machine (add registers, compare two values, complement a number, access data at an address in memory, etc.) may be defined in from 40 to 400 or so instruction codes that can be defined with a combination of unique bit patterns in a byte or word of storage and a systematic positioning of the *arguments* (e.g., memory addresses or items of data in CPU registers). For easier human memorization and readability in program listings of these codes, they usually allow octal or hexadecimal notation in place of binary numbers. Experienced machine language programmers memorize these and work with them; in the early days of computing, these codes were about all they had for a programming language.

The following is an example of a machine language instruction for a Motorola 6800 microprocessor, which is a simpler 8/16-bit ancestor of the 32-bit 68000 series microprocessors widely used in microcomputer systems today:

	Instruction code	Argument
In binary	10001011	00000101
In hexadecimal	8B	05

This instruction means to "add immediate" (code 10001011, or 8B, stored in first byte) the constant 00000101 (or 05, stored in second byte) to the contents of Accumulator A and store the result in place of what was originally in Accumulator A. Printouts of programs written this way could run on for pages of dense numbers; even experts could spend weeks trying to locate errors causing such programs to misbehave. Today programming at this level would be enormously tedious and expensive, but it is useful for the professionals called *system programmers* to be able to read output that appears in this form. For the purposes of the construction professional, however, it is sufficient just to know that this is the most rudimentary level of programming. One can take solace in the knowledge that even most computer professionals cannot understand it either. They also look with awe at the system programming wizards who do.

ASSEMBLY LANGUAGE. The next level of programming down from computer machine code and toward human understanding is ***assembly language***. Assembly language is still specific to a particular machine-hardware design, but it uses mnemonic codes made up of a few letters and possibly some symbols (e.g., #, %) to express the equivalent numeric machine codes. For example, 8B in the preceding accumulator addition has the mnemonic code *ADDA*, which is much clearer to a human, on the Motorola 6800.

For the most part, mnemonic instructions in assembly language match their numeric equivalents in machine code one for one. If programmers are good at it, they can also write the fastest and most memory-efficient programs this way—as well as can be done directly in machine language because of the one-to-one mapping to machine code. Because it is still specific to a particular hardware, however, software written for one type of machine must be totally rewritten in a different assembly language if it is to be reimplemented for another type. In computer terms, we say that such software is not *portable*. Also note that the computer cannot directly execute the assembly language as written by the programmer (called ***source code***). First, a program called an *assembler* must check the programmer's writing for correct ***syntax***, then translate the mnemonic assembly language ***statements*** into their equivalent machine language codes. If the programmer has referenced any previously written segments of code (such as a print routine), the program called a ***linker*** must then combine the various units of code into a single machine-executable program. Thus, having an assembly language that makes it easier for the programmer moves us one or two steps away from the code that is understood by the computer.

An incremental step beyond assembler language is a ***macro assembler***. It works with the same mnemonic codes as the basic assembler and similarly produces machine-specific code. However, it enables programmers to create *named* ***procedures*** containing sections of assembly language that might be called repeatedly. Having done so just once, the programmer can then refer to this procedure by its name, perhaps with appropriate arguments, whenever its function is needed again. Manufacturers supplying system software with macro assemblers usually include a well-written library of common macro procedures, such as mathematical functions, and sorting and print routines.

COMPILED PROCEDURAL LANGUAGES. In the late 1950s and early 1960s, the limitations of machine-specific programming techniques stimulated development of ***procedural languages,*** which can be implemented in forms to be either *compiled* (i.e., translated) to machine code or *interpreted* directly as the computer executes the program. This subsection will concentrate on languages normally designed to be compiled, and the next subsection will address languages normally designed to be interpreted.

Two of the first and most successful compiled procedural languages were *FORTRAN* (FORmula TRANslation) and *COBOL* (COmmon Business-Oriented Language). Important subsequent examples include *Pascal*, *C*, and *Ada*. These and other languages in this category are described in Chapter 8. The focus here will be more on the general nature of such languages. Basically, they offer key words and a syntax that enable a programmer to write in a way that looks more like the language of the application (mathematical formulas, English words, etc.) than that of the computer. Also, program listings are much shorter and easier to follow than if the program were written in assembly language. The programmer normally uses a text editor to create the source program and then uses a program called a *compiler* to translate the source code into machine code.

The compiler's function is somewhat like the assembler for assembly language; that is, it checks the source code for correct syntax and then translates it. However, there is not a line-for-line equivalence of the source code with the machine code. Indeed, one line of code in the procedural language (such as a trigonometric formula or a printer output statement) might translate into a dozen or more lines of machine code (or its assembly language equivalent). In assembly language the programmer might have to write his or her own routine even for a common operation like division if it is not available in the machine's instruction set, and certainly would have to do so (or copy from libraries of previously written routines) for more complex mathematical or logical functions. In procedural languages, however, most of the functions needed in applications for which the language is designed are built in and can be called upon with a short keyword and perhaps a list of arguments.

The machine-readable output of the compiler's translation is called *object code* or *object module*. This can be combined with other code using the same linker program mentioned for assembly language to produce the final machine-executable program that performs the intended application. Once successfully compiled and linked, these steps do not have to be repeated each time the program is run. The machine-executable module can be copied and run on other computers of the same type without needing the source code or the compiler and linker programs.

One of the major advantages of compiler-level languages is portability. Source code that adheres closely to recognized standards for a given language should need few if any changes when it is moved to another type of computer. Instead of reprogramming, one just recompiles the source code with the compiler designed for the new computer and its operating system, relinks with the new linker, and the program should run. If only that were the case! Although there are indeed precisely defined national standards for languages like FORTRAN and COBOL, manufacturers respond to programmers' desires for extensions that take advantage of characteristics of particular hardware, and portability is thus compromised. For newer languages, several dialects may be in use while discussions drag on about what features and syntax should become a standard. Even so, good compilers point out most of the differences when code from one machine is recompiled for another, enabling programmers to focus on the relatively small parts of the programs likely to need changing.

Because a given procedural language may not take advantage of all the hardware instructions available on a particular computer, or because a particular compiler for a certain machine may not translate the procedural language's statements into efficient machine code, programs written at this level usually do not run as fast as, and take more memory than, the same program if written in assembly language. Sometimes compiler programs are hastily and inefficiently written, perhaps for marketing reasons to expand the range of languages said to be available for a machine; their output may run 10 times slower than the equivalent programs well-written in assembly language. However, a very good compiler, one that optimizes well for the target hardware, often produces results that come close to assembly language written by an expert. For popular computers and languages, competition among third-party compiler developers also pushes products ahead for more efficient and cost-effective versions. They are thus shown with an overall efficiency factor of 2 in Table 4-1. Flexibility for implementing varied applications is also close to that of assembly language programming.

Because such compiler-based procedural languages (and the following interpretive versions) are written much closer to the human's understanding of the application problem, they are sometimes called *high-level* languages (implying, one hopes, that humans are still on top of things in a world being infiltrated by computers). They are also called *third-generation* languages, representing the third major advance in programming after machine code and assembly language. In terms of ease of use by normal human beings, however, they remain fairly difficult, so costs and development effort rank somewhere in the middle of Table 4-1.

INTERPRETED LANGUAGES. Interactive computers began evolving in the 1960s; with these computers the user works directly with the machine by using a terminal rather than by preparing a batch of cards in advance to be input at a later time. Two computer scientists at Dartmouth College then conceived the idea that an interactive programming language designed to take advantage of the characteristics of interactive computers would work well for teaching. Professors John Kemeny and Thomas Kurtz thus developed *BASIC* (Beginner's All-purpose Symbolic Instruction Code). Other languages that have evolved since include a cryptic mathematical one called *APL* (A Programming Language), a language with a small but dedicated following called *Forth*, a graphically-oriented teaching language called *Logo*, and some versions of the two non-procedural languages used in artificial intelligence—*LISP* and *Prolog*. We will return to some of these in Part IV, and will concentrate here on just general characteristics of interpretive languages.

Interpretive languages offer several advantages to developers. First, the text editor used for creating the code is usually built in. If the programmer makes a syntax error when typing a line of code, the interpreter points it out immediately. Second, if the programmer wants to see what happens when the code typed thus far executes, he or she can do this immediately. In contrast, when one is using a compiler the errors are not flagged until the program code—itself usually prepared with a separate text editor—is complete enough to make sense when compiled, nor can the operation of the program be tested until it has been compiled and linked. Since these are separate and often time-consuming steps, one does not invoke them more often than necessary. With interpretive languages, however, there is more of an incentive to explore alternatives—to use trial and error, if you wish—and one learns quickly from mistakes. This feature facilitates teaching a new language to novices.

With an interpretive language, the source program is never fully translated into a contiguous set of machine instructions. Rather, program statements and data serve as input to the ***interpreter*** program, which in turn carries out the specified functions in the logical order of the user's program. Lines of source code are interpreted and executed only when encountered; they are not precompiled. If, perhaps in an iterative loop through a section of a program, statements are encountered again, they are reinterpreted every time they occur. Furthermore, interpreter software must always be present to run programs written for an interpretive environment. In contrast, compiler-based programs are free of the software used for development once they have been compiled and linked successfully.

The obvious advantage of interpreters over compilers is that development proceeds more rapidly and easily, so in Table 4-1 there is a correspondingly improved index of development cost and difficulty. The penalty, however, is that the code usually runs much more slowly, perhaps by a factor of 10 to 100 or so, than precompiled machine code. This slowdown occurs mainly because of the repeated and temporary translation of each line of code every time it is encountered and because the interpreter is the program actually running at the machine level; the application program only runs indirectly through the interpreter. This factor makes interpreted languages much less suitable for time-sensitive applications (e.g., those involving dynamic graphic CRT displays, process control, etc.) or for those handling large amounts of data or complex calculations. Memory requirements, however, can go either way, depending on the combined size of the interpreter and program code versus the size of the programs and all the library routines that are linked into a compiled program. Flexibility for developing a variety of applications is not inherently limited by the fact that a language is interpreted (except for time-sensitive applications), but some interpreted languages are indeed limited in flexibility.

Before leaving procedural languages of both the interpreted and compiled varieties, note that the name and syntax of the language do not necessarily limit the way it executes—in compiled or interpreted form—on the computer. There are indeed compilers for languages like BASIC and APL, and there are interpreted (or partially interpreted) versions of languages like Pascal. Indeed, one can have the best of both worlds. For example, as the language features of BASIC evolved to handle major applications in science, engineering, and business, BASIC compilers were developed to speed up the execution of the increasingly large and complex programs. The programmer today can have the ease and speed of code development and testing in the interpretive environment, and then compile and link the completed program for efficient and standalone execution (without giving away the source code). Developers of numerous custom business programs work in just this way.

GENERIC PROBLEM-ORIENTED DEVELOPMENT ENVIRONMENTS. Numerous tools fall under *generic **development environments***. These are self-contained software development and implementation systems oriented toward certain categories of problems. They include the spreadsheet and database programs that have become familiar in almost every office environment and also problem-oriented languages for simulation, engineering applications, business data processing, and others. Certain artificial intelligence software and expert-system shells for developing qualitative and judgment-oriented applications could also be included here.

Generic tools are well suited for ease of software development and utilization in their intended areas of application, but they are less flexible than procedural languages for developing a wide range of applications. For example, spreadsheets serve well the types of tabular modeling, computations, and reporting found in many business and engineering applications, but, though usable in some cases, they are clumsy at word processing, database management, or project scheduling. Database programs are useful in creating applications designed for managing large files of information and for preparing reports based on the information contained therein, but most have limited computational abilities. Generic simulation software can model a wide range of real-world phenomena but are of limited use for routine calculations or for information management. Similar limitations apply to most expert-system development shells.

The appropriate applications and limitations of spreadsheets, databases, and expert-system software are explored in some detail in Chapters 9 through 11 of Part IV, so we will not continue with them here. Before we leave development tools, however, it is worth mentioning a few other types that have many practical applications but that will not receive separate attention later in this book. These include program generators, problem-oriented languages, and mathematical modeling environments.

Program generators are a type of development software used in a manner whereby the programmer describes at a fairly high level the type of information files the application will require; the computations, sorting and searching to be performed with that information; and the appearance of input screens and printer reports. Examples include RPG and BRADS, developed by IBM for smaller business computers back in the 1970s, and Borland's ObjectVision for PC-based Windows. As an example of high-level interaction for program specification, a report might be described, in effect, by moving a cursor around on the screen, typing in the text for headings and lines to divide rows and columns, and showing where various types of computed output fields are to be displayed. The structure of information to be stored in files could be described in a similar form. Desired computations could be expressed in terms of simple mathematical or English-like statements. The program generator software then takes all of these high-level descriptive specifications as input and literally translates them into a programming language, such as business BASIC. That code, in turn, can execute as an application program on the host computer, be modified at the level of the procedural language, or be compiled and duplicated for distribution to various users in an organization.

Problem-oriented languages provide the user with a limited set of keywords and a syntax for writing programs that focus on certain application areas. The keywords usually reflect the vocabulary of the intended application, and the syntax is simple enough for users who know their application area to use with little added knowledge of computer programming. Perhaps the most familiar example related to construction is the Integrated Civil Engineering System (ICES), and its various subcomponents such as COGO (for surveying and earthwork computations), PROJECT-II (for CPM schedule calculations), ROADS (for highway engineering), and STRUDL (for structural design). ICES was originally developed at MIT in the 1960s, but several of its components have evolved and are still available from commercial software firms. MicroCyclone, a problem-oriented language for simulation, will be described in Chapter 15.

Mathematical modeling environments enable not only mathematicians but also scientists, engineers, and business analysts to build complex mathematical computer models almost as if they were writing the equations on paper. The software solves not only numerical problems but also those dealing with symbolic algebra, trigonometry, calculus, and statistics. Examples include Wolfram's Mathematica and MathCAD by MathSoft. Most software in this category also has excellent graphical capabilities, showing, for example, three-dimensional contour plots of complex mathematical functions. This capability makes it much easier to explore and reason about what otherwise might be impenetrably difficult and complex model behavior.

What all of these generic problem-oriented tools have in common is that they make it much easier to create applications, build models, and solve problems in their intended areas of application than it would be to do so with traditional programming languages. Indeed, in most cases users probably would not even try to tackle these problems if it were not for the availability of these new tools; it would be too difficult, expensive, or time-consuming relative to the value of the results. With the new tools, users are moving into all kinds of new areas with a degree of analytical and computational ability that was previously unknown on such a broad level. There already have been enormous benefits to business, industry, and the professions.

The main disadvantage of such tools is their limited scope, but taken as a whole they are tackling most areas of importance. Other disadvantages relate to compatibility between tools when one tries to build applications that involve two or more of them (e.g., a spreadsheet, a database, and a simulation program), but developers are working to provide better linkages at least among the more popular tools. Portability is also limited to the computers on which particular tools are available. The speed and efficiency of applications relative to procedural languages are variable. Some are very slow or take up enormous amounts of memory even for smaller problems, but in many cases they are at least as good as interpreters, and some can approach the speed of compiler-level languages. In fact, compilers have been developed to produce direct machine-executable code for applications developed with popular spreadsheet and database packages. Thus the index of efficiency in Table 4-1 becomes very case-dependent as you near the bottom.

Support Software for Program Development

At this point it is appropriate to digress from the hierarchy of software development tools to examine some other software that supports the development process. Although a construction crew could build a house with a few basic hand tools, work goes faster and is more effective if they also skillfully employ power tools and specialty tools to get the job done. Similarly, when one is developing software with assembly language or compiler-level procedural languages, there are numerous auxiliary tools whose availability and skilled application significantly influence the programmer's productivity and quality of work. Although few construction professionals need to know how to use such tools, it is important to be aware of their value and be sure that they are available—and used well—if hiring programmers to implement custom applications for a construction company.

As a minimum, traditional program development software includes text editors that enable a programmer to type a program or data into a computer for initial development and modification as well as compilers and linkers that translate the source program into the machine code that is understandable and executable by the computer. Even at this basic level there is a range of quality. Some compilers and assemblers produce compact and efficient code, and others do 10 times worse. The basic text editors included with many operating systems are often limited in functionality and inefficient to use beyond a few lines of input. Better *programming environments* include powerful editors that not only make it easier to create, search, and manipulate text but might also automatically format code, according to conventions of the programming language in use, and perhaps even do some checking for syntax errors. A good programming environment also keeps track of which program modules have been modified since they were last compiled, and avoids recompiling before relinking modules when it is not necessary.

Some compilers and assemblers make available varying degrees of *code optimization,* usually with a trade-off in speed of recompilation. While code is under development, optimization will be left off to speed development time, but as the code nears completion the availability of optimization can make it faster and/or more memory-efficient. Similarly, linkers may offer features such as the ability to *overlay* portions of the application code from separate modules kept on disk, which enables larger programs to run where memory size might otherwise be insufficient.

Beyond the basics, other useful program development tools include *libraries* of common **subroutines** and **functions** for operations such as input, output, and mathematical calculations. In quality systems, these library functions and subroutines implement the best-known methods for performing their respective tasks. Application programmers who know what they are and how to use them can avoid the time and cost associated with "reinventing" this code. Examples include not only mathematical and statistical functions but also procedures for **sorting** and merging data, complex file management, and many others. Related to this tool is *librarian software* that enables the programmer systematically to create and access libraries of his or her own program routines that have been successful and that could be useful again in other programs.

Diagnostic software* can help programmers locate arithmetic and logical errors in the program code (a process called *debugging*) and verify that it runs correctly. Lacking such software, the programmer is left to read long listings in search of errors and use trial and error to track down some problems, both an expensive waste of resources. Diagnostic software can offer many functions, such as displaying the values of program variables at various stages of execution and producing traces of the execution logic. Diagnostic software can also enable the programmer to single-step through portions of the program in exactly the sequence each line executes to spot unexpected changes in values or branches in logic just when they occur.

Other software supports the important function of program **documentation**, that is, explaining how the program works so that others can understand how to fix it, expand it, or convert it to another computer in the future. Important information provided by documentation software can include tables of cross references to all variables and subroutines used in the program and *flowcharts* of program logic.

Cross-compilers* and *cross-assemblers* enable a programmer working in one hardware and/or operating system environment—typically one that is powerful and offers many programming conveniences—to develop programs and convert them into machine-executable modules for another environment. For example, a programmer working on a workstation might develop code to be stored in ROM memory for a microprocessor to be used in monitoring or controlling some functions in a crane. The latter computer would probably have no software development facilities at all of its own.

One last category of program development aids to be mentioned here is *code optimizers*. These programs monitor and analyze the execution activity of programs that are under development and near completion. They report to the programmer how much time the program spends in various parts of its code, they spot sections of code that repeatedly execute unnecessarily (such as arithmetic with constants contained within an iterative loop), and they provide other information that suggests to the programmer where to make modifications or best spend time fine-tuning the code for better execution. Good assemblers and compilers also do some of this optimization (such as taking arithmetic with constants out of loops), but optimizer software usually goes much further and provides far more statistical information on the behavior of the program when it is executed with its real-world application data.

Although this brief overview of software tools for application development is now complete, considerably more information on several of the tools is given in the four chapters of Part IV. Also important, given a particular type of application need and the organizational resources available to meet it, is how to choose the appropriate development tools from the huge variety available. Part III, especially in Chapter 6, has more to say on this subject.

FILE STRUCTURES FOR INFORMATION ORGANIZATION AND STORAGE

In most computer applications, the associated data are kept separately from the programs that process it. Physical locations of the files of information can be *online*, meaning that the device containing the information is continuously attached to the computer and ready for access, and *offline*, meaning that the information may be contained on tape cassettes or disk packs stored away from the drives on which they can be mounted. Online storage is used for information needed on a continuous basis, such as the files for an airline reservation system, whereas offline storage serves for archiving annual records, backing up online files, and so forth.

There are many ways of organizing and storing information; the most suitable in a particular case depends upon the nature of the application. Two very general categories include simpler *data files* that contain records of information, each similar in format, and more complex *databases* that might provide integrated access to interrelated blocks of information that can be viewed from different perspectives. This section addresses some of the basic file structures and introduces terminology common to database systems as well. The next section will focus separately on databases themselves.

File processing and database management systems can be built into software that is specifically designed for custom applications, such as a file for a scheduling program or a database for an estimating system. Examples of files and databases are included in Part V of this book. Alternatively, one can acquire generic file processing and database management *package software* and use their development tools to custom-tailor files, control menus, input screens, and output reports to fit a specific application. Chapter 10 shows how such generic tools can work for specific construction applications.

The Elements of File Structure

At the lowest level in a given application, we identify the various *data items* with which the application works. For example, in Fig. 4-1 are some items that might occur in a critical path scheduling (CPM) application. These include inputs to the schedule calculations, such as the activity labels, descriptions, durations, and the labels of their following activities, and also the outputs of CPM calculations,

FIGURE 4-1
Elements of data in CPM scheduling.

such as the early and late starts and finishes for each activity (ES, EF, LS, and LF) and their total and free floats (TF and FF).

As a first step in organizing the data items into a file, the application designer will specify how a collection of related items for a single entry could be combined into a *record*. In Fig. 4-2 we see how the preceding CPM data items might be placed in a record for a single CPM activity. Each data item is put into a separate *field*. In addition to specifying the *sequence* of the data fields, the programmer must decide the *size* of each field (e.g., number of characters or magnitude of a number it can hold), whether that size is *fixed* (e.g., 24 characters for the activity description) or *variable* (as many characters as desired, perhaps with some special character designating where the description ends), the *type* of data stored in the field (e.g., text, integer or floating point numbers, or a true/false value), and even whether there is to be a fixed or variable number of fields for all or certain types of data (e.g., for the labels of the following activities—limit them for example to 5, whether needed or not, or somehow put a marker at the end of the last follower, no matter how many). The file designer may decide not to store some items at all, and just compute them on demand instead. For example, part of the CPM output results might be computed when needed, especially the free float, which is just a simple subtraction between two other elements that already might be stored (LS − ES or LF − EF).

The next level up in the information organization is a *file*, which is a collection of related records organized in a manner whereby the computer software can recognize that they belong together. In the example, it could be the collection of all of the records defining a particular critical path schedule, such as that shown schematically in Fig. 4-3. In addition to the activity records, there might also be one or more *header records* specifying the name of the file and other descriptive information pertinent to the application (e.g., project name and location for this schedule) and perhaps the date of file creation or last modification and the number of records it presently contains.

The tabular nature of such simple file structures has led many people to use the term *flat file* to describe them. They look like rows and columns laid out on a piece of paper of an arbitrarily large enough size to accommodate the width of the records horizontally and the number of records vertically. Internally, on the

Label	Duration	Description	Fol1	Fol2	Fol3	Fol4	ES	LS	EF	LF	TF	FF

FIGURE 4-2
Record structure for a CPM activity.

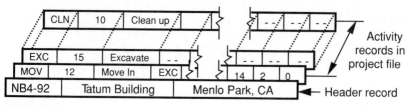

FIGURE 4-3
File of CPM activities for a project schedule.

electronic media of a disk, the sequence of records might differ radically from the neat tabular order that flat file implies.

File Organization and Utilization

Considerable thought goes into designing a file organization most suitable to the anticipated use of the application software. The organization helps determine the means for creating, processing, selecting, and extracting information, and it can also facilitate or frustrate file maintenance.

Certain fields containing data elements of the same type might be designated as *keys* for sorting or searching a file. In the example, the activity labels and the early and late start and finish times might be thus specified and perhaps also the total floats. The labels could be used to sort records into alphabetical order. Sorts according to the early and late start and finish times may be needed by the program's specified calculation procedures. Total float could help sort out all of the critical activities (those with zero float). The keys and the other elements of the records also constitute *attributes* of the records, the names of which might be thought of as column headings if the records were organized into a table.

When the organization of the file is known, programs can be written to perform actions with the data with varying degrees of efficiency, depending on the organization. For example, a telephone book has a name, address, and telephone number as attributes in each of its records. Most telephone books are organized alphabetically, first by last name, then by first name and perhaps middle initial. Addresses and telephone numbers, though they might be what you are looking for, are carried as non-key attributes. If the telephone book were available as an electronic file, it would be easy to write a program that could search rapidly through the file for a given name and report the address or telephone number for that person. However, it would be much harder to take a given address or telephone number and find the corresponding person; the file is not well organized for that purpose. For use by the telephone company and emergency services, inverted telephone books do exist where one can do what we have just said is difficult, but in doing so they have used another of the attributes as the primary sort and search key. Finding the telephone number from a switchboard connection, they can quickly locate, for example, the address from which a distraught child is trying to report an emergency. One can achieve an equivalent function electronically by creating separate *indexes* to the same file information, where the computer program looks first to the index to locate the particular record it desires.

In routine business data processing, certain names are given to files that serve different purposes. In an ongoing application such as payroll or accounts payable, there will most certainly be a *master file* that keeps an updated record of the current status of each employee or supplier. In payroll, each record would have a name, address, social security number, and year-to-date earnings, taxes, and deductions. For interim information processing related to each master record, a *transaction file* will be accumulated until it is again time to update the master file. In the payroll case, transactions would be timecards for each employee in a given pay period; these would be used not only to produce a paycheck but also to update the master file. The transactions might be retained online for a short period in case questions arise and then be discarded or written to an *archive file* for offline storage. The master file, however, remains online for access to up-to-date information when needed. A *report file* is usually very transient in nature and retains the output of a program only until the comparatively slow mechanical printer finishes printing it. Similarly, a *sort file* holds information from an interim stage of data processing while records are rearranged into a new order for searching or reporting.

Within conventional file systems, data can be organized **sequentially** or for **direct access**. *Sequential* means that, to get to a particular item of information in the file, it is necessary to search through information that precedes it. Data stored on tape is inherently sequential, but disk files can also be organized this way. *Direct access* means that the computer programs can instruct the storage device to go directly, or perhaps in just a few steps, to the location that contains the desired information. Applications that need quick reference to items of information in an unpredictable order need direct access files. For example, an equipment-dispatching system needs to be able to respond promptly to any foreman or superintendent making an inquiry about any machine. Sequential systems better suit well-organized files of information processed on a periodic basis, such as a weekly payroll. There are also compromises between sequential and direct access methods, such as the *indexed sequential access method* (ISAM).

SEQUENTIAL FILE PROCESSING. An example of traditional sequential file processing appears in Fig. 4-4. Think of it as a weekly payroll run. First, a program checks the week's transactions for data input errors (e.g., an employee badge number does not match a name, a carpenter coded on a pipefitter wage scale, or 400 hours of work entered instead of 40), and sends the rejects to an error file or directly to the clerk's screen for correction or reentry. Once the records are correct, they are next sorted based upon some key, most likely one that corresponds to the record sequence in the master file. The payroll processing program then matches transaction records against corresponding master records (perhaps by social security number); where there is a match the appropriate calculations update the record in the master file. Information for printing a paycheck and other weekly reports will probably also be sent to a print file at this time. There is not necessarily a transaction record for every record in the master file. In payroll, employees will be retained in the master file for at least a full year for income tax reporting, but, owing to the frequent hiring and layoffs in the industry, certain

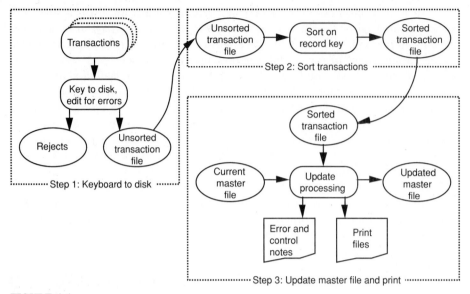

FIGURE 4-4
Sequential file processing.

employees may not have been working (and thus creating transaction data) in the past week. However, in most applications a new record in the master file should be created before any transactions can take place.

Advantages of such sequential file organization are that it is simple to design and maintain, it is efficient for handling many transactions together in a *batch* mode, and it is suitable for low-cost tape media. Depending on the way the media are used, the previous tapes or disks can also serve as automatic backups in case the current master file is damaged. Then one only needs to reprocess the recently sorted transactions against the previous master file to recreate the current master file. Disadvantages are that one normally must search the whole file up to the point where the desired record is located, transactions must first be sorted before processing, and the master file is only up-to-date to the last period when the transactions were run. It can be difficult in such systems, for example, to extract the needed information quickly to produce a paycheck if an employee is terminated in the middle of a pay period, or to find out how a given cost account is doing this week.

DIRECT ACCESS FILE PROCESSING. Direct access files have become increasingly prevalent in business data processing. Decreasing costs of high-speed disk storage and of powerful computer processors have made the efficiency advantages of sequential processing less of an issue; numerous advantages of direct access files offset these anyway. They do need a type of storage medium that facilitates direct access to selected records; most common are magnetic and optical disks, but some types of nonvolatile memory devices also serve, such as the credit-card-size memory banks available for some portable computers.

Many schemes exist for storing and determining the location of records on disk, but most involve some form of *index* that associates at least the beginning of a series of records with a physical location on the media, and perhaps *pointers* from one record to those that relate closely to it. The records need not be contiguous, and indeed the files do become *fragmented* as new records are added and old ones are modified or deleted. Where efficiency of a file organization scheme is enhanced by a physical order to its records (e.g., to do sequential processing or to accommodate the mechanical characteristics of disk drives), then either automatically or upon request the application program or a utility might periodically *resequence* and *compress* (i.e., reclaim the small fragments of disk space where records were deleted) the contents of the file.

Figure 4-5 is a schematic example of direct access file processing. The master file is online all the time and is available for inquiries from terminals or other programs. Transactions can either be handled immediately (such as the aforementioned employee termination) or be batched for more efficient periodic processing. Compared with sequential processing, these systems stay more up to date, and inquiries and reports reflect the latest information. Another advantage is immediate access to any desired data. Disadvantages are that programs without good error checking on inputs can allow users inadvertently or deliberately to corrupt the information in the files. Backup must be planned and scheduled carefully because it is not likely to be an automatic by-product of this method of file processing. These systems can be more complex to program, the files usually take more space than sequential equivalents, and applications best suited to sequential processing may run less efficiently this way. Also, as business procedures become more and more dependent on such online information systems, things grind to a halt when

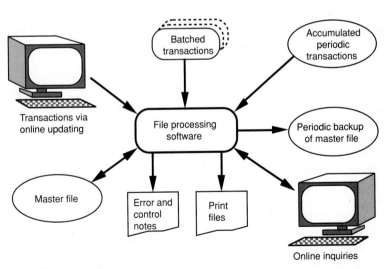

FIGURE 4-5
Direct access file processing.

the systems are unavailable. The most common bureaucratic excuse of the late twentieth century seems to be something like, "I'd really like to help you, but the computer is down."

INFORMATION ORGANIZATION AND STORAGE IN DATABASES

The term *database* sometimes is used loosely to mean the collection of files contained at a particular computer installation. However, in the data processing profession today, ***database*** refers to a complex system of software and hardware that integrates the various files for a number of different applications into one commonly accessible pool of information. This ***integration*** makes it possible to reduce the duplication of items, such as the names of people needed by both personnel and payroll applications, and also to provide for direct ties between different applications, such as materials management and accounts payable.

The evolution of the need for database systems parallels that of organizations. At the start of a business enterprise, all of its pertinent information conveniently resides in one place: the mind of the entrepreneur-founder. As a business grows and adds more people, it divides into specialty functions and departments, and these may develop local information systems to serve their own needs. However, inconsistencies in information content and differing priorities can lead to misunderstandings and even conflict among the departments and people, and management may have difficulty pulling together information needed to make the best decisions for the overall good of the enterprise. There will also be duplication of data entry and reporting functions if they are handled in separate file systems because there are many overlaps in needs. For example, construction equipment information is needed in estimating, field operations, and cost control as well as in the equipment department. By moving beyond independent file systems to an integrated database, we can overcome many of these problems and restore some of the closer working relationships, even in larger organizations, that existed when they were smaller.

Designing a database system concurrently with rethinking the allocation of tasks performed in an organization leads to several advantages. First, although responsibilities for various elements of data entry may be ***distributed***, any one type of data item need only be entered once. Second, consistency and synchronization of data in various applications become automatic when data are shared across the various applications whose update status might otherwise be out of phase. Providing a reasonably consistent user interface for interacting with various applications in a company facilitates training and productivity. Finally, since all of the relevant data for an organization are in one system of interrelated files, which may be logically independent of the applications accessing them, it can be easier to add a new application, to make inquiries, or to produce hybrid reports that draw from several different departments.

Such information concentration has disadvantages, of course. Erroneous data input simultaneously impacts the information used by several applications, not just that of the department responsible for it. With integrated database information systems much more attention must be paid to the software that screens and checks input data to be sure that they are reasonable and accurate. Similarly, in the face of deliberate fraud, the most vulnerable systems (e.g., payroll and accounts payable, which print checks) could become accessible to more personnel than those who are screened to work in these areas. More attention thus must go into implementing software and procedural security systems regarding who has access to what programs and files so as to exclude those who are not authorized. Finally, with "all the eggs in one basket," when the computer information is down, all departments are affected simultaneously. Backup procedures become critical for the whole database and its related software, including off-site storage, and contingency plans may even be needed for using an alternative computer elsewhere to keep the organization going in case a fire or other catastrophe causes a long-term disruption to the primary system.

The underlying concepts and technology of database systems introduce some complexities beyond those of files. For example, one must consider the ways that elements of data in various applications might be interrelated. Standard relationships recognized in most database systems from one type of data (or record or file) to another include the following:

one-to-one: In this case there is a singular relationship between one item and another. For example, a specific *employee name* will be associated with one and only one *social security number*.

one-to-many: In this case one data item might link to one or more of another type of data item. A specific *vendor* may have submitted many separate *invoices*, but each of these invoices refers only to that one vendor.

many-to-one: A many-to-one relationship occurs when two or more of one type of data are associated with only one item of another type. In a shop maintenance file, many separate *repair records* might relate to just one *crane* in the equipment inventory.

many-to-many: A many-to-many relationship indicates that one or more of one type of data item can be associated with one or more of another. For example, in a database supporting both cost control and payroll functions, one *employee* might have worked on several *cost items* last week, but each active cost item also had several employees.

Bubble diagrams are a graphical technique used by database designers in laying out the relationships between elements of information in designing a database. The notation used in these diagrams is summarized in Fig. 4-6, which provides an example for construction equipment management. The "bubbles" in this case are the divided boxes, where the top half is the generic description of what is

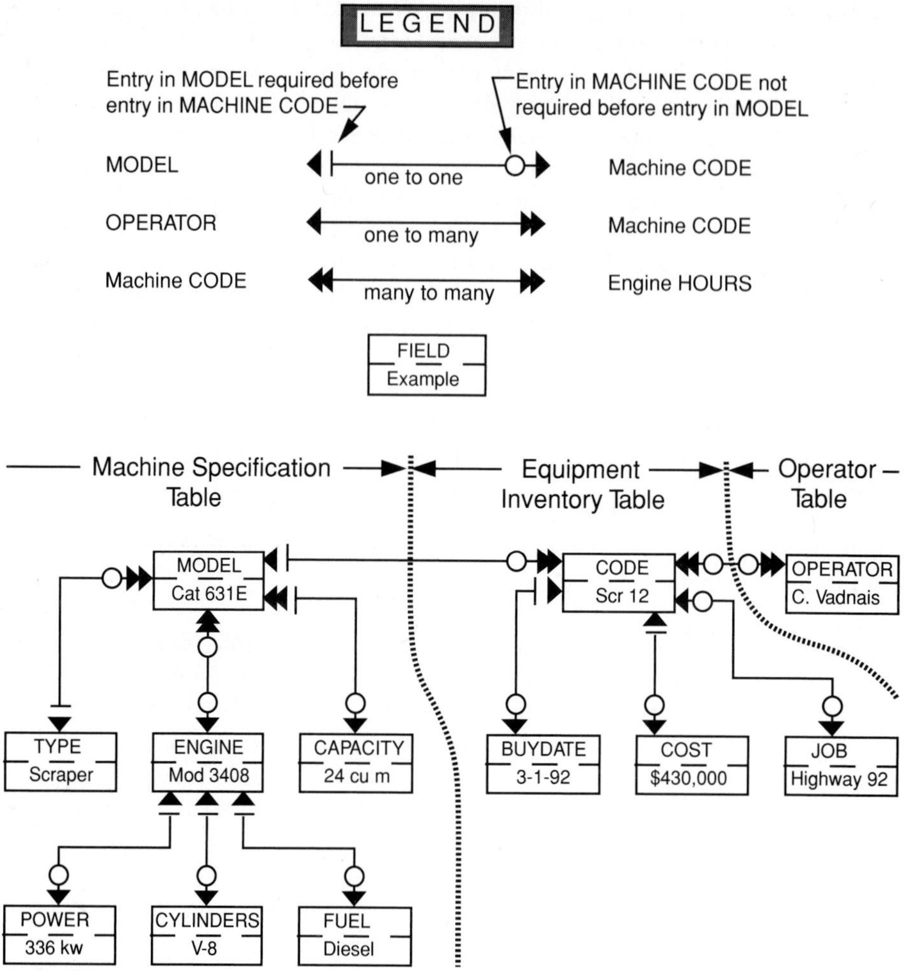

FIGURE 4-6
Bubble diagram as a graphical expression of relationships.

contained in the data field, and the bottom half gives a specific instance. The significance of the single-directed arrows and double-headed directed arrows is given in the legend of the figure. The bar across the line just before the arrow head means that the indicated item in the pair connected by the arrow must be entered in the database before its counterpart can be entered. The open circle near the arrow means that the item need not be entered before its counterpart item. A bar at each end would be an unacceptable "chicken-and-egg" or "deadlock" situation, but circles at each end mean that either end of the relationship can be entered first, whichever is feasible. This example of a database model will be developed in more detail in Chapter 10.

We can subdivide database technologies into three types: network, hierarchical, and relational. These are summarized in Fig. 4-7, which illustrates an

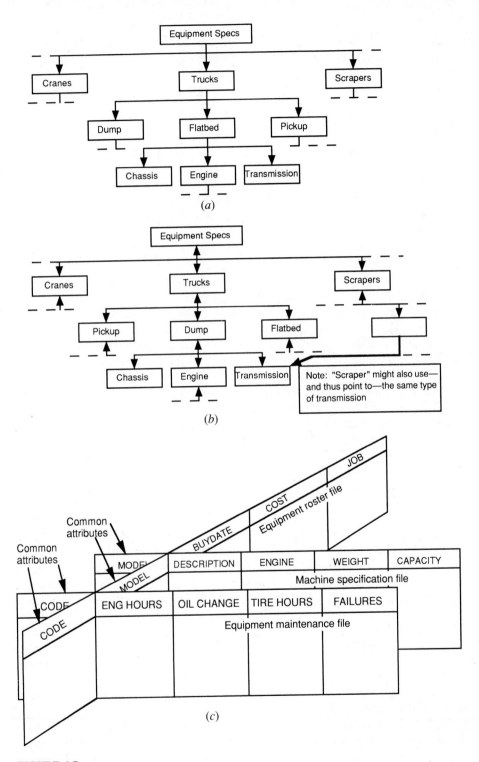

FIGURE 4-7
An equipment application of alternative database technologies: (*a*) hierarchical, (*b*) network, (*c*) relational.

equipment application. Briefly, the *hierarchical database* (or multifile) approach is the closest to conventional flat file systems. Its index system facilitates searches and cross-correlations among the elements of data stored in different files. To get to an item at a lower level in the hierarchy, such as an engine specification, the user still has to come in from the top of the hierarchy and work down step by step. *Network databases* somewhat resemble hierarchical ones, but inquiries can start from many points, not just the top. They use more complex indexing schemes and have more pointers and other structures to add data access flexibility. ***Relational databases*** conceptually seem the simplest for many users yet are among the most flexible for designing complex database systems; they are also among the most demanding for computers. Basically, they link information in various ***tables*** (you might think of each table as a flat file) via attributes that are common to them. Only one attribute is needed to link two tables. If, say, the first table is linked to the second by one attribute, and the second table is linked by another attribute to a third table, one can then seek combinations of information from the first and third, even though there might not be an attribute common to them.

Developers impose a conceptual structure on a database via a concept called a *schema*. This is the overall description for the database and its utilization, including the access paths that the user will have available to get at information stored in the database. Software can search and manipulate the database according to rules contained in the schema. Related to this is the idea of a *subschema*, which would be a description of how certain users or applications can work with a subset of a database, perhaps using security procedures to limit the extent to which they can search, modify, or copy information.

Many databases also provide a ***query language***—or the ability for a developer to custom-build one—that will enable a regular user to access and manipulate the database using high-level, nontechnical command terms, menu-based procedures, or some other method. In one extreme form of development, these query languages are called *natural language processors* if they enable the user to invoke some subset of a human language such as English. For example, the user might type (or perhaps even speak) something like, "Find me all of the bulldozers that need an oil change before Labor Day." The natural language *parser* (i.e., software that analyzes text and extracts its significant content) will probably have a predefined set of words that enables it to understand "Find," "all," and "before." But it may need to interact with the user several times to define "bulldozer" (perhaps it equates this to "tractor"), Labor Day (it might work in a specific calendar format by which it can somehow find the first Monday in September), and so on. If the parser has learning capability, these new terms will be stored and be available for future use without having to ask the user to define them again.

Finally, most database software provides a *report writer* capability to enable the user to design custom reports based on the information contained in the database. Advanced versions offer considerable flexibility in formats, graphics, fonts, and the like.

It is beyond the scope of this book to go into detail for each database technology. Chapter 10, however, presents an example of an application using a relational database, which is the most common type used on microcomputers beyond simple file systems.

SUMMARY

This chapter introduced basic concepts of computer system software, development software, data files, and database technologies. These provide computer hardware with a kind of intellectual capacity to enable the combined hardware/software system to perform useful functions.

Underlying any computer is its operating system software, which is the set of programs that supervise (1) the allocation of its resources (CPU, memory, I/O channels, mass-storage devices, etc.) to other programs, such as applications and utilities, and their users and (2) the operation of these resources. Operating systems automatically perform many functions that once required hours of tedious, error-prone, manual effort for users of earlier generations of computers. They range from simple single-user, single-task, disk-based operating systems (DOS) through multitasking and time-sharing operating systems, on up to those with real-time management and virtual memory capabilities, and are now evolving to handle computers built with dozens or hundreds of parallel processors. Closely associated with operating systems are device driver programs that interface the computer to a variety of hardware peripheral devices.

Some utility programs come with the operating system. These usually include file handling and disk backup programs, batch command processors, text editors, debugging software, and program linkers. Optional utilities start with enhanced versions of the basic system utilities and go on to screen savers, electronic mail and terminal emulation programs, advanced diagnostic software, and many others.

There is a whole hierarchy of programming tools for developing new computer application software. They range from machine-specific assembly language through compiler-based and interpreter-based procedural languages such as Pascal and C, to generic problem-oriented development environments for applications such as spreadsheets and databases. In general, the further up this hierarchy you go, the easier (and thus less expensive) programming becomes, but there can be compromises in terms of program efficiency and development flexibility. Many of these tools are supported by a variety of auxiliary software for code optimization, diagnosing errors, cross-compiling for other computers, and so forth. Libraries of modules to perform common functions are also available so that programmers can avoid "reinventing the wheel."

Basic file systems for single applications, also called flat files, can be either offline for batch processing or online for direct access processing. The files themselves are built up from data elements that are collected into the systematically organized fields of a record for a single entry; a group of records, in turn, makes a file. Data fields used for sorting and searching are called keys, whereas the others are carried along as attributes of the record. One or more indexes can be used to facilitate different kinds of access to a file. Functional types of files include master files, which are maintained and updated over an extended period of time; transaction files, which collect entries until they can be updated into the master file; archive files, which provide a long-term offline backup for information that is not presently needed but may be required for an audit or other need; report files, which hold output until it is printed; and sort files, which are created as working storage while a file is reorganized and then deleted when done. Some common methods of file access include sequential, direct access, and indexed sequential.

Databases can hold the information equivalent of several application files but do so in a way that provides integrated access to and processing of the information. Databases help improve consistency, minimize the duplication of data entry, and enable users to make inquiries and produce reports that draw pieces of information from several different application areas. However, with so much information in one place, they require much higher standards for data entry screening, authorization to access and modify data, and backup so that the potential losses in the case of problems can be minimized. Several types of logical relationships can be established among the elements of a database, including one-to-one, one-to-many, and many-to-many. Different methods for implementing database software are hierarchical, network, and relational. Concepts associated with database applications include the schemas and subschemas that describe the way the information is organized, accessed, and processed; query languages to make it easier for users to work with a database; and report writers for customizing output.

REVIEW QUESTIONS

The following questions address concepts and terminology related to computer software and information organization and storage.

1. Assume that a time-shared minicomputer installation uses an operating system with virtual memory capabilities. Which choice describes what this assumption means?
   (*a*) Its primary memory (RAM) resides on its Winchester hard disk and is paged into the CPU only as needed by various users sharing the system.
   (*b*) Its operating system is more virtuous than those used by microcomputers.
   (*c*) Its virtual memory is transient; that is, it is lost when power is turned off.
   (*d*) A programmer does not have to be concerned about the maximum amount of physical RAM memory available to run a program.
2. What software utility capability permits the automatic invocation of a series of operating system procedures or programs ?
3. What is the device driver?
   (*a*) The operator of a computer processor or storage device
   (*b*) The electronic interface module between the computer's CPU and a peripheral storage device such as a disk or tape drive
   (*c*) The computer program that interfaces a computer processor and a peripheral
4. In the following list, which functions are performed by a single-user disk-based operating system such as those used on personal computers? Briefly justify those you select. Also, note the reasons for rejecting those you do not select.

(a) Provide ready and convenient access to program development tools such as a text editor, compiler, and linker

(b) Enable the CPU to execute directly from high-level source-language program code such as FORTRAN or Pascal

(c) Enable automatic, unattended compilation, linkage, and execution of programs

(d) Enable concurrent execution of several application programs by time-sharing the CPU in allocating fractions of a second to each of the application programs

(e) Enable virtual memory page swapping so that large programs can run in a limited physical memory

5. True or false: Unlike a BASIC language *interpreter*, a FORTRAN *compiler* makes a direct translation of each line of source code into a single corresponding instruction in machine code that, once all source code is translated, can be directly executed by the computer.

6. With what does a linker combine the translated source code to make up the executable load module?

7. True or false: A *source program* is one that is designed to read in the input data for an application, whereas the *object program* then operates on the data and produces the output results.

8. State four main ways by which the software development procedure is made easier for a program developer using a modern computer with a time-shared, disk-based operating system and additional software utilities such as a batch command file processor.

9. Briefly describe the difference between a multiprogramming and a time-sharing operating system.

10. Effective and productive application program development with compiled procedural languages is greatly facilitated by good program development software. Minimum requirements are an editor, a compiler, and a linker. List four other useful program development software tools and briefly describe what each does.

11. State three main differences between the following two types of disk-based operating system concepts:

(a) Multitask batch

(b) Real-time, multiuser, multitask

12. Briefly state the most appropriate purpose or function of each of the following types of computer software:

(a) A FORTRAN program

(b) A BASIC interpreter

(c) A real-time, multiuser operating system

(d) A peripheral device driver

(e) A COBOL compiler

13. Briefly describe the physical storage characteristics of magnetic tape and magnetic disks (see Chapter 3), and then compare their relative advantages and disadvantages for storing data in the following kinds of file structures:

(a) Sequential transaction file

(b) Direct access file

14. Explain the terms *data item, record, file, sort key, record attribute*, and *header* in the context of construction payroll information kept on a magnetic disk storage unit in a computer system. Where relevant, explain how the physical organization and

operating characteristics of the disk drive and the storage medium relate to the terms described.

15. Briefly list four of the most important ways in which a database structure for electronic information storage differs from an application-specific file-oriented structure.

SUGGESTIONS FOR FURTHER READING

Aho, Alfred V., and Jeffrey D. Ullman. *Foundations of Computer Science*. New York: Computer Science Press, 1992. An authoritative and well-written 765-page introduction that follows the guidelines of the Association for Computing Machinery (ACM) for a first course in computer science.

Brookshear, J. Glenn. *Computer Science: An Overview,* 3rd ed. New York: Benjamin-Cummings, 1991. A good text for an introduction to computer science.

Stallings, William. *Operating Systems*. New York: Macmillan, 1992. Provides a good overview of operating systems, including several now in use. Author's experience provides a practice-oriented focus.

Tanenbaum, Andrew S. *Modern Operating Systems*. Englewood Cliffs, N.J.: Prentice Hall, 1992. Provides a good introduction to various types of operating systems and how they work. It also covers related system software.

Warford, J. S. *Computer Science*. Lexington, Mass.: Heath, 1991. A good text for a first course on computer science. Could be read for self-teaching.

Wilson, Leslie B., and Robert G. Clark. *Comparative Programming Languages,* 2nd ed. Menlo Park, Calif.: Addison-Wesley, 1993. A good background, classification, and overview of the types of computer languages, their evolution and interrelationships, and their applications and limitations. It describes all the major languages and many others.

PART

III

APPLICATION PLANNING, DEVELOPMENT, AND MANAGEMENT

There is nothing more difficult to plan, more doubtful of success, nor more dangerous to manage than the creation of a new system. For the initiator has the enmity of all who would profit by the preservation of the old institutions and merely lukewarm defenders in those who would gain by the new ones. The hesitation of the latter arises in part from the fear of their adversaries, who have the laws on their side, and in part from the general skepticism of mankind which does not really believe an innovation until experience proves its value. So it happens that whenever his enemies have occasion to attack the innovation they do so with the passion of partisans while the others defend him sluggishly, so that the innovator and his party are alike vulnerable.

Nicholas Machiavelli*

*The Prince, 1513.

103

Implementing change is not easy, yet most institutions, consciously or not, make the job even tougher than it needs to be. Anyone who has tried to implement in a construction organization major changes involving computer technology knows what Machiavelli meant. Certainly there will be technical problems when dealing with a technology advancing as rapidly as computers, but ironically most of the problems in computer initiatives arise because the proponents ignore the very principles of organization and management that work so well in successful construction projects.

Planning, delegating, monitoring, and decision making are just as important—and maybe even more so—in ensuring the success of new computer applications as they are in construction management. Although today's design and construction professionals would not even think of building a house, let alone a commercial building or industrial facility, on the basis of just a few ideas sketched on the back of an envelope, it is amazing how many of these same people will start acquiring computer hardware and software vital to the needs of their businesses with little more planning than that sketch.

People and business organizations respond to changes involving computers as they do to other types of change. Nobody likes being blindsided by a new development for which he or she is unprepared, even if it turns out to be beneficial in the long run. Successful design and construction professionals understand this in the context of a construction project, and they are careful in managing changes once work begins. Thus it is all the more surprising how many of them have allowed changes in computer technology to happen almost randomly—or at least without significant consultation and coordination—within their firms.

The three chapters that make up Part III do not break new ground in terms of planning and management. Indeed, they show how many of the well-proven principles and practices of managing construction organizations and construction projects can apply just as well in improving the chances of success for computer-related initiatives. Chapter 5 concentrates on the planning stage, including the feasibility study and analysis that should precede acquisition or development of a new computer-based application. Chapter 6 explains details of software and system design and provides guidance for managing software development and documentation. It also discusses human factors that influence individual human-machine interactions. Chapter 7 examines procurement of commercial systems as an alternative to custom development, compares alternative procedures for implementing new systems, and provides guidelines for ongoing computer applications management.

Beyond reviewing these basic principles, we cannot adequately describe here the underlying problem of how to integrate technological change into the culture of modern construction organizations, let alone solve it. Little has changed in this regard since Machiavelli's time. But those firms that find the answers and apply what they learn will be well on their way to moving civil engineering and construction back into the mainstream of technological progress.

CHAPTER
5

FEASIBILITY STUDY AND ANALYSIS

CLEANUP RULE: There is more than one way to solve a computer problem. The first solution is almost always to buy more hardware, which is almost always the wrong solution.

William E. Perry*

There is a remarkable parallel between the process of planning, designing, and building a home, school, highway, or industrial plant and that of planning, designing, and acquiring or developing a major computer application. Using the respective terminologies of the computer and construction fields, Table 5-1 is an illustration of the close correlation between the various stages. Its third column also provides a rough indication of the approximate time that might be allocated to each stage in successful projects of either type. Note that about half of the overall time goes into careful thought and analysis, making clear the objectives and the design for achieving them, before either construction or programming begins. A good design minimizes the changes and delays that might otherwise occur during construction or software development, helps keep the overall costs and schedule under control, and better ensures that the owner or user will be pleased with the resulting product.

At each major stage of the process in any significant application development or acquisition effort—especially after the feasibility study, analysis, and design stages and before implementation—there should be a pause for management and users to review and approve what has been accomplished and recommended to that point before going on. In this way misunderstandings in intentions or interpretation can be caught and corrected before being "cast in electronic concrete," and

---

Cleaning Up the Computer Mess, 1986.

TABLE 5-1
Chronological stages in computer and construction projects

Stage in development or acquisition process for:		
Computer application	**Construction project**	**Time**
Users recognize need for application	Owner has need for structure	Before

Feasibility study and analysis
  Scope, alternatives, costs, benefits
  Data gathering about present systems
  General hardware and software types
  Criteria for design or purchase
	Concept and feasibility study
	  Scope, alternatives, costs, benefits
	  Study appearance and space use
	  Process, structural concept, etc.
	  Develop design criteria
		25%

System and software design
  Input and output forms
  File structure and contents
  Programming and testing methods
  Hardware to be acquired
  Operating procedures
  User and system documentation
  Training procedures
	Detailed design and specifications
	  Site work and foundations
	  Structure
	  Electrical and mechanical systems
	  Enclosure
	  Roof
	  Interior partitions
	  Finish materials
		30%

Programming and documentation
  Top-down approach
  Schedule sequence and coordination
  of modules
  Coordination of variables, data
  passing, files, etc.
  Testing and reliability
	Construction
	  Schedule and budget
	  Procurement and materials management
	  Methods and equipment
	  Labor and supervision
	  Codes and inspections
	  Punch-lists and/or start-up tests
		40%

Implementation
  Installation and start-up
  User training
	Move-in or turnover
	  Logistics of move or transition
	  Learning and adjustments
		5%

Operation and utilization
  Corrections and adjustments
  Hardware and software maintenance
  Enhancements and additions
	Owner occupancy or operation
	  Warranty follow-up
	  Maintenance and utilization
	  Remodeling and additions
		There-
		after

management can be sure that the scope, schedule, and budget for the application remain under control.

This chapter concentrates on the earlier stages of developing computer applications—the feasibility study and analysis. Its concepts apply whether one intends simply to go out and buy a computer and a package program to handle a particular application need or to undertake a full custom development project in-house. Chapter 5 also places emphasis on the organizational context in which the process takes place. The remaining stages in Table 5-1 will be covered in Chapters 6 and 7.

Another important concept related to application development efforts is that the quality and effectiveness of the result, as well as its cost and schedule, are determined predominantly by activities and decisions made in the earlier stages shown in Table 5-1. This concept is generally well recognized in project management and is sometimes described in a *level-of-influence diagram* such as Figure

5-1. The chronological schedule along the time axis summarizes the key stages from Table 5-1. Reading from the left vertical axis, we see that the curve descending across the graph from the upper-left corner represents how the degree of influence on the outcome declines steadily over time. The solid line going from lower left to upper right shows the cumulative effect of the actions and decisions on costs, schedule, system effectiveness, or other relevant measure of performance, as plotted against the right vertical axis.

The biggest early decision is whether to conduct the feasibility study at all and thus commit to a path leading to a new application. Other early influential activities include defining the scope and developing the general criteria according to which the application will be developed or acquired. Much more time and effort may actually go into system design, and even more into software development, but options here are constrained increasingly by successive decisions and actions as they are taken. Backing up at this stage to reconsider the plan or to redesign adds to the costs and delays the schedule. Similarly, once the step has been taken

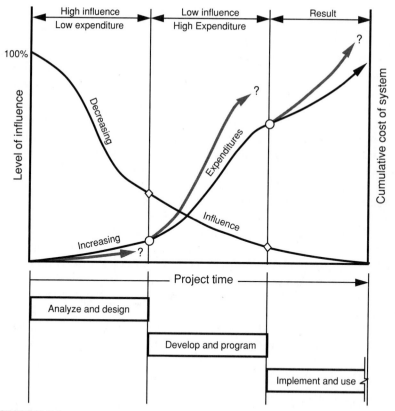

FIGURE 5-1
Level of influence on project outcome. (Adapted from Boyd C. Paulson, Jr.,
Designing to Reduce Construction Costs, *Journal of the Construction Division*,
ASCE, vol. 102, no. CO4, December 1976, p. 588. Reproduced with permission
from ASCE.)

to acquire a particular system or application, most subsequent decisions must be made within the narrow confines of its capabilities. One can still influence the outcome by deciding how to implement the new system and how well to prepare and train the potential users, but again these actions are largely determined by the system that has been made available through design and development or by purchase. Once the system is in operation, changes can be enormously expensive and disruptive relative to taking time to make them at the feasibility study and design stages, so the users may have to live with the results of all that has gone before—for better or worse.

OVERVIEW OF FEASIBILITY STUDY AND ANALYSIS

In construction, the feasibility study for a computer application typically starts when some group thinks it has a task or operation that could be improved using a computer and new software. Methods for communicating this need range from informal discussions over a period of time, perhaps with a supervisor, to a written request sent to a computer administrator or planning committee.

For purposes of illustration, let us say that such a need has arisen in an estimating department that finds itself overwhelmed in a booming market by all the work management would like to bid. Owing to fatigue, haste, and the sheer volume of work, they have made a few errors lately, causing a couple of low bids with too much money "left on the table" and the loss of a few second-low bids for no other reason than clerical error. They have asked management for approval to acquire microcomputers and application software to help improve their productivity and accuracy. Recognizing the load on the department, management has given tentative approval, subject to a feasibility study.

Lacking both the time and the expertise to study the computer market carefully, the department enlists help from a computer consultant with expertise in construction—a bright young person who already has successfully helped other contractors with similar needs. (In this chapter we will now use the term *consultant* to stand for whatever person or group provides the technical expertise needed for a feasibility study; we will also assume that this consultant is the coordinator for people involved in the study.)

The extent of the feasibility study and analysis should be appropriate to the scope of the application problem. With acquisition plans of this magnitude, the first step should be to plan the study itself. Although the level of detail will vary with the magnitude of the application, some things to be done are as follows:

- Investigate and define the general *scope and objectives* of the application
- Conduct a rigorous study of the methods used by the *present application*
- *Evaluate business issues*, such as the security, confidentiality, public image, and externally imposed requirements
- Develop a more precise statement of needs and objectives and *establish a specific set of criteria* for evaluating proposed solutions

- *Explore alternatives* in procedures and technology
- *Consult other users* with similar applications
- *Assess and compare the costs* of each alternative
- *Assess the benefits* of each alternative
- *Evaluate each of the main alternatives* with respect to the criteria, costs, and benefits
- *Select one of the alternative solutions* for acquisition or development
- *Prepare a schedule, budget, and plan* for acquisition or development, followed by implementation of the selected alternative
- Summarize the findings and recommendations in a *report to management and users*; seek authorization to proceed before going on

The quality of the investigation, analysis and thinking that goes into these steps will directly influence the success of the application. Before we describe this procedure in more detail in the remaining sections of this chapter, let us first review some of the important organizational principles that support them.

IDENTIFY ORGANIZATIONAL FACTORS

Those who would succeed in implementing changes in an organization will do well (1) to recognize the sources of conflict that may arise, anticipate the forms of subtle and overt resistance that are common in individuals and groups, and understand the reasons that may account for them and then (2) to follow well-established management guidelines for reducing the conflicts and responding to the fears and uncertainties that may cause resistance.

Sources of Conflict

In developing and implementing new computer applications, some conflicts arise inadvertently through *misunderstandings in communication*. Although many construction people have difficulty understanding the concepts and terminology used by computer professionals, the converse is also true. Both sides should be considerate of each other and take a positive attitude toward their mutual education. Minimizing jargon, patiently explaining technical terms and concepts, and offering open, noncritical responses to questions will all help.

Other conflicts arise from *differences in priorities and objectives*. Some technically inclined people may want to implement the most advanced, feature-rich systems available, whereas others undoubtedly want proven and straightforward systems that get their job done with the minimum of difficulty. Management may be especially concerned to keep the budget under control. All sides must work together and reach a mutual understanding as early as they can on common priorities and objectives.

In computer-related development efforts, *cost and schedule overruns* have often been a source of conflict. In part, this type of conflict can be reduced by establishing realistic expectations to begin with, since the computer industry has been notoriously over-optimistic in this respect. Before starting, there should be agreed budgets and schedules, and project monitoring will help reduce the chances of unanticipated surprises and disappointments.

Forms of Resistance

Resistance may show up in many ways. First, in the early stages of planning discussed in this chapter, certain people may deliberately *withhold information* that could be useful in designing a better application. Second, once implementation begins there may be people who stubbornly *persist in doing things the old way*, not only causing duplication but also reducing the benefits of the new system. Third, even with the new system in use, it can be subverted if people *distort or falsify the input information* that it needs to be useful. A well-known example is the common tendency of field supervisors to juggle numbers between accounts on cost reports to make them come out close to the budget.

Most of the reasons for such resistance will be found in a basic understanding of human nature and social behavior. For example, a new and different computer application may create in some people the fear that they will not be able to master it, and thus they may see it as a threat to job security. Furthermore, to make the most of some new computer applications, there may be good reasons to reorganize departments and working groups. But organizations are social entities as well as productive ones, so such reorganizations may disrupt friendships and familiar relationships. Finally, a new system may be a threat to the ego of people whose self-esteem and respect from others is based in part on high skill at procedures about to be made obsolete.

Guidelines for Implementing Change in Organizations

This is not the place for a minicourse on human and organizational behavior. However, it is worth outlining a few guidelines that will be helpful in carrying out the feasibility study and analysis procedure that is described in the rest of this chapter.

- *Maintain openness and honesty* throughout the planning, design, development, and implementation process. The rumors that otherwise circulate usually tend to be worse than the reality. Keep people informed about forthcoming opportunities in the new system.
- *Encourage participatory planning* in defining goals and objectives and in influencing the design or procurement of the new system.
- *Managerial support and involvement* should be evident from the beginning to the end of the planning and implementation process.
- The *goals for the change* should be understood and viewed positively by all concerned.

- There should be an effort to *coordinate goals* of this new system with those of other goals in the organization, and to *maximize overall benefits*. One must be careful not to suboptimize systems to fit the needs of a particular group, for example, while compromising the success of the larger organization within which the group works.
- There must be ample opportunities for *education and training* on using the new system, and positive incentives for it. The environment for learning should be an encouraging and supportive one.
- The organization and content of the new system must be *designed for the people who will use it*. We can do little to redesign people to suit computers, but a great deal can be done to improve computer software to suit the needs of people.

Certainly there still will be problems and failures in trying to implement computer applications in construction. But the fact that some organizations are succeeding as a matter of course proves it can be done well. Experience has shown that these organizations are the ones who spend the bulk of their efforts on the organizational and human factors rather than just the technical facets of design and implementation. With this background, we can now move on to describe a systematic procedure through which these principles can be applied.

ESTABLISH SCOPE AND OBJECTIVES

The scope and objectives of computer applications for construction are less tangible than for a construction project but defining them is no less important. To illustrate the problem of defining scope, we recall the following list of possibilities that was discussed in more detail in the section on estimating in Chapter 2:

- Maintenance and preparation of master checklists
- Quantity take-off
- Productivity and cost analyses
- Compilation and organization of summary and backup computations
- Compilation and analysis of subcontractor bids
- Distribution of overhead and indirect costs
- Analysis of alternatives and contingencies in determining markup
- Preparation and delivery of the bid or proposal

The consultant must find out which of these are highest priority in this department and get some idea of where problems are occurring. Is it more important to handle or even increase the sheer volume of bid work just now? Or would it be better to have some help in selecting the most promising projects on which to concentrate? Is accuracy a normal problem in this department—perhaps suggesting a need for better forms and procedures and tighter checks—or has it just been a recent casualty? How is this estimating department run anyway? Do superintendents and project managers participate in making the estimates, or are all parts of a bid prepared by a full-time staff? What is the range of project types this company bids? How do they deal with bids from subcontractors?

With questions like these, the consultant will soon get a feel for the scope of the problem. He or she will then be able to define objectives and set tentative priorities. The consultant should abstain at this stage from recommending a specific solution. There is much more thinking and analysis to be done first.

STUDY CURRENT SYSTEM AND NEEDS

Having an initial grasp on the scope and objectives, the consultant will seek more detailed information to improve his or her understanding of the application's needs and problem areas, and obtain more data that can support the selection or design of a new system that best meets the needs. Typical activities in this study include the following:

- Interview all key parties involved in the present and proposed application
- Gather examples of printed documents related to the application
- Identify and understand any computer systems that may currently be in use
- Establish functional flow of information
- Establish costs and durations of various steps and procedures in the system

The consultant may already know many of the users by now, but it is important to return to *interview* them for more detailed information. In a small group, such as the estimating staff for a mid-sized contractor, each person should be interviewed. If it is a large group, care should be exercised at least to get a representative sampling of job functions and opinions. Managers as well as people who provide input to the group (e.g., payroll and equipment departments supply costs) and who depend on its output (e.g., superintendents) also should be contacted. In conducting interviews, the consultant will try to assess the skills and knowledge of the present and potential users of the system, listen to their complaints about current obstacles and frustrations, and get their opinions on steps that will improve the system. The consultant should also identify creative and articulate users who might most usefully contribute to the selection or design of the new system.

Over time, organizations accumulate various *documents* that can help one to understand the system to which they apply. Examples include forms, reports, policy and procedure manuals, organization charts, job descriptions, mission statements, progress reports, and so forth. In most organizations, there will be differences between what is in these documents and the way the interviews reveal that things are actually done; but there is knowledge even in these differences, since the unwritten procedures worked out by those doing the job may have deviated for good reason from what was originally planned. A new system should recognize these factors and not reimplement shortcomings.

If a *current system* is employed in the application, the consultant should find out about it (e.g., examine its manuals and reports and observe how it is used).

The consultant should concentrate on the shortcomings of the system that have in part led to the desire for a new system, but also find out what is good about it. What things will people miss and criticize if they are not somehow carried forward or improved upon in the new system?

The consultant should identify the major and minor *components of information* processed in the system, and establish its *functional flow*. For example, a materials requisition document, used by field supervisors to request supplies for their projects, might have three carbon copies (pink, yellow, and green). How well do they understand these documents? Who approves and signs a requisition before it becomes a purchase order to an outside vendor? How long does this take? Are there money limits on how much can be requested at various levels of signature approval? The consultant might literally follow the paper trail to see what happens to the original plus each of the pink, yellow, and green carbon copies of the requisition. After a procedure has been around for many years, people may forget the origin and purpose of some of its details. The consultant may well discover that the pink copy, for example, ends with a clerk who dutifully files it in a large cabinet with thousands of others and updates an index of what is in there. When asked what they are then used for, the clerk replies that nobody has ever asked since he or she has worked there, but the pink copies are in the file just in case. In case of what? Who knows! Analysis of the requirements of any established system of significant scope usually turns up procedural dead ends of this type, and there is no point carrying them into the new system.

As a result of this much analysis, the consultant should be able to prepare lists of the people involved; describe the information input, processing, and output; and make *functional flow charts* to help understand what is going on. Figure 5-2 is an example of one such flow chart. Things like the pink copy will show up as dead ends, and less conspicuous inefficiencies may emerge as well. Whether or not a new computer application comes about, cleaner and more efficient procedures and a tighter organization should result from this analysis.

With this level of detail in the information collected, the costs and durations of each step can be computed so that at least the direct costs of the application become known. It is often sobering to realize just how much time and money go into some forms and procedures—perhaps three hours and $80 just to process a single requisition, whether it be for a $25,000 compressor or a $25 keg of nails. In the latter case, the analysis might lead to a simpler procedure, but simplification could increase the company's exposure to abuse, so there are trade-offs in such decisions. This stage of the study provides real facts to guide the design of the new system and to measure its success against the current one.

IDENTIFY BUSINESS ISSUES AND REQUIREMENTS

One must be mindful of the broader context of the system and how it affects the relationships of its users and even its organization with the world outside. We will call this step the analysis of business issues and requirements. It identifies

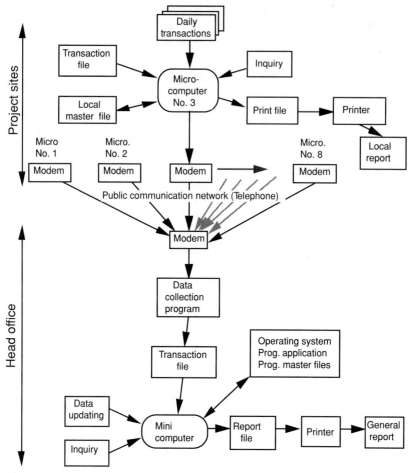

FIGURE 5-2
Information flow diagram. (Adapted from a figure submitted in a term project by Mr. Takeshi Shibata for a course at Stanford University on computer applications in construction, February 1986.)

more general criteria that the new system must meet. There are many possibilities, depending on the application. Those to be addressed here include some of the secondary purposes of the information or products produced by the system, issues of confidentiality and security, timeliness, and reliability, and satisfaction of externally imposed requirements.

It may seem that the *purpose of a system* should be obvious by its nature: a cost system should report costs; a CAD system should produce drawings; a payroll system should produce checks. But there are often more subtle issues involved. For example, if a cost report is intended only for internal consumption, then only its content, accuracy, and timeliness may be important. One can provide in-house training on how to use the report. But if it also goes out to clients, then its style and appearance may be important to impress them favorably. Form designers should give special thought to documents that request money or information from outside parties; forms need to be courteous, readily comprehended, and easy to respond to.

When *confidentiality* is an issue, as it usually is in systems dealing with money and personnel and probably with a company's productivity figures as well, then means for controlling access to related information systems become an issue. *Security* is also important with systems that print checks (notably payroll and accounts payable) but also may apply to materials and equipment management systems and even to controlling access to certain areas on job sites. Systems of this type might be designed with controls via *hardware* (e.g., dial-up telephone modems that connect only if the party calling is one that appears on a small list of those authorized), *software* (e.g., **passwords**), and *procedures* (e.g., require that two people from different departments sign checks to make them valid).

The importance of *timeliness*, *accuracy*, and *reliability* seems readily apparent, but it can vary significantly with the application. Financial systems put high stress on accuracy, but their timeliness may be satisfied with monthly cycles for printing income statements and balance sheets. Payroll systems must be deadly accurate *and* timely *and* reliable; payroll thus can be the most demanding of all construction business systems. Cost-control systems usually favor timeliness over accuracy—the value of information deteriorates rapidly over time for operations in progress; reliability and accuracy are also desirable here, of course.

The requirements from *external parties* vary with the application. Clear examples include payroll systems and their associated taxes and benefits, and accounting systems for tax purposes and for full auditability in public corporations. Bonding companies and banks also may have a right to accurate information if they have loans or bonds at stake. Even scheduling and CAD systems may have to produce results that conform to standards imposed by contracts with owners.

All such business issues and requirements should be identified. Those that are indeed important must be specified as criteria to be met in acquiring or developing systems within their domains of influence.

REFINE SCOPE AND OBJECTIVES INTO CRITERIA

The detailed study of the current system and the identification of business issues and requirements will lead to revisions and clarifications in the scope and objectives that were identified earlier in the feasibility study. These revisions plus other new information will lead to a detailed set of *criteria* for the proposed application. Some of the categories to be addressed typically include those shown below:

- The amount and type of information to be processed
- Skill levels required
- Input methods
- Output reports

- Amount of, and limits on access to, information in files
- Controls and procedures for confidentiality and security
- Training and support from developers or vendors

There must be some *justification* for each criterion at this stage, since everything has a price. Long wish lists from the initial scope study and investigation of the current system may need to be pared down to what is technically feasible and affordable. The criteria can be ranked, of course, with some deemed essential and others only desirable. Ranking is helpful in making choices among alternatives in package software and also in deciding on how far to go in custom software development. A numerical weighting for each criterion can also further refine the evaluation process. Once developed, these criteria will help evaluate alternatives to find the one that comes closest to meeting the needs and objectives of the individual, group or organization concerned and to satisfying the requirements of those with whom they interact.

EXPLORE ALTERNATIVES

Several alternatives for meeting the application needs should be explored at this stage, preferably with significant input from the users to complement the technical and market expertise of the consultant. These should not be limited to solutions involving computers. Quite to the contrary, all kinds should be considered. The breadth at this stage might encompass the following:

- Computer versus noncomputer
- In-house versus external sources for development of software or even for operation
- Package (with compromises) versus custom (at higher cost) system
- Centralized versus decentralized

Information on *package solutions* can be sought from software directories put out by manufacturers, trade associations, or information services focused on construction. Reviews in trade magazines and recommendations from construction user groups should be explored. Computer and construction industry conferences and trade shows with manufacturer and vendor exhibits provide good opportunities to see many alternative systems in a short time. Initial inquiries like these can lead to follow-ups directly to the developers and vendors that appear to meet at least some of the identified needs.

If promising package solutions do not appear to be available, in-house or consultant-based *custom software development* becomes more attractive, although it can be expensive. Even here, there is a wide range of alternatives, including many of the application development tools to be introduced in Part IV. Procedural languages, spreadsheets, databases, and expert system software all can be—and have been—applied to create software useful for estimating. One must be very careful, however, in finding those that best suit the needs at hand, and be even more careful in choosing people to use the tools for development.

CONSULT OTHER USERS

As part of the feasibility study, it is useful to talk to users in other firms (or other divisions of the same firm) who have applied computers in applications similar to those being considered. Some formal avenues for discussion include both national and local contractor associations and professional societies, which often have committees or user groups whose meetings focus on computer applications in construction. From these sources, patterns may emerge, such as the wide adoption of certain types of software by contractors of your type, or the scornful consensus that a certain potentially interesting package is fraught with bugs or has poor vendor support. Such information can help a firm avoid making costly errors and can make attendance at the meetings well worthwhile. These contacts and others may lead to opportunities to visit firms using computer applications of the type being considered. Developers or vendors may also suggest a list of clients that you might visit, although this could be a biased sample. Take good notes when visiting users and be considerate—so that you can call back with more detailed questions that will inevitably arise if you later consider this software more seriously.

ASSESS COSTS

It should now be possible to focus on the most promising alternatives to get a better estimate of costs and benefits. How this is done depends on the information gathered thus far and on the types of alternatives being considered (e.g., package software versus custom development). This section looks at a typical approach to estimating application costs, and the next examines benefits.

It is difficult enough to make conceptual cost estimates in construction, where the technology advances slowly and the products still are mostly tangible. It is much harder at this early stage to estimate costs for computer applications, where the products are less tangible, the parameters are more abstract, and technology advances rapidly.

Some computer costs are easy to estimate, of course, but these can be misleading. If a department needs two new microcomputers and some package software, one can almost get their costs out of a catalog. But hardware and software costs are only a fraction of those associated with even routine applications. For example, user training often costs more than the software to which it applies.

An important underlying principle here is that personnel costs in computer applications usually far exceed computer hardware and software costs. Application development or acquisition efforts must thus put their focus on the people who will be using the systems and on the results of their work, and not worry so much about lesser differences between the costs of competing hardware and software products.

To begin, we can list general categories of costs:

- Hardware (whether new or existing)
- Package software acquisition
- Custom software development
- Setup and data conversion
- Support for software and hardware
- Documentation
- Training (both formal and on-the-job)
- Communications
- Supplies and accessories
- Space and furnishings

If one is starting from scratch, new **computer hardware** might be required for people using the application. This might include microcomputers, monitors, printers, modems, and cables. Costs can come from catalogs, manufacturers, retail stores, or consultants. Two concepts to keep in mind here are (1) hardware can be capitalized and depreciated over several years and (2) it might be acquired incrementally over several months or years. Therefore, it may be inappropriate to lump all of the costs into the first year.

Software costs can differ radically depending on whether package software or custom development is expected, so we will look at them separately. **Package software** such as spreadsheets, databases, and scheduling programs may seem inexpensive at first, and their prices are readily obtained. Bear in mind, however, that copyright laws generally require that either one package or license be acquired per anticipated user or workstation or site licenses for multiple users be obtained. Package software aimed at more limited markets like construction estimating tends to be more expensive than generic packages and often is only offered for an annual license fee.

Estimating the time and costs for *custom software development* is difficult even for experts. A cynic might say, "Make your best estimate, double it, and you will be low by only 50 percent—*if* things go well!" A famous and useful book on the subject of managing computer software development is Frederick P. Brooks, Jr.'s appropriately titled *The Mythical Man-Month*. The development time required will vary significantly with the choice of tools to be used (recall the range of effort levels in Table 4-1) and with the skill, creativity, and productivity of the programmer (one in the eightieth percentile might be five times as good as one in the twentieth percentile).

As a rule of thumb, it is tempting to say that if good packages are available at reasonable cost, then there must be an extraordinary justification for developing custom software. The risks of development costs, schedules and product quality being worse than expected have proven historically to be very great; users and management must be made well aware of these risks. Nevertheless, if an application is to involve several people using computers extensively, productivity gains from custom software can indeed provide a competitive edge, and this option should not be ruled out. Furthermore, in any construction organization there still are lots of worthwhile applications for which quality package solutions are not available; numerous contractors have improved their competitiveness by getting there first with custom software.

Whether generic or custom, significant costs of such a system may be incurred in *setup* to meet the needs of the application and in **conversion** to get the data loaded and ready to use. Although a word processor may be ready to go almost right out of the box for the very next letter to be typed, screen and report layouts and programs written in "macros" may be needed first to get spreadsheet or database applications ready for intended users, and it may take weeks of data entry to get cost control and estimating systems ready to use.

Support for hardware and software is essential for successful applications, but it has its costs. For example, one must allow for annual maintenance agreements on hardware and for acquiring updates to software, both of which typically run 10 percent to 20 percent per year based on original purchase costs. For more complex systems, such as for CAD or construction estimating, support can also include onsite installation and the right to telephone the developer to ask questions about capabilities or seek help in case of problems. Such support is typically included with the purchase cost for the first few months but may go on a fee basis thereafter. Even for internally developed systems, the need for support must be recognized. Inevitably, somebody in an office becomes known as a local expert, and more and more of this person's time is consumed in setting up new workstations and software, answering questions, and solving problems. If there is a network or central file server, somebody has to be responsible for that, too. Eventually the position evolves into a formal one, such as a "microcomputer administrator," "network manager," or similar title. In any case, some fraction of salary and overhead related to such in-house expertise should be recognized as a support cost.

Historically, **documentation** has been a weak link in computer applications. Although the software may have been fine, the manuals were so poor that users simply could not figure out how best to use the software; thousands of hours have been wasted as a result. In the case of commercial package software, the situation has vastly improved; the quality of writing, organization, and presentation in manuals has become an important competitive factor. However, the old problems persist in more specialized software sectors, including construction, in spite of the fact that one may pay much more for such products. Particularly in the case of custom software, good documentation is likely to be neglected altogether unless it is specifically budgeted and measured against strong criteria set forth at the outset.

There are many options for *training*. Traditional forms include in-house or external seminars provided by developers, vendors, consultants, or educational institutes. Costs for these should include not only the fees for the days or weeks involved but also employee time and travel, and possibly the costs of temporary replacements. Videotapes and other self-study aids are sold to explain popular software such as spreadsheets and databases, and to introduce popular computer systems. Commercial packages often include tutorial manuals and online educational software, some of which can be very effective for independent and motivated employees. Probably the worst "training" method of all—unfortunately, one that is all too common—is just to let users puzzle away at their desks, slowly learning a subset of software capabilities over many months and feeling stupid when they ask questions of busy office experts. This is not only unproductive but also demoralizing, and thus it can undermine what might have been a good system.

Given the decentralized nature of construction, *communications* costs can also be significant. For example, microcomputers in the field may need to exchange cost, payroll, equipment, and materials information with systems in the main office. Although modems and related devices can be included with the costs of the computer equipment, telephone charges need to be recognized as well. Depending on the volume of data, a dedicated line may be needed or perhaps be shared with a fax machine, itself integrated into a computer.

Supplies and accessories include disk and tape storage media, paper, ribbons, laser printer cartridges, and miscellaneous gadgetry such as mouse pads, stands for CRTs, and bins for printers. Within a given application area, their costs should not vary significantly from one alternative to another. A rule of thumb such as "15 percent to 30 percent of initial hardware costs annually" is normally sufficient at this stage.

In some cases one must also allow for *office space*, *furniture*, and *special wiring*. If software is being added to an existing system, these costs may not be incurred. But new hardware often requires specialized furniture, and the space needed by workers who both have regular desk duties and use computers exceeds that traditionally allocated in office planning. To keep power cords and network cables from accumulating hazardously on the floor, technicians may need to install special ducts and trays. Furniture, space, and cable installation all can be priced— construction firms already do this well.

There will still be uncertainties in these costs, and ranges should be noted where appropriate. Even where costs cannot be estimated accurately, it is still worth applying intuition or judgment to suggest a range in which they may fall. For example, if five estimators will be using a new software package, one might make an educated guess that each will spend 50 to 100 hours between formal training and just "getting up to speed." One could not only multiply this by their salaries and benefits but also estimate the loss of output resulting from their inattention to their primary jobs. These costs should later be offset by benefits if a new application proves justified. Even if people involved in the study at this stage absolutely cannot estimate costs for certain items, they should still make a list of these intangibles to discuss later with management or others; perhaps these people will have a better idea of the costs. However, by now it should be possible to make reasonable estimates in most categories; the chances are that some errors will offset each other, so the total should be fairly reliable. As time goes on, more

TABLE 5-2
Preliminary budget*

Item	Unit Price	Quantity	Amount
Hard-disk-based head-office minicomputer system	$15,000.00	1	$15,000
Field microcomputer systems	2,000.00	8	12,000
Laser printer	1,000.00	9	9,000
Modem	200.00	9	1,800
Installation cost (telephone, cables, etc.)			2,000
Software package acquisition cost (including license fee)			
Spreadsheet	400.00	9	3,600
Database	500.00	9	4,500
Communications	200.00	9	1,800
Miscellaneous utilities			2,500
Computer supplies			
Floppy disks			500
Printer paper			700
Others			300
Contract programming (160 hours at head office)	80.00	160	12,800
In-house educational seminar (at head office)			
Fee for instructor (from outside of the company)	800.00	2	1,600
Transportation cost (8 employees from the sites)	500.00	8	4,000
Lodging and meals (8 employees from the sites)	150.00	8	1,200
Materials and supplies			500
Telephone charge (charged to head office)			
Long-distance area (4 sites × 10 min/day × 52 weeks)	0.50	10,400	5,200
Regional area (4 sites × 10 min/day × 52 weeks)	0.25	10,400	2,600
Miscellaneous			5,000
Total			$86,600

* Adapted from a table submitted in a term project by Mr. Takeshi Shibata for a course at Stanford University on a computer applications in construction.

precise figures can be plugged into this framework, eventually leading to a budget like that shown in Table 5-2.

ASSESS BENEFITS

The in-depth analysis of the present system, combined with the better knowledge of the narrower range of application alternatives now seriously being considered, should make it possible to improve the estimate of benefits. The general goal is to provide input for a good relative evaluation of benefits that can be compared with each other and with the corresponding costs, and to be sure that there is enough substance in the benefits category to comfortably offset the costs. The more uncertain the cost and schedule estimates (such as with custom software development), the higher should be the perceived benefits.

But if estimating costs is difficult, estimating benefits is more so. Nevertheless, at least for business investment decisions, we need to compare benefits to costs, so the effort is warranted. Since benefits analysis is very situation specific, this section will be confined to the following general categories in which benefits can be evaluated, and examples of cases in construction will be presented.

- Productivity enhancement
- Marketing
- Improved accuracy and consistency
- Improved quality
- Improvement for employee jobs

Productivity benefits have many forms. In the earlier days of computing, some such "benefits" were described in terms of the number of people who could be replaced by automation. Productivity benefits now more commonly refer to improvements in the amount, quality, and value of work performed and thus are a measure of the ability of a firm to expand capacities and markets. With retraining and reorganization, employees often find the quality of their jobs improves as well. These latter benefits, however, are difficult to quantify, especially if the people concerned do not yet have experience with what the computer might do for them.

Both productivity and *marketing* benefits accrued when an excavating and grading firm introduced a computer-based quantity take-off system at a cost of about $20,000 per workstation. The direct output per estimator for this task increased by a factor of four, and quality and consistency improved as well. Rather than lay off estimators, the firm was able to bid more work in an increasingly competitive market. Additional benefits came in negotiated work, where the ease and productivity of the system enabled the firm to explore money-saving alternatives to site designers' original plans; discussing such alternatives with potential clients not only enhanced the firm's technical and business image but also won contracts as well. On the basis only of direct costs and estimating output, each workstation paid for itself in less than six months, and the marketing benefits, though difficult to quantify, were very real extras.

Design consultants once debated whether the cost of CAD workstations could offer enough gains in drafting output to justify replacing manual methods (for a decade evidence seemed evenly split, although declining system costs and improved capabilities have since made the debate moot). Their focus was on labor productivity. At the time, a large design-construction firm contemplated switching from a two-dimensional (2-D) to a three-dimensional CAD system. They had had mixed experience with 2-D and were apprehensive about the even higher costs of 3-D systems. They did move ahead, and the substantial benefits that arose in the *accuracy*, *consistency*, and *quality* of work produced have since made the 3-D decision seem truly prophetic. In 2-D one is just speeding up drafting for the most part—a labor versus capital tradeoff. Moving to 3-D, however, they soon found that interference checking as facility components were being designed by various disciplines working on an integrated model had real benefits for construction as well as design. Before 3-D, field construction rework attributable to design errors accounted for about 5 percent of industrial process plant costs; many of these resulted from design coordination failures where, for example, an electrical designer ran a cable tray through the same space that a piping designer located a pipe or a structural designer placed a beam. It is difficult to catch these problems in piles of 2-D drawings separated by disciplines or in 2-D CAD. With integrated 3-D CAD, more and more such errors were caught early in the design stage, and construction rework for the resulting projects started dropping under 1 percent. The 4 percent difference paid for the 3-D CAD systems many times over.

But this case also illustrates the context-sensitive nature of evaluating the benefits of new technologies. The frequency of design coordination errors actually may differ little between either consulting or design-construct firms. However, a design-construct firm can more readily capitalize on such savings and thus incorporate them into the benefits analysis. On the other hand, architectural or engineering consulting firms—which do not perform physical construction—would have a much harder time here because, in effect, they would be asking the owner to pay to avoid mistakes that the owner (naively or otherwise) would say should not be made in the first place; for such consultants, this sort of benefit is embarrassing even if it could be recovered.

Benefits to *employee jobs* increasingly are being recognized. Employees who have suffered years of drudgery in such tasks as quantity take-off, transcribing time cards, preparing cost reports, updating drawing submittal logs, and other such tasks can readily appreciate a good computer application that reduces the tedium and improves their productivity and accuracy. Similarly, interviewers visiting colleges in search of young new construction professionals have been reporting that students now ask about what computer facilities they can expect to use in their jobs, and they look down on companies not perceived to be up-to-date. They know how computers have helped them through college and resist going back to earlier methods. Progressive firms have capitalized on this trend.

The results of the benefits analysis can be summarized in a table with monetary estimates, as was done for costs in Table 5-2. Some examples of quantifiable benefits have been mentioned in the preceding examples. More probably, however, the benefits analysis will have to rely much more upon lists of qualitative benefits, organized into categories of related items. Such lists still can serve as a good basis for discussion and decision making.

EVALUATE ALTERNATIVES

There should be enough information now on hand to narrow the choices, using a mixture of the expert's experience and users' input on what seems most feasible to them. Discussions with people in trade associations, user groups, and other firms may have lent credence to some alternatives and discounted others. With good criteria established and information about costs and benefits, the study can proceed to a more conclusive evaluation of the promising alternatives.

Each viable alternative should be examined in light of the criteria to see, first, if it has significant omissions or shortcomings in capabilities (e.g., many microcomputer accounting systems, being designed for retail or manufacturing firms, do not serve the project-based accounting needs of contractors) and, second, if it has special strengths (e.g., a package designed almost exactly as if customized for the application at hand). It may be obvious by now that there really is not much substance behind some alternatives (e.g., the software industry is notorious for putting out tempting advertisements long before products actually materialize, if ever). More systematic considerations for this evaluation might include the following:

- Availability of commercial application software suitable for company needs
- Adaptability of available programs to specific in-house needs
- Extent to which company is committed to its current systems, and effectiveness of current systems versus advantages of new alternatives
- Interest and capability of employees in supporting new applications
- Technical and managerial expertise to do in-house custom development or to supervise consulting developers in this process
- Costs and benefits compared among alternatives

Generally, given the significant risks involved for in-house development, a firm is wise to examine all possible alternatives before embarking on a custom software effort. Decisions most often come down to choices among third-party systems, usually involving commercial packages or consulting developers. Available software and systems will have different strengths and weaknesses, depending upon the needs of the case under consideration.

In large measure, this evaluation is mostly technical and functional, just to be sure which alternatives are still feasible. Then, for those that look promising, one should look more closely at costs and benefits. There should be enough information available by now to begin developing a ranking; perhaps at least one alternative looks promising against most of the criteria, while others drop away. As an example of such a ranking scheme, Table 5-3 takes a set of weighted criteria and shows how they might be applied to fictitious alternatives A, B, and C. If two or three contenders still look like good possibilities, the study may again iterate through their functional merits and refine the analyses of costs and benefits.

Having some idea of both costs and benefits, one is tempted to compute benefit/cost ratios or run cash flow analyses to rank alternatives, but in most cases it is probably too early to apply such methods in other than a tentative way. Many of the relevant costs and benefits may be inadequately quantified. The seeming objectivity of such financial methods thus may be misleading, given that the inputs are so tentative. Real and important qualitative factors usually get played down in such efforts, and benefits tend to suffer more in this computation than costs. But if there is sufficient quantification of the benefits or revenues that might be expected as well as the costs, and if their timing can be determined, then one

TABLE 5-3
Quantitative comparison of alternatives*

NEW COMPUTER SYSTEM EVALUATION AND RANKING				
Criteria for evaluation	Possible points	--Proposals--		
		A	B	C
1. System performance (42%)				
a. Hardware performance	60	60	41	42
b. Software performance	60	39	43	31
c. Communications performance	30	21	25	15
d. Expansion capabilities	25	20	23	15
2. Vendor capabilities (25%)				
a. Vendor performance	40	25	32	21
b. Maintenance and backup	20	14	18	13
c. Installation support	20	12	18	12
d. Staff preference	25	15	20	5
3. Cost (33%)				
a. Purchase or lease price	50	35	23	40
b. Maintenance	20	17	5	18
c. Ongoing educational cost	25	21	9	17
d. Backup and recovery	20	15	10	14
e. One-time system & ed. costs	20	18	11	14
	415	312	278	257

Source: Gordon B. Davis and Margrethe H. Olson. *Management Information Systems*, 2nd ed. New York: McGraw-Hill, 1985, Table 19-3, p. 622. Reproduced with permission of McGraw-Hill, Inc.

can run a present-worth analysis based on a forecasted cash flow discounted at an appropriate interest rate. If, based on these quantifiable entities alone, the return is satisfactory, then one can reasonably hope that the unknown and intangible benefits will clearly make the investment worthwhile. The 3-D CAD example discussed earlier was of this type. In other cases, early analyses or even experience may show that the payback is so quick, as in the quantity take-off case mentioned before, that it will take little time at all to justify switching the whole operation to the new method. A detailed cash flow analysis to prove this may not be worth the effort if it is done only to justify the obvious decision.

SELECT BEST ALTERNATIVE AND DEVELOP PLAN FOR IMPLEMENTATION

The scope and objectives for the application should be clear by now, and a dozen or more possible alternatives for meeting them may have been explored. One alternative should have emerged as the obvious—or at least the preferred—choice. It may well be a compromise, such as a package system combined with some customization of reports, or even the integration of two or more package systems that may require some custom programming to "glue" them together. The decision will be based in part on rankings such as those in Table 5-3 and in part on the evaluation of benefits and costs. Often it will involve a compromise between users (who want the best system with the most features for their needs) and management (who may say that the second choice better suits their budget constraints). If significant conflicts of opinion remain, especially among the users upon whom a successful implementation will depend, then it is best to continue in the earlier phases of the study a little longer until a more satisfactory solution can be found. In any case, it is best not to proceed toward implementation in the face of significant opposition. An expert who tries to force onto the users what appears to be the best system from his or her own perspective will learn why human beings are noted for their ability to prove experts to be wrong—even when they are right!

Assuming that the chosen system is likely to have broad support, the consultant should pause, consolidate the findings and analyses that have been achieved thus far, support the recommended solution, and prepare a plan for its acquisition or development. The plan should include a budget such as that shown in Table 5-2 and also a tentative schedule such as that in Fig. 5-3.

PRESENT RECOMMENDATIONS

We now reach the end of the feasibility study; it is time to report our findings and conclusions so that management and users can decide whether and how to continue. The manner and content of the report may range from a brief telephone call to a formal written report and oral presentation. If the feasibility study has turned up several possible good solutions, perhaps two to a half dozen will be described in this report, pointing out their pros and cons and—qualitatively, at least—their relative costs and benefits. If the findings have led to a recommendation to pursue a computer-based application, then management approval and

Activity \ Weeks:	1	2	3	4	5	6	7	8	9	10	11	12	13	14	15	16
Feasibility study and analysis	▦	▦														
System and software design			▦	▦												
Programming						▦	▦	▦	▦							
Testing and corrections										▦	▦					
Hardware procurement, installation						▦			▦							
Documentation										▦	▦	▦	▦			
Introductory seminar														▦		
Implementation and data entry															▦	▦

FIGURE 5-3
Preliminary schedule.

support should be sought and be forthcoming before moving on—support mani-
fested in a commitment of resources to implement the plan. With this support and
approval, we are now ready to proceed to the detailed design and development
of the system (Chapter 6) or to acquisition and installation of a package system
(Chapter 7).

SUMMARY

There are numerous approaches to bringing computer applications to construction,
including package microcomputer systems with all the software for onsite planning
and control, specialty programs for applications such as estimating in the home
office and for scheduling and resource allocation, and access to time-sharing
systems for advanced programs that are only occasionally needed in the company.
Since the needs of construction companies vary so widely, no one vendor can best
suit them all. Rather, for a particular company or application, a study should be
made of several alternative approaches, and careful selection criteria should be
developed to choose the one best suited to the needs. Given the time and cost
involved in developing quality software, one should consider all available vendor
alternatives before embarking on an in-house development effort for any major
application.

The key stages in development or procurement of a computer-based system
for a typical construction application of significant magnitude include (1) recognition of the need and a request; (2) a feasibility study and analysis of the scope,
alternatives, costs, and benefits of the proposed application; (3) system and software design; (4) procurement and/or program development and documentation;
(5) training and implementation; and (6) operation and utilization. The earlier
stages in this process, which usually are smaller in cost, have a disproportionately
high level of influence on the success of the eventual system, so they should
receive appropriate management attention.

The feasibility study explores the scope and objectives of the application,
gathers information from users and other sources about the present and desired
system, and suggests possible alternative solutions. The analysis compares benefits and costs and leads to a recommended approach, along with a preliminary
schedule, budget, and plan. The quality of this study and analysis will be much
improved if those involved have a basic understanding of the types of human and
organizational issues that are likely to arise in moving to a new system. As the
study nears completion, its results should be documented and reported to users
and management, and approval should be obtained before proceeding to system
development or procurement.

REVIEW QUESTIONS

1. Assume that, as an engineer for a medium-size building construction contractor, you
   have the responsibility of developing a new estimating system. The company currently does about $50,000,000 of work per year. Of this, 60 percent is represented by

subcontractor costs. At present there are no computers being used by the estimating department. Approximate expenditures now associated with the estimating department are as follows:

- Quantity take-off   6 people at \$30,000/yr
- Item productivity and cost analysis   4 people at \$40,000/yr
- Bid summarization and preparation $\frac{1}{2}$ person equivalent, at \$48,000/yr

Assume that the costs or benefits of the various estimating functions could be improved through computer applications. The general cost of each application, and the expected savings, are as follows:

Application	Cost/month	Savings
(1) Quantity take-off and data entry	\$2,000	Either cut 4 staff or triple capacity
(2) Productivity and cost analysis	\$4,000	No staff savings, but increase company profits by 0.5 percent of volume through better estimates
(3) Bid summarization and preparation:		No staff savings, but improve ability to evaluate last-minute
if added to software in (2)	\$2,000	subcontractor bids for 1 percent expected reduction in average sub costs
if acquired to stand alone	\$4,000	

(a) Briefly classify the costs and benefits into "economically quantifiable" and "significant but intangible qualitative" categories, with at least four items of costs and four items of benefits in each category.

(b) Complete the following table, showing, for each application, its annual cost, its annual savings, and the benefit/cost ratio.

Application	Annual cost	Annual savings	B/C ratio
(1) Quantity take-off			
(2) Productivity and cost			
(3) Bid summarization and preparation:			
with item (2)			
without item (2)			

(c) Assuming that this was a skeptical contractor who would like to see at least a 3:1 return on money invested in computer applications in order to become interested, and assuming that the given cost and benefit figures were reasonably accurate, what would be your recommendations, if any, with respect to (1) choice of applications and (2) sequence of implementation, if computers are to be used for estimating applications in this company? Justify your recommendations. You may wish to include nonfinancial considerations in your reasoning, such as the practicality of implementation, chances of acceptance by users, and impact on the contractor's growth potential.

2. Assume that, as an engineer for a medium-size heavy and highway construction contractor, you have been given the responsibility of developing a new equipment information system. The company currently does about $50,000,000 of work per year, and equipment is its major asset. At present there are no computers being used by the equipment department. Approximate *annual* expenditures now associated with equipment are as follows:

Ownership	
Depreciation	$5,000,000
Major repairs, overhauls, etc.	$3,000,000
Interest, insurance, taxes, etc.	$2,000,000
Operation	
Fuel, oil, grease, tires, field repairs	$6,000,000
Labor	$8,000,000
Total	$24,000,000

Assume that the costs or benefits of the various equipment cost functions can be improved through computer applications. On the basis of a preliminary feasibility study, the expected costs and savings of each of three possible applications are as follows:

Application	Cost/month	Savings
(1) Equipment records	$1,000	1% of ownership costs, not including repairs (inventory, costs, etc.)
(2) Equipment maintenance	$1,800	3% of major repairs, plus 2% of nonlabor operating costs
(3) Dispatching	$2,500	1.5% of total ownership and operating costs

(a) Briefly classify the costs and benefits into "economically quantifiable" and "significant but intangible qualitative" categories, with at least four items of costs and four items of benefits in each category.

(b) Complete the following table, showing, for each application, its annual cost, its annual savings, and the benefit/cost ratio.

Application	Annual cost	Annual savings	B/C ratio
(1) Equipment records			
(2) Equipment maintenance			
(3) Dispatching			

(c) Assuming that this was a skeptical contractor who would like to see at least a 6:1 return on money invested in computer applications in order to become interested, that the given cost and benefit figures were reasonably accurate, and that the annual budget for any combination of these applications should not exceed $40,000, what would be your recommendations with respect to (1) choice of applications, and (2) sequence of implementation (if any), if computers are to be used for equipment applications in this company? Justify your recommendations, particularly for any departures from the guidelines given here.

(*d*) Consider the nature of each of the three applications mentioned here, the probable difficulty of implementation, and the likely interest and training of the people who might be the users. Briefly mention some of the key issues involved and indicate how, if at all, these considerations might alter the financially based answers to part (*c*).

SUGGESTIONS FOR FURTHER READING

Brooks, Frederick P., Jr. *The Mythical Man-Month*. Reading, Mass.: Addison-Wesley, 1982. The classic book on the difficulty of managing software projects, with good ideas on how to overcome inherent problems.

Davis, Gordon B., and Margrethe H. Olson. *Management Information Systems: Conceptual Foundations, Structure, and Development*, 2d ed. New York: McGraw-Hill, 1985. A standard text on the analysis, design, and implementation of computer applications. The scope and depth are designed more for a business school-based course on information systems.

Lucas, Henry C., Jr. *The Analysis, Design, and Implementation of Information Systems*. New York: McGraw-Hill, 1985. Another good, practice-oriented text with useful guidelines for successfully developing computer-based applications. Also more of a business-school focus.

Martin, James. *Information Engineering, Book I, Introduction*. Englewood Cliffs, N.J.: Prentice Hall, 1989. This and the next reference are the first and second volumes of a three-book series by one of the most popular authors and seminar leaders in the field. Focused on industry practitioners.

Martin, James. *Information Engineering, Book II, Planning and Analysis*. Englewood Cliffs, N.J.: Prentice Hall, 1990.

McLeod, Raymond, Jr. *Management Information Systems*, 5th ed. New York: Macmillan, 1993. The most recent version of a popular text in this field. Quite readable for a construction professional.

CHAPTER
6

DESIGN AND DEVELOPMENT

[Computer pioneer and U.S. Navy Rear Admiral Grace Murray Hopper] enjoyed chiding her computer colleagues for playing it safe. The most dangerous phrase employed by computer people, Hopper would say, was "But we've always done it this way." To stress that, she kept a clock in her office that operated in counterclockwise fashion. After a while people realized that they could learn to tell time that way, and there was no reason why clocks had to run clockwise. She encouraged people to be innovative. One of her favorite pieces of advice was, It is easier to apologize than to get permission.

Robert Slater*

Design and development for a new computer application are analogous to design and construction for a new building. Design proceeds on the basis of information and criteria made available in the feasibility study and analysis, and draws upon the knowledge and experience of professionals, to set forth detailed plans and specifications for an end product. The production phase—whether for a construction project or a computer-based system—further enlists the knowledge and skills of professionals and artisans to produce a work that meets or exceeds the specifications. Both of these phases are creative activities, and they yield a unique, custom product. Success requires not only professional skill but also continued effective management to marshal the necessary resources toward the desired objective. There may be easier alternatives—leasing existing space in a building or buying a packaged computer system, for example—but they involve compromises deemed unacceptable to those embarking on original design and development.

Portraits in Silicon, 1987.

This chapter presents the design and development process in five main sections:

- Human factors to consider in application design
- Elements of designing computer applications
- Software development
- Testing and verification
- Documentation

For purposes of written organization, they are introduced sequentially, but in reality there are substantial overlaps. Human factors should be kept in mind throughout. Testing takes place during software development as well as at the end. Programming and documentation can and often should proceed concurrently, especially for documenting the code itself.

System design should also anticipate procedures for training, implementation, and managing ongoing operations. But as separate topics these will be deferred to Chapter 7, where they will be considered following an examination of direct procurement of a finished product as an alternative to original design and development.

HUMAN FACTORS IN SYSTEM DESIGN

The term *human factors* can encompass many things. At a high level it concerns organizational issues such as those addressed in Chapter 5. It might refer to the personal traits and needs of people doing the design and development work, particularly in regard to managing them successfully as personnel. But in this section we use the term to describe (1) the understanding of the physiological and psychological nature of different human beings as related to their abilities to work with computers and (2) the design of computer applications to accommodate human needs and limitations.

Relationship of System Characteristics to Human Behavior

There are some general characteristics of any system that influence the way people use it. For example, ***response time*** tells how long it takes for the computer to acknowledge a human action such as clicking a mouse button or pressing a "Return" key to select a menu item. Human sensitivity to delay depends on the context. In typing on a keyboard or drawing a line on the screen using a mouse, it is disconcerting if the delay between action and result approaches even one second; imagine trying to play a piano with one-second delays after playing the keys; think how much harder it is to carry on an international telephone conversation where transmission via a geosynchronous satellite causes words to lag by a second or so. In such cases, where actions and feedback are intimately linked, human thinking is confused even by momentary delays. Quick psychological feedback at this level very much contributes to a lively and responsive feel in a computer application, and lack of quick feedback makes an application seem sluggish and awkward.

In other cases, such as selecting menu item "Recalculate" for a large spreadsheet model, or entering a command that invokes an extensive sort or search procedure in a database, the actions cause *anticipated* delays; having initiated the action, one's mind can relax while it is carried out. Even here, however, we expect some *consistency* in response. Extensive searches or complex calculations are expected to take longer than short and simple ones. Procedures of similar complexity are expected to take similar times. If the computer replies too quickly, our surprise soon turns to disappointment if this means our action failed prematurely. If it takes too long, we become apprehensive that it is not only failing but also may be significantly damaging our earlier work as well. We feel uncertain about if and when to abort the action, which might be about to succeed given a few more seconds. Unpleasant surprise, disappointment, apprehension, uncertainty, and anxiety are all emotions that we usually would prefer to do without in computer applications—at least in professional life if not in videogames. Software can be designed to provide visual cues (e.g., filling in a bar graph scaled to the length of the process) and other types of feedback that reassure us about what is going on during longer procedures.

The widely used concept of *chunking*, sometimes expressed as the heuristic rule of "seven plus or minus two," also applies to the design of computer applications for human users. One theory here is that our short-term memories typically can only retain about seven discrete items of data while in the course of performing a task. For example, we can usually look up a seven-digit telephone number and dial it without rechecking, but many of us have to look again if we also must include an area code, and especially if we have to append an accounting charge code. Applied to computer software, the concept of chunking influences the syntax for command languages, the layout of menus and forms, information presentation in reports, and the hierarchical structure of subprograms.

The need for reassurance also is important when we design for easy *recovery from failures*—whether self-inflicted or attributable to faults in hardware and software. Anyone who has spent an hour or so of creative enjoyment writing or drawing with a computer program—only to lose his or her work accidentally by choosing "cut" instead of "save" in a command, by kicking out the power cord, or by deleting the file once it was saved—knows that feeling of despair not just for the time lost but of disbelief that doing it over will have the same magic as the first time around. Software can be designed to help protect us from such hazards of computer usage: "Undo" commands, messages like "Are you sure?", and automatic saving of work strengthen quality software today. Systems that fail often and impede recovery from errors quickly discourage people from using them at all if they can be avoided, or at least leave them wary.

System *accessibility* becomes less of an issue as computers become cheaper and more widely available, but administrators and budgets still impose limits, and some of these can be counterproductive. Placing microcomputers and terminals in central pools rather than at individual desks discourages routine usage. Just as some people will circle a parking lot for 10 minutes looking for a closer space to save a minute of walking, others will work by hand at their desks rather than walk down the hall to work quickly and accurately on a computer. Administrators may be delighted with records that show pool computers are well utilized, but high utilization may imply that contention for scarce resources is pushing other people to work less productively. Similarly, some companies place the latest and fastest computers with employees who have the highest titles, letting older models trickle down over time to those who actually use computers the most.

Methods of *accounting* for the costs of computing resources can also influence their usage. There is an analogy here with long-distance versus local telephone calls. If local calls are included in the base rate for having a telephone, one makes them casually, with no consideration to costs. In long-distance calling, people think more about what they will say before they call, and may even reschedule calls to times with cheaper rates. Similarly with computers, if there is no time-based charge, then we will use them casually whenever they seem to be useful in getting our work done. But if we are expected to keep track of time, disk capacity, and other measures of resource usage—perhaps so these costs can be billed to projects or departments—then we are more likely to be mindful of cost in deciding whether to use a computer, or do something the hard way, or not at all. There are merits to different cost-recovery schemes, but sometimes administrators become too preoccupied with tangible computer costs at the expense of the direct and indirect benefits for productivity, quality, and employee morale.

Human Factors in Device and Software Design

Computer devices and software should include concepts of *ergonomics* in their design. This term characterizes our physiological interaction with artifacts around us—such as the position of a driver relative to the controls in a car, a pilot in an aircraft, or an operator in a batch plant. But it also has implications in psychology, such as how well we respond to different colors and positions of words and objects in a graphic layout. Eyestrain, mental fatigue, and even headaches can be caused by flickering screens, fuzzy images, glare, heat, vibration, and high-frequency noises. They should be minimized or avoided in selecting and locating CRT display screens, disk drives, printers, and other devices. Poor seating, awkward keyboard positions, and even the tactile feel of keys themselves can cause muscle fatigue, backaches and ligament problems, and they further can contribute to errors while slowing productivity. Manufacturer and employer ignorance and neglect of such ergonomic factors have caused governments in some countries and states to start imposing safety and health regulations that set minimum ergonomic standards for computers.

Software can be designed to work with numerous alternative devices for input and output. In addition to keyboards of various types for input, we can choose a joystick, mouse, digitizing tablet, touch screen, light pen, bar code reader, or even speech-input device. For output, there is great variety just within CRT displays (size, color, resolution, text-oriented, graphics-oriented, etc.), and we also can choose audio speakers, printers of many kinds, plotters, and others. The choice of hardware must be made early so that software can be designed to suit it. Choices in a particular case depend greatly on the characteristics of the intended users. (Do they type? How well do they read? How often will they use the application? Can we train them?) The answers will differ for estimating in an office, designing with CAD, controlling materials in a warehouse, operating a batch plant, or running a motor grader with computer-automated blade control. For example, a computer-based crane instrumentation package has a liquid-crystal text display and an analog dial for output; its user inputs, in addition to numerous sensors, include a four-position switch to select French, English, German, or Spanish as the display language, and a few buttons to choose which parameters to show. It is a complete and valuable construction computer application but has no CRT screen, keyboard, or printer.

User Characteristics and Circumstances

Designers must be careful to match the structure and presentation of the application to the needs, skills, and knowledge of the users and to the frequency with which they are likely to use the system. To illustrate, in Fig. 6-1 are plots of learning curves for two different approaches to the same general application. In system A, the user quickly masters the capabilities that the easy-to-use software provides but over time finds the options limited, thus flattening the eventual level of proficiency. This design approach could be good for a system that has lots of new users or is used infrequently by most. In system B, on the other hand, the initial rate of

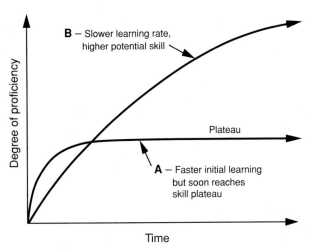

FIGURE 6-1
Rate of learning versus eventual proficiency in applications.

learning is slower, yet eventual productivity is higher. This approach could better suit users whose time and training to develop proficiency will be repaid by higher productivity in the long run.

Special consideration should be given to new users, whether for infrequent use or eventually to become proficient. Learners require explanations and prompting to help them choose appropriate responses; instructions could be embedded in a program introduction and elsewhere. Possible choices to a response could be illustrated or a typical response shown. Programs to be used regularly after learning could have both novice and expert modes, perhaps by querying the user initially and then suppressing verbose explanations for experts. A *help* option should remain available for further information. The syntax and grammar (i.e., the rules for the way things must be entered into the computer) should be kept simple. Dialogue methods particularly helpful to learners are menus of choices, form-filling, and question-answer formats. Mnemonic command languages with key words and parameters required to be in a certain strict order are much more difficult to learn but may be preferred by experts. Where used, vocabulary should be drawn from that of the user's familiar application, not from computerese. Useful redundancy can provide several ways to do something, as in today's window-oriented computer programs, which often provide keystroke equivalents for frequently used menu commands. The menus themselves remain as an aid to learning new commands. The designer also should be attentive to the aesthetics of screen design and the context they create, to help the users remember where they are in the program. Errors need to be anticipated, explained, and processed smoothly.

Designers must consider how much of the user's attention will be focused on using the computer software itself versus its application. For example, a construction dispatcher, like an airline reservation agent, will most likely be focused on the field superintendent at the other end of a telephone or two-way radio conversation while using a computer database to see if the desired equipment is available. Simultaneously, the dispatcher might initiate a second call to a rental agency to see what is available there. The program itself had better be easy enough to operate almost subconsciously. Similar circumstances apply with an estimating bid-sheet program used to analyze and select reliable low bids from the supplier and subcontractor quotes that come pouring in during the last hour before the proposal is due. Mistakes here can be very expensive, so the computer program must not add to the chaos. On the other hand, a scheduler should be able to focus more intensively on a scheduling program, and can take time to learn the large range of features and capabilities that modern scheduling software offers. In any case, the designer must be well aware of the circumstances in which the application will be used and design accordingly.

Improving User Understanding of an Application

What makes some software applications seem natural or intuitive while others are difficult and confusing? What makes some interesting while others are boring? Logical sequencing and layout of operations within programs help a great deal.

Graphical metaphors for familiar objects such as windows, desk tops, desk accessories, and the like make it seem natural to confine different applications to different rectangular areas (windows) on a screen, which itself resembles a glass viewport into the information contained in the computer. With a mouse we can attach to or "pick up" objects on the screen (such as tabs or arrows in scroll bars) to move windows around or move text or drawings within windows. In writing, we can "cut" and "paste," and in drawing we can have enough "tools" to make an artist envious. Similarly, designers can provide variety in interaction, both to enable a user to customize an application to taste and to change it over time.

In our metaphors and attempts at logic, however, we must be careful not to add to a user's misunderstanding of what really goes on in a computer. To some extent it is good to have a conceptual understanding of what happens behind the scenes—to have a semantic as well as a syntactic understanding of an application. *Syntactic understanding* implies rote memorization of the various rules and procedures that make it possible to use a given tool, be it an automobile or a computer program. "Let's see. The book says to turn the key all the way to the right to start the car. Wow! It must be that turning the key is what makes cars start." Later, when a low battery makes the engine barely turn over, this person tries turning the key harder, and breaks it off in the lock! *Semantic understanding* includes the concepts or principles behind a technology or method, not just what to do according to prescriptive rules—the why and how as well as the what and when. Somebody who understood that the key just turned a switch that tripped a relay to allow electric current to flow from the battery to the starter motor, which in turn rotates the engine fast enough to take in air and compress it with fuel that is ignited and starts the engine going, which in turn runs an alternator that charges up the battery for future starts. . . . In any case, somebody with a semantic understanding of how a car works would not have broken the key off in the lock.

Most computer applications rely too extensively on syntactic explanations to teach users to work with them. Users often report being "mystified," "baffled," "confused," and "ready to scream" when a certain combination of syntax produces an unexpected result: "It just quit on me!"; "All of a sudden the beginning text was at the end!"; and "When I went back, my file was gone! I know I saved it." The first user might have accidentally typed CTRL-Z instead of CTRL-something else (and the software developer likely forgot to disable this common system-level command in the program); the second may have done an inverse sort (which the programmer put in the cut-and-paste menu for some reason); and the third's file might have been alive and usable in a different directory, had the user even realized that there was a hierarchical directory structure on the computer. Software and its documentation should be designed to provide a better semantic understanding of its underlying concepts.

DESIGN

It seems reasonable that a skilled and experienced craftsperson could build some shelves or even a garden shed using only a rough sketch for guidance, tools from the cabinet in the garage, and a tentative shopping list of materials to take to the lumber store. But one would not start to build a house, let alone an industrial plant or hydroelectric project, without detailed plans and specifications. The scope of the design effort should be commensurate with the magnitude of the project. Similarly, a knowledgeable user might quickly whip up a spreadsheet to do a bit of calculation and reporting for a specific one-shot need, and do so working directly at the computer, thinking it out along the way, revising as necessary. But this would be a foolhardy way to start developing a large and complex construction application, such as a construction estimating system, a materials management system, a payroll system, or an equipment management system. Following a thorough feasibility study and analysis, we next need a detailed design.

This section will look at the components of application software design in the following main categories:

- Output reports and procedures
- Input forms and methods
- File and database type, structure, and content
- Programming methodology, structure, and procedures
- Testing and verification standards and procedures
- Equipment procurement specifications
- Documentation requirements and standards
- Training methods and requirements
- Implementation and conversion plan
- Operations procedures
- Revising cost and schedule estimates
- Consolidating and reporting on design

Starting with the output and then working backward for awhile to programming may at first seem odd, but this sequence is analogous to coming up with the functional and spatial layout for a building before designing the structure to support it. Also, designing reports and input forms before any programming has been done on the system that will produce them is quite easy. Therefore, we will start with the desired results and work back from there.

Output Reports

The design of output reports includes not only their layout and content but also the procedures that are associated with their use. Specific points to consider in designing a particular report include the following:

- Intended audience: external versus internal, trained versus casual, and so on
- Content, format, and number of copies (if paper)
- Method or media for presentation (e.g., CRT display, printed, graphic plotter)
- Access restrictions and circulation priorities

- Security while in use and for disposal
- Procedures and schedule for production and distribution

There are several useful tools and methods for designing reports. For designing paper reports, one can obtain preprinted blank forms that contain a faint grid in a matrix the height and width of a typical printer page (e.g., 132 characters wide by 56 lines long). With no more than a pencil and an eraser, one can lay out title blocks, column headings, and sample output data fields for a typical report. If, in discussions with users, it makes sense to revise or relocate the title or headings, this change can be made readily with the eraser and pencil—no special computer skills required! Similarly, a standard word processor or drawing program can be used to design CRT reports and simulate how they might appear on the screen. Rather than having to change and recompile a computer program to make adjustments in the appearance of the report, one can do it quickly and conveniently as one normally uses the program, even with a potential user watching nearby, just by moving the cursor, typing, inserting, and deleting characters and lines. Furthermore, by jumping from one screen design to another, one can partially simulate the sequence in which screens would appear while interacting with the application once it was completed, yet still incorporate recommended changes before programming even begins.

Special software has been created to help developers design and simulate the behavior of a program in a similar but somewhat more sophisticated manner than that just described. For example, the co-inventor of the original VisiCalc spreadsheet program subsequently published a developer tool called Dan Bricklin's Demo Program. Apart from not processing real data, it indeed simulated the behavior of typical business application programs. More recent products of this type include Demo II (Intersolv), ShowPartner (Brightbill-Roberts), and Proteus (Genus).

With reports thus designed via paper drafts or screen simulations, the programmer can concentrate on writing code that directly implements the desired result. This approach is far superior to doing the draft in computer code, and then going through a series of draft edits and code changes to get to the same result; this would be analogous to having carpenters build and tear down walls until the doors and windows end up in the right place.

The following three figures illustrate specifications for some reports designed for an equipment information system. The intended users' location, such as project sites or head office, and the contents and type for all reports in the system are shown in Table 6-1. Figure 6-2 is a tabular printer report, and Fig. 6-3 is an illustration of a graphical report.

Input Methods and Forms

The previous section on human factors mentioned a wide range of input methods available to suit different applications and people. These ranged from keyboard or

TABLE 6-1
Summary specifications for output reports[*]

Report and use	Contents	Type
	Project sites	
Project progress report	Planned and actual progress comparison—in terms of whole project, job, machine	Graph
Productivity report	Weekly and monthly production rate and unit cost calculation—in terms of job, machine, operator	Table, graph
	Head office	
Project progress report	Planned and actual progress comparison—in terms of whole company and projects	Graph
Productivity report	Weekly and monthly production rate and unit cost calculation—in terms of project, machine	Table, graph
	Whole company's historical production rate and unit cost analysis	Table, graph
Dispatching report	Possible dispatching selections for machines and operators, or their combination	Table
	Cost comparison of in-house and lease machine	Table
Maintenance report	Maintenance record of each machine	Table
	Periodic maintenance necessary in the next period	Table
Inventory report	Historical record and specification of each machine	Table
	Installment to be paid in the next period	Table
	Depreciation calculation for tax purposes	Table

Source: Adapted from a table prepared by Mr. Takeshi Shibata for a course at Stanford University on computer applications in construction, 1986.

*All output forms can be displayed both in CRT and hard copy.

Monthly Machine Production Rate			Machine Type: Scraper					Month: July 1992	
Proj.	State	Type	Machine	No.	Prod (CY)	Op hrs	Prod rate	Hourly cost	Unit price
A	N.CA	HW	CAT 631D	2	25,200	250	100.80	$80	$0.79
B	N.CA	HW	CAT 637D	2	31,860	226	140.97	$120	$0.85
C	N.CA	DEVP	CAT 637D	5	81,485	542	150.34	$120	$0.80
D	N.CA	DEVP	CAT 657E	3	80,940	349	231.92	$180	$0.78
E	S.CA	HW	CAT 631D	1	12,510	87	143.79	$80	$0.56
F	S.CA	HW	CAT 631D	2	29,770	196	151.89	$80	$0.53
G	S.CA	DEVL	CAT 637D	3	54,800	309	177.35	$120	$0.68
H	NEV	HW	CAT 631D	2	41,355	240	172.31	$80	$0.46
I	NEV	Other	CAT 621B	1	8,790	102	86.18	$70	$0.81
J	ARIZ	HW	CAT 621B	1	15,380	125	123.04	$70	$0.57
K	ORE	HW	CAT 631D	2	28,200	180	156.67	$80	$0.51
L	ORE	DEVL	CAT 657E	4	74,650	360	207.36	$180	$0.87

FIGURE 6-2
Example of printed report design. (Adapted from a figure prepared by Mr. Takeshi Shibata for a course at Stanford University on computer applications in contruction, 1986.)

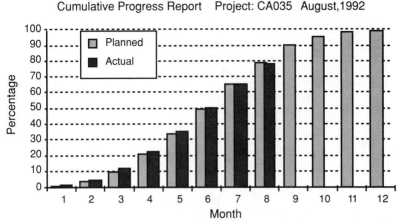

FIGURE 6-3

Example of graphical report design. (Adapted from a figure prepared by Mr. Takeshi Shibata for a course at Stanford University on computer applications in construction, 1986.)

mouse input to touch screens and bar code readers. Some of these are specialized and thus are beyond the scope of our general overview. Most applications use a keyboard or a mouse to make selections, answer questions, and fill in forms displayed on a CRT screen. Closely related to screen design is the design of paper input forms that someone in the field fills in by hand and then submits for clerical staff to enter into a computer as part of a batch of transactions. In either method, some of the following considerations should influence the design:

- Document type, format, and content
- Data entry schedule and procedures
- Data checking (typing, logic, and errors in reasonableness)
- Error correction procedures (e.g., online error messages)
- Batch input by clerical staff versus online input by application users
- Technology (e.g., CRT, OCR scanner, bar code, etc.)

Tools similar to those used for designing reports can also help design menus, screen forms, and messages encountered during data entry. These include preprinted blank forms for drafts, standard word processor or drawing programs to design CRT displays, and design-demonstration programs. Again the designer can simulate how data-entry interaction might take place before actually writing program code to implement the designs.

Table 6-2 is a summary of specifications for input forms to be used in the aforementioned equipment information system. Figure 6-4 contains designs for three typical input forms that might be printed on paper or appear on a CRT screen.

TABLE 6-2
Summary of specifications for input forms

Input form and use	Contents	Method
	Project sites	
Job card	Date, number, project code, job code, operator name, operation hours, production quantity, comments	Form filling
Field maintenance card	Date, project code, machine code, maintenance code, quantity of parts, fuel code, gallons, comments	Form filling
Project schedule change	Date, job code, start date, finish date, quantity, machine code	Q & A
Machine list updating	Date, machine name, machine code, expected period (until _____)—add, delete	Q & A
Operator list updating	Date, operator name, operator code, expected period (until _____)—add, delete	Q & A
Dispatching request	Date, project code, machine code, operator code, expected period (from _____ to _____), comments	Form filling
	Head office	
Shop maintenance card	Date, machine code, shop code, maintenance code, quantity of parts, charge, comments	Form filling
Machine registration	Date, machine name, machine code, specs, source (name, tel, contact), installment schedule depreciation plan, comments—add, change, delete	Form filling
Operator registration	Date, operator name, operator code, salary, personal record (age, address, etc.)—add, change, delete	Form filling
Project list updating	Date, project name, project code, place, scale, duration, budget, progress schedule, comments —add, change, delete	Form filling
Machine cost	Date, machine code, purchase price, lease price, region factor, information source—change	Q & A
Installment payment	Date, machine code, invoice number, year, month —add	Q & A
Maintenance cost	Date, maintenance code, price, unit, source —change	Q & A

Source: Shibata, 1986.

File and Database Design

Chapter 4 provided information pertinent to the design of files and databases. Record layout such as that shown in Figs. 4-2 and 4-3 and bubble diagrams such as that in Fig. 4-6 apply well in file and database design. As a general overview, however, the following broader range of factors should be considered:

- Type, content, usage, media, and so on
- General structure (sequential, direct access, database, etc.)

```
***** Job Card Input *****

Date               _____
Project code       ___ - _____
Job code           ___ - _____
Machine code       __ - ___ - ____ - __
Operator name      _____
Operation hours    Start    Finish
                   ____ - ____
                   ____ - ____
                   ____ - ____
                   ____ - ____
Production unit        ____
Production quantity    _____
Comments _____
(up to 30  _____
   chars)
                 Press ? for help
```

```
***** Dispatching Request Input *****

Date               _____
Project code    ___ - _____
Machine code    __ - ___ - _____ - _
Operator code   ___, ___, ___, ___
Expected period  from _____
                   to   _____
Comments_____
(up to 100 _____
   chars)  _____
           _____
                 Press ? for help
```

```
***** Machine Registration Input *****

Date  _____
Name  _____
Code  __ - ___ - _____ - _
Specs Flywheel power     _____ HP
      Operation weight   _____ lb
      Top Speed (loaded) _____mph
      G.V.W.             _____ lb
Purchase source _____
  Tel ___ - _____ Contact _____
Installment  Down payment $ _____
             Monthly charge $ _____
Depreciation  Method    _____
              Period  ____ years
Comments _____
(up to 100 _____
   chars)  _____
           _____
                 Press ? for help
```

FIGURE 6-4
Examples of input form designs. (Adapted from Shibata, 1986.)

- Volume and frequency of transactions
- Reliability, security, backup, restoration
- Procedures and controls for data entry, modification, and access
- Interface to files of other applications

The intended storage media and the technology for files (e.g., indexed sequential, direct access) or databases (indexed, relational, etc.) depend on the nature of the application. Design details will include data types, field sizes and numbers of fields within records, number of records and their interrelationships, methods for adding, searching, and retrieving records, the relationships between files in a database, and so forth. The specification should include overall file or database capacities, target search and retrieval times, and outlines of file contents such as

TABLE 6-3
Summary of file specifications

Jobsite transaction file	
Data structure:	Sequential
Data elements:	Job cards (date, project code, job code, machine code, operator name, operation hours, production, comments)
	Field maintenance cards (date, project code, machine code, maintenance code, quantity, fuel record, comments)
	Request (date, project code, machine code, operator code, period)
Jobsite master file	
Data structure:	Database
Data elements:	Accumulated job and field maintenance records for a period
	Project progress schedule (in terms of production and cost)
	Current list of machines and operators used in the project
Headquarters transaction file	
Data structure:	Sequential
Data elements:	Accumulated data transferred from each project site in one day
Headquarters master file	
Data structure:	Database
Data elements:	Sorted data from all projects (job and maintenance records)
	Machine and operator inventory (machine code, name, spec, history /operator code, name, personal records)
	Project list (project code, name, progress schedule, budget)
	Market price and lease cost of machines
	Installment and tax payment records
	Dispatching request list from the projects
	Standard maintenance cost table (tire, fuel, oil, filter, etc.)
	Company's historical performance data

Source: Shibata, 1986.

that shown in Table 6-3. More detailed specifications for each file would also note field widths, data types, pointers, and such.

Program Design

For the forms, reports, and data files mentioned in this section, several procedural languages or database development environments could produce applications that, superficially at least, were indistinguishable by the user. We must choose a language or environment that is likely to suit the application as well as the skills available to implement it. On the assumption we have made a choice, the following considerations should go into a specification for how the application software should be developed:

- Hierarchical structure of components; define bounds on modules; interfaces
- Standards for declaring local and global *variable* types, names, and sizes

- Methods and standards for passing variables among subprocedures (e.g., global declaration, argument lists, interim files, spreadsheet links)
- Input: extent of data screening and checking, reliability, interface to user, assumptions about user, use of forms
- Processing: what must be done; algorithms and standards
- Output: methods of selecting reports; display or printing options
- Audit trails (e.g., to trace path of data through financial accounting programs)
- Speed and efficiency
- Contingencies: failure, interim restart, exceptions, and so forth
- Portability and maintainability (within limits of chosen tool)

Chapter 4 gave some idea of the comparative advantages and disadvantages of various programming tools to satisfy these criteria, and Part IV will describe several tools, so we will stop with the preceding list for purposes of design.

For our example, Fig. 6-5 includes the hierarchical relationships among the equipment system's modules. Table 6-4 is a brief summary of a specification for application software to handle the equipment information system. Much more detail would be required for the final design. For example, flowcharts for specific processing routines will be introduced later in this chapter.

Testing and Verification

The nature of testing depends on the software development methodology to be used. A few general points to keep in mind at the design stage are as follows:

- Allow for testing modules and partial modules during development
- Design procedures to successively integrate and test components of software, files, and hardware up to and including the final system
- Test both for exceptional conditions and for realistic production volumes
- Some cases will require approved standard procedures for verification of results

We will return to testing later in this chapter in the sequence where it normally takes place in the development process. At this point we will just recognize that designers should anticipate the need for testing and prepare specifications for the extent, methods, and results that will satisfactorily assure users that the application software works correctly.

Equipment Specifications

In the design stage we also need to develop specifications for the computer equipment that will run the planned application. That way, some can be ordered and delivered for use in software development, and the rest can arrive in time for training and implementation. Chapter 3 discussed hardware, and Chapter 7 will

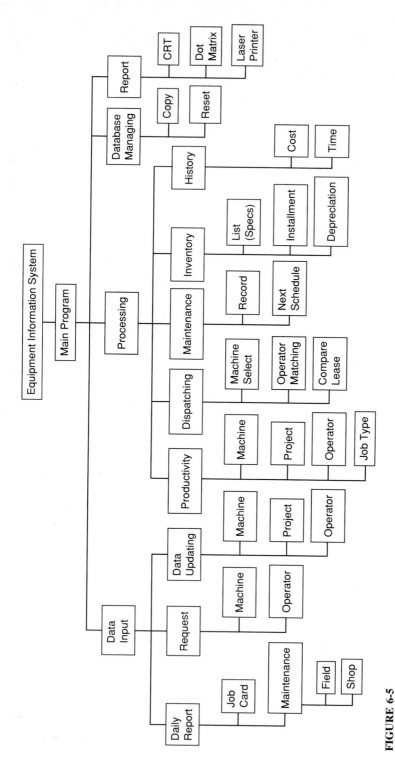

FIGURE 6-5
An example of system hierarchy. (Adapted from Shibata, 1986.)

TABLE 6-4
Program specification

Program on jobsite microcomputer
Local use, independent from the main system except for data transfer

Subroutines
    Data input: Daily input of job cards and field maintenance cards
                Updating of the project progress schedule
                Updating of the machine and operator lists (add, delete, change)
                Request for machine and operator dispatching
    Processing: Project progress analysis
                Production rate and cost analysis
    Database manage: Copy database to floppy disk and reset database contents
    Report: Weekly and monthly project report printout
    Data transfer: Send daily inputs (transaction file) to the head office
Development software
    Programmable generic database: for Processing and Report menus
    Programmable communications package: for Data transfer

Program on equipment department microcomputer
Central use at the head office as the main system

Subroutines
    Data input: Updating the machine and operator inventory (add, delete, change)
                Updating the current project list and related data
                Updating the market price and lease cost of machine
                Input monthly installment and tax payment transactions
                Updating standard maintenance costs
    Processing: Production rate (in terms of machine, project, operator, and job)
                Dispatching (machine selection, operator matching, economic com-
                    parison with lease)
                Maintenance (record of each machine, schedule for the next period)
                Inventory (listing of machines with specifications, installment
                    schedule for the next period, depreciation calculation)
                Historical analysis (project progress, company performance)
    Database manage: Copy database to floppy disk and reset database contents
    Data receive: Mode to receive the files transferred from the project sites
Development software
    Pascal procedural language: for Processing and Report menus
    Programmable generic database: for Database construction purpose
    Programmable communications package: for Data receiving

Source: Shibata, 1986.

describe procedures for its procurement. Here we will just summarize general considerations for this process.

- Identify any excess capacity that may be available on existing equipment
- Consider new or upgraded processors, larger disks, extra terminals, and special peripherals to enhance capabilities of existing equipment
- Determine types and amounts of new equipment that will be needed
- Develop requirements and specifications for the new equipment

- Define vendor evaluation criteria
- Prepare requests for proposals (RFP) or requests for quotations (RFQ)
- Solicit and evaluate proposals from reputable vendors
- Order equipment to be delivered according to the desired schedule

Chapter 7 discusses requests for proposals, vendor relationships, and purchase contract legalities in more detail.

Specifications for Documentation

The types needed and standards required for documentation should be specified in advance, although details will depend on the software that is developed. Specifications should deal with the following aspects of the documentation process and thus make clear the standards that are to be expected:

- Describe the content of each type of documentation desired (e.g., what types of charts and figures to put in the programmer's manual)
- State the quality of writing and organization (e.g., whether or not technical writers should be involved, or perhaps specify the use of automatic spelling, grammar, and style checkers)
- Request the inclusion of tables of contents and reference indexes
- Describe the way manuals should be typed, printed, and bound (e.g., laser-quality output using specified fonts, style, spacing, and layout—perhaps attaching an example style guide; xerox or offset printing; labeled tab dividers between sections; and spiral ring binding so manuals will open flat)

One could also cite example documents that illustrate the standards desired, but it would be unreasonable to expect manuals and tutorials of commercial package quality for a custom application. Specifications should require that detailed outlines be submitted and approved before developers proceed to write manuals. We will say more about what goes into each type of documentation later in this chapter.

Training Methods and Requirements

This section of the design should identify the type and number of people who will need training, estimate the extent and timing of that training, state the preferred location, and indicate the methods that are desired. Alternatives might include the following:

- Training in full system capabilities for some users; concepts only for others
- Training to take place in-house, at the vendor's site, or at a third-party venue
- Possible methods, such as in-house courses, on-the-job tutoring, self-study books, audio cassettes, videotapes, online tutorials (possibly multimedia), and the like
- The type and amount of training and reference materials given to users

The specific approach will depend on the number of people to be trained, the complexity of the application software, and other factors. If only a few users are involved, then perhaps the developer can work with them informally for a short time, providing on-the-job training to get them started, and then be available by telephone or on follow-up visits to answer questions that arise. On the other hand, if dozens of people need training, seminars and in-house courses become the likely mode. Only if there are hundreds of people, or if the application involves a widely used package program, do self-study books, videotapes, and online tutorials come into the picture.

Chapter 7 will further explore the subject of training as a phase in implementation, but it should still be anticipated and tentatively planned at the design stage.

Implementation and Conversion

Implementation is the process of introducing the new application to its target users, and conversion is the process by which data and procedures from the old system transfer to the new one. General goals are to minimize disruption, avoid risks of failure, and create a positive climate for the new system. The design should specify the following:

- Implementation strategy
- Activities (e.g., install hardware and software, convert old data, etc.)
- Responsibilities (to plan, to supervise, to assist)
- Resources required (e.g., temporary personnel, consultants, etc.)
- Schedule (lead-in, conversion, follow-up)

Chapter 7 will discuss four main alternative strategies: total conversion, parallel operations, multiphased, and pilot project. Each has its advantages and disadvantages, depending on the type of application. Designers should think ahead to this stage. Perhaps provisions can be made in software development, file design, equipment acquisition, training, or otherwise to facilitate the implementation and conversion process.

Operating Procedures

The design should be influenced not only by how the application will be used but also by the operating procedures that will make it more effective and dependable. These include considerations such as the following:

- Staffing and responsibilities
- Input and output procedures
- Procedures for processing under various design conditions
- Procedures for backup and security
- Procedures for dealing with exceptions, errors, and failures

This subject shades into overall computer system management, which will be covered in more detail in Chapter 7. Table 6-5 is an example of a procedure specification that focuses on the equipment information system.

Revising Cost and Schedule Estimates

On the basis of the design for each of the system components described in this section, one should now be able to prepare revised assessments of costs, benefits, resources, and schedules. In particular, hardware to be acquired can be priced more accurately than was done in the feasibility study, since it has now been specified in some detail. Knowing the approach to software development and having a better idea of what is involved should help in that regard. Documentation, training, conversion, and operating procedures should provide a better idea of the people involved and their time commitments, which should further refine costs and schedules. Benefits may be only a little more clear, but in any case they can be weighed against the better picture of costs.

Consolidating and Reporting on Design

At the end of this second major stage in the planning and development process, it is time to consolidate findings and recommendations produced thus far and report them to management and users. Specifications similar to the examples presented in this section will be attached to the report. Approval and support should be forthcoming before proceeding to write programs, prepare documentation, or order equipment.

We will now assume that we have such approval and support and continue with development and implementation. The remaining sections of this chapter cover those steps that apply to developing custom applications: programming, testing, and preparing documentation. Steps common to both custom development and direct procurement of package systems will be deferred to Chapter 7. These include equipment and software acquisition, training, implementation, and ongoing system management.

PROGRAM DEVELOPMENT

In a computer application development project, programming is analogous to the construction stage of a building project. Skilled professionals work with their tools and materials to create the product that has been designed. As with construction, strong organization and management provide the framework in which the resources can best be applied, and without them the project can descend into chaos and end in failure.

The underlying complexity of software development is illustrated by the iceberg analogy in Fig. 6-6. On the surface, one sees the program under development. Just below are the editor and language translators or the generic development software used most by the programmer. Deeper beneath the surface, however, is a

TABLE 6-5
Operations procedures

Step	Time	Procedure (needed time)	Responsibility
		Project sites	
1	16:45	Fill out job cards manually about what has been done with production in that day (15 min)	Job Superintendent
2	16:45	Fill out field maintenance cards manually about what has been done with equipment in that day (10 min)	Superintendent (Equipment)
3a	Any time	Input project schedule change data directly into microcomputer (15 min)	Project Manager
3b	Any time	Fill out dispatching request card if necessary (5 min)	Project Manager
3c	Any time	Input machine and operation list updating data directly into microcomputer (10 min)	Superintendent (Equipment)
4	9:30	Type the last day's filled-out cards into microcomputer and save them as a transaction file (30 min)	Clerk or Secretary at the site office
5	10:00	Check data errors in the transaction file (5 min)	Field Engineer
6	10:10–12:00	Send the corrected transaction file to the head office computer whenever telephone line is open before noon and incorporate the data into the local master file (10 min)	Field Engineer
7	Any time	Do processing work and print out weekly and monthly project report of which information is limited to the locally obtained data (depends on work)	Project Manager or Field Engineer
		Head office	
8	In morning	Same procedures as above are taken in the equipment department at the head office regarding shop maintenance cards filled out last day (45 min)	Clerk of Equipment Department
9	13:00	Check data errors in the transaction file, which consists of the transferred data from each project site, and incorporate them into the master file (30 min)	Computer staff
10a	Any time	Input machine registration data directly if needed	Equipment Director
10b	Any time	Input operator registration data directly if needed	Personnel Director
10c	Any time	Input project list updating data directly if needed	Operation Director
10d	Any time	Input machine cost and maintenance cost updating data directly if needed	Clerk of Equipment Department
10e	Any time	Input installment payment data directly if needed	Accounting Director
11	Any time	Do processing work and print out weekly and monthly reports in the provided formats (depends on work)	Equipment Director, Computer staff

Source: Shibata, 1986.

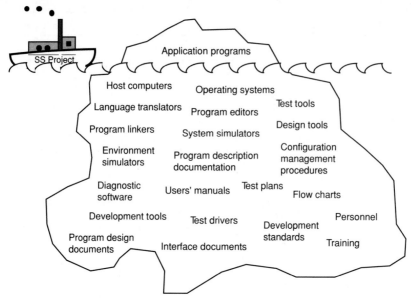

FIGURE 6-6
Software to support application development. (Adapted from H. Mark Grove, "DOD
Policy for Acquisition of Embedded Computer Resources," *Concepts, The Journal of
Defense Systems Acquisition Management*, vol. 5, no. 4, Autumn 1982, Figure 6, p. 19.)

substructure of other software systems and tools, and their quality and capabilities
also contribute to the end result. Many of these were explained in Chapter 4.

This section will discuss principles that can enhance application programming efforts. These include the following:

- Freeze scope before starting
- Apply project management techniques
- Select the appropriate tools
- Use proven programming methods for software development
- Maintain quality and accuracy in programming

Programming methodologies themselves are not described at this point. That subject is deferred to Part IV of this book.

Freeze Scope before Starting

Construction project managers know how frequent change orders can wreak havoc
with schedules and budgets once construction gets under way. The same problem
impacts software development, which is even more vulnerable; the ramifications
of change are much more clear when something physical is being built of steel,
concrete, and lumber. To laypersons, it seems so easy to ask the programmer just
to make a few changes here and there. They do not appreciate how changes to an

earlier module upon which days or months of subsequent work have been based can cause much of that effort to be scrapped.

Assuming that programming can start with a good design, management should freeze the scope at the outset so that the computer professionals have a chance of holding to their schedule and budget. No doubt various users and even programmers themselves will think up new ideas to improve work in progress, but such tendencies must be channeled constructively. Most suggestions can simply be collected and held for later consideration when the software undergoes its next major revision. However, some changes will indeed be compelling—perhaps to correct a flaw in the design or to add something subsequently deemed essential. Formal procedures should be established for submitting these change requests, evaluating their impacts and benefits, and obtaining the appropriate approvals—with additional resources and time extensions if needed—before allowing these changes to be included in the scope of current work.

Project Management

Project management for software is similar to that for construction: set objectives, define tasks, delegate assignments and responsibilities, monitor schedules and budgets, assess performance, and make corrections to keep the project on track. Indeed, the very same computer-based project management tools that assist construction project managers—CPM programs, cost-control systems, and such—are often used by their counterparts in the computer profession.

What differs, of course, is the nature of the work being managed. Whereas construction projects involve interdependencies between foundations, structural work, mechanical and electrical systems, finishes, and the like, computer programmers worry about interdependencies in program logic, integration of modules in a certain sequence, and having files and databases ready to test. Quality control involves various levels of testing and debugging, and construction's "punch list" has its counterpart in final system verification. Libraries of program modules are similar to prefabrication, and offsite work in specialty shops may be similar to cross-development on more sophisticated computer systems. Just as constructors worry about satisfying building codes, computer professionals try to adhere to established programming standards, especially if portability to other computers is a design goal.

Select the Appropriate Tools

A central consideration in application development is the choice of tools to use, particularly the programming language or generic application program. The following types of factors go into the decision:

- The nature of the application itself: Will it have large or complex data files? Is execution speed important? Does it need a graphical interface?
- Programming skills of those who will develop and maintain the application

- The degree of structure, internal documentation, and standardization needed
- Capabilities and quality of the tool that implements the chosen language
- Quality and reputation of the firm that developed the tool
- The tool's capabilities for debugging, testing, and verification
- Vendor support such as new releases, telephone consulting, and training
- Conformance to recognized standards (restrictions and extensions)
- Ability to interface to other tools (e.g., procedural language access to a database)
- color graphics CRT, analog input from instrument, etc.)
- Availability of libraries of subroutines (e.g., graphics, statistics)

This list is by no means exhaustive, but it gives some indication of the range of things to consider. The choice will be the one that best suits the needs of the application, the skills of its developers, and the performance desired by its users.

Where transferability among different types of computers is of major importance, high-level languages adhering to recognized standards should be used; otherwise, translating from one language into another or even from one dialect into another can take an enormous amount of time. Most implementations of even a standardized language have extensions made by the particular manufacturer; these should be avoided if the program is to be implemented on other machines. If machine-dependent features are necessary, they should be isolated in the program and be well documented.

Sometimes a *two-stage development process* is appropriate, each stage using entirely different kinds of tools. In the first, a prototype is built employing tools that are relatively easy to use, such as programmed procedures using a macro language built into a generic spreadsheet or database development package, or perhaps a high-level expert system shell. The result of this effort could be tried for a while, perhaps by a pilot test group such as a specific construction site office. Problems could be corrected and recommended enhancements could be implemented during the trial period. In the second stage, following an evaluation to confirm that the prototype has been successful, the design it embodies could then be reimplemented in, say, a compiler-based procedural language to gain speed or capacity. The resulting product could then be disseminated and integrated into all of the company's applicable operations.

Programming Methods

Programmers take overall guidance from the design documents mentioned earlier in this chapter. Before programming, they sometimes further refine and detail the design using techniques such as *flowcharts*, at least for some of the more complex or critical modules. Figure 6-7 is an illustration of a plastic *template* containing symbols used to draw such flowcharts, and the equivalent result can be achieved using computer-aided-design software intended for this purpose. Figure 6-8 is a demonstration of a simplified flowchart for a prime number calculation. Other techniques employed by programmers to describe the design more explicitly are *decision tables* and *pseudo code*, the former being a formal expression of logic and

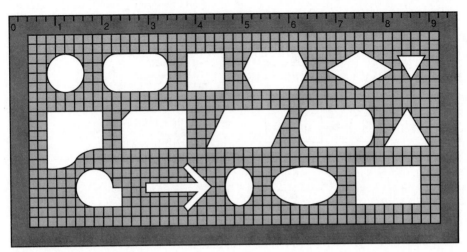

FIGURE 6-7
Template of flowchart symbols.

the latter a general way of writing procedures that can then be readily translated into the specific syntax of a given procedural language. Pseudo code enables the programmer to concentrate on *algorithms* rather than language syntax while working out the details of the programs and procedures.

How the program is written is as important as the language in which it is written. Logical and clear *flow of control* is critical to the readability of a program by someone who did not write it (and sometimes even to the one who did). Recommended control structures are sequential execution of statements, forward skipping upon the results of a stipulated test, and iteration. Variables and subprocedures should be defined before they are used in subsequent sections of the code. Variables should be isolated in submodules to the extent possible rather than exist globally throughout large sections of a program; this allows them to be independently validated before they are integrated.

The program should be organized into sections or *modules* that perform particular tasks. These modules are typically called from a main control section. Each module should be restricted to a size that can be easily understood. The separation of a program into modules enhances its readability and provides reusable code for other programs. Libraries of such general modules can often significantly reduce programming effort. Another benefit of modularity is that it eases the implementation of large programs on small computers by overlaying segments of code as the program executes.

In recent decades, programmers have debated the philosophy and specific techniques for *structured programming*. Traditionally, design proceeded *top-down*, but program modules were implemented *bottom-up*. Most now argue that the top-down approach should also be used for program development, with dummy routines temporarily holding the place of subordinate modules to be developed later. The bottom-up approach sometimes results in inconsistencies and incompatibilities as submodules are combined into bigger sections, resulting in

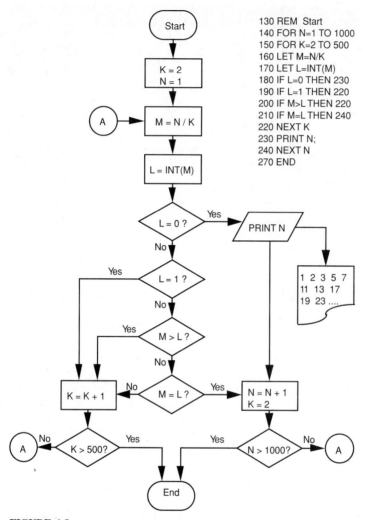

```
130 REM  Start
140 FOR N=1 TO 1000
150 FOR K=2 TO 500
160 LET M=N/K
170 LET L=INT(M)
180 IF L=0 THEN 230
190 IF L=1 THEN 220
200 IF M>L THEN 220
210 IF M=L THEN 240
220 NEXT K
230 PRINT N;
240 NEXT N
270 END
```

FIGURE 6-8

Example flowchart for calculating prime numbers to 1000; equivalent
BASIC code shown at upper right.

considerable rework to resolve the problems between them. A thorough design
and rigorous specifications help to minimize this problem, but so does top-down
programming.

The importance of thoroughly debugging a program before serving it up
before a group of demanding users need not be emphasized to those who have
done it. It is equally important for program modules under development. Minimum
and maximum ranges of input variables should be tested for their effect upon the
program. In general, the integrity of each module should be verified before it is
integrated with other parts of the program. The next section has more to say on
this subject.

TESTING AND VERIFICATION

Organizations should not implement programs that have not been tested by a live audience, at least not without a clearly specified caution. Good programming practice suggests that programs receive extensive pilot tests with real users before they are released. The following paragraphs amplify some of these ideas.

As individual program modules are developed, it should be possible to test them even though the modules with which they will be interdependent are not yet ready. For example, to test one of the computational modules in a system, a text editor might be used to create files of interim input test data—data that eventually would be produced by the data-entry module of the system. Similarly, one might write simplified routines to perform a function eventually to be handled by a module that will interface to the one being tested.

As development proceeds, there should be a planned sequence in which modules will be successively combined and tested for their interaction as well as their individual performance. If development is being done according to rigorous program design standards, this testing should go well. But if each module developer is working too independently and with insufficient criteria, integrated testing will reveal numerous incompatibilities and inconsistencies that will have to be resolved.

Developers naturally tend to think in terms of *extremes* and *exceptions* in terms of the normal and erroneous types of data entry and processing that might be encountered. The problem is how to know if one has allowed for all possible conditions that the program may encounter when implemented in practice—anticipating "Murphy's Law," which says that anything that can go wrong will go wrong. There is a field of computer science that mathematically tries to prove that all conditions have been tested, but such analytical methods tend to bog down when faced with large and complex systems. In practice, programmers tend more to rely on their imaginations to think up parameters and conditions for interim testing. Once software nears completion, they use formal procedures called *alpha testing* (enlisting a set of people to test the program who themselves are quite knowledgeable about computers—perhaps working colleagues) and *beta testing* (trying it out on a set of users who are representative of the target market for the program). Testers report back about errors encountered, procedures that seemed slow or awkward, and so forth. Alpha testing and beta testing depend more on accumulating empirical experience than mathematical analysis, but they are the main practical alternative for programs of significant size and complexity. There may be several iterations of such testing.

An area of testing that tends to be neglected is to test software with realistic *volumes* of data. Creating large files of distinct data is tedious and expensive for developers, and they may not have a feel for what volumes users eventually will be dealing with anyway. Such testing is important, however. For example, one might find that the performance time increases exponentially rather than linearly with volume and soon becomes unacceptable when one moves much beyond sample file sizes used by developers.

In some cases, software is required to go through formal verification procedures. For example, programs for structural analysis or CPM scheduling might be tested with a known set of standard data that a standards organization or professional society has developed and made available for this purpose. Programs that pass the tests can then be certified for a certain level of performance and/or accuracy.

Finally, after an application has been in use for several months, it is a good idea to conduct a postaudit or retrospective evaluation. Compare the use in practice to the original goals: How well were they met? What were the costs and schedules compared with the estimates? What were the shortcomings? How might they be corrected? This type of feedback is especially useful in making better feasibility studies and designs in the future and can give a good idea of how well we do in estimating costs and schedules.

DOCUMENTATION

Documentation for a transferable, enduring program must be especially thorough. It may be divided into these four components:

- Programmer's comments internal to the program source code
- A programmer's manual describing technical aspects of how the program works
- A user's reference manual
- Educational materials for user training

The following sections provide guidelines for each of these.

Internal Program Documentation

Internal documentation should begin with the first lines of code. Every program should have an introductory section that describes the program in an abstract form, the programmer, the institution, the target computer, and the date. Modules of the program and even individual lines of code should have comments explaining the function of the module, the variables, the specifications and sources of formulas or algorithms, and any tricky or nonobvious coding. The readability of the program will be improved by indentations that show the various levels of logic flow. Variables should have meaningful names to suggest the real entities they represent in formulas or objects. Any machine-dependent code should be marked clearly and be isolated where possible from the mainstream of the code.

Technical Documentation External to the Program

The following documents should be included with each program to provide a technical description of how it works:

- Abstract of the program: what it does, who wrote it, where it comes from, when it was written, and for what computer it was written
- Program listing
- Flowchart or other graphical illustration of program modules
- Description of all variables used for input, processing, and output
- Description of all files created and/or used
- Description of all mathematical formulas, algorithms, or tables of values used
- Transcript of a typical run with both interaction steps taken by the user and sample output reports

Together, these items provide a record of the software's production. They will be invaluable to others who need to understand, modify, or interface other programs to the one that is described.

User's Reference Materials

In addition to the technical documentation both internal and external to the program, application software must have user's reference materials written with the aims of introducing the program to someone who knows nothing about it and explaining its full capabilities. The user's manual should include a description of the theories, principles, and methodologies that are used by the program. It is very important to make explicit any required background knowledge on the part of the user. If the program deviates from standards in its area of application (e.g., simplification of concepts), it is important to point this out. The manual should tell the user how to handle errors and exceptions. Further reading and reference materials should be listed. Documentation can also be incorporated into files accessible to the program so that the user can retrieve it online with a help command.

User Training Materials

Training materials should provide an introductory explanation of the program content, then lead the user through a sample interaction. This sample should be chosen to illustrate a typical but fairly simple use of the program. Finally, the training guide could contain supplemental examples and possibly exercises to solve, additional areas of application, and related topics.

SUMMARY

Design, programming, testing, and documentation are central to the production of software applications. All, however, focus on making the applications more useful for human users, so they should be built upon a good understanding of underlying human factors that relate to computers.

At one level, system characteristics directly influence human behavior. Examples include response time, consistency, chunking, recovery from failures, accessibility, and procedures for accounting for the use of system resources. At a more fundamental level are the ergonomic characteristics of devices used for input and output. Anticipated learning curve rates for different kinds of users in different situations should also be recognized in the design, development, and documentation of applications. More subtle objectives are to give users a semantic understanding as well as a syntactic one of applications, at least for those with which they will work closely over an extended period of time.

There are many approaches, tools, and methods available for designing systems. This chapter mainly focused on the components to be designed. These included output reports and procedures, input forms and methods, files and databases, programming methods, testing and verification standards and procedures, equipment procurement specifications, documentation type and content, training requirements, plans for implementation, and operations procedures.

Program development should not proceed until the design is sufficiently well advanced that an initial scope can be approved and frozen. Otherwise, as with construction, change requests will wreak havoc with programming schedules and budgets. Once under way, project management techniques similar to those we use in construction can help guide the programming phase to successful completion. Skilled and experienced programmers should have available and be able to apply a wide variety of general and specialty tools that can speed their work and help produce high-quality results. They will also use recommended programming methods such as flowcharts to detail the design of important sections of code, modularization, and top-down programming.

Testing procedures are extensively used even during the programming process. Individual modules are tested and verified before they are assembled into larger subsystems, and the overall system should be thoroughly tested when ready. Test data should address extremes and exceptions to normal processing and should also provide realistic volumes of transactions that represent what will happen in actual applications. More formal procedures can include alpha and beta test phases, standard certification tests, and a follow-up audit.

Several kinds of documentation can be developed to serve different users and purposes. These include internal and external program documentation intended mainly to describe the workings of the code to other programmers, and also training and reference materials for the application's end users.

REVIEW QUESTIONS

1. Briefly explain the relationship of each of the following four characteristics of a commercial time-sharing computer system to the behavior of problem-solving users interacting with the system:

   - Response time
   - System accessibility

- Recovery from system failure
- Allocation of charges for system resources

2. Assume that, as a new engineer for a medium-size building contractor, you have been given the responsibility of *designing* a new estimating system. At present there are no computers being used by the estimating department. From the feasibility study and analysis you may assume that the company had found it best not to automate quantity take-off at this stage, but they do want to use spreadsheet software—for bid item productivity and cost computations and for final bid spreadsheet preparation—and relational database software—for maintaining historical cost files from past and current projects; for current labor, equipment and materials costs; and for lists of potential subcontractors. The hardware environment is to be a local area network with five desktop personal computers and a network file server (i.e., a hard disk accessible by each of the PCs).

 (*a*) Briefly describe three of the most effective methods or techniques you would use, before any programming takes place, to document and communicate the design to the estimators who will be using the system so that they can clearly understand the scope of what the application will do and get some feel for how it will appear and interact when they run it. These communications methods should be flexible so that the estimators can suggest modifications and otherwise interact with you periodically as the design takes shape.

 (*b*) Briefly describe three of the most effective and accurate methods or techniques you would use to document the design specifications in a form that could be sent by mail to a freelance spreadsheet/database programmer located in a distant city who has been retained to develop the software. (Note that this question does not ask you to prepare the design itself; just say how you would do it.)

3. Two major alternatives to design were presented in this chapter: top-down and bottom-up. Assume that, as a new engineer for a medium-size civil works construction contractor, you have been given the responsibility of *designing* a new equipment management system. At present there are no computers being used by the equipment department. To focus your thinking, consider the specific task of designing a maintenance subsystem.

 (*a*) How would you organize the subsystem from a top-down perspective? Show the breakdown with a hierarchical chart with two or three levels. Your answer need not be comprehensive in terms of the application's terminology or detail but should give the general idea of your approach.

 (*b*) In a bottom-up approach, what types of information would you consider? Illustrate with two or three examples of what you mean.

4. The following list includes some general concepts that apply in the design and programming of a construction computer application such as a payroll system, cost-control system, and the like. Briefly explain each and, where appropriate, suggest limitations, guidelines, applications, advantages, or disadvantages.

 (*a*) System flow charts

 (*b*) Modular structure in programming

 (*c*) Top-down programming

 (*d*) The use of superefficient and complex programming techniques that take advantage of the hardware characteristics of a particular machine

5. Among the recommended guidelines for the design and programming of computer applications are the following:

- Clearly and completely define the problem
- Plan and design carefully before programming
- Proceed top-down when programming

Now assume that, on the basis of an earlier feasibility study and analysis, you have chosen to implement the maintenance module of an equipment management system.

(a) Considering the guidelines, list five of the most important factors that might best define to a programmer the scope of this particular application.

(b) What are the implications of the second guideline in allocating your time between the design and programming phases of this application? Give an approximate percentage breakdown of your recommendation.

(c) Outline a top-down approach for this application, with particular emphasis on the structure and interfaces for the main program and subprograms. (E.g., draw a hierarchical block diagram such as that shown in Fig. 6-8.)

6. For a system of the type described in Question 5, describe the purpose and content of each of the following types of documentation:

- Training manual
- User's reference manual
- Programmer's manual

SUGGESTIONS FOR FURTHER READING

Humphrey, W. S. *Managing the Software Process*. Reading, Mass.: Addison-Wesley, 1989. An excellent book on bringing software projects to a successful completion and implementation. Covers a variety of techniques well proven in practice.

Macintosh Human Interface Guidelines. Menlo Park, Calif.: Addison-Wesley, 1992. Provides excellent insights into design techniques that have made Apple's Macintosh software so successful and widely imitated. Well-illustrated and easy to read.

Martin, James. *Information Engineering, Book III, Design & Construction*. Englewood Cliffs, N.J.: Prentice Hall, 1990. The third book in the series mentioned in the list in Chapter 5 of suggested readings. Good management orientation. Includes specific techniques and systems approaches.

Metzger, Philip W. *Managing a Programming Project*, 2d ed. Englewood Cliffs, N.J.: Prentice Hall, 1981. Good book with a mixture of management and computer-oriented perspectives.

Pfleeger, Shari Lawrence. *Software Engineering: The Production of Quality Software*, 2d. ed. New York: Macmillan, 1991. Excellent book. Particularly useful to construction readers for its analogies to managing a construction project.

PROCUREMENT, TRAINING, IMPLEMENTATION, AND SYSTEM MANAGEMENT

There are two lies in the computer business. The first is the tailor's lie—you'll have it tomorrow. . . . The second lie is even more widespread. It can be expressed in two words, which are both a slogan and a myth: "It's compatible."

Jean-Louis Gassée*

Initially, this chapter parallels the design and development processes reviewed in Chapter 6, exploring procurement of commercial systems as an alternative to creating custom software. It then moves on to discuss topics common to both approaches, including training, implementation, and system management.

PROCUREMENT

Even the procurement of a commercial system should be preceded by a feasibility study and analysis of the type described in Chapter 5. Indeed, the design phase is valuable for evaluating commercial systems if a group decides to take this route instead of custom software. This section thus emphasizes aspects that relate specifically to the procurement alternative, including the following:

- Gathering information from published sources
- Alternatives available in computer services and products

The Third Apple: Personal Computers and the Cultural Revolution, 1987.

- Procedures for soliciting and evaluating proposals
- Contractual and financial arrangements

Several of these topics also apply to procuring hardware and other components for custom applications.

Sources of Information

There are numerous sources of information for commercial systems. Some of the more common ones are listed here.

- *Published* **information services** typically are offered on a periodic or annual basis. An example is the *DataPro Directory*, which continually surveys the computer business for new hardware and software and issues frequent updates of technical specifications and evaluations. This makes it very easy to compare alternative vendors within categories.
- *Computer database services* enable subscribers to use a microcomputer and a modem to telephone the service's remote computer, link into the desired database, and search for abstracts and even the full text of descriptive articles, reviews, and so on. An example is *Dialog Information Services* (a Knight-Ridder company), which has several databases dealing with computers. Others are *CompuServe* (a division of H&R Block), *The Source* (owned by Reader's Digest), and *The Dow Jones News/Retrieval Service.*
- The *magazines and journals published by contractor associations and professional societies* often contain articles and reviews about computers and advertisements from computer hardware and software vendors offering products for construction. Independent construction trade magazines such as *Engineering News-Record* also carry information about computer applications.
- Numerous *computer magazines, journals, and tabloid newspapers* are evident in any large bookstore or library. There are quarterlies, monthlies, weeklies, and even dailies. Some are narrowly and technically oriented to scientists and technicians working in arcane branches of computer science, but others are comprehensible to readers with little computer background. An example that covers several aspects of the microcomputer field is *Byte Magazine* (a McGraw-Hill publication), whereas others are focused on specific types of computers and operating systems. The problem with such sources is that one must sift through an enormous amount of information to come across products, such as scheduling and simulation programs, that apply directly to construction. Many of the major publications are abstracted in the aforementioned computer information databases, so a search through databases in areas of specific interest can help one more rapidly locate issues that contain relevant articles.
- Most metropolitan areas contain numerous *computer stores* where one can see products in person and ask sales personnel for demonstrations. They are useful for learning about general-purpose software such as word processors, spread-

sheets, and databases, but few such stores carry applications designed for construction, and typically the sales staff knows little about where to locate such packages. Indeed, a problem with such stores is that a reasonably knowledgeable customer often knows more about computers than the salespeople, which makes information-gathering frustrating at times.

- One finds *trade shows* in the computer industry, in the construction industry, and as exhibits associated with conventions of professional societies. Almost all contain relevant information for computer applications in construction, and one can acquire a bewildering quantity of information in a just a few days at such shows. Since manufacturers and developers often send top technical and managerial people to important shows, they can also provide opportunities for face-to-face discussions with those who really can answer questions knowledgeably. The annual *AEC Expo,* well focused on construction, also attracts a number of computer-oriented conferences that are held nearby and almost concurrently, making it a good place to gather information quickly.

- *User groups* often exist when there are several people using systems of a similar type in a given locale; some user groups are national, and both types often put out informative publications. Such groups also exist for commercial packages like databases, spreadsheets, and CAD systems. More useful to construction, however, are those within local contractor associations and professional groups.

- Once one has examined sources such as those previously mentioned to find out who produces systems for a given application, then the next obvious source is contacting the *developers and vendors* themselves. These range from individuals working out of their homes to major corporations. Some are divisions of large engineering contractors and construction equipment manufacturers. Occasionally one finds useful programs offered by government agencies, such as the U.S. Army Corps of Engineers, or from universities. You can write or call these sources for more detailed information and demonstrations, and the quality of the response may give some clues about the quality of the vendor and its product support.

By screening such sources, one should obtain information about several products that might meet the needs of the planned application.

Alternatives in the Marketplace

In parallel with the search for information, one can consider commercial alternatives for achieving a similar application goal. This section briefly touches on some of the main ones, including third-party services and hardware and software purchase options.

THIRD-PARTY SERVICES. Third-party services are available for almost every aspect of computer applications and operations. Some of the more common ones are mentioned here.

- Computer *service bureaus* exist in many areas to provide contractor payroll and accounting services and sometimes other applications. Such services, offered by independent firms and some banks, have their own computers and the appropriate software. The contractor can either submit paper forms (e.g., timecards) for input by the service or possibly have remote data entry and printing at its own site. The service keeps the software up-to-date, maintains security, and otherwise relieves the contractor of the worries of maintaining an in-house computer system. The costs are often reasonable, owing to the economies of scale available to service bureaus, and many contractors use them.

- *Time-sharing* services typically are nationally centralized complexes of computers that are accessible via local calls into national networks. They sometimes offer powerful project management software and some engineering analysis and design programs that may be needed only intermittently. Related to these are the aforementioned information services that provide access to a variety of databases on a wide range of topics. Some firms that traditionally published estimating cost data and materials and equipment specifications in book form have also begun offering such information by computer time-sharing or on optical discs.

- In the 1960s the concept of third-party *facility management* services developed in firms like Ross Perot's EDS, which eventually was acquired by General Motors. Basically they offer technical and managerial staff that can work as a type of subcontractor to manage and operate a computer facility and its applications right on the client's premises. Although such arrangements are popular with some banks, manufacturers, and institutions, they have not been used much in construction.

- *Consulting services* for computer-related feasibility studies, analyses, designs, procurement, and implementation are offered by several public accounting firms and more specialized management consultants. Some have considerable expertise in construction and have found a good market in the industry.

- In any metropolitan area, numerous firms and independent freelancers offer computer *programming, documentation,* and *training* services. There are often construction specialists within these categories, and they typically make their services known in the activities and publications of local contractor associations. Some specialize in certain tools (languages, spreadsheets, databases, etc.) and others in certain applications (e.g., accounting and cost control). Others go in more for training via seminars and workshops.

- Although most computer owners obtain *maintenance and repair services* from manufacturers or local dealers, there are also third-party firms that try to compete by offering better service at lower cost. They exist both as local repair shops, much like independent auto repair firms, and as on-site technicians who will come out in response to a call.

HARDWARE AND SOFTWARE. Alternatives for purchasing hardware and software include manufacturers, retailers, brokers, mail-order firms, value-added resellers, and others. They are discussed briefly here.

- Some manufacturers and developers maintain a *direct sales force,* whereas others work exclusively through dealers. Some of those who sell directly do so only for the largest types of clients—the top 500 businesses, major institutions, and government agencies—for whom they offer substantial volume discounts. For smaller sales, where available, there may be no financial advantage in dealing directly. Specialized construction software often is sold directly by developers to customers of all sizes—sometimes only this way—and users can establish closer working relationships with such developers.

- *Retailers* mostly consist of storefront operations. Like other types of retailers, they range from full-price boutiques to high-volume discounters. There are national and regional chains and franchises as well as independent stores. Sometimes computers are just part of the retailer's much wider product line, such as consumer electronics. Even department stores sell computers. One would hope that the higher prices some firms charge might entitle the customer to better service and support, but in practice there is only a loose correlation in this regard.

- *Brokers* typically operate between manufacturers and the retail market, often in used equipment. They sometimes help dispose of surplus equipment that is about to be superseded, and to locate peripherals that are compatible with older systems. Similarly, some customers have access to *wholesalers* and *distributors,* which operate much as they do for other products used in construction.

- One cannot look at a popular computer magazine without noticing advertisements for *mail-order firms.* They range from reputable and efficient firms that provide excellent delivery, service, and value down to those that just take your check and disappear. Particularly for smaller orders, especially for software and accessories, many customers consider the reputable mail-order firms to be an excellent option. Even the largest name-brand computer manufacturers have started this type of direct marketing. Retailers complain that low-overhead mail-order firms do not provide face-to-face customer support, but the argument would seem more valid if retailers did better at this themselves.

- ***Value-added resellers*** (VARs) acquire computers and system software at a discount from manufacturers, develop their own software and possibly specialized hardware (e.g., CAD or instrumentation) on top of those products, then integrate them all into working systems that provide added functionality for the customer. The customer may not actually see the separate prices of the computer hardware and software, although manufacturers do usually allow the VARs to pass along manufacturers' warranty and service arrangements to the customers.

In all such purchase arrangements, the buyer must think about compatibility of hardware and software, potential for expansion, performance, and longer-term vendor relationships. Large construction firms with diverse geographic markets, or those located far from major metropolitan areas, may have to concentrate on products from larger manufacturers and on vendors that are likely to have service widely available.

Procedures for Soliciting and Evaluating Proposals

Before soliciting proposals for hardware or software, we first need specific criteria for the type of specification that would result from the analysis and design phases discussed in Chapters 5 and 6, although not necessarily at the same level of detail. For equipment, these would include general hardware capacity and performance specifications, the type of operating system the computer should be able to run, and possibly brand names of acceptable manufacturers. However, one could leave it to the dealer to decide on which disk drives, CRT displays, printers, memory modules, interface boards, and other items to integrate with the processor to meet the desired specifications. Costs can often be reduced this way, but the dealer should then guarantee and support the systems that integrate components from multiple manufacturers. Similarly, one might specify generic and application software packages or, alternatively, provide detailed specifications for what the software should accomplish and leave it to the vendors to recommend solutions.

For smaller requests, a vendor may not be able to invest much time responding to complex specifications. Rather, it will probably have informal discussions with the buyer that soon lead to a proposal or quotation from the vendor. This offer should be made in writing and remain valid for a few weeks or months while a comparative evaluation takes place. For larger purchases, there will often be ongoing discussions and exchanges of specifications and proposed solutions, sometimes over a period of weeks or months. The process can be a formal one with development of a written *request for proposal* (RFP) that is sent to potential vendors. It would include not only technical specifications but also the desired business and contractual terms, much like an owner's solicitation for bids in construction. Regardless of the size or complexity of the purchase, one should always insist on a thorough demonstration of the capabilities of the proposed hardware and software. Preferably, the demonstration should allow hands-on access for the user, and even let the user try sample input files containing his or her own data, if possible.

Eventually some alternative proposals should be available for consideration. At this stage a quantitative tabular matrix evaluation such as that shown in Table 5-3 could help in sorting and ranking the alternatives. The evaluation should consider not only the technical and functional characteristics of the system and the economic terms but also factors such as support for installation, training, and maintenance. One should check references from other users for the proposed items and for the reliability and reputation of the vendors. Depending on the terms offered by the vendors, one might be able to combine components from two or more sources for a lower overall cost, but the buyer then assumes a greater risk for ensuring that the components will successfully integrate into a working system. In evaluating the economics of alternative proposals, remember what Chapter 5 said about the overall costs of a system: people costs and other indirect costs can considerably outweigh those of hardware and software.

Contractual and Financial Arrangements

After selecting the vendor, we continue to firm up financial and contractual arrangements for acquisition and support. In some ways, acquiring computers is similar to acquiring construction equipment. Options range from lump-sum purchase through various financing packages to rentals, leases, and lease-purchase agreements. The one that works best depends on the market and the financial situation of the contractor at the time concerned. However, there are a few dissimilarities. For example, computer equipment becomes obsolete far more rapidly than most construction equipment, so realistic depreciation schedules can run faster than the law allows for maximum tax benefits. This rapid obsolescence also means that cumulative rental or lease payments can exceed the full purchase cost in little more than a year, making these less attractive options than they might be for construction machinery.

The final dissimilarity is that software is often made available on a licensed basis in which ownership remains with the developer, and the buyer legally pays only for the right to use a copy under terms specified in the license agreement. The fee may have to be renewed annually, and if the licensee fails in this or other terms of the agreement (such as making illicit extra copies to use on other computers), the developer may have the right to take back the program and even sue for damages. If you look carefully at the fine print contained even on mass-produced generic packages, you will often see terms like these.

Ongoing contractual arrangements deal with warranties for hardware and software, training included or available at extra cost, the right to updates to software and sometimes to hardware for a certain period of time, technical support for questions related to system installation and utilization, and ongoing service and repair agreements. Warranties used to be very weak in the computer business, but market competition has brought improvements. Note, however, that warranties almost always exclude consequential damages (e.g., lost files of information crucial to a business, errors in bid calculations, etc.) that may occur as a result of hardware and software failures. As in any business, read the fine print, and caveat emptor.

TRAINING

Training was considered briefly in Chapter 6 as part of the design. Training activities typically begin shortly before the system is ready to be installed and then continue as needed for new users and for changes and enhancements to the system. Things to be considered here include selection of participants, curriculum, methods and materials, training personnel, and schedule.

In participant selection, if there are just a few users all can be trained. The main consideration, when using methods that take time away from the job, is to rotate them in a manner that keeps their work going. For larger groups, a good strategy is to pick motivated, quick learners who work well with other people first; they can then assist in training their coworkers. We should also consider different levels and focuses of training for people working with different aspects of the system.

Curriculum design depends mainly on the nature of the application and the needs of the trainees. Normally it should include the semantic concepts behind the system as well as the practical details of how to use it. For a complex system, it is best to cover the main capabilities of the application first and then have advanced sessions for more difficult, less used, or optional capabilities. The curriculum should almost always include exercises to provide hands-on experience, even with approaches based on seminars, workbooks, videotapes, or other "offline" methods. These exercises can either be supervised in a learning laboratory or workshop setting or be handled by allowing employees to do them back at their offices if computers and tutorial assistance are available there.

Chapter 6 listed some options in training methodology and materials. These and other possibilities include the following:

- Training seminars and workshops
- On-the-job training using reference sources and tutorial assistance
- Interactive *multimedia*-based training systems
- Online interactive tutorial software
- Audio and video cassettes and/or slides
- Tutorial manuals
- Programmed-learning workbooks

Each has its advantages and disadvantages, depending on the size of the group, the nature of the application, and the skill and motivation of the users.

Classroom and tutorial approaches to training will require selecting people to do the teaching. This can be accomplished by sending employees to courses offered by vendors or independent firms, by hiring people to teach in-house, or by having in-house employees take on a teaching role. In any of these approaches, teachers should be selected for their teaching abilities as well as for their technical knowledge of the application system. Asking a knowledgeable but introverted programmer to mumble through a presentation aided by a series of overly technical transparencies is likely to do more harm than good.

Alternative approaches, including self-study workbooks, online tutorials, and videotapes, should be carefully reviewed and tested before they are used for training. Computer professionals and knowledgeable users often skip these and thus do not know what they contain. They just assume they will be good for new users. However, too many materials of this type turn out to be little more than marketing demonstrations and do little to teach practical skills. Others may be sufficiently detailed but boring. Some may be based on earlier and obsolete versions of the software and thus cause confusion when their explanations conflict with or fall short of the version of the software that is being installed. Where competing training materials are available—as is the case for popular generic software packages—the best should be chosen following a careful evaluation.

Whether training should be scheduled en masse or on a rotating basis depends on the nature and complexity of the application and on the impact of taking employees from their work. Training that takes a week or more may have to be rotated, whereas simpler systems may be introduced to the whole group in a few hours. For centralized software, the schedule should be set to have all key users trained before the new system goes into operation. For example, changing a company's accounting system requires that all affected employees be ready to work with the system when the conversion is made. Individual standalone programs like spreadsheets and word processors and project-oriented software like scheduling can be introduced and learned incrementally.

Some training options may be included with the purchase price of a commercial system. Expensive systems often do include seminars or courses for a limited number of employees, but even here travel and living expenses may have to be added. Low-cost packages are more likely to provide just tutorial manuals and perhaps online training software. Beyond this, any number of options could be available for purchase. They should be evaluated and budgeted like any other component of the application development, procurement, and implementation process.

IMPLEMENTATION

Preparation for implementation should begin before a new system is actually installed and the old application is converted over to the new. Even as early as the feasibility study and analysis, and certainly when development or procurement is under way, developers should prepare users to be realistic about what to expect when the new system comes, including what steps to anticipate in training and implementation.

Management should consider the advantages and disadvantages of the following four possible implementation strategies and select the one that best serves the needs of the application and its people:

- Total conversion at a fixed date
- Parallel operations with a gradual transition
- Phased implementation of separate modules
- Pilot implementation

Figure 7-1 is a conceptual schedule for each of these four methods.

Total conversion, sometimes called "going cold turkey," means that the old system stops and the new one starts on a fixed date, with little or no overlap. Although it sounds risky, if it is well planned, follows a proven system, and involves users who are trained and ready to go, total conversion can be the most effective and least stressful alternative. Although the implementation itself is a very busy time, it avoids duplication of efforts and makes sure that all users are on board for the new system. For centralized systems, such as payroll and

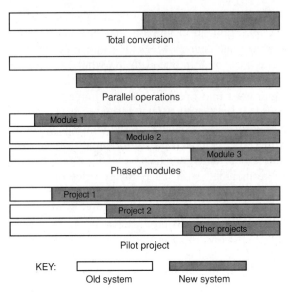

FIGURE 7-1
Alternative implementation strategies.

accounting, this approach is almost essential; such systems do not lend themselves to working the old and new ways simultaneously. The main potential risk is that, if the new system is faulty or the implementation is poorly planned, chaos may indeed ensue, demoralizing the users whose commitment is needed for success.

The strategy of *parallel operations with a gradual transition* seems cautious and prudent since one can check new results against old results to be sure that all is going well. Such caution may be appropriate for a new and unproved system, but there are at least two major disadvantages to this approach. First, it takes duplicate efforts to keep two systems going simultaneously. Unless temporary staff are added and trained sufficiently in advance of implementation, existing staff may become so stressed with the added complexity and long working hours required to run two systems that the effectiveness of both systems deteriorates. Second, given the option of continuing with old ways, some people will stall as long as possible to avoid learning the new system. Implementation thus can drag out much longer than might reasonably have been anticipated, prolonging the stress on the staff who have to support both systems.

Some applications lend themselves to the *phased implementation* of separate modules that will eventually make up an integrated system. For example, a new accounting system might start with installing a general ledger, followed soon after by a payroll system. Later the firm could add accounts payable, accounts receivable, materials management, and job-costing modules. Advantages of this approach are that the training and implementation workload can be distributed over a period of time, and corrections and adjustments can be made before committing to the full implementation. The main disadvantage is the possible difficulties of bridging incompatibilities between the components of the old and new systems during the phased transition period.

A *pilot implementation* can work well for certain types of applications. For example, project-oriented software, such as that for scheduling or cost control, might first be introduced on a project that has interested and capable construction professionals who have the motivation and initiative to try to make it succeed. If they are successful, their experience will help convince potential users on other sites. If the pilot test fails or is otherwise unsatisfactory, the "damage" is confined to one site; the people there still may be constructive in offering practical advice for helping similar attempts succeed in the future. This approach can also test concurrently two or more alternative packages tentatively being considered, but we must be careful about creating rival camps of supporters who force the company into a situation of maintaining multiple standards. The main disadvantage of the pilot approach is that it is not suitable for centralized systems such as accounting.

Taking these alternatives into consideration and choosing the approach that best suits the application and users at hand, management can develop detailed plans to facilitate implementation and ensure its success. As implementation time nears, management and staff must have ready to go not only the hardware and software but also the forms and procedures to be used. If they are converting from an old system, there must be a plan for readying its data files for input into the new system. Perhaps this can be done with automatic conversion software, or it may require hiring temporary data-entry people who laboriously type data into the new system. For periodic systems like payroll and accounting, the timing of such steps is critical to reducing the stress and chances for problems in conversion. People who will be involved in the implementation and subsequent use of the system should meet in advance to identify and assign the tasks to be performed.

A fairly detailed plan and schedule will facilitate the implementation of large or complex systems or those that affect many people. Activities should be broken down to a daily or even an hourly level of detail, depending on the nature of the application.

SYSTEM MANAGEMENT

As computers become more and more essential in construction business activities, computer system management becomes vital to our organizations. The planning, development, and implementation stages discussed thus far in Chapters 5, 6, and 7 are part of this management responsibility, but related activities continue into daily operations once the systems are in place. This section briefly examines key facets of ongoing system management, including maintaining good relations with developers and vendors, providing service and support, and managing the people and activities involved in daily operations.

Vendor Relations

Computer systems are composed of complex technical products, and vendor relations in this area are not unlike those for construction equipment. A contractor should maintain technical as well as business relationships with its primary dealers and suppliers, who in turn should have some understanding of the needs of the contractor. A good dealer will have knowledgeable and responsive sales-people and technicians and will update the contractor's computer manager with current product literature and announcements about seminars, trade shows, and other learning opportunities. Both sides should be able to respect and trust each other and to keep long-term needs and relationships in mind in their short-term interactions. A good dealer performs a kind of consulting role as well as a sales function; the dealer helps the contractor (1) find the best products for its needs, (2) plan capacity and compatibility for future growth, and (3) avoid bad decisions. Such dealers will also "go the extra mile" in providing service and support and in administering warranty and return policies, and the contractor can reciprocate with ongoing business. Needless to say, such relationships are what one often can only hope for in an ideal commercial world. Where a contractor finds dealers who aspire to these ideals, the relationships are worth cultivating and maintaining for the long term.

There are a number of pitfalls in maintaining such relationships. For example, a contractor's computer system may have products obtained from multiple vendors, and certainly from multiple manufacturers, yet, by and large, they must work as an integrated system. When errors and failures occur, "finger pointing" exercises may arise in which each manufacturer claims that the other's component is at fault, leaving the contractor's staff to isolate the offending component. This situation is particularly difficult when the problem is not with any one component but rather with conflicts and incompatibilities in the way the components work together. A dealer who takes overall system responsibility can be very helpful in troubleshooting such problems.

In remote areas one may have to choose between trading with local dealers, who may be lacking in product range, technical ability, and competitive pricing, and having long-distance relationships with dealers in metropolitan areas, who may be unable to provide responsive service and support. Sometimes sticking with the products of major manufacturers and with dealers who are part of chains that have good central technical staffs to back up their dealers can help in this situation.

Other issues relate to limits on vendor liability and to ambiguities about where responsibility falls between manufacturers and vendors. Some manufacturers will only warrant their products if they are obtained through manufacturer-authorized vendors; yet in these cases there is still a so-called *gray market* of dealers who obtain products through other channels and often sell them at an attractive discount. Unknowing buyers can then be stuck without a warranty unless the dealer explicitly takes this responsibility. Finally, dealers as well as manufacturers try to avoid all liability for consequential damages.

Service and Support

Service and support often come from different sources for hardware and software. There are also numerous trade-offs to consider in optimizing the balance between costs and levels of service. This section briefly explores some of the components and options.

The first consideration in either software or hardware is how much support comes with the initial purchase. Local dealers will often come out and set up the hardware and install an initial complement of software when a system is purchased. They may also spend some time with the new user to explain how to start using the system. Most microcomputers plug together even more easily than component stereo systems, at least at the hardware level, and common software packages are also easy to install. New users can gain confidence by doing this job themselves. However, installing workstations and larger computers and some types of networks can take skilled technicians. Similarly, just installing the system software for some of the more complex machines can take an expert 10 hours or more. Setting up complex databases and business information systems can also be complex. These tasks may become a never-ending nightmare if unprepared buyers attempt them, so again technical support may be essential.

Next consider warranty coverage. The minimum coverage for hardware is 90 days, and increasingly vendors offer one to three years. Rights to return software may expire as soon as the package is opened, but today the terms are usually better. One should also consider whether coverage is for on-site or take-in service or consultation, and whether backup systems will be provided while a system is in for repair or service. Even under warranty, the levels of service vary as widely as those of the maintenance agreements discussed later.

HARDWARE SERVICE AND SUPPORT. After the warranty has expired, dealers, manufacturers, and third parties offer hardware service agreements at varying costs. Where possible, obtain integrated coverage from a single vendor even when several manufacturers' products are in the system. Manufacturers of some commonly used computers may service the more widely used products of third-party vendors that are installed in their systems. Regardless of source, typical levels can range as follows:

- 24-hours/day availability, respond within two hours
- 8-hours/day, 5-day week, respond in one day
- Take-in depot support
- Technical training, parts, and supplies for user self-maintenance

The first type is the most expensive and is normally used only for central mainframes or large minicomputer installations that run on a multishift basis, and where even a few hours' disruption would cause major problems for the functions they support. This level may also apply to computers used in real-time process-control applications for which there are significant economic or safety consequences for prolonged failure. The second level is most typical for business minicomputers and microcomputers. Usually if the client telephones in with a problem, a service technician will be on-site later the same day or at most by the next day. Technicians can often isolate problems and swap components to get the system running again soon after they arrive. The third level requires that the client take defective or suspect equipment to the vendor's site for repair. Although there may be a lower cost for this level of service, one should not underestimate the costs of employee time to perform the errands and of the potentially longer interruption to the user whose system is in the shop. The fourth level is designed more for clients who employ their own in-house technical staff but need access to vendor parts and technician training to maintain these in-house capabilities. This level applies more to firms that own large numbers of systems where the in-house option may become more economical.

Other factors that affect hardware service include remote diagnosis, the quality of service technicians, and the extent of coverage. *Remote diagnosis* refers to the capability of service technicians to connect to the client's computer by telephone and run diagnostic software to try to isolate problems before sending out a technician to the site. Many problems turn out to be purely related to software and can be corrected by reloading the needed components over the telephone line. Others turn out not to be system problems at all and simply are the result of the user's misunderstanding. With the remote link, the support person can guide the user through the correct operating procedures.

The importance of quality service personnel seems self-evident given the complex nature of computer technology. Nevertheless, some service vendors do provide narrowly or inadequately trained technicians who may hit their limits on some service calls, causing prolonged delays until more capable people are brought on-site. Even worse, poor technicians can cause more problems than existed initially, such as shorting out whole circuit boards when all that was needed was a new chip. The buyer must beware of service firms, dealers, and manufacturers who have poor reputations. Even if all hardware costs are covered by the service agreement, it almost never covers the costs of unnecessary or prolonged disruption to the client's business.

Full coverage includes all parts, labor, supplies, and shipping costs needed to make repairs and may include periodic maintenance and diagnostics as well. Other agreements may cover only some of these costs. The buyer must carefully read the service agreements to be sure of the coverage that is provided.

SOFTWARE SERVICE AND SUPPORT. Software service and support are often fragmented among the various vendors that supply the different programs that go into a particular computer system. The issues are somewhat different, but the basic principles are much the same as for hardware. Again, there are various levels of support available, such as the following:

- On-site and/or telephone consulting and troubleshooting
- Newsletters and notifications of updates and error corrections
- Software supported and tested by vendor, but installation and maintenance are the user's responsibility
- As-is, no support (e.g., software developed by user groups)

At the first and highest level, vendors provide ongoing services to answer questions and fix problems related to the use of their software product. The better ones have toll-free numbers, provide enough staff to minimize the waiting time before offering help, and have well-trained people who can indeed answer most questions (even when poorly communicated by the user) and who somehow remain polite and professional even when users call up to blame all their own shortcomings (like never opening the manual) on the product and vendor.

At the next level, the vendor maintains a mailing list of those who have bought its products and sends out notices when updates and corrections have been made available. Some also send out newsletters with helpful tips and solutions to problems other users have encountered. Although these things certainly do have a marketing function—especially when high prices are charged for "upgrades" that are mainly *bug*-fixes—the better vendors make these services available at reasonable cost and sometimes even for free.

The next level is software that is commercially developed and tested and that stays up-to-date with an ongoing development program. However, obtaining updates may be the user's responsibility, and significant fees may be charged for new releases.

Finally, there is a vast amount of unsupported software, some of which is distributed informally as *freeware*. Most of it is available from user groups and network services. These range from games and utilities developed by hobbyists on up to major packages resulting from university or government research. Sometimes voluntary or modest fees are requested under the term *shareware,* and a few dedicated developers in this category continue to put out upgrades for their loyal following. In general, however, all risks and responsibilities are with the user, and in most cases the products are worth about as much as you pay for them. They are tried once and discarded. For some people and applications, however, there are some real bargains in unsupported software, and the better products become known to those for whom they are useful.

COSTS AND TRADE-OFFS. Service agreements are somewhat like insurance. If nothing goes wrong and you are satisfied with the current level of performance, you may wonder why you spent the money. If you have a major and otherwise costly failure, or you get a major upgrade at low cost, you are glad you bought the agreement. As in anything, however, there are trade-offs, and these can be analyzed to develop good policies.

Hardware and software support agreements often end up costing between 10 percent and 20 percent of the original purchase costs per year. However, computer systems tend to be reliable enough that, without an agreement, most installations will spend much less than half of this on actual repairs and upgrades. The difference may buy security and peace of mind for the few times when things really go wrong and do so at the worst possible time.

In the case of centralized systems, such as mainframes, minicomputers, or network servers—where a large fraction of an organization will be disrupted by a failure—it is good practice to have a service agreement in place. This agreement can also be good when an office has only a few microcomputers, particularly if none of the users has any aptitude for troubleshooting problems that can arise.

However, where there is considerable redundancy and backup, such as an organization that has numerous microcomputers of a similar type, only a few of which might be heavily used, it can make sense to accommodate a failure by shifting priorities or doubling up while a system is taken in for repair on a fee basis. As the numbers grow larger, it can also make sense to have an in-house person who at least has enough skill to distinguish user errors from genuine computer failures, and to run basic diagnostic software to isolate components that have failed. Keeping on hand a small backup supply of memory chips, video boards, monitors, disk drives, cables, and other failure-prone components usually can help this person get systems running again in little time. Only in a few cases will it be necessary to take failed machines in to the vendor.

A problem designed to let the reader analyze service costs and trade-offs of the type discussed in this section is included at the end of this chapter (see review question 4). It provides deeper insight into some of the advantages and disadvantages of different policies in different situations.

Operations Management

Operations management involves numerous routine activities, staffing and personnel issues, and security. This section touches on some key elements in each.

ROUTINE OPERATIONS AND MAINTENANCE. Typical routine operational duties include those listed here. They must be planned, staffed, and monitored to ensure the efficient and secure operation of systems used by an organization.

- Periodic maintenance for hardware (cleaning, adjusting)
- Procurement and installation of supplies (paper, disks, tapes, ribbons, etc.)
- Mounting and dismounting of tapes and disks
- Backing up, maintaining, and cleaning up of files on disks
- Running of centralized programs on a demand or scheduled basis
- Start-up, shutdown, and reconfiguration of systems
- Centralized data entry
- Collation and distribution of reports

Most computer hardware needs little routine maintenance, particularly computer processors, disk drives, and monitors. However, especially in dusty areas, air filters occasionally need cleaning or changing, and dust can be vacuumed or blown out (with special nonmetallic, static-free tools) to facilitate cooling and prevent shorts. Printers, particularly impact types, generate paper particles that need to be cleaned out periodically. Screens and keyboards should also be cleaned occasionally.

Supplies are noticed most when they run out, so there should be a clear delegation of responsibility to be sure that adequate amounts are procured and on hand. Central printers and servers should have a person responsible for installing disks, ribbons, paper, and so forth. Users can handle supplies for their own systems, but some controls on obtaining these supplies may be necessary to reduce waste and pilferage.

Central systems and servers need to have **operators** responsible for mounting tapes and disks used for backups and other periodic or on-demand operations. Backup itself is a vital function and should be carefully planned and monitored. Typically there will be rotational schedules, with some media periodically rotated to an off-site location in case of major catastrophe. Where all users keep their files on a central computer or server, this function can be achieved with reasonable assurance. Applications that depend on users to back up their own files require continuing education and reminders, and even occasional inspections to be sure that key company files are indeed being backed up.

For some operations the central staff has a responsibility to run programs such as payroll and accounting. This responsibility might belong to the computer staff or relevant user department. In either case, schedules and procedures should be planned and monitored. Centralized and network-based systems may require certain procedures for start-up, shutdown and reconfiguration (for example, to add new devices). Responsibilities and schedules should be made clear, and unauthorized persons should not feel free to do these tasks on their own. There can also be centralized clerical functions, such as data entry from forms like time cards, and the collation and distribution of reports. These can be handled by a computer staff or by the associated departments.

Where there are enough operational functions to be performed, a full-time manager and a staff may be needed. These will be discussed next.

STAFFING AND PERSONNEL. Staffing needs for computer operations vary widely depending on the size of organization and the extent of its computer applications. Smaller firms may run without full-time computer professionals, or they may develop skills in-house in people who have other jobs. Larger organizations will have professional staffs.

Professional staffs for larger computer installations often divide into administrative, clerical, hardware operations, and program development functions. Administrative people can come from either the organization's primary functional areas or a computer background. Clerical staffs performing functions like data entry or report distribution may be derived similarly. Operations staffs (starting, stopping, and maintaining hardware; mounting tapes and disks; changing printer ribbons and forms; performing backups; etc.) tend to be vocationally trained for these tasks, and they may know little about the construction business in which they work. Those developing and maintaining computer software can come from either a professional or a vocational background, and they also tend to be aligned more with computers than construction as a career endeavor. There thus can be a considerable management challenge to integrate all of these people into a construction organization, but success in meeting this challenge can improve the computer operation for all concerned.

Like others in an organization, employees whose work focuses on computer functions need job descriptions, professional opportunities, and support to maintain their skills and career development paths; and means are needed to evaluate their performance and recommend their advancement. In light of the wide range of skills that are possible in various computer jobs, particularly those involving software development and support, it can be difficult for smaller organizations to have the ability to evaluate employees, so outside assistance may be needed. Hiring the right manager will certainly help. With the rapid growth of computer applications in most organizations, there should be ample opportunities to manage these people and respond to their development needs. In other industries, the computer-based information functions have even provided paths into general management, and this development should be anticipated in construction.

SECURITY. Both operations and staffing involve the important subject of security in computer operations management. As more and more information and procedures vital to the success of business organizations are maintained in computer systems, these systems have become the most critical aspects of many businesses. Airlines, banks, insurance companies, government agencies, and others could not function without their computer systems, and that situation is fast approaching in construction. Key security issues include physical security, prevention of unauthorized electronic access, and personnel.

Physical security includes mitigating the potential damage from fire, earthquakes, weather, and other natural hazards and restricting unauthorized access to sensitive areas. Larger installations include sensors, fire suppression systems, earthquake-resistant floors and structures, and identity-card access systems, as well as procedures that go with these. In any case, there should be good off-site backup of essential computer files and contingency plans for establishing operations on alternative computers in case a major catastrophe disables the primary site.

Physical security can also involve computer hardware design and support installations. For example, the Tandem Computer Company specializes in redundant processors whereby a second machine always shadows the primary one and kicks in immediately if the primary one goes down. They are used in stock exchanges, reservation systems, hospitals, and other time-critical applications. A variety of backup power installations are available, ranging from batteries for microcomputers to diesel generator sets for mainframe installations.

Software can also be designed to check and double-check the integrity of data and to be continuously backing up files as applications run. Both software and hardware can be designed to minimize the chances of unauthorized persons illegally "breaking in" to sensitive installations via networks or telephone lines, without ever coming near the installation itself. Passwords and procedures, data encryption schemes, and other means are available to improve system security via software.

Personnel aspects remain among the most difficult problems in security. People make mistakes, get sloppy with procedures, or deliberately try to compromise systems. Sometimes those closest to systems are the greatest problem. For example, although passwords are designed to prevent unauthorized access, they also pose a human memory problem. It is thus not uncommon to find passwords taped to the desk or computer of people who are authorized to use the system. This human failing has been exploited repeatedly by those who know this and look for the posted notes.

More difficult is deliberate fraud or mischief perpetrated by dishonest or disgruntled employees with access to a system. Dissatisfied employees have been known to use magnets to destroy data kept on disks and tapes and to appropriate sensitive information that might be used to embarrass management in labor negotiations. Procedures can be designed to make it difficult to print checks or authorize disbursements without a certain number of checks and balances, but nothing is foolproof. One of the nation's major finance thefts was perpetrated by a programmer who designed a program subroutine to take the difference between rounding and truncating the bits below one cent in all transactions and deposit that difference in an account he set up under a different name. Although each deposit was under a penny, over time they added up to millions of dollars. He would probably never have been caught except that eventually he could not resist boasting about the technical aspects of his achievement. The best defense, of course, is to hire trustworthy employees and treat them well, but no system is invulnerable to determined criminals.

SUMMARY

Procuring computer applications is usually a better alternative than developing application software in-house. The likelihood of successful procurement will improve, however, if good procedures are followed. Many of the analysis procedures discussed in Chapter 5 apply here, as do some of the design methods presented in Chapter 6. But these will be augmented with a careful search through a variety of sources of information to identify several different types of solutions and alternative products available within them. Common alternatives to acquiring package systems in construction include computer service bureaus and time-sharing services, as well as third-party developers and consultants.

Sources of commercial hardware and software include manufacturers' direct sales forces, retail stores, brokers, mail-order firms, and value-added resellers. Depending on the scope of the application, formal requests for proposals can be developed and issued to potential vendors. A tabular matrix or other means can help to objectively compare and evaluate proposals that may be submitted. Once a vendor is selected, most of the usual contractual and financial arrangements can formalize the purchase, but special consideration should be given to the rapid obsolescence of computer equipment and to the fact that most commercial software is licensed rather than sold outright.

Approaches to training depend on the scope and complexity of the application and the number and characteristics of its users. Some training options may be included with the purchase of an application package, and others may be available from the developer, vendor, or third-party firms. On-the-job training, seminars and workshops, and tutorial manuals are the most common, particularly when the number of users is small, but some of the more widely used packages can also be supported by polished video-based courses, online tutorial software, and interactive multimedia systems. In any case, special attention must be given to acquiring quality instruction and materials and to matching the level of teaching to the capabilities and needs of the users.

A carefully planned implementation strategy for a new system is critical to its acceptance and success. Major alternatives are total conversion, parallel operations, phased modules, and pilot implementation. Each has its advantages and disadvantages, and different kinds of applications suit themselves to different strategies.

Ongoing system management is facilitated by building and maintaining good relations with vendors who can provide sound technical and business advice. But there are numerous pitfalls in the computer market, so it is also essential to have in-house expertise or reliable consultants to assist in major decisions. Key areas on which system management should focus include ongoing service and support for both hardware and software. Several technical and business alternatives are available for each. Continuing tasks include routine operations and maintenance, staffing and personnel, and security, among others.

REVIEW QUESTIONS

1. Assume that you are planning to acquire a network-based system of five microcomputers for use in the home office of a construction company.

   (a) Outline in general an evaluation and selection procedure that would ensure that, for a given amount of money, you obtain the best system to serve the needs of the company.

   (b) What would be a few of the comparative advantages and disadvantages of leasing versus buying the equipment you need?

2. This question assumes that you are using the categorical tabular point method (e.g., Table 5-3) for ranking vendor proposals for a new computer hardware and software system for a contractor's financial and administrative applications. For this question assume you are the systems analyst and consultant for the acquisition study; you are the most computer-savvy person in the company; and you have had good cooperation in developing the selection criteria and in carrying out the evaluation. How should you respond if, out of a possible total of 400 points, the system you initially favored, and still think has the best technical merits, wound up in second place at 290, some 50 points below the 340 points of the system that has emerged as the company favorite?

3. Four alternative approaches to implementation of a new system are (1) total conversion at a specified date; (2) parallel operation/gradual transition; (3) phased introduction of related modules; and (4) pilot implementation.

   (a) State the single most important advantage and the single most important disadvantage of each method for a construction company.

   (b) State and briefly justify the method you would select for (1) a major change in the company's financial accounting system, and (2) a change from manual bar charts to microcomputer-based CPM for field project scheduling.

4. This problem uses information about operating and maintaining microcomputers to develop policies for maintenance contracts and staffing. First, consider three alternative

ways of providing maintenance: (1) assume no in-house expertise, consider putting all machines on a carry-in depot maintenance agreement, and take them in whenever there is a failure; (2) allocate some time from an in-house engineering employee who is more knowledgeable and interested in computers than anyone else in the office, and consider combining this with a carry-in agreement; and (3) hire an experienced microcomputer professional to be responsible for the upkeep and routine maintenance and repair of microcomputers; this also could be combined with a carry-in agreement if it is cost-effective to do so.

Assume that a manufacturer maintenance agreement costs $300 per year per microcomputer. On the average, assume that a micro fails about twice a year, and if there is no maintenance agreement, the average repair cost will be $200 per trip to the repair shop if taken to the vendor. Whether or not there is an agreement, the average depot repair takes the machine away for three days. We will also account for another $100 for the total cost of the courier to take the machine to the depot and pick it up. In addition, there is a $50 per hour impact cost for the first two hours of user disruption if the machine fails but can be fixed on-site within two hours, and $200 total if the machine is taken away to the depot and the user has to move his or her backup files and share another employee's computer.

As already stated, all failures under policy (1) require a trip to the depot. In policy (2), assume our interested amateur can fix the problem 60 percent of the time with an average investment of one hour. Assume that he will try for no more than two hours before giving up and sending the machine to the depot. With benefits and overhead, his time is worth $60 per hour to the company, not allowing for the disruption of his primary task. Only time actually used is charged to this maintenance and repair function. In policy (3), assume that the microcomputer professional keeps a stock of spare boards and parts and can fix the problem 90 percent of the time within one hour, and that she also will try no more than two hours before sending it in for repair. Her time, with overhead and benefits, is worth $80 per hour to the company. She will be hired to perform other microcomputer-related duties as well, but assume that at least 20 percent of her time must be budgeted to the maintenance and repair function, even if all of this time is not used. All personnel work a 40-hour week.

(a) Compute the average annual cost per micro for the following conditions. In policy (3) for this question, charge for the professional's time only on a unit repair basis. Do not try to include the 20 percent annual allocation.

Policy (1), with and without the annual maintenance agreement
Policy (2), with and without the annual maintenance agreement
Policy (3), with and without the annual maintenance agreement

(b) In light of the 20 percent annual allocation in policy (3) for the professional, what break-even number of microcomputers would justify moving from policy (2) to policy (3)?

5. A project manager wishes to choose between a microcomputer and a terminal, which must be online to the remote computer whenever in use, to use for transmission of cost and payroll data from the site office to the home office computer for central processing. Some relevant information is given below:

- Job duration is 52 weeks.
- All hardware costs must be charged to this job (no salvage assumed).

- The terminal costs $500 to buy, plus $20/month maintenance.
- The microcomputer costs $2000 to buy, plus $50 per month maintenance.
- A data-entry person can type an average of 4 characters per second.
- A 480 character-per-second modem can be purchased for $200.
- The weekly payroll entry volume is 100 records at 300 characters per record.
- The entry volume for the weekly cost report is 200 records at 120 characters per record.
- Telephone charges to the home office are $0.75 per minute.

. Assume that the staff time, log on time, home office computer costs, and other costs not specifically priced here will be the same for either alternative.

  (a) Assuming that this is the only application used to justify the purchase, determine the most economical solution—between the microcomputer and the terminal—for this application. Show the production and cost calculations that support your answer.

  (b) Comment on the implications for people costs and communications costs compared to those of the computer or terminal. What is the effect of the extra hardware investment on overall costs?

SUGGESTIONS FOR FURTHER READING

The following books are particularly suitable for managers responsible for planning and supervising business computer systems, including modern distributed network systems. They also have sections pertinent to the material covered in Chapters 5 and 6.

Frenzel, Carroll W. *Management of Information Technology.* Boston, Mass.: Boyd and Fraser, 1992.

Hussain, Donna S., and K. M. Hussain. *Information Management.* Englewood Cliffs, N.J.: Prentice Hall, 1992.

McNurlin, Barbara C., and Ralph H. Sprague, Jr. *Information Systems Management in Practice.* Englewood Cliffs, N.J.: Prentice Hall, 1989.

APPLICATION
DEVELOPMENT
TOOLS

Give me a place to stand, and I will move the earth.

Archimedes*

C hapter 6 discussed the process of creating original computer-based systems custom-designed to handle applications in construction. Chapters 8 to 11 now introduce four categories of application development software that can support this process:

- Procedural languages
- Spreadsheets
- Databases
- Artificial intelligence (AI) and knowledge-based systems

On the Lever.

These are not the only types of development software, but they are the main ones being used for construction applications. Others, mentioned in Chapter 4, included assembly language, program generators, problem-oriented languages, and mathematical modeling environments.

Explaining in detail how to use each language, generic package, or AI approach is well beyond the scope of these chapters. Take a language out of Chapter 8 and you will find numerous books purporting to cover the fundamentals, and other whole books exploring intermediate and advanced techniques or specialized applications of the language. The same proliferation of reading material has been happening with methods and tools for artificial intelligence and knowledge-based systems. A good bookstore will have shelves full of books about each of the more common spreadsheet and database packages, again at beginning, intermediate, and specialized levels. In this text each chapter makes some recommendations for reading in this literature, but we will not attempt to perform the function of all those shelves of books.

Although not explaining software development tools in detail, each chapter in Part IV first introduces the general *concepts* of each tool and describes some of its major capabilities. It also mentions some of the differences to look for among specific tools within the broad category. It then describes in a general way the types of construction applications for which tools in the category are most suitable. Finally, each chapter develops two examples that show, first, distinctly different types of construction applications within its category and, second, the application of real development products to these examples. This approach gives added insight into how the tools themselves work in practice. In each chapter, one example involves an MS-DOS/IBM PC and compatible system environment and the other will be on an Apple Macintosh.

CHAPTER
8

PROCEDURAL LANGUAGES

Therefore is the name of it called Babel; because the Lord did there confound the language of all the earth....

Genesis 11:9

Since the late 1950s, procedural languages have been the primary vehicle for developing computer application software. Even though generic spreadsheets, databases, expert system shells, and other alternatives are handling more and more construction applications, procedural languages will be with us for some time. Although most construction engineers and managers will not be directly involved in such programming, it is nevertheless useful for them to understand some basic concepts and differences among various procedural languages so that they can better influence and guide the selection or use of such software for their companies and projects.

Chapter 4 defined procedural languages in general terms and described supporting development tools such as text editors, compilers, linkers, interpreters, libraries of common subroutines, and tools that assist in debugging new programs under development. Chapter 6 presented some criteria and procedures to guide the selection of the tools that are most suitable for a given organization and application. This chapter first presents some fundamental concepts of procedural languages. Next, it describes some of the languages that are available, then their advantages and limitations in typical construction applications. The key parts of two simplified construction examples, developed in two different languages, are presented next to convey an idea of what programs written in a procedural language look like. We chose the languages BASIC and Pascal for these examples because they are two of the most commonly used teaching languages and are widely available on microcomputers in a form that is more easily accessible to novices than some of the other languages used for professional software development.

BASIC CONCEPTS OF PROCEDURAL LANGUAGES

Procedural languages provide rules of syntax; data definitions and data structures; execution control mechanisms; mnemonic keywords; unary, relational, equality, arithmetic, assignment, and logical operators; input and output capabilities; and prewritten libraries of common functions and subroutines. They remain the most flexible and powerful of the application development tools and are still the most widely used approach to creating new applications.

Procedural languages are built up from fundamental components that work together according to precisely defined sets of rules. At the elementary level, they provide some way of representing constants and variables. A *constant* is a fixed piece of data or code, such as the *integer* 45, the decimal number 24.374, a Boolean value TRUE, the set of characters (called a *string*) XF-42J, a hexadecimal or octal number, and so on. They can either be read into programs from input sources or be permanently embedded in code by a programmer, such as an *integer* to define the number of times a piece of code should be executed, or a *real* (i.e., decimal) number to represent the constant π. Languages vary somewhat in the range of values they permit for numeric constants (e.g., from -32768 to $+32767$ if integer constants are limited to 16 bits of storage), precision (e.g., floating-point numbers limited to eight significant decimal digits if 24 bits contain the mantissa in their internal scientific notation), and the specific characters that can represent alphanumeric data (e.g., strings may not be allowed to start with numbers, or certain characters like $ may define special names). Furthermore, these limits and conventions may vary with different implementations (i.e., different compilers and interpreters running on different hardware) of a given language. You should always consult the programmer's manual to determine these implementation-specific limits and conventions. The way that constants are represented (e.g., strings within single quotation marks, octal numbers designated with an initial # character, etc.) is normally standardized for a given language.

Variables, in effect, are temporary holding places, usually in memory, where a series of changeable values can reside temporarily. For example, they may hold data being read in from a disk file, interim results of computations or character manipulations, and so on. They are usually named with a set of characters that bears some resemblance to the purpose of the variable (e.g., Haul_Time for a component of a truck's production cycle, where the _ character joins two words into a single string). Some languages impose tight limits on the number of characters that can be used in variable names (e.g., six in FORTRAN), which can cause some obscure abbreviations for more descriptive words (e.g., HLTIME). Most languages require that programmers *declare* variables with respect to type (e.g., integer, real, complex, character, logical, etc.) and possibly in size (see following discussion of arrays) before using them in a program (e.g., with the key word INTEGER followed by the names of variables of this type). Some languages have *implicit* naming conventions (e.g., unless declared otherwise, variable names starting with the letters I, J, K, L, M, N in FORTRAN are of the integer type; BASIC variable names ending in $ hold character strings). Generally, however, one should declare all variables. Variables can also be defined as *local*, where they are known only within the portion of a program (i.e., in a procedure or subroutine) where they have been declared, or *global*, where they are known and accessible throughout the program.

All languages allow the imposition of some types of *structure* on data used in a program. As a minimum, programmers can declare variables to represent *arrays* of data, from one, two, or three dimensions to many more in some cases. For example, the declaration REAL Mydata(6,10,2) might represent a matrix of real numbers, with the name Mydata, 6 columns wide, 10 rows high, and 2 layers deep, for a total of 120 storage places for numbers ($6 \times 10 \times 2$). These arrays can facilitate efficient iterative processing of similar data elements. In earlier languages and implementations, these declared allocations of memory were fixed at the time the program was compiled and loaded into memory (in the example, 120 memory locations would be blocked out and reserved for Mydata) whether they were needed or not. In such implementations, programmers must ensure the arrays are large enough for whatever quantities of data the program may need to handle. In other cases, language implementations allow for the *dynamic allocation* of storage; that is, storage can expand and contract as needed and be traded off with other variables that might use the space at different times. Finally, some languages allow *hierarchical* data structures (e.g., at the lowest level might be the elements of CPM scheduling, the next level up might be all of those elements that relate to a particular activity, etc.).

Most languages provide arithmetic and logical *operators*. Examples of *arithmetic operators* found in almost all languages are + for addition and − for subtraction. However, almost no language uses $\times$ for multiplication and $\div$ for division. Instead, many use * for multiplication and / for division, and some use the words DIV, MULT, and so on. The symbols for exponentiation vary even more. Some use the double asterisk (for example, 3**2 equals 9), whereas others need a keyboard with the special character ↑, or perhaps just the ^ found on most keyboards. There can also be integer operators such as DIV and MOD, which return the integer quotient and remainder, respectively, from an integer division. *Logical operators* include AND, OR, NOT, and so forth, but some languages enclose them in other characters, such as .AND., .OR., and the like, or even use special character symbols. There are also operators used in tests of inequalities, such as EQ, NE, LE, LT, GT, GE, or the equivalent symbols $=, \neq, \leq, <, >, \geq$. Again, whereas many if not all of these operators are provided in most procedural languages, the symbols for these operators and the rules for their usage vary significantly from one language to another. Constants, variables, and operators can combine to form *expressions* [e.g., PI*(RADIUS^2)*HEIGHT].

The functional capabilities or vocabulary of a programming language is defined in a set of *keywords* or *reserved words*. Examples from BASIC include GOTO, PRINT, IF *some test* THEN, OPEN, and END. These must be typed exactly as specified in the language, and normally the same sets of characters cannot be used as variable names or for any other purpose.

These words can combine with logical and arithmetic operators and with variables and constants to form **statements**. Statements are analogous to sentences in a verbal language and usually express a specific task to be performed. Each language will define a **syntax** for constructing its statements, which amounts to the rules of grammar for the programming language. Most programming languages are extremely particular about spelling, capitalization, punctuation, and the order and location of elements within statements. The syntax of standard BASIC requires line numbers in front of each line; the syntax of earlier versions of FORTRAN, which dates from the days of punched cards, specifies exactly which columns can be used for the comment symbol (Column 1), statement labels (Columns 2 to 5), line continuation marks (Column 6), and the statements themselves (Columns 7 to 72). Pascal and several others are noted for their use of punctuation marks (; and .). The syntax can vary so much from one language to another that some seem entirely backward of others. Some, such as Forth, *post-fix* their operators at the end of their statements, whereas most others *in-fix* them algebraically among the variables and constants. Programmers must carefully learn and use the syntax specific to a given language, and they must take care not to confuse it with others. There can be *declarative statements* (e.g., to specify variable data types and define the size of arrays); *procedural statements,* to define the beginning and end of procedures and functions; *assignment statements,* to specify an action to be performed (such as arithmetic or a string manipulation); *control statements* (e.g., to state how many times a given series of statements should be performed); *conditional tests* (e.g., IF *condition* DO *something* ELSE DO *something else*); *input and output statements;* and statements describing the *format* or appearance of screens and reports.

Most languages specify a minimum number of library **functions, procedures,** and/or **subroutines**. Specific implementations of any given language usually add more than the minimum specified in the language standard, but using these can lead to incompatibilities unless informally agreed-upon *extended standards* have evolved. For example, it is common to find logarithmic (natural and base-10 logs and exponentials), statistical (averages, standard deviations, etc.), and trigonometric (sine, cosine, tangent, and often their inverses and others as well) functions, particularly in languages designed for science and engineering. Others have functions that deal with decimal currency (e.g., round, truncate, percentages, interest calculations, etc.) and character strings (extracting, inserting, combining, etc.), particularly those languages designed for business purposes. Some come with good sorting and searching procedures.

Computer programming would have limited applicability if languages merely provided the ability to list a series of statements and execute them sequentially, from top to bottom, like reading a book. The power of computer programming resides in the *control structures* that languages provide and in the ability to iteratively repeat sections of code and procedures tens, hundreds, and thousands of times. As a minimum, computer languages provide the means to transfer control directly or conditionally (subject to an arithmetical or logical test) from one part of a program to another. For example, a GOTO *label* statement, when encountered, will transfer control to the statement with the corresponding *label* rather than continue to the statement after the GOTO. If the statement is of the form IF *some test* THEN GOTO *label,* it establishes a branch, where execution will either continue to the next statement or transfer to *label,* depending on the result of *some test.* More complex control structures include iterative loops, which are pieces of code that will execute repeatedly either for a specified number of times or until some test condition is satisfied. At a higher level, most languages allow us to divide code into relatively independent procedures that can be tested and reused in other programs, and which enable programmers to isolate (make "local") certain variables, constants, and logical routines. These elements of structured programming enable programmers to build complex systems from well-tested modules under conditions that can make the overall system very reliable.

Some languages are weak in structure. For example, although one can use GOSUB to transfer to and RETURN to come back from pieces of code in standard BASIC, all parts of the program are actually accessible at any time from any other part, and all variables are in effect global, even though the programmer lays the statements out in separate areas. Thus, a standard BASIC program is really one single program, and too often it has transfers all over the place that make it logically like a tangle of spaghetti (indeed, programmers sometimes use the term *spaghetti code* to describe such programs). Other languages allow structure but make it too easy for programmers, out of ignorance or expediency, to violate good structured programming practices (e.g., via COMMON blocks in FORTRAN, or GOTO statements in BASIC). Languages like Pascal, Ada, and Modula-2 almost force good structured programming, although the restrictions they impose sometimes rankle certain programmers. We provide a Pascal example later in this chapter.

Finally, programming languages usually provide *input and output (I/O)* capabilities to receive input from terminals, networks, files, instruments, and so on and to send output data to display screens, printers, files, networks, and automatically controlled devices. Some languages are limited in capability (handling keyboards, screens, and disk files, but little else), whereas others can reach almost anything that can be connected to the computer. The standard definition of Pascal has almost no provision for I/O, although numerous implementations over time have evolved to a kind of standard. Common implementations of BASIC have rather simple I/O capabilities, but these have provided a surprising amount of flexibility in the applications that have been developed, including real-time data acquisition and process control. FORTRAN's standard I/O capabilities are similarly lacking, but its rich scientific and engineering heritage has created a large number of public and proprietary library routines for advanced graphics, process control, reporting, and almost any other application that can be done on computers. C is much admired by advanced programmers for its low-level capabilities that provide experts with a tremendous amount of flexibility in I/O programming. In recent years, the complex demands of the advanced graphical user interfaces (GUIs) (i.e., those fancy windows, icons, and mice that we take for granted today) have required a much higher level of sophistication in the I/O programming capabilities of languages that support application development for these environments.

In addition to the languages themselves, there is a plethora of software that constitutes the *programming environment* surrounding the language. For further information, we refer you to the editors, compilers, debuggers, and so on described in Chapter 4. Suffice it to say that some language developers do an excellent job of integrating these tools with their language, and this greatly facilitates the ease, productivity, and reliability of program development.

SOME COMMONLY USED PROCEDURAL LANGUAGES

This section briefly reviews some of the major procedural programming languages and, for some, provides short segments of program code to illustrate their differences. The code segments perform a simple crane computation, as shown in Fig. 8-1. They take as inputs from the user the length L and angle A of a boom, use these to compute the horizontal distance R from the boom hinge point to the point under the load, and then print the result on the screen.

Two of the first and most successful procedural languages implemented for compilation were FORTRAN and COBOL. FORTRAN source code looks a lot like the mathematical language of scientists and engineers and remains a mainstay of their professions. The name of the language itself means FORmula TRANslation. The sequence of mathematical formulas and the logical controls on the program's execution constitute a *procedure*. Figure 8-2 is a listing of the FORTRAN code that performs the required operations.

BASIC, which stands for Beginners' All-purpose Symbolic Instruction Code, is a useful and flexible language that is often included or available at low cost with personal computers. Although in its standard form it is lacking in the procedural modularization, rigorous control structures, strong data typing, I/O flexibility, and some other features of more advanced languages, its simple

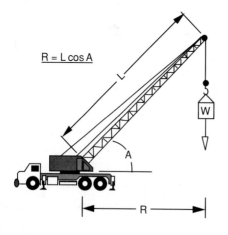

$$R = L \cos A$$

FIGURE 8-1
Crane problem.

```
C --- EXAMPLE CRANE CALCULATION
      REAL ANGLE, LENGTH, PI, RADIUS
      PI = 3.1415926
C
      READ (1,10) ANGLE
  10  FORMAT ('Input the boom length (meters): ', F5.1)
      READ (1,20) LENGTH
  20  FORMAT ('Input the boom angle (degrees): ', F5.1)
C
      RADIUS = LENGTH * COS (ANGLE * PI/180)
C
      WRITE (2,*) 'The horizontal projection is: ', RADIUS, ' meters'
      RETURN
      END
```

FIGURE 8-2
Example of FORTRAN source code.

mnemonics, straightforward syntax, and typical interpretive implementation make it easy for novices to learn and use for programming. In Fig. 8-3 we see how BASIC tackles the crane problem. More advanced versions of the language borrow structured programming concepts and a richer array of functions from other languages and employ compilers to generate fast code, both of which make these versions suitable for professional program development.

COBOL, which stands for COmmon Business-Oriented Language, was developed mainly for business applications such as accounting, financial reporting, personnel records, and marketing. It offers flexible constructs for information storage, sorting, and searching, and it precisely formats business reports. The language also uses English-like words, sentences, and paragraphs to express its formulas and procedures; this characteristic makes it somewhat self-documenting for long-term development and maintenance by a series of programmers in the business environment. Micro Focus COBOL by Micro Focus of Palo Alto, California (also marketed by Microsoft), is a widely used microcomputer version.

```
100   REM  **** Example Crane Calculation ****
110   REM
120   REM - Input the length and boom angle:
130       INPUT "Input the boom length (meters) : ", LENGTH
140       INPUT "Input the boom angle (degrees) : ", ANGLE
150   REM
160   REM - Compute the radius:
170       PI = 3.1415926#
180       RADIUS = LENGTH * COS (ANGLE * PI/180)
190   REM
200   REM - Print the radius:
210       PRINT "For a boom length ", LENGTH," and angle ", ANGLE
220       PRINT "The horizontal projection is: ", RADIUS, " meters."
230   END
```

FIGURE 8-3
Example of BASIC source code.

Figure 8-4 is a COBOL program that performs the equivalent function to the FORTRAN code shown in Fig. 8-2. It includes the four main divisions (identification, environment, data, and procedure) required by COBOL programs. Note that the lack of trigonometric functions in COBOL requires us to define explicitly an approximate trigonometric table in the data division and to limit the input angles in the procedure division for this technically oriented problem. Business programs rarely if ever need such functions.

In the 1960s there was also a reasonably successful effort to combine advantages of FORTRAN and COBOL into a more complex language called PL-1.

```
IDENTIFICATION DIVISION.
PROGRAM-ID. CRANE.
*   EXAMPLE CRANE CALCULATION

ENVIRONMENT DIVISION.
SOURCE-COMPUTER. IBM-PC.
OBJECT-COMPUTER.  IBM-PC.

DATA DIVISION.
WORKING-STORAGE SECTION.
01 PI              PIC 9V9(7)      VALUE 3.1415927.
01 COS_15          PIC V9(4)       VALUE 0.9659.
01 COS_30          PIC V9(4)       VALUE 0.8660.
01 COS_45          PIC V9(4)       VALUE 0.7071.
01 COS_60          PIC V9(4)       VALUE 0.5000.
01 COS_75          PIC V9(4)       VALUE 0.2588.
01 LENGTH          PIC 99V99       VALUE ZERO.
01 RADIUS          PIC 99V99       VALUE ZERO.
01 ANGLE           PIC 99          VALUE ZERO.

PROCEDURE DIVISION.
*   INPUT THE LENGTH AND ANGLE
    DISPLAY "INPUT THE BOOM LENGTH (METERS):          ".
    ACCEPT LENGTH.
    DISPLAY "INPUT THE ANGLE (15, 30, 45, 60, 75 DEGREES):      ".
    ACCEPT ANGLE.

*   COMPUTE THE RADIUS
    IF ANGLE = 15
         COMPUTE RADIUS = COS_15 * LENGTH
    ELSE
         IF ANGLE = 30
              COMPUTE RADIUS = COS_30 * LENGTH
         ELSE
              IF ANGLE = 45
              COMPUTE RADIUS = COS_45 * LENGTH
            ELSE
              IF ANGLE = 60
                   COMPUTE RADIUS = COS_60 * LENGTH
              ELSE
                   COMPUTE RADIUS = COS_75 * LENGTH.
*     PRINT RADIUS
      DISPLAY "THE HORIZONTAL PROJECTION IS:", RADIUS.
      STOP RUN.
```

FIGURE 8-4
Example of COBOL source code (from Mohan Manavazhi).

Although it was widely used for many years, one seldom hears of it anymore. Even today most engineering and scientific programs have been written in FORTRAN, and most mainframe business programs in large companies and institutions have been written in COBOL.

Over the years many other languages have come and several have gone. Some have been designed to implement evolving principles of computer science better and to promote good programming practice. An early example developed in Europe, which was a contemporary of FORTRAN and COBOL, was ALGOL. It laid the groundwork for some of the better languages that have followed, including Pascal, Modula-2, and Ada.

A Swiss professor named Nicholas Wirth originally developed Pascal around 1970 to teach sound programming practices to college students, and it has become the dominant teaching language in the United States as well (although an increasing number of schools have shifted to C). Early versions of the language did not even have all of the functionality (such as input and output) that was needed in practical applications. However, Pascal's strong data typing and top-down procedural structures promote good programming methods, and commercial implementations of the language have evolved that can implement the most complex of business, scientific, and engineering applications. Indeed, many of the spreadsheets, word processors, video games, and other popular software in the market today were first written in Pascal. Figure 8-5 is a section of Pascal code that performs the crane calculation.

```
program Crane;
{ EXAMPLE CRANE CALCULATION }

  const
   pi = 3.1415927;
  var
   length, angle, radius: REAL;

begin

{ Input the length and angle }
  write(output, 'Input the boom length (meters): ');
  readln(input, Length);
  write(output, 'Input the boom angle (degrees): ');
  readln(input, Angle);

{ Compute the radius }
  radius := length * COS(angle * pi / 180);

{ Print the radius }
  writeln;
  writeln(output,'The horizontal projection is:',radius:5:0,' meters')

end.
```

FIGURE 8-5
Example of Pascal source code.

Professor Wirth also developed a more advanced language called Modula-2, which has various commercial implementations. However, perhaps because Pascal itself has evolved sufficiently to serve programmers' needs, Modula-2 has been slower to catch on. Its code appears similar to that of Pascal.

The Ada language, which initially was promoted and used especially in the defense sector, has also evolved from some of the principles set forth in Pascal. It has some objectives in common with PL-1, particularly to provide a full complement of capabilities that can support a wide range of management, engineering, and scientific applications. Consequently, the language is large and complex and is used mainly by professionals working on major application development projects.

C, which was originally developed at Bell Laboratories along with the UNIX operating system, offers some of the machine-specific efficiency and flexibility advantages of assembly language programming, and it has grown in popularity with professional, scientific, and engineering programmers. There has been a trend toward using C to develop system software as well as application programs, and today it is probably the most widely used language for commercial software development, at least for application packages used on microcomputers and workstations. Chapter 11 mentions an object-oriented derivative called C++. Figure 8-6 is a C program that solves the example of the crane problem.

There are many other languages, including APL, Forth, Logo, RPG, and others. Some remain important in certain areas, others serve the interests of enthusiasts, and a few are declining from times when they enjoyed more popularity.

```
/* EXAMPLE CRANE PROGRAM */

#include <stdio.h>
#include <math.h>

#define PI 3.1415927

main()
{
   double length, radius, angle;

   /* Input the length and angle */
   printf("Input the boom length (meters) : ");
   scanf("%lf", &length);
   printf("Input the boom angle (degrees) : ");
   scanf("%lf", &angle);

   /* Compute the radius */
   radius = length * cos(angle * PI/180);

   /* Print the radius */
   printf("The horizontal projection is: %5.1f meters\n", radius);
}
```

FIGURE 8-6
Example of C source code (from Mohan Manavazhi).

Few if any of these are used much in construction, so they have not been described here.

One also sometimes hears of other computer languages such as Prolog and LISP. However, these fall in a somewhat different category from the procedural languages that have been described in this chapter and we will return to them in Chapter 11.

ADVANTAGES AND LIMITATIONS

Most procedural languages are suitable for a wide range of applications, a range as broad as almost any that are likely to be needed in construction. Nevertheless, languages such as FORTRAN and C are more commonly used for mathematical and engineering applications, whereas COBOL and some of the more advanced versions of BASIC are more likely to be found in custom business applications. Pascal, Modula-2, and Ada have not been used so much in construction, but they have potential for either type of application. ALGOL and PL-1 are not used much anymore, although some large construction firms once had a significant amount of software written in PL-1 for IBM mainframes.

Apart from these generalizations, availability of qualified programmers and compatibility with the existing software environment remain some of the main criteria when choosing a language. In almost any metropolitan area are many programmers who work well in FORTRAN, COBOL, and BASIC; and those who know C and Pascal are also widely available, particularly in academic and technical settings.

Where complex data file structures and flexible input and reporting are required, COBOL and PL-1 have historically been preferred, but Ada is a growing force in this area, particularly as a result of the U.S. government requirements supporting this language. Pascal and C have also become increasingly important for commercial applications, especially in microcomputer and workstation environments. FORTRAN has long been dominant in scientific and engineering computing, but C is taking a larger share of this market, and BASIC continues to be used for simpler programs.

Languages such as Pascal and Ada tend to enforce the most rigorous programmer discipline, whereas C is considered more vulnerable to unexpected behavior if not used with care. Although easy, BASIC is considered by some to preclude good programming practices. Judged by modern standards, FORTRAN and COBOL have been considered backward, but later versions of these languages have borrowed features and capabilities from the newer languages to stay competitive. Indeed, one wag said that although he does not know precisely what the language of engineering computation will look like in the twenty-first century, we can be pretty sure that it will still be called FORTRAN.

In general, it is difficult to set limits on what construction applications can be implemented with procedural languages. They can handle virtually anything that can be done with computers. But we have indicated before that this can be an expensive and time-consuming route to some kinds of application software development. After reading Chapters 9 through 11, you may conclude that we should fall back on procedural languages mainly in cases where more specialized development environments are inappropriate or too limited.

CONCRETE STRENGTH TEST CALCULATIONS IMPLEMENTED IN BASIC

The first example, to be implemented in the BASIC language, is from concrete quality control. The purpose is to have a program that will input from a disk file the concrete cylinder test results from all the pours completed in a week's time, make some statistical computations, and print a report of the results. The input records for each pour contain the following data:

- Pour number (integer)
- Pour location (alphabetic)
- Design strength (integer)
- Number of cylinders (integer)
- Strength of each cylinder (integer)

The program should compute the following:

- The mean strength for each pour
- The mean strength for all tests of cylinders for concrete of a given design strength
- The standard deviation for all cylinders of a given design strength

The program should then print out a neat report with a heading and the results of the tests and computations. One line for each pour should give, in successive columns, the pour number, location, design strength, and the mean of the cylinder test results. A double asterisk (**) should be placed by the mean if it is below the design strength. Below these results should be summary computations for all cylinders of a given design strength that show, in successive columns, the design strength (only for those design strengths used that week), the number of pours using that design strength, the total number of cylinders tested for those pours, the mean of those tests, and their standard deviation.

Figure 8-7 is a listing of a program for this application. It is written in Microsoft GW-BASIC to run on an MS-DOS-based microcomputer. Comments included in the program designate its main sections. This simple, relatively unstructured program proceeds sequentially through the steps of defining variables (statement numbers 130–240), reading in a week's data from a disk file (statement numbers 420–550), computing the mean of each pour's cylinder breaks and printing the results for each pour (statement numbers 620–760), finding the number of

```
100 REM *** CONCRETE CYLINDER TEST ANALYSIS ***
110 '
120 '      Define variables:
130 DIM POURNO(100)        'Pour number
140 DIM POURLOC$(100)      'Pour location
150 DIM DESNSTR(100)       'Design strength
160 DIM NUMCYL(100)        'Number of cylinders in test
170 DIM TESTS(100,6)       'Test results
180 DIM POURMEAN(100)      'Mean strength for each pour
190 DIM DESNUSED(10)       'Design strengths used this week
200 DIM DESNMEAN(10)       'Mean of test results for each design
210 DIM DESNSTDV(10)       'Standard deviation for each mix design
220 DIM DESNSUM(10)        'Sum of test results for given design strength
230 DIM TOTDCYL(10)        'Total number of cylinders for given design
240 DIM NUMPOUR(10)        'Total number of pours using given design
300 '
310 CLS          '       Clear screen and print program title:
320 PRINT TAB(8);"*** CYLTEST -- THE CONCRETE TEST ANALYSIS PROGRAM ***"
330 PRINT TAB(8);"---------------------------------------------------------"
400 '
410 '      Input the data file:
420 PRINT : INPUT "    --- Enter name of test data file: ", FILENAME$
430 OPEN FILENAME$ FOR INPUT ACCESS READ AS #1
440 PRINT
450 PRINT "    --- The data in the input file are being read in..."
460 PRINT
470 FOR I = 1 TO 100
480   INPUT# 1, POURNO(I)
490   IF POURNO(I) = 9999 THEN GOTO 540
500   RECS = I
510   INPUT# 1,POURLOC$(I),DESNSTR(I),NUMCYL(I)
520   FOR J = 1 TO NUMCYL(I) :  INPUT# 1, TESTS(I,J)  :  NEXT J
530 NEXT I
540 CLOSE #1
550 PRINT "    ---"; RECS; " records have been input."
560 PRINT : INPUT "    --- Press RETURN to continue:", ANYKEY$
600 '
610 '      Compute and print data for each pour:
620 CLS
630 PRINT " ***** SUMMARY OF POUR TEST RESULTS *****" : PRINT
640 PRINT "Pour  Location                 Design N    Mean"
650 PRINT "----- -------------------- ----- -   -----"
660 FOR I = 1 TO RECS
670   TESTSUM = 0
680   DBLSTAR$ = "  "
690   FOR J = 1 TO NUMCYL(I)
700     TESTSUM = TESTSUM + TESTS(I,J)
710   NEXT J
720   POURMEAN(I) = TESTSUM / NUMCYL(I)
730   IF POURMEAN(I) < DESNSTR(I) THEN DBLSTAR$ = "**"
740   PRINT USING "##### & ##### ##   ##### &";
      POURNO(I),POURLOC$(I),DESNSTR(I),NUMCYL(I),POURMEAN(I),DBLSTAR$
750 NEXT I
760 PRINT : INPUT "    --- Press RETURN to continue:", ANYKEY$
```

FIGURE 8-7
Cylinder test program.

```
800 '
810 '_____Determine number of design strengths used this week:
820 CLS
830 PRINT: PRINT: PRINT "--- Statistical calculations in progress . . ."
840 DESNUSED(1) = DESNSTR(1)
850 NUMDESN = 1
860 FOR I = 1 TO RECS-1
870   FOR J = I+1 TO RECS
880     IF DESNSTR(I) = DESNSTR(J) GOTO 950
890     FOR K = 1 TO NUMDESN
900       IF DESNUSED(K) = DESNSTR(J) THEN GOTO 950
910     NEXT K
920     NUMDESN = NUMDESN + 1
930     DESNUSED(NUMDESN) = DESNSTR(J)
940     GOTO 960
950   NEXT J
960 NEXT I
1000 '
1010 CLS : PRINT : PRINT
1020 PRINT "    **** RESULTS OF STATISTICAL CALCULATIONS ****" : PRINT
1030 PRINT "Desn Str   Num Pours   Num Cyls   Mean Str   Std Dev"
1040 PRINT "--------   ---------   --------   --------   -------"
1050 FOR I = 1 TO NUMDESN
1060   DESNSUM(I)  = 0
1070   TOTDCYL(I)  = 0
1080   NUMPOUR(I)  = 0
1090   DESNMEAN(I) = 0
1100   DESNSTDV(I) = 0
1110 '_____Compute mean of test results for given design:
1120   FOR J = 1 TO RECS
1130     IF DESNUSED(I) <> DESNSTR(J) THEN GOTO 1190
1140     NUMPOUR(I) = NUMPOUR(I) + 1
1150     FOR K = 1 TO NUMCYL(J)
1160       TOTDCYL(I) = TOTDCYL(I) + 1
1170       DESNSUM(I) = DESNSUM(I) + TESTS(J,K)
1180     NEXT K
1190   NEXT J
1200   DESNMEAN(I) = DESNSUM(I) / TOTDCYL(I)
1210 '_____Compute standard deviation:
1220   FOR J2 = 1 TO RECS
1230     IF DESNUSED(I) <> DESNSTR(J2) THEN GOTO 1270
1240     FOR K = 1 TO NUMCYL(J2)
1250       DESNSTDV(I) = DESNSTDV(I) + (TESTS(J2,K) - DESNMEAN(I))^2
1260     NEXT K
1270   NEXT J2
1280   DESNSTDV(I) = SQR (DESNSTDV(I)/TOTDCYL(I))
1290   PRINT USING" #####        ##        ##        #####      ####";
          DESNUSED(I),NUMPOUR(I),TOTDCYL(I),DESNMEAN(I),DESNSTDV(I)
1300 NEXT I
1310 PRINT "--------------------------------------------------" : PRINT
1400 END
```

FIGURE 8-7 (*continued*)

different design strengths used that week (statement numbers 840–960), and finding the mean (statement numbers 1120–1200) and standard deviation (statement numbers 1220–1280) for all the cylinder breaks for all of the tests corresponding to each design strength contained in that week's data file. Except for the request for a file name in statement number 420 and the request to press a key in statement numbers 560 and 760, this is basically a **batch** rather than an **interactive program**. As presently written, the program directs reports to the user's screen via PRINT statements. The output could be redirected to a printer by using BASIC's LPRINT keyword in place of each PRINT for the statements that create reports.

Figure 8-8 is an example set of input data for the program. Such a file could be prepared with a text editor. Although the file looks fairly structured, the main requirements of the program are to have a separate line for the information for each pour and to leave at least one space between the data fields in each record. The column headings are not part of the file and are included here only for clarity. N refers to the number of cylinder test breaks that follow on the same line for each pour.

Figure 8-9 is the first screen that appears when the program is run. Its main purpose is to ask the user to identify the name of the file that contains the test results and confirm that the data have been read in. The user's input (cyltest.dat) appears in this figure in boldface.

Figures 8-10 and 8-11 are the two output reports produced by this program. They follow the specifications of the problem given earlier in this section.

```
Pour   Location                 Design
Numbr  of pour                  strength N Test results (up to 6)
-----  --------------------     ---- -  ----------------------------
21579  "BLOCK 1 FOOTING      "  2500 4  2500 2400 2600 2550
21779  "MAT SLAB, NW QUARTER"  2500 6  2300 2600 2400 2350 2400 2500
21879  "RETAINING WALL 1     "  3000 4  3040 3200 3100 2950
22079  "ABUTMENT 1           "  3000 5  3040 3200 3150 3400 3200
22279  "PRECAST DECK SLAB 1  "  4000 6  4200 3980 4150 4300 4230 4001
22479  "COLUMNS, ROW A       "  4000 4  4100 4200 4300 4050
22679  "BLOCK 2 FOOTING      "  2500 4  2560 2700 2670 2560
22879  "LATERAL WALL         "  3000 2  2600 2750
30279  "PRECAST DECK SLAB 2  "  4000 6  4500 4800 5200 4400 4700 4444
30579  "MAT SLAB, NE QUARTER"  2500 3  2600 2600 2600
30879  "ABUTMENT NO. 2       "  3000 3  3400 3560 3420 2800
31079  "COLUMNS IN ROW B     "  4000 6  4300 4200 4300 4321 4567 4353
31279  "MAT SLAB, SW QUADRAN"  2500 4  2300 2100 2600 2400
31579  "ABUTMENT NO. 3       "  3000 2  3278 3098
31779  "DECK SLAB NO. 3      "  4000 6  4444 4333 4222 4111 4533 4440
32079  "BLOCK 3 FOOTING      "  2500 3  2560 2650 2600
32279  "INVERTED WALL        "  3000 4  3090 3080 3070 3060
32479  "COLUMNS, ROW C       "  4000 6  4300 4567 5432 4321 4567 4356
 9999
```

FIGURE 8-8
Cylinder test data file.

```
        **** CYLTEST -- THE CONCRETE TEST ANALYSIS PROGRAM ****
        --------------------------------------------------------

    --- Enter name of test data file: cyltest.dat

    --- The data in the input file are being read in...

    --- 18  records have been input.

    --- Press RETURN to continue:
```

FIGURE 8-9
First screen displayed by CYLTEST program.

```
       ***** SUMMARY OF POUR TEST RESULTS *****

       Pour   Location              Design N    Mean
       -----  --------------------  -----  -    -----
       21579 BLOCK 1 FOOTING         2500  4    2513
       21779 MAT SLAB, NW QUARTER    2500  6    2425 **
       21879 RETAINING WALL 1        3000  4    3073
       22079 ABUTMENT 1              3000  5    3198
       22279 PRECAST DECK SLAB 1     4000  6    4144
       22479 COLUMNS, ROW A          4000  4    4163
       22679 BLOCK 2 FOOTING         2500  4    2623
       22879 LATERAL WALL            3000  2    2675 **
       30279 PRECAST DECK SLAB 2     4000  6    4674
       30579 MAT SLAB, NE QUARTER    2500  3    2600
       30879 ABUTMENT NO. 2          3000  4    3295
       31079 COLUMNS IN ROW B        4000  6    4340
       31279 MAT SLAB, SW QUADRAN    2500  4    2350 **
       31579 ABUTMENT NO. 3          3000  2    3188
       31779 DECK SLAB NO. 3         4000  6    4347
       32079 BLOCK 3 FOOTING         2500  3    2603
       32279 INVERTED WALL           3000  4    3075
       32479 COLUMNS, ROW C          4000  6    4591

           --- Press RETURN to continue:
```

FIGURE 8-10
Report by pour number.

```
       **** RESULTS OF STATISTICAL CALCULATIONS ****

       Desn Str   Num Pours   Num Cyls   Mean Str   Std Dev
       --------   ---------   --------   --------   -------
           2500          6         24       2504        141
           3000          6         21       3118        223
           4000          6         34       4389        298
       ----------------------------------------------------
```

FIGURE 8-11
Report by design strength.

EQUIPMENT CYCLE TIME COMPUTATION IMPLEMENTED IN PASCAL

This section describes a Pascal program to do calculations of cycle times for earthmoving equipment using some basic principles of mechanics combined with machine power and weight characteristics and haul-road characteristics. The application involves the analysis of the four main components of a hauler's cycle time, as shown in Fig. 8-12. This example shows a truck loaded by a front-end loader.

In this program the time for the loading operation will be given as input data, not computed. Similarly, the time for dumping will be given. The time calculations will thus focus on haul and return times. For the haul and return times, we could look up the manufacturer's rimpull and braking curves, but instead we will use the equation for a hyperbola, which can make a good approximation. Specifically, for engine power we will say that

$$V(\text{kph}) = E \times \left(\frac{P}{F}\right) \times k$$

where
- V is the machine's speed in kilometers per hour
- E is the efficiency factor for the drive train (usually about 80%)
- P is the engine power in kilowatts
- F is the drawbar pull or rimpull in kilograms
- k is a constant to convert the parameters into consistent units

The force F is determined by the combination of the slope of the road and the rolling resistance as applied to the weight of the machine and its load (if any). As long as F is positive, power is applied. If F goes negative (i.e., where the gravitational force from a downgrade exceeds the rolling resistance), then we are

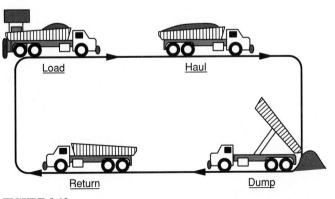

FIGURE 8-12
Diagram showing a truck's load, haul, dump, and return cycle components.

in a braking situation, and the velocity equation is slightly modified, as follows:

$$V(\text{kph}) = \frac{1}{E} \times \left(\frac{R}{F} \right) \times k$$

In this case, note that the drive train inefficiencies help retard the machine, so we take $1/E$. R is the retarding force (also in kilowatts) from the brakes, engine compression, or other means used to control the speed of the vehicle. Since a hyperbola has asymptotes that go to infinity, we also need to impose upper bounds (i.e., speed limits) on these equations for cases where the computed speeds would otherwise exceed the maximum design speeds of the machine.

For purposes of simplicity, this will be a straight deterministic calculation. No provision will be made for queuing (waiting in line for a pusher or loader), variations in times or load amounts, or other uncertainties such as breakdowns. Those topics will be left to the simulation programs in Chapter 15. Similarly, we will assume a simple one-segment haul road and will come back on that same road for the return. Our focus here is more on showing how Pascal can be used to solve a small construction problem, not on showing how complex a problem it can solve. However, we will include one extra check to handle cases where the forces balance to zero.

The program takes the following data as input:

Weight:	the empty weight of the machine (in metric tonnes)
Load:	the weight of the load that the machine carries (in metric tonnes)
Max_Speed:	the maximum speed permitted for the haul or return (in kph)
Power:	the maximum engine power (in kilowatts)
Retarder:	the maximum retarding capability (in kilowatts)
Efficiency:	the drive train efficiency (0 < Efficiency < 1)
Length:	the length of the haul road (in meters)
Grade:	the slope of the road in the haul direction (in percent)
Roll_Resist:	the rolling resistance of the haul (in kg per tonne of machine + load)
Load_Time:	the load time in minutes (including acceleration and maneuver in pit)
Dump_Time:	the time to dump (including acceleration and maneuver at dump)

As variables for values computed in the program, we will also define the following:

Rimpull:	the rimpull or drawbar pull force (in kilonewtons)
Haul_Speed:	the haul speed (in kph)
Return_Speed:	the return speed (in kph)
Cycle_Time:	the total of the load + haul + dump + return times

Figure 8-13 is a partial listing of a program for this application, called **Hauler**. It was programmed using Symantec's development system called THINK Pascal on a Macintosh microcomputer. Comments included within braces in the program designate its main sections and explain certain computations. It is

```
program Hauler;       { Calculates the cycle time for a hauler }

uses
    Vehicle_Calcs;                  { Calculates the haul and return velocity }

var
    Weight, Load, Max_Speed: real;                    {input vehicle parameters}
    Power, Retarder, Efficiency: real;                {input engine parameters}
    Length, Grade, Rolling_Resistance : real;         {input road parameters}
    Load_Time, Dump_Time: real;                       {input cycle components}
    Rimpull, Haul_Speed, Return_Speed, Cycle_Time: real;   {computed values}

begin

{ Print an introductory header }
    writeln(output, '     *** RIMPULL AND CYCLE TIME CALCULATIONS ***');
    writeln;
    writeln(output, 'In the data entry section, type decimal numbers and hit the Return key.');
    writeln;

{ Ask the user for the input parameters }

    writeln(output, 'The following section requests vehicle characteristics:');
    writeln;

    write(output, 'Input the vehicle empty weight (kg): ');
    readln(input, Weight);
    write(output, 'Input the load capacity (kg): ');
    readln(input, Load);
    write(output, 'Input maximum safe speed (kph): ');
    readln(input, Max_Speed);
    writeln;

    writeln(output, 'The following section requests engine characteristics:');
    write(output, 'Input the engine power (kw): ');
    readln(input, Power);
    write(output, 'Input the retarder power (kw): ');
    readln(input, Retarder);
    write(output, 'Input the percentage drive train efficiency (%): ');
    readln(input, Efficiency);
    writeln;

    writeln(output, 'The following section requests road characteristics:');
    write(output, 'Input the length of the haul road (meters): ');
    readln(input, Length);
    write(output, 'Input the grade of the haul (%): ');
    readln(input, Grade);
    write(output, 'Input the rolling resistance (kg/tonne): ');
    readln(input, Rolling_Resistance);
    writeln;
```

FIGURE 8-13
Hauler program.

```
    writeln(output, 'The following section requests fixed cycle components:');
    write(output, 'Input the load time (minutes): ');
    readln(input, Load_Time);
    write(output, 'Input the haul time (minutes): ');
    readln(input, Dump_Time);
    writeln;

{ Call Velocity function to calculate the haul speed }
    Haul_Speed := Velocity(Weight, Load, Max_Speed, Power, Retarder, Efficiency, Grade,
                        Rolling_Resistance);

{ Call Velocity function to calculate the return speed }
    Grade := -Grade;                            { Reverse the grade for same  road upon return }
    Load := 0.0;                                { Load = 0 when empty upon return }
    Return_Speed := Velocity(Weight, Load, Max_Speed, Power, Retarder, Efficiency, Grade,
                        Rolling_Resistance);

{ Calculate the cycle time }
    Cycle_Time := Load_Time + Dump_Time +
                        ((Length / Haul_Speed) + (Length / Return_Speed)) * 60 / 1000;

{ Print a summary report }
    writeln;
    writeln(output, ' --- Report of Cycle Time Results ---');
    writeln;
    writeln(output, 'The haul speed is:  ', Haul_Speed : 8 : 1, ' km/hr');
    writeln(output, 'The return speed is: ', Return_Speed : 8 : 1, ' km/hr');
    writeln(output, 'The cycle time is:  ', Cycle_Time : 8 : 1, ' min');
    writeln;

end.
```

FIGURE 8-13(*continued*)

a simple program and, with the comments, should be mostly self-explanatory. **Vehicle_Calcs**, mentioned after the **uses** keyword in the third line, refers to a *function* subroutine that is presented in Fig. 8-14. It is followed by the declaration of the variables, all of which are type **real** in this case. Next comes a set of interactive I/O statements using the keyboard and screen.

A function subroutine called **Velocity** handles the calculations for **Haul_Speed** and **Return_Speed**. It is contained in a separate file within a program unit named Vehicle_Calcs that in turn is cited by the main program, Hauler. The same routine is called for both the haul and return time, thus saving duplication of this code. Before calling it the second time, **Grade** is set to the opposite sign from its input value, since the return is on the same road as the haul. **Load** is set to zero for the return.

Once Velocity has computed both the haul and return times, the four components of the cycle time (including the user's input values for load and dump times) are added to compute the total cycle time. The last part of the main program outputs a summary of the computed results to the user's screen.

Figure 8-14 is a list of the file containing the unit Vehicle_Calcs and its function called Velocity. A *function* is a subroutine that returns a value to the calling

```
unit Vehicle_Calcs;

interface

   function Velocity (Wt, Load, Max_Spd, Pwr, Retard, Eff, Grade, RollResist: real): real;

implementation
{ Calculate the haul or return velocity }
   function Velocity (Wt, Load, Max_Spd, Pwr, Retard, Eff, Grade, RollResist: real): real;

   var
      Rimpull, Speed: real;                  { Temporary variables used within this function }

   begin

{ Calculate the rimpull }
   Rimpull := (Wt + Load) * (Grade / 100.0 + RollResist / 1000.0) * 9.8;

{ Calculate the speed ------ }
{ Check first if Rimpull would be close to zero divide; set Speed to Max_Spd if so }
{ Else check if positive power is needed; computer speed if it is }
{ Else retarder is needed if down grade exceeds rolling resistance }
      if (abs(Rimpull) < 1) then
         Speed := Max_Spd
      else if (Rimpull >= 1) then
         Speed := (((Pwr * 1000) * (Eff / 100.0)) / Rimpull) * (3600 / 1000)
      else if (Rimpull <= -1) then
         Speed := ((Retard * 1000) / (Eff * (-Rimpull) / 100.0)) * (3600 / 1000);

{ Check to be sure maximum speed is not exceeded; limit to max speed if it is }
      if (Speed > Max_Spd) then
         Speed := Max_Spd;

      Velocity := Speed;             { Assign result to name of function }

   end;

end.
```

FIGURE 8-14
Calculation function called by the Hauler program.

program when its name is called. In this case, the Velocity call passes a set of eight *arguments* to the function. The definition of the function uses abbreviated versions of the variable names defined earlier. This demonstrates that the names in the function need not be the same as those in the calling program and that they are just placeholders for the actual values that are passed to the function called in the calling program (Hauler in this case). The top half of the listing contains the interface conventions used by THINK Pascal. The actual calculations come after the **begin** keyword. The program starts by computing the **Rimpull** needed to propel the vehicle against a combined rolling and grade resistance if this is positive, or to brake the vehicle if a downgrade produces a force greater than the rolling resistance; in the former instance, power is applied, whereas in the latter the retarder is applied. Note that rolling resistance is always positive, whereas the grade resistance depends on whether the vehicle is going up or down the road. The **Speed** calculation contains three options implemented via an **if** . . . **then** . . . **else** statement. The first option will catch a divide-by-zero condition (in this case, if Rimpull is less than 1). The value of Speed would be so large even at 1 that the **Max_Spd** should be taken as the limit. The second branch handles the case where combined resistance is positive and engine power is needed. The third branch handles the case where net resistance is negative, thus requiring the use of the retarder to hold the vehicle at a safe velocity on the downgrade. In this case, drive train inefficiencies are assumed to work in our favor, so **Eff** moves to the denominator. After computing Speed, the program checks to see if the result of the second or third branches exceeds Max_Spd, and the latter is substituted if the result is too high. Finally, the function assigns the value of Speed to its name, Velocity, so that the value returns to the calling program, Hauler.

In Fig. 8-15 is shown a typical view of a Macintosh screen containing four windows that might be open while creating a program in the THINK Pascal development environment. THINK Pascal includes an "intelligent" built-in editor that automatically checks syntax and formats the program text (boldfaces keywords, indents, etc.) while the programmer is typing. Powerful built-in debugging tools help the programmer detect logical errors that might have crept into the program. Once all or part of a program is ready for testing, the programmer can open the **Observe** window in order to select any of the variables to observe their sequential changes while the program is executing step by step. The programmer can choose to run all or part of a program in single steps and to set breakpoints (shown as stop signs in the margin) by any statement in the program. The **project** window keeps track of all program modules and libraries that make up the total program being developed, allows the programmer to turn on any of several options to specify the amount of code to be included for debugging purposes in prototype modules, and handles the object code automatically and transparently. These are but a few of the many powerful capabilities available in a modern procedural language development environment like THINK Pascal.

This particular program uses standard Pascal code and does not create a graphical user interface (e.g., with its own pull-down menus, recognition of mouse input, etc.) that is characteristic of a graphically oriented computer such as a Macintosh. Thus it should be possible to type in this program, compile it, and run it on almost any computer with few if any changes. However, THINK Pascal does indeed support the development of genuine Macintosh applications, desk accessories, and other software using its many extensions to Pascal, and its rich Macintosh-specific libraries of code. It has been widely used to create many of the commercial Macintosh programs now found in the software marketplace.

When THINK Pascal is used to build a standard text-oriented Pascal program such as our example, its user interaction takes place in a *Text* window. This behaves similarly to the main screen of a character-oriented computer such as a DOS-based PC. In Fig. 8-16 is an example of the dialog of the user interaction with the Hauler program. The user's inputs are shown in bold text. The output report of results appears on the same screen.

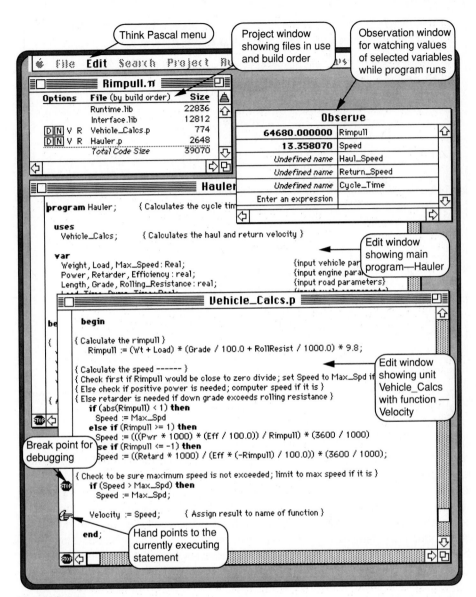

FIGURE 8-15
THINK Pascal being used to build the Hauler program.

With a procedural language like Pascal, a developer can create almost any type of construction application that a computer can run. Using a powerful programming environment such as THINK Pascal makes this task far easier to do today than it was with similar languages even a decade ago. Back in the days of punched cards, making a few changes to a program and finding the results took hours and even days. Even with the interactive environments of minicom-

```
          *** RIMPULL AND CYCLE TIME CALCULATIONS ***

In the data entry section, type decimal numbers and hit the Return key.

The following section requests vehicle characteristics:

Input the vehicle empty weight (kg): 30000
Input the load capacity (kg): 25000
Input maximum safe speed (kph): 80

The following section requests engine characteristics:
Input the engine power (kw): 300
Input the retarder power (kw): 350
Input the percentage drive train efficiency (%): 80

The following section requests road characteristics:
Input the length of the haul road (meters): 4000
Input the grade of the haul (%): 6
Input the rolling resistance (kg/tonne): 40

The following section requests fixed cycle components:
Input the load time (minutes): 2
Input the haul time (minutes): 1


 ---- Report of Cycle Time Results ----

The haul speed is:      16.0 km/hr
The return speed is:    80.0 km/hr
The cycle time is:      21.0 min
```

FIGURE 8-16
Example of dialog and results of the Hauler program.

puters and earlier microcomputers, the process of editing, compiling, linking, and debugging was tedious and time-consuming. But with THINK Pascal and similar systems, the time required to see the results of changes is reduced to minutes and even seconds, and most of the tedious agony of programming has almost been eliminated. Although spreadsheets, databases, and other high-level tools may be easier and even more suitable in their special application domains, the productivity gains achieved in modern, highly integrated procedural language development environments certainly make these languages viable alternatives for creating computer applications in construction.

SUMMARY

Procedural languages have long been the standard tools for application software development, and they have evolved to be excellent tools for use on microcomputers. They offer more power, flexibility, and variety for creating new applications than any other development tool.

Procedural languages build programs from many elements—including constants, variables, data structures, arithmetic and logical operators—and with keywords that define their vocabulary. These elements combine to form expressions and statements using a syntax or grammar that further defines the nature and capability of the language. The statements can declare characteristics of variables, define the beginning and end of procedures, evaluate expressions and assign their results to variables, control the execution sequence of sections of code within a program, perform conditional tests, and format and direct the input and output to and from the computer. Structure, reliability, and ease of modification can be enhanced by modularizing programs into subprocedures and functions, containing logically contiguous sections of code, that meet well-defined objectives.

The environment in which programs are developed in a given language can also include the text editors to create source code, interpreters or compilers to translate the source statements to object or machine code, linkers to combine separately developed sections of code and to join them with code from libraries of prewritten routines that might have been included with or added to the language product, and various utilities to evaluate and debug code while it is under development. Figure 8-15 from the Hauler example is an illustration of a modern procedural language development environment.

There exist a tremendous variety of procedural languages. Some are more suited to engineering and scientific application development, others focus more on business applications, and a few try to do both well. This chapter provided comparative code segments from FORTRAN, COBOL, BASIC, Pascal, and C and also suggested where PL-1, ALGOL, Modula-2, and Ada fit into this spectrum.

The main advantages of procedural languages are their flexibility and breadth for creating almost any application that can run on a computer, the speed with which their code can run, and their portability across a wide variety of computer hardware platforms. But for novices they can be difficult to use and, in general, applications developed in this way can take longer and cost more to produce. In their appropriate application domains, the spreadsheet, database, and expert system development environments presented in the next three chapters can be more effective and economical. But a construction professional is still better off having a working knowledge of at least one procedural language, and for many useful construction applications they are the only way to go.

REVIEW QUESTIONS

1. Give five distinctly different examples of constants that might be defined in a computer program.
2. How does the name of a variable differ from its content at any given time?
3. What is meant by an implicit variable type declaration?
4. How does an array differ from a hierarchy in defining a data structure?
5. Which of the following are arithmetic operators commonly used in computer programming languages?

- +
- −
- >
- ≠
- ÷
- AND
- *
- LE

6. Which of the following could be placed in programming language statements?

• Constants	• Syntax
• Variables	• Reserved words
• Operators	• Conditional tests
• Strings	• Control transfers
• Expressions	• I/O actions

7. What is the general purpose of a library of procedures or functions?

8. In the control structure for a computer program, how does a conditional test enable the programmer to provide alternative branches in the sequence in which the code is executed?

9. How does the control structure of a structured modular language like Pascal differ from that of standard BASIC?

10. Give two distinctly different examples of input functions that might be performed by a computer program.

11. Give two distinctly different examples of output functions that might be performed by a computer program.

12. What was the first major procedural language that became widely used for scientific and engineering applications? Briefly state how the name of the language derives from its function.

13. What was the first major procedural language that became widely used for business applications? Briefly state how the name of the language derives from its function.

14. How do the origin and initial purpose of the BASIC language compare to those of the Pascal language? (You may wish to refer to Chapter 4 for part of the answer to this question.)

15. Historically, which of the early procedural languages was the ancestor of Pascal, Modula-2, and Ada?

16. What was the initial primary area of applications for the Ada language?

17. What is the trend in the use of the C language in software development today?

18. Relative to higher-level development tools like spreadsheets and databases, what are two of the primary advantages of using procedural languages for the development of construction application software?

19. Which of the two example application programs (concrete cylinder test analysis and hauler cycle time calculation) illustrated how to access a data file from a computer program?

20. Which of the two programs used a modularized structure?

DEVELOPMENT PROBLEMS

1. Extend the Cylinder Test Analysis Program

In the language of your choice, reimplement the cylinder test program and then make either or both of the following extensions:

(*a*) The code section in lines 800 to 960 (see Fig. 8-7) determines the number of design strengths used in the week so that these strengths can be used for the statistical summary in lines 1010 to 1310. But this determination is actually rather awkward. First sorting the records in ascending order of design strength and then doing the statistical calculations directly on each group of records with the same strength could be a better way to go about it. Try reimplementing that part of the program to work this way.

(*b*) As presented, the program proceeds linearly with little modular structure. Divide it into subprocedures with a main control program and the appropriate calls to subprocedures or functions.

2.   Extend the Hauler Cycle Time Program

In the language of your choice, reimplement the hauler cycle time program and then make either or both of the following extensions:

(*a*) Create a file of haulers (e.g., using data in a recent *Caterpillar Performance Handbook*). Then, instead of having the user type in a machine's weight, power, and so on, simply provide a listing of machines on file, ask the user to select the one to be used for the present calculation, and then read in the relevant data from the file.

(*b*) Add information to the file (or input section) giving the percentage of the machine's weight on the drive wheels when it is (1) empty and (2) full. In the input section for the road characteristics, add a parameter giving the coefficient of traction between the tires and the road (such data are also tabulated in the *Caterpillar Performance Handbook*). In the **Velocity** function, add the drive wheel load percentages and the coefficient of traction to the argument list of the procedure call. Then, after calculating the rimpull, perform a check to see if the coefficient of traction between the drive wheels and the road will actually permit the necessary rimpull force to be transmitted from the engine and drive train to the road without spinning the drive wheels. Add the result of this calculation to the summary report and print an appropriate message if traction is insufficient.

3.   Compactor Production Program

In this problem you are encouraged to create a new program from scratch. The application is to select and then compute the production of a high-speed compactor working on an earth-fill structure.

BACKGROUND. Construction soil compaction involves the interaction of engineering specifications, materials properties, standard test methods, equipment types and procedures, and environmental conditions (weather). The purpose can be to increase strength, reduce permeability, or reduce settlement. Part of the problem is selecting the right machine for a given set of soil conditions, but we will return to this in an expert system in Chapter 11. For the moment we will focus on production, where the parameters can include (1) lift height, (2) number of passes, (3) width of the roller(s), and (4) machine speed. Production is also affected by soil moisture content, density requirements, and plasticity, but we ignore these here. Thus a production calculation can be simplified to the following formula:

$$\text{Production, m}^3/\text{h} = \frac{W \times S \times L \times E}{P}$$

where W = the width of the roller(s) in meters
         S = machine speed in kilometers per hour
         L = lift height in millimeters
         E = an efficiency factor (e.g., a 45-minute hour $\rightarrow$ 0.75)
         P = the number of passes to achieve the specified density

EXAMPLE. Estimate production for a compactor, traveling at 6 kph, whose rollers are 1 m wide. The lift thickness is 200 mm, four passes are needed, and we assume the machine works a 50-minute efficiency hour.

$$\text{Production} = \frac{1 \text{ m} \times 2 \text{ wheels wide} \times 6 \text{ kph} \times 200 \text{ mm} \times 50/60}{4 \text{ passes}}$$

$$= 500 \text{ m}^3/\text{h}$$

PROGRAM REQUIREMENTS. Your program should implement a well-designed application that will enable a user to answer some common compaction production questions, such as (1) how many cubic meters per hour can be compacted for a given set of inputs; (2) how long will it take to complete a given volume with one compactor; and (3) how many compactors will it take to keep up with a loading and hauling fleet that is delivering material at a given rate. The design is up to you, but, in addition to providing a convenient way to input data and answer the preceding questions, you might include the following capabilities:

(*a*) Provide for selection from a file of predefined machines, such as:

Machine model	Weight	Power	Drum width
Cat 815B	20.0 tonne	161 kW	2×0.98 m
Cat 825C	32.4 tonne	235 kW	2×1.12 m
⋮			

(*b*) Provide an interactive input section whereby the user would specify the machine to be used, the number of passes, the speed, the lift height, and the efficiency factor (as a percent or as an N-minute hour).

(*c*) Provide a menu whereby the user can choose to find the answer to the following types of questions. Implement each as a separate subprocedure or function.
   (*1*) Compactor production
   (*2*) Time to compact a given volume (ask user at this point for that volume)
   (*3*) Number of machines to keep up with a given production rate (ask user at this point for that given rate)

(*d*) Isolate the compaction formula in a single subprocedure or function, and call it out wherever else it is needed in the program.

(*e*) Design and provide the program code to print summary reports of results.

You should select and use the procedural language that you prefer. This fairly simple program could be done in almost any language.

TEST CASES. In addition to input test data that you may wish to make up yourself, use the following test cases for your program:

Case 1: What is the production of a Cat 815B where the speed is 10 kph, the lift height is 200 mm, the number of passes is 4, and it works a 40-minute efficiency hour?

Case 2: Given a Cat 825C, a speed of 12 kph, a lift height of 250 mm, 5 passes, and a 45-minute efficiency hour, how long will it take one machine to compact 2000 cubic meters of material?

Case 3: Given a scraper fleet that is delivering material at a rate of 1200 m^3 per hour, and one or more Cat 815B compactors that are working on a fill requiring 150-mm lifts, 6 passes at 8 kph, and working a 50-minute efficiency hour, how many compactors will it take to keep up with the scraper fleet?

SUGGESTIONS FOR FURTHER READING

Although it may seem that there are a large number of references here, there are only one to three for any given language. Focus on those for the languages that interest you. Also note that the book titled *Comparative Programming Languages,* by Leslie B. Wilson and Robert G. Clark, listed at the end of Chapter 4, has an overview of computer languages.

Adams, Jeanne C., Walter S. Brainerd, Jeanne T. Martin, Brian T. Smith, and Jerrold L. Wagener. *FORTRAN 90 Handbook.* New York: McGraw-Hill, 1992. A large and comprehensive text and reference book on the latest version of this widely used engineering and scientific programming language.

Barnes, J. G. P. *Programming in ADA,* 3d ed. Menlo Park, Calif.: Addison-Wesley, 1989. Good standard text for this large, complex, and flexible language.

Cooper, Douglas. *Oh! Pascal,* 3d ed. New York: W. W. Norton, 1993. Widely used college text noted for its clear and easily understood writing. Includes disk with examples.

Deitel, H. M., and P. J. Deitel. *C: How to Program.* Englewood Cliffs, N.J.: Prentice Hall, 1992. Well-written 600-page book intended as a text for a first course in computer programming.

Ellis, T. M. R. *FORTRAN 77 Programming,* 2d. ed. Menlo Park, Calif.: Addison-Wesley, 1990. Good FORTRAN text. Includes information on the FORTRAN 90 standard.

Gardner, J. *From C to C: An Introduction to ANSI Standard C.* New York: John Wiley & Sons, 1989. A popular college text for an introduction to programming with C.

Jatich, Alida, and Peter Nowak. *Micro Focus Workbench: Developing Mainframe Programs on the PC.* New York: John Wiley & Sons, 1992. Micro Focus COBOL brought mainframe COBOL applications to the PC world. This book is a way to learn about the system before buying the software.

Jensen, K., N. Wirth, A. B. Mickel, and J. F. Miner. *Pascal User Manual and Report,* ISO Pascal Standard, 4th ed. New York: Springer-Verlag, 1991. The standard reference for the definition of the Pascal language.

Kernighan, Brian W., and Dennis M. Ritchie. *The C Programming Language,* 2d ed. Englewood Cliffs, N.J.: Prentice Hall, 1988. The standard reference for the definition of the C language.

McCracken, D. D., and D. G. Golden. *A Simplified Guide to Structured COBOL Programming.* New York: John Wiley and Sons, 1988. McCracken has long been respected for his excellent introductions to programming languages, and his COBOL book is in this tradition.

McNally, Clayton L., Jr., and Peter Molchan, Jr. *Micro Focus COBOL Workbench*. Boston, Mass.: QED Publishing Group, 1993. Another good introduction to a PC version of COBOL.

Microsoft GW-BASIC Interpreter. Redmond, Wash.: Microsoft Corp., 1988. The language most commonly included with MS-DOS-based PC computers. Antiquated but widely used text.

Nelson, Ross P. *Running Visual BASIC for Windows*. Redmond, Wash.: Microsoft Press, 1993. Visual BASIC has modern features of structured and object-oriented programming and has rapidly become popular for developing business applications. This is but one example of a large number of books that have recently capitalized on this trend.

Riddle, Douglas F. *Programming in Pascal*. San Francisco, Calif.: Dellen, 1991. Designed to follow the Association for Computing Machinery (ACM) guidelines for a first course in computer programming, this 508-page book provides a good introduction to the Pascal language as well.

Terry, P. D. *An Introduction to Programming with Modula-2*. Menlo Park, Calif.: Addison-Wesley, 1987. Recommended for an introduction to programming in this language.

THINK Pascal User Manual. Cupertino, Calif.: Symantec Corp., 1991. An excellent example of a modern software development environment. The company makes a similar environment for C and C++.

Waite, Mitchell, Robert Arinson, Christy Gemmell, and Harry Henderson. *Microsoft QuickBASIC Bible*. Redmond, Wash.: Microsoft Press, 1990. Microsoft's QuickBASIC is intended as an enhanced replacement for the old DOS BASIC, and this book is the text from that company's press. It has many worthy competitors, so this is only an example if you are interested in QuickBASIC.

Wirth, Niklaus. *Programming in Modula-2*, 4th ed. New York: Springer-Verlag, 1988. The language developer's own book for the definition of Modula-2.

CHAPTER
9

DEVELOPING APPLICATIONS WITH SPREADSHEETS

As far as the laws of mathematics refer to reality, they are not certain, and as far as they are certain, they do not refer to reality.

Albert Einstein*

Electronic spreadsheets are a fairly recent development in computing, yet they have become an almost indispensable tool in construction engineering and management. For years contractors used paper spreadsheets and calculators for estimating, accounting, finance, and production computations, so they quite naturally took to the electronic version. Electronic spreadsheets have taken the drudgery out of computations and have improved accuracy and presentation standards. They have also added powerful and convenient analytical and graphical capabilities that were infeasible or too expensive with manual methods.

The first widely used electronic spreadsheet was VisiCalc, developed by Dan Bricklin and Bob Frankston in 1979 for the Apple II and other Motorola 6502-based microcomputers and later for the CP/M operating system used on some Intel 8080-based microcomputers. It was considered a brilliant program and by itself justified the purchase of thousands of early microcomputers. It answered the question, "What practical value is there to a personal computer?" VisiCalc was soon emulated by Sorcim's SuperCalc, Microsoft's MultiPlan, and several others. The one that really expanded the application of spreadsheets, however, was Lotus 1-2-3, whose 1982 introduction followed soon after that of the IBM PC. The combination of the PC and Lotus 1-2-3 made personal computers an

---

*In *The Tao of Physics*, by F. Capra, Ch. 2, 1975.

accepted mainstay in the corporate world. It was also Lotus 1-2-3 that introduced programmable *macro* commands to spreadsheets and thus made spreadsheet software into a development tool capable of providing standalone applications to non-computer-oriented users.

Lotus 1-2-3 itself has been the model for a number of clones, that is, programs that, while offering their own distinctive advantages and features, basically stick closely to the original's standards and techniques. The most enduring of these in this highly competitive market has been Borland's Quattro Pro. Some major design alternatives have also evolved, such as Microsoft's Excel and Claris' Resolve, and even an innovative new Lotus product called Improv. These in turn have influenced succeeding generations of Lotus 1-2-3 and its earlier contemporaries. Such is the competitive nature of the software marketplace, most of which works to the consumer's advantage.

This chapter first examines some of the basic concepts underlying computer-based *spreadsheet programs* and indicates how one works with them. It then mentions more advanced concepts and features that have become available in spreadsheet programs, such as online help, graphics, programmed macro commands, added dimensions, and integration with other applications. The chapter then discusses the advantages and limitations of spreadsheets for various types of applications and emphasizes those for which spreadsheets are most appropriate. The final two sections present examples. The first is a single spreadsheet implemented in Lotus 1-2-3 for scraper load-growth optimization. It illustrates not only basic spreadsheet design but also the use of programmed macros and graphics. The second example is programmed in Microsoft's Excel and shows how multiple spreadsheets can be interlinked to form various components of a construction estimating system.

BASIC CONCEPTS OF SPREADSHEETS

Although some people occasionally use spreadsheets to type letters or do other text applications, and most spreadsheets have limited file-management capabilities, they are intended primarily for applications involving computations, particularly those that can be organized into rows and columns of text and numbers. Figure 9-1 is a simplified spreadsheet for the summary page of a bid estimate. Each row corresponds to a major bid item in the project. In each row, the labor, materials, and equipment are summed to form the total in the last column. Also, column totals show the total cost for labor, materials, equipment, and the overall direct costs. Just below the direct costs is overhead, which adds 8 percent to total direct costs. Below that is a contingency, which is 10 percent of total labor only. Adding a profit of 5 percent of the total cost and contingency gives a total bid of $23,520.

In Fig. 9-1, we have labeled the columns alphabetically, and the rows numerically. The intersection of a row and a column we will call a *cell,* which we name by the relevant column-row designators. For example, the total direct cost is in cell E8. Using this type of designation, we can write general formulas for

	A	B	C	D	E
1		Estimate for Bid			
2					
3	Item	Labor	Materials	Equipment	Item Total
4	Excavate	$1,000	$0	$2,000	$3,000
5	Build Foundation	2,000	1,000	1,000	4,000
6	Build Structure	3,000	3,000	1,000	7,000
7	Finish Work	2,000	4,000	0	6,000
8	Direct cost totals:	$8,000	$8,000	$4,000	$20,000
9	Overhead:				1,600
10	Contingency:				800
11	Profit:				1,120
12	Total bid:				$23,520

FIGURE 9-1
Simple spreadsheet.

what takes place in cells where calculations are done. For example, the total cost for excavation can be written as

$$E4 = B4 + C4 + D4$$

As an alternative way of expressing a sum, the formula for the total labor cost could be written using a function (SUM) and a form of notation called a *range* as follows:

$$B8 = SUM (B4..B7)$$

Here, B4 through B7 (B4..B7) are the cells that make up this particular one-dimensional range, but a range can be multidimensional as well (e.g., range B4..D7 contains the direct-cost components). Thus, using ranges, the total contained in cell E8 can be computed in any of the following ways:

$$E8 = SUM (E4..E7)$$
$$E8 = SUM (B8..D8)$$
$$E8 = SUM (B4..D7)$$

The formulas for the overhead, contingency, profit, and bid total can be written as

$$E9 \;\; = E8 * 0.08$$
$$E10 = B8 * 0.10$$
$$E11 = SUM (E8..E10) * 0.05$$
$$E12 = SUM (E8..E11).$$

These formulas illustrate the types of *relationships*, which in spreadsheet terminology are usually called *references*, that can be established between cells in an electronic spreadsheet. But the power of an electronic spreadsheet is far greater than simple arithmetic, as we shall see.

First, we need to introduce another concept. The size of a computer screen is finite and two-dimensional, yet spreadsheets can be created that are far larger and even three-dimensional. The problem of large spreadsheets is solved by imagining the display screen to be a window that can be moved around the spreadsheet beneath it, as shown in Fig. 9-2. The window usually is maneuvered about either with the directional arrow keys on a keyboard or by using a mouse to manipulate horizontal and vertical graphical scroll bars directly on the screen. The screen can also be split into two or more windows in order to examine cells in different parts of the spreadsheet simultaneously.

The spreadsheet not only performs computations but also can serve concurrently as the means for both input of information and output of results. This feature is quite different from most procedural programming languages, where the program code is separate from its input and output. For example, to change the foundation labor cost in Fig. 9-1, the user would simply type the new number into cell B5, and computations in all cells that refer to cell B5 or depend on it indirectly will be updated in an instant. The new appearance of the screen constitutes the new report, and this in turn could be printed.

The contents of each cell can be a simple numeric or string constant or a formula such as those in Fig. 9-1. Cells can also contain logical tests and other control mechanisms mentioned for procedural languages. Like many procedural languages, spreadsheet systems contain a large number of predefined functions,

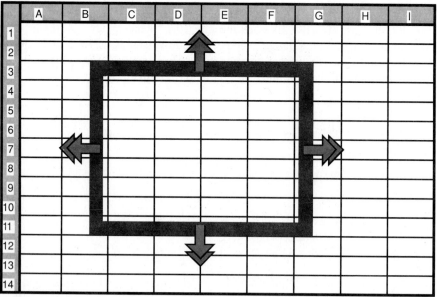

FIGURE 9-2
A window view into a large spreadsheet.

such as those for trigonometry, statistics, string manipulation, and others. Table 9-1 is a partial set of the functions contained in a typical spreadsheet package, divided into major categories. These functions can be mixed in with other parts of formulas contained within cells. We include several examples in the two cases presented later in this chapter.

WORKING WITH SPREADSHEET PROGRAMS

In the main field of the spreadsheet, one moves a *pointer* (typically shown by boldly outlining the cell, showing it in reverse video, or using a special cursor symbol) from cell to cell using directional arrow keys found on a keyboard or, in some cases, a mouse to manipulate a graphical user interface. The cell under the pointer at any moment is the *current cell* or *active cell*. When a cell is active, text, numbers, or a formula can be entered (if it was blank), replaced (by typing over what was there), or edited. Also while a cell is active, its contents appear in an *entry* or editing *line* or a *formula bar*, usually at the top or bottom of the screen, as shown in Fig. 9-3. Text or a number will appear similarly in both the cell and in the formula bar. However, in the case of a formula, the *formula* itself will appear (and be available for editing) in the formula bar, whereas the *result* of its computation normally will appear in the cell on the spreadsheet. This is the case shown for cell E11 on Fig. 9-3.

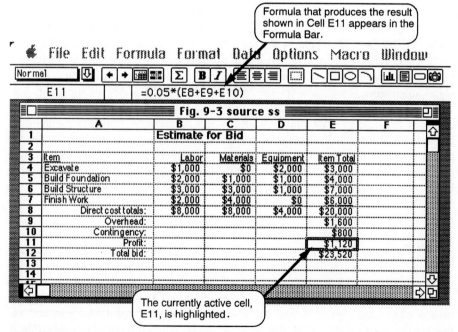

FIGURE 9-3
Working within a cell on the main field of a spreadsheet.

While one is working with a spreadsheet program, ***commands*** are available to perform a wide variety of functions. These commands normally are invoked using a hierarchical menu system. The menu choices typically fall into the following main categories:

- File manipulation (start new file, open old file, save file, close file, etc.)
- Editing (cut, copy, insert, delete, etc.) both within cells and for groups of cells
- Selecting and inserting functions and formulas (e.g., see Table 9-1)
- Formatting (forms of representing numbers, fonts, text styles, alignment, borders, column widths, etc.)
- Data manipulation (sorting data in ranges, defining databases, etc.)
- Display options (defining print area, setting headings, turning automatic recalculation on or off, turning grid lines on and off, etc.)
- Write, record, or use programmed macro routines

The new user can become productive knowing just a subset of the available capabilities, and the advanced user has the resources to build very complex spreadsheet models that would be enormously difficult or expensive to produce with other types of application development tools.

To implement the hierarchical menu system for keyboard-oriented computer systems, the VisiCalc and subsequent Lotus 1-2-3 packages used function keys and a horizontal text menu accessed by typing a special character such as a slash (/). Once the menu is invoked, the left and right arrow keys enable the user to move horizontally across the current line on the menu bar, and the current choice within the line is highlighted in reverse video. The desired selection can be made using the Enter or Return key. This might produce an immediate action or just take the user to a further set of choices at the next level down in the menu hierarchy. One can move a level back up the hierarchy by pressing the Escape (Esc) key. Figure 9-4 is an illustration of a Lotus menu at the second level down where the user has requested **/ File** and is about to go to **Save**. A short explanation of the options or the effect that would be produced by the action currently under the cursor appears on a line beneath the menu. As an alternative to moving around the menu with the arrow keys, one usually has the option simply to type the first letter (or other unique letter designated for this purpose) of the words in the menu bars. Thus the sequence of keystrokes **/FS** (for **F**ile **S**ave) would produce the same effect as the menu selection process just mentioned.

The main alternative to keyboard manipulation is a graphical pull-down menu manipulated with a mouse. In Fig. 9-5 is a similar (File Save) function requested using the mouse option in Microsoft Excel. Shown horizontally across the top of the screen is the first level of the hierarchy; the choices under each main menu option would appear in windows that pop down below the main menu as one moves the pointer across the main menu. At any point, one can pull the pointer down to select one of the items in the pull-down menu; choosing **File** in this particular case would produce the pull-down menu that is currently displayed.

TABLE 9-1
Typical built-in spreadsheet functions common to Lotus 1-2-3 and Excel

Database		Text (String)	
DAVG	average specified db values	CHAR	return character for ASCII code
DCOUNT	count nonblank cells	FIND	locate substring within string
DMAX	find greatest value in field	LEFT	extract leftmost n characters
DMIN	find minimum value in field	LENGTH	compute length of string
DSTDEV	standard deviation of entries	MID	extract n midstring characters
DSUM	add values in field	REPLACE	replace characters in string
DVAR	variance of selected entries	VALUE	convert characters to number

Mathematical		Trigonometric	
ABS	absolute value	ACOS	arc cosine
EXP	raise e to given power	ASIN	arc sine
INT	integer part of number	ATAN	arc tangent
LN	natural logarithm	COS	cosine
LOG	common logarithm	SIN	sine
MOD	remainder of a division	TAN	tangent
PI	value of π		
RAND	generate random number		
SQRT	square root of number		

Logical		Statistical	
AND	true if arguments true	AVG	average list of values
IF	perform logical T/F test	COUNT	count number of entries
NOT	reverse logic of argument	MAX	find greatest value in list
		STD	standard deviation of list
		SUM	add numbers in list

Financial		Lookup	
DDB	double declining balance	CHOOSE	choose value from list
FV	future value of investment	HLOOKUP	find value of cell in row
IRR	internal rate of return	INDEX	use index to choose value
NPV	net present value of series	LOOKUP	find values in an array
PMT	periodic loan payments	VLOOKUP	find value of cell in column

Date and time		Special	
DATE	return given date as number	CELL	find specs of cell contents
DAY	calculate day of the month	COLS	count columns in range
MONTH	calculate the month number	NA	mark cells with info not available
YEAR	calculate number of year	ROWS	count rows in range

Notes: Lotus 1-2-3 precedes function names with @.
Minor variations of names in Lotus and Excel are mixed in here.

If further subordinate choices or information are needed, either a side window, dialog box, or other mechanism will appear on the screen. Choosing **Save** here the first time would cause Excel to display a pop-up window to allow the user to select the disk and directory (folder) in which to place the file and to permit the user to choose among several alternative format options (e.g., Lotus 1-2-3). Such

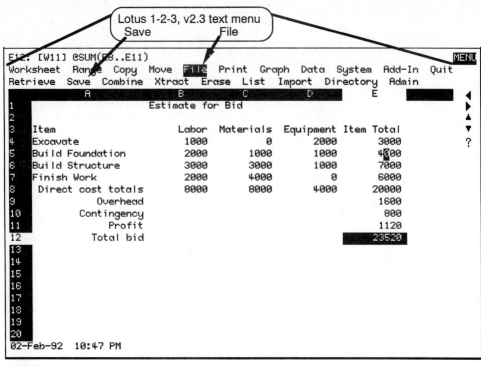

FIGURE 9-4
An example of a horizontal text menu.

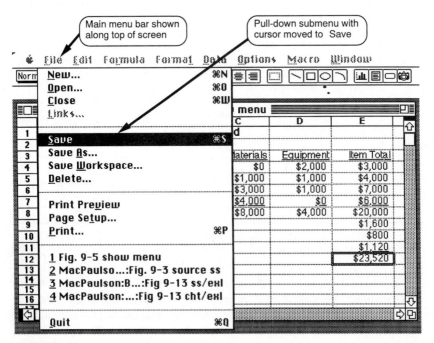

FIGURE 9-5
An example of a graphical pull-down menu.

graphical menu systems also commonly have keyboard alternatives for frequently used commands, such as Command-S as an alternative to **File Save** in this case. Once memorized, keyboard commands are convenient shortcuts for some users.

Spreadsheets offer numerous facilities to speed the development of large models. For example, the power to *copy* and *replicate* existing work is very helpful. In the example from Fig. 9-1, the formula in cell E4 is similar in form to that in cells E5, E6, and E7. Once E4 has been entered in an electronic spreadsheet, it can simply be replicated downward and the spreadsheet software automatically takes care of making each cell refer to the data in the correctly corresponding rows. Similarly, the formula in cell B8 can be replicated to the right into cells C8, D8, and even E8, since each just adds the column of numbers above it. Even where the formulas and range references do not precisely correspond, it is often easier to copy a formula that is similar, and make editing changes to make the copied one correct, than it is to type in the new formula from scratch.

Copying and replication are just a few of the numerous techniques that experienced spreadsheet users learn to exploit to make model development quicker and easier. It is beyond the scope of this chapter to teach such detailed techniques, but the results of using some of them will appear in the case examples later on.

Equally important, spreadsheets offer the advantage that, once the model has been developed and tested, making changes and seeing the results are easy, quick, and reliable. This facility has encouraged the use of *"What if? analysis"* to test out the impact of changing different parameters on the model as a whole. For example, if the labor cost in cell B5 of the spreadsheet in Fig. 9-1 should change, in a manual approach calculations would have to be redone for the column addition at cell B8, the row addition at cell E5, the column addition at cell E8, and the math in each of cells E9 through E12. The chances of making an arithmetic error, particularly under time pressure, are significant when working manually. With an electronic spreadsheet, one would only enter the change in cell B5, and everything else would be automatic, accurate, and instantaneous. Thus one could explore various possible outcomes before making the contingency allowance, determining the profit, and submitting the final bid.

ADVANCED SPREADSHEET CONCEPTS
AND FEATURES

The concepts and capabilities described in the preceding two sections are rudimentary to even the simplest and least costly spreadsheet software. Most programs go far beyond what has been described. This section briefly describes some of the extras that are commonly offered. They include added features and functions for the spreadsheet itself, protection and security, online help, report form definition, graphical reporting, database manipulation and reporting, a third dimension for cell organization, linkages among spreadsheets, programmed macros, and integration with other types of software.

Added Spreadsheet Features and Functions

As with any type of commercially successful software, spreadsheet developers continuously add to their products to make them more competitive in the marketplace. Common additions include extensions to the list of functions shown in Table 9-1, new editing capabilities, more options in the menus, enhanced formatting (text sizes, types and styles, better control of row and column sizes, etc.), enhanced display and reporting options, improved documentation and training materials, and so on. Some of these enhancements add real value, but others are obscure or rarely needed and just add complexity and bulk. Most user needs are well served by the simpler and less expensive spreadsheet packages. The quality, effectiveness, and ease with which the basic functionality is implemented are much more important for most people than long lists of features, Some of the more complex packages offer the option to switch between two levels of functionality. Thus the same package, perhaps adopted as an organizational standard, can serve the needs of both elementary and advanced users.

Protection and Security

A major advantage of spreadsheets is that it is very easy to make modifications to alter or enhance the capability of a model. But this can also mean that it is easy for a user inadvertently to alter or delete a formula, a range, or even the whole model. From the beginning, spreadsheet software has offered the ability to *protect* or *lock* designated cells or ranges of cells while leaving other cells available for modification. This is usually accomplished simply by designating the cell or range and selecting **Protect** from the menu. Protected cells cannot be modified unless one first takes the step to switch off the protection status. For example, once a spreadsheet has been developed and tested, one would want to protect cells containing headings, formulas, and other parts intended to stay unaltered. In Fig. 9-1, all cells except the actual cost data entry cells in range B4..D7 should be protected before turning the spreadsheet over for routine use (assuming that all estimates used the same four bid items and markups). Although the results appearing in protected formula cells will change when input data change, the formulas themselves stay the same.

A similar capability available in some spreadsheet programs is the option to *hide* cells or ranges of cells. This action might be taken with parts of the spreadsheet that contain tables of data or formulas that the user does not really need to see while using the model. This option can also make it possible to present a cleaner appearance in the design of the spreadsheet parts that the user does see, while keeping the closely related calculations or data conveniently nearby for the developer to see when needed.

Additional security is made available in some spreadsheet packages in the form of *password protection* or the equivalent. Access could thus be denied altogether, or just be made available at the data-entry and reporting level, or at the developer level. At the data-entry level, the user would have the password to use the system but not the one that would provide the ability to change the protection status or modify the spreadsheet itself.

Online Help

A major spreadsheet advance introduced with the Lotus 1-2-3 program was the incorporation of *online, **context-sensitive help*** for learning and working with the spreadsheet. It also contained an excellent online tutorial that set a standard for tutorials in a wide variety of other microcomputer programs that have followed. The idea of context-sensitive help is that you can be in the middle of some sequence of steps—perhaps at the second level down in the menu or about to insert a statistical function in a formula—yet you can press the function key for help if you are uncertain about how to do the next step or what its effect will be. The program in effect takes you directly to the pertinent explanation in the online manual. After reading it, you can either use a single keystroke to return where you left off or continue in the help system to read closely related material, or branch to anywhere else in the help files, as desired. This context-sensitive help not only is convenient but also is an excellent learning tool, since people learn best when they are trying to solve a specific problem. Some similar capability now exists in most spreadsheet packages, although the depth and level of effectiveness differ considerably from one package to another.

Report Form Definition

Initially, spreadsheets were strictly character-oriented packages, with limited font sizes and types. Borders on form designs were done with dash and vertical bar characters and a few others. With today's graphically-oriented computers, all kinds of font types, sizes, and styles are available, together with lines and shading abilities to design some very professional-looking report forms.

Particularly when spreadsheet software includes database capabilities, they also might offer the ability to design separate output forms for information selected and sorted from the database, much like the file-management packages to be discussed in the next chapter.

Graphics

Most packages now offer some type of business graphics. These are usually produced as separate reports, but now some of the more advanced spreadsheet packages offer a relatively free-form ability to integrate the graphs onto the same screen window or output page as the tabular spreadsheet from which they get the data to plot.

Graphics capability allows one to take data from one to several columns or rows of numbers and use them to plot a bar chart, pie chart, line graph, or, in some packages, even plot X-Y graphs correlating two sets of data. Although these capabilities are fairly simple compared with the graphics in symbolic math packages and dedicated business graphics software, they are quite handy for a wide variety of business reporting. Different packages vary somewhat in the types of graphs they offer, and especially in the richness and flexibility in the graphical standards they use for axes and grids, shading for bars and pie segments, symbols for data points and line types on line graphs, styles for showing keys or legends, and font types, sizes, and positions for titles and labels.

The two graphs in Fig. 9-6 are examples of the different capabilities that are available. The first (Fig. 9-6a) is a bar graph that takes two sets of data and plots them against the same vertical axis. The horizontal axis corresponds to the number of items (data cells) to be plotted, whereas the vertical axis scales automatically to accommodate the largest value. Figure 9-6b is an X-Y plot of one value (time) against another (load) for three different sets of data. It was produced by an Excel spreadsheet application similar to the Lotus 1-2-3 example in Fig. 9-13.

Database

Most current spreadsheet programs include what the issuers usually call a *database capability,* but it is usually very limited compared with the standalone database applications described in the next chapter. Conceptually, spreadsheet database commands allow a user to designate a two-dimensional portion of a spreadsheet to contain a table of data that can be accessed and manipulated in a manner similar to a flat file. Its rows become records; and columns headed by field names can serve as criteria for sorting, searching, and selection. This table can be either on the same sheet as another application or on a separate worksheet that can be linked to applications needing to access its data.

Entering data into the database portion of a worksheet is basically the same as entering data into any other part of a spreadsheet. Once the information is available, it can be designated as a database. Typical operations one can then perform include the following: queries or searches to find data that meet specific criteria (e.g., equal to, less than, greater than, or not equal to, a given name or number), multikeyed sorts by rows or columns in ascending or descending order, extraction or deletion of subsets of data that conform to user-defined criteria, statistical calculations (e.g., total count, sum, average, maximum, minimum, standard deviation, and variance, for numbers in a designated range, or for just those in a subset of records fitting specified criteria within that range), and display or printing of reports. Some programs offer flexible capabilities for designing attractive input and output forms that need not conform to the rigid row-and-column layout of the database itself.

The spreadsheet database capabilities described thus far are similar to those of simple file programs to be described in Chapter 10. Chapter 10 also presents a file program example that could also be handled with a spreadsheet's database capability. Thus we will not focus on the internal database aspects of spreadsheets in this chapter.

One useful database-related development for spreadsheets is the recent addition to some packages of remote database access capabilities. For example, a

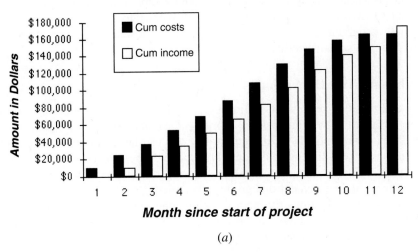

(a)

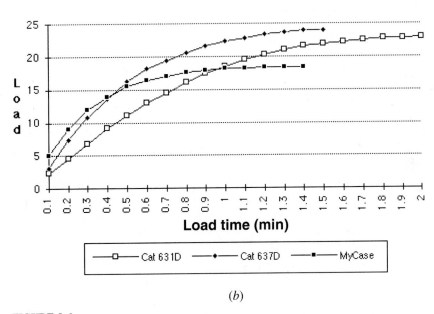

(b)

FIGURE 9-6
Two examples of spreadsheet graphics: (a) Bar graph; (b) X-Y line graph.

developer can now write macro commands that enable a user to network or telephone into a database that might be running on a corporate mainframe, department
minicomputer, or an information service bureau. The macros can implement commands that imitate the **structured query language (SQL)** of the host database to

extract from that database information to be downloaded to the microcomputer-based spreadsheet for use in further analysis. An example might be to find current data about construction materials costs and pull them into an estimating spreadsheet. With programmed macros, a developer can substantially automate such functions for the spreadsheet user.

Going beyond Two Dimensions

The original types of spreadsheet software, as well as today's simpler ones, limit a single model to a two-dimensional worksheet. If one needs to integrate two or more spreadsheet models of different types, one can lay them out on different parts of the same two-dimensional plane. For example, the spreadsheet calculation shown in Fig. 9-1 might have backup calculations for each line item (e.g., for the labor, material, and equipment that go into the structure), and these in turn might have subordinate calculations for labor or equipment productivity, the quantity of material, and so forth. Although not revealing detail, Fig. 9-7a shows how these various spreadsheets might be laid out.

There are, however, several problems with this arrangement. Two or more of the submodels that make up the same spreadsheet may have to share the same rows or columns, as do the "Bid Summary" and "Labor" submodels in Fig. 9-7a. If the developer later needs to insert an additional row or column into the "Bid Summary" submodel, then a blank row gets inserted into the "Labor" submodel, or any others that might be intersected. One way to avoid this is to lay out the submodels along a diagonal, say from upper left to lower right, as shown in Fig. 9-7b. However, in some packages this rapidly consumes computer memory and in any case makes for some long reaches to navigate around the spreadsheet territory.

One alternative is to add a third dimension to spreadsheet models. This approach is used in version 3.0 of Lotus 1-2-3, which was first marketed in 1991, and in other spreadsheet packages. To illustrate the basic concept, we will consider a bid recapitulation sheet that summarizes the bid quantity, unit price, and total price for each line item. Each unit price and total price contains labor, materials, and equipment components. One could add these as six extra columns to the main spreadsheet. But an alternative is to imagine that these components are behind each of the summary numbers, in a third dimension behind the spreadsheet, as illustrated in Fig. 9-8. In this type of spreadsheet, commands, keystrokes, or mouse *clicks* move the user in and out of the third dimension, in a manner similar to up and down, left and right in 2-D; the user sees each successive layer appear on the screen as if a section were being taken through the 3-D figure. In some spreadsheet programs, one can also imagine rotating the model to view it from the side or top. For example, the process of rotating Fig. 9-8 sideways 90 degrees (right) and going into the layer that represents total costs (now at back) gives a model similar to Fig. 9-1, where the components of labor, materials, and equipment are emphasized. Looking down and moving in from the top, one could get a detailed view of both unit costs and total costs of the resources—the labor, materials, and equipment, successively—for a given bid item. The ability to view data in a model from such extra perspectives literally adds a new dimension

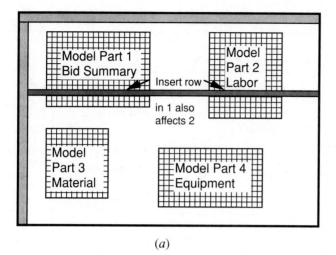

(*a*)

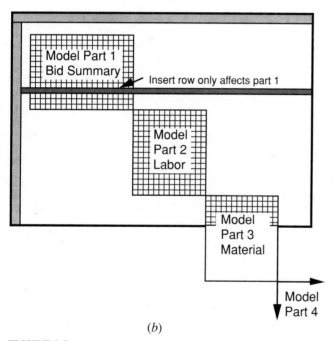

(*b*)

FIGURE 9-7
Multiple tables on one two-dimensional spreadsheet: (*a*) Adjacent
spreadsheet components; (*b*) diagonal spreadsheet components.

to one's power to model and analyze information. It is also a more compact
representation than can be achieved by laying out the equivalent of the various
layers in different subtables on a single two-dimensional surface.

Another powerful way to build models with interlinked components is
to establish linkages between specific cells and ranges of multiple spreadsheets.

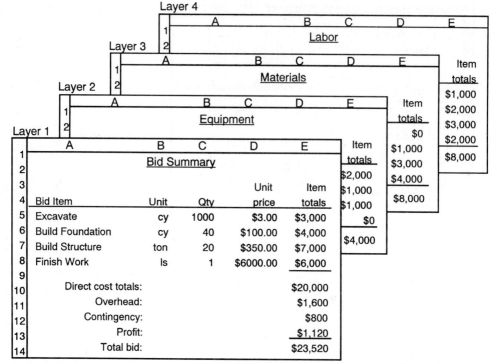

FIGURE 9-8
Three-dimensional spreadsheet.

This approach was introduced by Microsoft in 1984 in a product called MultiPlan, the first spreadsheet available for the then new Apple Macintosh microcomputer. The concept has been carried forward and advanced in that company's Excel spreadsheet package and is available in others. The basic concept is similar to that of establishing relationships among the cells of a single spreadsheet, except that some cells can be on other spreadsheets. Making a change to data in a cell or range on one sheet causes updates to any other spreadsheet cells or ranges that directly or indirectly depend on the data in that changed cell or range. Each sheet can be quite different in shape and design; it is through the cells and linkages that the sheets interact. For example, there might be worksheets containing labor wage rates for each craft, ownership and operating costs for machines, materials price tables, tables of productivity data, and so forth, and these could feed into item productivity and cost computations that draw upon those resources. Changing the fuel consumption on a machine worksheet or updating a wage for a craft would automatically reflect through all components of a bid that draw upon those resources. We will leave the explanation of linked spreadsheets at this stage for now, because the second case example in this chapter uses Excel in this way.

Programmable Macro Commands

Programmable macros are available in many advanced spreadsheet packages to enable developers to create preprogrammed routines that, when invoked by the user or another macro in a model, automatically execute sequences of actions of the type that the user would otherwise do directly with the mouse and/or keyboard. More advanced versions further add logical control capabilities (e.g., IF *condition* THEN *do this* ELSE *do that*) that automatically make some of the decisions that the user would probably make anyway if confronted with a similar set of interim results.

Although the behavior of macros is similar to that of programs written in a procedural language, their appearance is quite different. Essentially they look like transcriptions of the sequence of keystroke commands, menu choices, cursor or mouse movements, and query replies that the user would make when doing the equivalent routine directly on the computer. Indeed, one method some systems provide to create macros is for the user to turn on the "record" or "learn" mode in the menu, execute the series of desired commands, then save the "recording" for later "playback." The recorded information may indeed be the actual keystrokes, or it may be some text to describe what was done (e.g., {down} for one press of the down-arrow cursor key). As an alternative to the record mode, one can simply type in this text and test it when the program is complete, or incrementally, much as one would do in writing a BASIC program using an interpreter. In some systems the macro commands are placed directly on a portion of the spreadsheet that they control. In others they go into a separate file that is linked to the spreadsheet. Both types will be illustrated in the Lotus and Excel examples later in this chapter. As a brief illustration, the two columns of Fig. 9-9 contain, respectively, Lotus and Excel macros to sort a table of subcontractor bids, originally in alphabetical order by name, into ascending order of price.

At one level, macros are a convenience; an experienced user can make keystroke equivalents of a sequence of frequently used commands. The most wonderful thing about macros, however, is that they enable developers to create custom spreadsheet software much as they would do with other programming tools; test, debug, and protect the software; and then turn it over to users who know little about spreadsheets. These users can work directly with the application that was developed and not worry about learning much about the spreadsheet software behind it. This capability has led to an aftermarket of custom spreadsheet application templates, to be discussed shortly.

Integrated Software

Although we have now touched upon the basic concepts of spreadsheets as such, it is important to mention two primary mechanisms by which they can become part of *integrated software*. The first is as an integrated package, and the second is by designing a standalone spreadsheet package so that its applications can link directly to other software.

Integrated packages have been around since a few years after spreadsheets were invented. Basically, they take applications such as a spreadsheet, a word

	A	B	C
3	Company	Phone	Price
4			
5	Advanced Painting	325-3399	$12,847
6	Bay Area Painting	327-9342	$13,984
7	Bob Davies Painting	922-1811	$11,783
8	Empire Paint Co.	723-3923	$12,118
9	House Doctors	911-0000	$11,644
10	Master Painters	364-9935	$12,998
11	Peninsula Painters	243-1355	$11,875
12	Steves's Painting Co.	984-8045	$12,086
13	Yan-Go Artistic Painting	359-4000	$16,922

Before macro executes

After macro executes

	A	B	C
3	Company	Phone	Price
4			
5	House Doctors	911-0000	$11,644
6	Bob Davies Painting	922-1811	$11,783
7	Peninsula Painters	243-1355	$11,875
8	Steves's Painting Co.	984-8045	$12,086
9	Empire Paint Co.	723-3923	$12,118
10	Advanced Painting	325-3399	$12,847
11	Master Painters	364-9935	$12,998
12	Bay Area Painting	327-9342	$13,984
13	Yan-Go Artistic Painting	359-4000	$16,922

Lotus 1-2-3 Macro

```
\s      /ds{EDIT}dA5..C13~
        pC5..C5~
        ~g
```

Translation:

Macro named " \s "
   (activated with Alt-s)
/ **d**ata **s**ort ; **edit**; **d**ata range **A5..C13**
primary sort on column C
go (i.e., do the requested sort)

Excel Macro

```
PainterSort (s)
=SELECT("R5C1:R13C3")
=SORT(1,"R5C3",1)
=RETURN()
```

Translation:

Macro named PainterSort
   (activated with Opt-Cmd-s)
Select the range Row5Col1:Row13Col3
Primary sort on column 3
Return to spreadsheet

FIGURE 9-9
Two types of spreadsheet macros.

processor, a communications program, a file-management program, and perhaps an enhanced business graphics program and combine them into one package with several distinct modules or modes corresponding to these different applications. An early attempt from the company that developed VisiCalc was VisiOn, but it was slow to market and never caught on. The first really successful one was a Lotus product called Symphony, which combines spreadsheet, word processing, graphics, file management, and communications capabilities, though the spreadsheet is certainly the most advanced of its modules relative to corresponding standalone products. Another success was Ashton-Tate's Framework. With the development of portable and laptop microcomputers, some simpler integrated packages became especially popular to provide the basic functionality that one would want when traveling. Two PC examples are Microsoft's Works and Spinnaker's PFS:WindowWorks, each of which combines a spreadsheet with a word processor, spelling checker, communications program, and file management program and can fit in the RAM disk storage of even the smaller laptop computers. Recent Macintosh examples are ClarisWorks and Symantec's GreatWorks, which also include drawing modules.

The main advantages of these programs, apart from the simplified portability just mentioned, are the smooth integration of the different modules and the consistent interface that each offers to the user. Upon starting the program, one usually sees a main menu that asks which module the user wants first, but once underway one can easily flip back and forth between the modules, perhaps working on a spreadsheet, then writing a form letter that incorporates the spreadsheet in its body, then linking to the database to output individually addressed letters to all or a selected subset of people contained in a mailing list file. One might even use the communications program to dial up a remote mainframe and send some of the letters by electronic mail. If, in the course of writing the letter, you decide to change something in the spreadsheet, the changes will reflect automatically in the copy of the spreadsheet contained in the letter. The main problem with integrated packages is that most of the modules are a compromise short of the capabilities of the equivalent combination of independent programs, but for many if not most users they are more than adequate.

The alternative approach to integration is to design linkage mechanisms into programs that can hook into applications developed in other types of software. Initially this was done for packages developed by the same firm (e.g., Microsoft Excel and Word can behave similarly to what was just described for Works), but over time some de facto standards have evolved that have made links between packages from different developers increasingly effective. When the links go beyond simple text copies into active updating, such that a change in a spreadsheet is automatically reflected in a report being prepared on a word processor, the connections are sometimes called *hot links*. Recent Microsoft Windows software-linking technologies now moving into application software are **Dynamic Data Exchange** (**DDE**) and Object Linking and Embedding (OLE). The main disadvantages of integrating separate package programs are that the sum of the costs of the independent modules usually exceeds by far the cost of an integrated package; memory demands are high when they run concurrently; and the user interfaces may vary significantly from one package to another, making learning even more difficult.

Templates

Programmable spreadsheets have encouraged the development and marketing of prewritten applications that are built upon commercial spreadsheet software. They are commonly called **templates**, the idea being that they are blank forms ready for the user to begin inserting data. Examples of applications include income tax preparation, personal finance, and construction estimating. Developers usually assume that the user already has the necessary spreadsheet package to run the template, but the availability of spreadsheet compilers and run-time licenses sometimes makes the application entirely self-contained.

APPLICATIONS AND LIMITATIONS

Since modern electronic spreadsheets are so flexible and powerful, and since their capabilities are growing continuously, it is hazardous to say what cannot be done with them. Indeed, one could write several books on what can be done. This section will thus confine itself to some general guidelines.

Spreadsheets are primarily intended for applications involving calculations and quantitative modeling. Suitable applications are found in most fields of endeavor, such as business and economics, engineering, science, mathematics, and statistics. Even complex numerical models involving finite elements and differential equations have been developed with spreadsheets. In construction, one can create financial models for land development projects, prepare estimates, design forms for concrete, compute labor and equipment productivity, analyze costs, and document claims. Indeed, spreadsheets can cover more construction applications than almost any other single software tool, short of making enormous investments to develop applications in procedural languages. The two examples given later in this chapter will offer just a glimpse at the possibilities.

Because the applications are so many and varied, perhaps it would be best to suggest some areas where spreadsheets may *not* be the best tool. These include many applications where other tools described elsewhere in this book may do the job better or more economically.

Spreadsheets are usually clumsy for writing letters, reports, and other predominantly text-oriented tasks. Although some developers have attempted to include improved text editing and even word processing features directly into spreadsheets, none have really come close to the capabilities of today's better word processors. If a letter, memo, or brief report consists mostly of tables of computations, then using the spreadsheet itself to add the needed narrative text and print it might be sufficient. Beyond this, it is usually better to prepare the main document with a word processor and then copy the spreadsheet material into that. Integrated software makes this task particularly easy, but it is not much harder with modern standalone programs.

Many spreadsheet programs contain limited database capabilities, and some users seem determined to use them for file and database management tasks simply because they are familiar with spreadsheet software and do not want to learn to use another program. However, applications that involve large or complex bodies of information are usually best left to file and database management programs, particularly if any computational aspects of the application can be handled by the limited capabilities of this type contained in such software. This choice should almost certainly be recommended if the application is mostly intended just for data entry, information storage and retrieval, and reporting. File and database programs work from disk storage, whereas most spreadsheet programs work only with what can be held in memory. That is, much larger amounts of information can be handled with files and databases. The sort and search capabilities of file and database management software are usually much faster and more versatile. Also, file and database software usually have much better report design capabilities. If multiple interrelated files are involved in a common application, then relational databases will almost certainly be better than spreadsheets. Therefore, spreadsheets should mainly be used for smaller files, especially for those that integrate directly into spreadsheet applications.

One sometimes hears the term *simulation* used in conjunction with spreadsheet applications, but this usually refers to the aforementioned "what-if?" type of analysis, where spreadsheets make it easy to test out the effect of input changes on the results. This usage should not be confused with the powerful type of cyclical system modeling and simulation discussed in Chapter 15, and it is usually not a good idea to try to implement such applications in spreadsheet software.

Spreadsheets are predominantly quantitative and computational tools and thus are very limited in handling the type of heuristic and qualitative applications that have lent themselves well to expert systems and artificial intelligence software. The tools described in Chapter 11 are intended for that purpose.

Spreadsheets may be even more limited when compared to application-specific software. Although many spreadsheets have some graphical capabilities, they are far removed from drawing and CAD software, and even advanced business graphics needs may be better served with software tools designed for that purpose. Similarly, whereas one can easily imagine developing tabular work schedules with spreadsheets, and a few people have even been determined enough to prove that you can use them for CPM schedule calculations, they do not begin to compete with scheduling programs intended for this purpose, particularly for preparing graphical output. Even in estimating, cost control, materials management, and other construction applications where spreadsheet applications have had and will continue to have a major and effective role, there are dedicated application programs that, for certain companies and needs, can do the job even better, although they may be expensive to acquire and update.

In a few cases, a spreadsheet application will become so successful and widely used in a company, yet limited in terms of speed or capacity, that it may be worth having the application reimplemented in a fast procedural language. The spreadsheet version served well as a design and test bed, but the application's success may have outgrown the tool that spawned it.

In summary, spreadsheets are wonderful tools, and they legitimately have a strong following. But you should evaluate each potential application in its own right, using the guidelines from Chapters 5 to 7, and not just plunge ahead to implement it with a spreadsheet only because it is a good tool and you know how to use it. Certainly, compromises in favor of familiarity at the expense of appropriateness can be in order, particularly for small applications or those with a short useful life. But even the compromises should be made with some consideration given to the alternatives.

EXAMPLE: OPTIMUM SCRAPER LOAD TIME IMPLEMENTED IN LOTUS 1-2-3 ON A PC

This first example deals with finding the optimum production cycle for a pusher-scraper operation. To do this we need to find the load, haul, dump, and return times for the scrapers, the cycle time for the pusher, the number of scrapers to be served by the pusher, and the overall system production and unit costs. The load time is particularly interesting here because of the nonlinear loading characteristics of a scraper. In this spreadsheet we will thus include some theory related to the optimum load time for scrapers that was developed by Caterpillar Inc.*

Optimum Load Time, Caterpillar Inc., Peoria, Illinois, 1968.

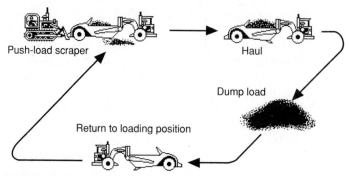

FIGURE 9-10
A scraper's load, haul, dump, and return cycle.

The basic production cycle is that shown in Fig. 9-10. To start, the operator of the scraper lowers its bowl so that the cutting edge goes into the ground, then the tractor pushes it forward so that a layer of earth peels off and builds up in the bowl. Once full, the pusher boosts the scraper on its way, and the scraper hauls the material to the dump site, unloads by forcing the earth back out with its ejector, and returns to the pusher. The duration of both the haul and return varies with their respective distances and speeds.

The pusher's part of the cycle consists of the following components: time to push a scraper forward to load, time needed to boost the load to get the scraper on its way, backtrack return time, and transfer time to get ready for the next scraper. Empirical values of 0.1 minute for boost time, 0.15 minute for transfer time, and a return time equivalent to 40 percent of the load time are suggested by Caterpillar based on past performance. Thus we have

$$\text{Pusher cycle time} = 0.25 \text{ min.} + 1.4 \times \text{load time}$$

The productive part of the pusher's output consists of the load that goes into each scraper; thus its load time for a given load is identical to that of the scraper. The pusher's cycle is shown in Fig. 9-11.

As a rough approximation, the number of scrapers that can be served by a pusher is computed as follows:

$$\text{Number of scrapers served} = \frac{\text{Scraper cycle time}}{\text{Pusher cycle time}}$$

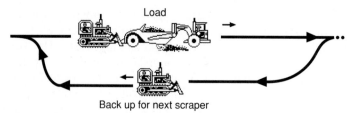

FIGURE 9-11
A pusher's cycle.

The haul and return times could be computed recognizing the grades and rolling resistance, and thus the required rimpull versus the gross weight and engine power, as was done for the second example in Chapter 8. However, for simplicity here we will assume the speeds as input data and divide them into the distances to get times. We will also assume the dump time as input. Strictly speaking, to get the pusher's cycle time, one must assume a load time, then check the cycle times it produces for pusher and scraper, compute the number of scrapers served, and iterate and converge to the optimum ratio of scrapers/pushers. For our purposes, we will assume the pusher's balance-of-cycle time (haul + dump + return) is given as input. The main focus here will be on determining the optimum load time and then the resulting production and costs.

The nonlinear characteristics of the load time result because of the mechanical process of loading a scraper. As more and more material enters the bowl, shoving more earth in becomes increasingly difficult, so the *rate* of loading decreases with time. Thus the first step is to develop the *load-growth curve*. This is usually done empirically by loading scrapers in a given material for 0.2 minute, 0.4 minute, and so on and by weighing the partial loads. The result is a curve like that shown on the right side of Fig. 9-12. The horizontal axis has the zero point at the base of the load-growth curve. To the right of this point is the load time. To the left is the balance-of-cycle time (i.e., total minus load time) for either scraper or pusher. Once determined for a given haul distance, the balance-of-cycle time is assumed to be constant. The vertical axis is the payload contained in the scraper, which varies with load time.

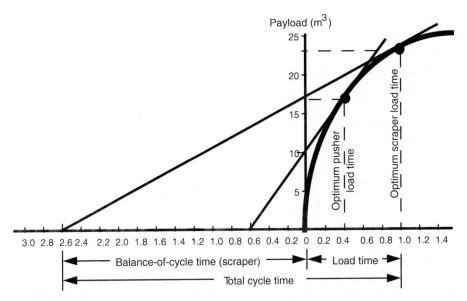

FIGURE 9-12
Scraper load-growth curve, with tangents to balance-of-cycle times for pusher and scraper. (Caterpillar Inc., *Optimum Load Time*, Peoria, Illinois, 1968, p. 11. Reproduced by permission of Caterpillar Inc.)

With the load-growth and balance-of-cycle times known, the graphical method to find the optimum load time for the scraper, shown by the longer straight line in Fig. 9-12, is to strike a tangent to the curve from the balance-of-cycle time for the scraper. The slope of the longer tangent shows the maximum production (y axis) per unit time (x axis) for a single scraper. Thus, rather than loading the scraper until it is nearly overflowing (perhaps at 1.4 minutes here), it is often better for overall throughput to load less, corresponding to the 1.0-minute point at the tangent in this case.

To obtain maximum production for a fleet of scrapers and pushers, however, one needs to consider the performance of the pusher as well. The shorter tangent intersects the time and scraper load that maximize the pusher's production. If there are plenty of scrapers so that the pusher is the limiting resource, then the pusher's time should be used to maximize system production even though the scrapers may be loaded light. If there is ample pusher time and scrapers are limiting, then the scraper's optimum time should be used. Times in between these two will generally also produce good results, but loading too long is unproductive. Generally, loading too short rather than too long is also better since it increases haul speeds and keeps down costs for tires and machine maintenance; this recommendation at first seems counter-intuitive, but it is proven in practice.

The spreadsheet for this application is written in Lotus 1-2-3 to run on an IBM PC or compatible with the MS-DOS operating system. It takes the following parameters as input:

- Scraper type and cost
- Pusher type and cost
- Distances (in meters) for the haul and return roads
- Average speeds (in kph) for the haul and return
- The dump time (in minutes)
- The scraper's load-growth curve, defined in m^3 at 0.1-minute intervals

The spreadsheet should plot the load-growth curve and compute and report the following:

- The haul and return times for the scraper (minutes)
- The optimum load times for scraper and pusher (minutes)
- The scrapers/pushers ratio at each load time
- The maximum production (m^3/hour) and corresponding fleet size

One possible solution for this case, together with a set of test data and results, is shown partially in Fig. 9-13 (macros and some off-sheet calculations do not appear here). In this spreadsheet, two columns (D and E) permit one to compare two different machines concurrently. To identify a machine and enter load, haul, dump, and return characteristics, the user can enter data in ranges D7..D16 and

```
A1: [W2]
Base Case    Your Case    Print    Quit    Manual
Modify base case
```

	A	B	C	D	E	F

```
 1
 2          Cycle Time Calculations With Load Growth Curve Considerations
 3
 4     INPUT SECTION: Use "Alt-R" to start    U.S.   <---Input Data--->
 5          Type "Enter" key after inputs    Units  Base Case Your Case
 6   ,------------------------------------- -------- -------- --------
 7   | Type of Scraper:                              Cat 631E  Test   |
 8   | Cost of Scraper:                      $/hr    $120.00  $140.00 |
 9   | Type of Pusher:                               Cat D9L  Cat D9  |
10   | Cost of Pusher:                       $/hr    $125.00  $150.00 |
11   | HAUL Distance:                        Meters   1200     900    |
12   | RETURN Distance:                      Meters   1500     900    |
13   | Average HAUL Speed:                   KPH       20       30    |
14   | Average RETURN Speed:                 KPH       35       45    |
15   | Dump Time:                            Min.     0.8      0.6    |
16   | Pusher Return Time:                   Min.     0.6      0.8    |
17   |                                                                |
18   | LOAD-GROWTH CURVE DATA --             Minute: Vol(c.m) Vol(c.m)|
19   |             Load at time:             0.1      2.4      5.0    |
20   |          "     "     "                0.2      4.6      9.0    |
21   |          "     "     "                0.3      6.8     12.0    |
22   |          "     "     "                0.4      9.2     14.0    |
23   |          "     "     "                0.5     11.2     15.6    |
24   |          "     "     "                0.6     13.0     16.6    |
25   |          "     "     "                0.7     14.6     16.8    |
26   |          "     "     "                0.8     16.2     17.6    |
27   |          "     "     "                0.9     17.5     18.0    |
28   |          "     "     "                1.0     18.6     18.2    |
29   |          "     "     "                1.1     19.6     18.3    |
30   |          "     "     "                1.2     20.4     18.3    |
31   |          "     "     "                1.3     21.0     18.4    |
32   |          "     "     "                1.4     21.6     18.4    |
33   |          "     "     "                1.5     22.0            |
34   |          "     "     "                1.6     22.3            |
35   |          "     "     "                1.7     22.6            |
36   |          "     "     "                1.8     22.8            |
37   |          "     "     "                1.9     22.9            |
38   |          "     "     "                2.0     23.0            |
39   `------------------------------------- -------- -------- --------`
40
41     RESULTS:
42   ,------------------------------------- -------- -------- --------,
43   | Haul Time: ---------------------->    Min.     3.6      1.8    |
44   | Dump Time: ---------------------->    Min.     0.8      0.6    |
45   | Return Time: -------------------->    Min.     2.6      1.2    |
46   |    Subtotal (Balance of cycle):       Min.     7.0      3.6    |
47   |                                                                |
48   | Determine Max Production Rate:        Cu.M/Min 2.61     4.00   |
49   | Lookup Corresponding Load Time:       Min.     1.70     0.80   |
50   | Lookup Corresponding Load Volume:     Cu.Mtrs. 22.60   17.60   |
51   | CycleTime (Load-Haul-Dump-Return)     Min.     8.67     4.40   |
52   |                                                                |
53   | Production (Per 60 Minute Hour)       Cu.M/Hr. 156.4    240.0  |
54   |                                                                |
55   `------------------------------------- -------- -------- --------`
```

FIGURE 9-13
Scraper spreadsheet.

E7..E16, respectively. Formulas in range D43..E53 then compute the cycle time and production data. A typical set is given below for range E43..E53:

E43: +E11/(E13*1000/60)   Haul time = haul dist./
                            (kph × 1000 m/km ÷ 60 min/h)

E44: +E15   Copy down entered dump time

E45: +E12/(E14*1000/60)   Return time = ret. dist./
                            (kph × 1000 m/km ÷ 60 min/h)

E46: @SUM(E43..E45)   Compute balance-of-cycle time

E48: @MAX(E62..E81)   Find max of
                            (production/cycle time)

E49: @VLOOKUP(E48,E62..G81,2) Match load in E48 to time
                               in G62..G81

E50: @VLOOKUP(E49,$C19..E38,2) Match time in E49 to load
                                in E19..E38

E51: +E46+E49   Haul-dump-return + load time

E53: +E50*60/E51   Load volume/cycle time

Not shown in Fig. 9-13 is a hidden calculation range in D62..G81. Formulas in D62..E81 compute the production rate (m^3 per minute) corresponding to every load and load time in D19..E38. Because the VLOOKUP Function expects to find returned values to the right of the LOOKUP value (e.g., look up a load and find the corresponding time), the load times from C19..C38 are duplicated in G62..G81 so that the formula (such as that in E49) can work. Although the hidden range is not shown in Fig. 9-13, the formulas below show typical entries in that range. These happen to be the ones that correspond to the optimum load time for the case in column E.

E69: +E26/(E$46+$C69)   Rate = load/
                        (load time + balance-of-cycle time)

G69: 0.8

In Fig. 9-14 we see the portion of the spreadsheet where the macros have been placed. The /xmH1 in cell G1 says to execute the macros starting in H1. Cells H1 to K1 contain the definition of the custom menu that is seen at the top of the main spreadsheet in Fig. 9-13. The row below that shows the explanation text displayed for each item in the menu as the cursor moves over the corresponding item (e.g., with cursor on **Base Case** the text reads **Modify base case**). Items under columns H to K starting in row 3 contain a series of keystrokes and commands that automatically execute to carry out the function of each menu selection. For example, the function in cell I3 does the equivalent of the keystrokes **/wtc** (/ worksheet-titles-clear) then **/re** (/ range-erase E7..E16) . . . and so on. The sequence **/xcH16** in cell G8 and elsewhere is actually a subroutine call to request the series of down-arrow keystrokes recorded as shown in cells H16..H25. The **/xr** in H26 is the equivalent of a return from subroutine and takes the sequence back

	G	H	I	J	K
1	/xmH1~	Base Case	Your Case	Print	Quit
2		Modify base case	Enter data for your case	Print results	Quit application
3		/reD19..D38~/wgrm	/wtc/reE7..E16~/reE19..E38~/wgrm	/pprA1..F55~	/wgra/wtc
4		{Home}{GoTo}D1~/wtv	{Home}{GoTo}D1~/wtv	oml3~mr80~mt2~mb2~	/QY
5		{GoTo}D4~	{GoTo}E4~	s{Bs}{Bs}{Bs}018~	
6		{Down}{Down}{Down}	{Down}{Down}{Down}	qgpq	
7		/xcL7~	/xcH16~	/wgra/wtc	
8		/xcH16~	{Down}{Down}	/xgSTART~	
9		/xcH16~	/xcH16~		
10		{GoTo}D40~{Calc}	/xcH16~		
11		{GoTo}D58~	{GoTo}E40~{Calc}		
12		/wgra/wtc	{GoTo}E58~		
13		/xgSTART~	/wgra		
14			/xgSTART~		
15					
16		{?}~{Down}			
17		{?}~{Down}			
18		{?}~{Down}			
19		{?}~{Down}			
20		{?}~{Down}			
21		{?}~{Down}			
22		{?}~{Down}			
23		{?}~{Down}			
24		{?}~{Down}			
25		{?}~{Down}			
26		/xr			

FIGURE 9-14

Selected scraper spreadsheet macros.

243

to the next step after the cell that made the call. Other referenced subroutine calls (e.g., /xcL7 in cell H7) are not reproduced here. It takes some familiarity with the syntax of the Lotus 1-2-3 command set to read the rest of these macros, but at least one should be able to get an idea about how they are written by studying this example. In most newer programs, one can automatically record a series of keystrokes to make a macro.

Figure 9-15 is an example of a graph that was produced by the standard X-Y graph mode in Lotus 1-2-3 using the data in cells C19..C38 for the x axis and the two load sets in cells D19..E38 for the load-growth curves plotted against the y axis. You may notice that they are not quite so fancy as the similar curves plotted in Fig. 9-6b, which were plotted by Excel on a Macintosh. The version in Fig. 9-15 is included to illustrate the more rudimentary form of spreadsheet graphics. Lotus 1-2-3 is also capable of higher-quality graphics, including labels, titles, legends, and graphic fill patterns, using its high Wysiwyg (what you see is what you get) mode.

The spreadsheet included here could be extended to include many more aspects of the scraper load-growth theory discussed earlier in this section. For example, no effort is made in this version to optimize on the basis of pusher production (i.e., cells D16..E16 are ignored). Similarly, we have not used the cost data that are provided for the scraper and pusher, which would be a simple enhancement. On a larger scale, one could imagine extending this example with a three-dimensional spreadsheet to maintain and be able to compare different types of scrapers and pushers with similar calculations on successive layers. The main input could be the distances, speeds, and material characteristics specific to a given application. The 3-D spreadsheet could have a summary page where these variables are entered, and the best equipment fleet would be pulled forward after the comparative computations take place. Even multiple load-growth curves

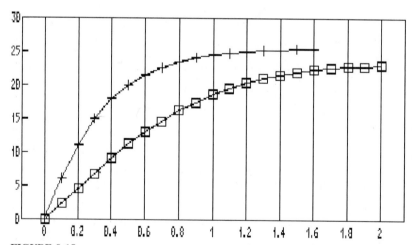

FIGURE 9-15
Scraper load-growth curve.

could be maintained for different machine and soil conditions, so that the overall application could concentrate on fleet selection for a given project.

It is typical of spreadsheets that one idea leads to another, and it is easy to let such an application grow over time. Eventually one hits limitations on implementing new ideas, and the cumulative user demand to overcome these limitations stimulates the developers to bring out spreadsheets with more and more features and capabilities.

EXAMPLE: ESTIMATING SYSTEM IMPLEMENTED IN MICROSOFT EXCEL ON A MACINTOSH

This second application is designed mainly to illustrate the linkage of multiple spreadsheets to create a more complex application. It is a simplified version of an estimating system for unit-price bidding that consists of the following components:

- Summary sheet
- Bid item calculation sheets (one example given)
- Resource cost summary tables
- Resource cost calculations (labor and equipment examples given)

The overall relationships among the spreadsheets are shown in Fig. 9-16. The resource cost calculations contain detailed labor and equipment cost elements and provide a total that is transferred to the appropriate resource cost tables. There

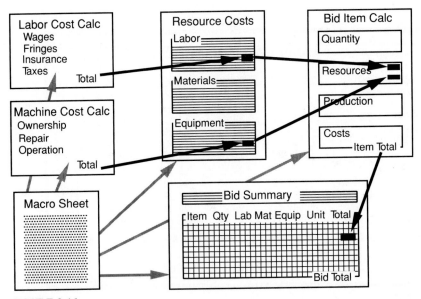

FIGURE 9-16
Interlinkages of spreadsheets.

could be either separate sheets for each craft or machine type or multiple columns for each major resource. The resource cost tables would normally be contained on separate sheets for materials, labor, and equipment but, to save space in this example, are consolidated here onto one sheet with just a few representative items of each type. The bid item calculation sheets each would contain a section for the item quantity estimate, crew composition and cost, productivity calculations, and the unit and total cost calculations, broken out for labor, materials, and equipment. The results from each of these sheets feed into the summary sheet, which adds the indirect costs, contingency, and profit and then distributes these back to the individual bid items for the unit and item prices to be used in the bid. Also shown is a macro sheet that partially automates the overall system.

The figures and text that follow introduce each of the spreadsheets and highlight their important features. They were developed in Microsoft Excel, version 3.0, on an Apple Macintosh microcomputer. They should be almost 100 percent compatible with versions of Excel that run on IBM PC and compatible microcomputers.

Simplified labor calculations for a few key crafts used on a building foundation project appear in Fig. 9-17. Basically there is a separate craft in each column, and similar formulas are used to add up the wages, benefits, payroll-based insurance, and taxes to develop an hourly straight-time labor rate. The sheet could easily be extended to show overtime rates and other variations on the basic cost, but it is kept simple here.

This sheet also displays a **Return to Menu** *button* at the bottom. This button is a *graphic object* that calls a small macro that is part of a set of menu-control macros that will be discussed later in this section.

	A	B	C	D	E	F
1		\$colspan Labor Cost Calculations				
2						
3	**Cost category**	*Carpenter*	*Cem. Msn*	*Ironworker*	*Laborer*	*Operator*
4	Base wage (s.t.)	$14.23	$16.42	$15.11	$12.57	$18.84
5	Fringes:					
6	Health & welfare	$1.55	$1.32	$1.67	$1.25	$1.74
7	Pension	$2.10	$1.96	$1.75	$1.50	$2.25
8	Vacation	$1.27	$1.36	$1.43	$1.10	$1.78
9	Training fund	$0.30	$0.21	$0.25	$0.05	$0.40
10	Total Fringes	$5.22	$4.85	$5.10	$3.90	$6.17
11	WC, PL & PD insurance	$1.99	$2.30	$2.12	$1.76	$2.64
12	FICA (So. Sec. & Med)	$1.19	$1.36	$1.27	$1.05	$1.58
13	SDI	$0.19	$0.22	$0.21	$0.17	$0.26
14	Unemployment Ins.	$1.09	$1.24	$1.16	$0.96	$1.44
15	**Total Labor Rate**	**$23.91**	**$26.40**	**$24.96**	**$20.40**	**$30.93**
16						
17						
18				Return to Menu		

FIGURE 9-17
Labor cost calculations for craft workers.

Figure 9-18 is similar, with a separate column for each of a few typical equipment types. Costs are broken down into subcategories under ownership and operating costs.

The first two spreadsheets are fairly simple. They show the type of information that would be maintained by the payroll and equipment departments, respectively. For our purposes, they mainly serve as input to the resource cost summary that is used as a reference for estimating purposes. This summary, Fig. 9-19, pulls together unit costs for materials and hourly costs for labor and equipment. It also includes a code for each resource type. This code will be used later for linking back to this sheet.

Note particularly the cells in ranges D5..D11 and D22..D29. If we look at the formulas that produce the unit costs for the operator and the crane, we will see the following:

```
D10: -> ='Fig 9-17 labor'!$F$15
D25: -> ='Fig 9-18 equip'!$C$17
```

These expressions in effect refer to the labor (cell F15) and equipment (cell C17) spreadsheets, which in this case are titled, respectively, "Fig 9-17 labor" and "Fig 9-18 equip," and request the contents of the indicated cell numbers. Any update on those source documents will be reflected automatically on the resource summary sheet.

Figure 9-20 is the backup calculation for an individual bid item. It is in a fairly general format that could apply to a variety of types of work. The example

	A	B	C	D	E	F
1			Equipment Cost Calculations			
2						
3	**Cost category**	*Backhoe*	*Crane*	*Dozer*	*Dmp Truck*	*Cnc. Pump*
4	*Ownership costs:*					
5	Capital cost	$185,000	$420,000	$225,000	$87,000	$280,000
6	Salvage value (%)	5.00%	15.00%	10.00%	10.00%	5.00%
7	Useful life (hours)	9,000	14,000	12,000	15,000	16,000
8	Hourly depreciation	$19.53	$25.50	$16.88	$5.22	$16.63
9	Interest, ins, & tax	$3.70	$5.40	$3.38	$1.04	$3.15
10	Total hourly own cost:	$23.23	$30.90	$20.25	$6.26	$19.78
11	*Operating costs:*					
12	Fuel	$9.75	$5.20	$14.32	$4.35	$6.94
13	Oil & lube	$1.95	$1.04	$2.86	$0.87	$1.39
14	Tires or tracks	$6.42	$2.35	$8.50	$5.26	$1.89
15	Repair & maint.	$8.22	$6.00	$7.50	$1.16	$5.25
16	Total hourly op costs:	$26.34	$14.59	$33.18	$11.64	$15.47
17	**Total Equip Rate**	**$49.57**	**$45.49**	**$53.43**	**$17.90**	**$35.24**
18						
19			Return to Menu			
20						

FIGURE 9-18
Machine cost calculations.

	A	B	C	D
1	Resource Summary			
2		*Code*	*Unit*	*Unit Cost*
3				
4	*Craft Labor Rates*			
5	Carpenter	C	hour	$23.91
6	Cement mason	CM	hour	$26.40
7	Foreman	GF	hour	$32.65
8	Ironworker	IW	hour	$24.96
9	Laborer	LB	hour	$20.40
10	Operator	OE	hour	$30.93
11	Teamster	TM	hour	$21.76
12				
13	*Materials Prices*			
14	Concrete		cu. yd.	$72.00
15	Forms		sq. ft.	$1.19
16	Lumber		fbm	$0.65
17	Reinforcing steel		ton	$624.00
18	Steel piles		lin. ft.	$19.25
19	Structural steel		ton	$944.00
20				
21	*Equipment Rates*			
22	Backhoe	BKH	hour	$49.57
23	Concrete pump	CNP	hour	$35.24
24	Compressor	CPR	hour	$8.25
25	Crane	CRN	hour	$45.49
26	Dozer-D8	D8L	hour	$53.43
27	Dump truck	DTK	hour	$17.90
28	Flat bed truck	FTK	hour	$15.65
29	Front-end loader	LDR	hour	$43.45
30				
31		Return to Menu		
32				

FIGURE 9-19
Resource cost summary.

shown here is for Item 1, the Basement Excavation. The first section is the quantity take-off calculation for the excavation volume. It shows a bank volume of 14,402 cubic yards and applies a swell factor of 25 percent to convert this to 18,003 loose cubic yards to use for loading and hauling production calculations.

The second section is more interesting. Here we indicate the resources that will be used. However, rather than just explicitly referring to the sheet name and cell labels that contain the needed cost information, the spreadsheet uses a VLOOKUP function. Here, just by typing in the resource code, it automatically goes to the correct section in the resource cost summary sheet and looks down the items available to find one that matches the resource code requested. Examples of spreadsheet formulas that handle this process for the laborer and backhoe are as follows:

```
F16: -> =VLOOKUP(C16,'Fig 9-19 Res sum'!$B$5:$D$11,3)
F24: -> =VLOOKUP(C24,'Fig 9-19 Res sum'!$B$22:$D$29,3)
```

	A	B	C	D	E	F	G	H	I
1									
2		Bid Item Calculation Sheet							
3									
4		**Bid Item Number:**		1		Description:	Basement excavation		
5									
6		*Quantity Take-Off*		Qty	Unit				
7		Basement excavation		13847	bcy				
8		Access ramp excavation		556	bcy				
9		Subtotal		14402	bcy				
10		Swell Factor		25%					
11		Production volume		18003	lcy				
12									
13		*Resources*	Code	Qty		Unit Rate		Total	
14		*Labor*							
15		Foreman	GF	1		$32.65	/hr	$32.65	
16		Laborer	LB	2		$20.40	/hr	$40.81	
17		Operator	OE	1		$30.93	/hr	$30.93	
18		Teamster	TM	3		$21.76	/hr	$65.27	
19		Labor crew total:						$169.65	
20		*Materials*							
21		None						$0.00	
22									
23		*Equipment*							
24		Backhoe	BKH	1		$49.57	/hr	$49.57	
25		Dump truck	DTK	3		$17.90	/hr	$53.71	
26		Equipment total:						$103.28	
27									
28		*Production*		Number		Prod. Rate		Output	Unit
29		Backhoe		1		120		120	lcy/hr
30		Dump truck		3		50		150	lcy/hr
31		Max output (minimum of backhoe, trucks)						120	lcy/hr
32									
33		Time required:		18,003 / 120			=	150	hours
34									
35		*Costs*		Cost / hr.		Duration		Total	
36		Labor		$169.65	x	150	=	$25,452	
37		Materials		$0.00	x	150	=	$0	
38		Equipment		$103.28	x	150	=	$15,495	
39						Item Total:		$40,947	
40						Pay Quantity:		14,402 bcy	
41		Return to Menu				Item Unit Price:		$2.84 /bcy	
42									

FIGURE 9-20
Bid item calculation for excavation.

The remaining sections of the bid item calculation in Fig. 9-20 include the production and cost calculations. One could also have the spreadsheet look up production rates in a table as it did the resource costs, but most estimators would rather exercise their own judgment in this matter. Thus we take the productivity rates as data input. The time and cost calculations for this item then follow automatically.

Figure 9-21 is the final bid summary. It is similar in concept to the original spreadsheet we introduced in Fig. 9-1, but it includes the quantity of the bid

	A	B	C	D	E	F	G	H	I
1									
2				Bid Item Summary Sheet					
3									
4							*Direct*	*Bid*	*Bid Unit*
5	*Bid Item*	*Qty*	*Unit*	*Labor*	*Mat'ls*	*Equip't*	*Cost*	*Total*	*Price*
6	Basement excavation	14,402	bcy	$25,452	$0	$15,495	$40,947	$49,190	$3.42
7	Footing excavation	284	bcy	$646	$0	$397	$1,043	$1,253	$4.41
8	Backfill & compact	3,556	ccy	$1,024	$0	$1,382	$2,406	$2,890	$0.81
9	Concrete footings	190	cy	$3,793	$15,929	$2,086	$21,807	$26,198	$138.15
10	Foundation slab	24,924	sf	$10,468	$45,417	$4,865	$60,750	$72,980	$2.93
11	Basement walls	107	cy	$24,000	$28,800	$3,876	$56,676	$68,086	$638.31
12	*Totals*			$65,383	$90,146	$28,101	$183,629	$220,597	
13	*Overhead*						$14,690		
14	*Contingency*						$7,846		
15	*Profit*						$14,432		
16	*Total bid*						$220,597		
17									
18					*Distribution Factor:*		1.2013		
19									
20					Return to Menu				
21									

FIGURE 9-21
Bid item summary and bid price calculation.

item and two extra columns for the bid item total and unit price (shown in box) because this form is for a unit-price bid. The direct costs in cells D6..F11 pull their numbers from the individual bid item calculation sheets, but only that for basement excavation is linked in this particular example. The main user inputs on this sheet are the overhead costs and the percentages to use for contingency and profit. There could also be a detailed cost form to back up the overhead cost calculation entered in cell G13, making even that automatic on this sheet, but it is not included here.

The total cost computed in cell G16 is divided by the direct cost from cell G12 to produce the distribution ratio shown in cell G18. All of the direct total costs are multiplied by this ratio to produce the total item bid costs in column H, and these in turn are divided by the quantities to produce the unit costs shown in column I. There is a check sum in cell H12, but because the spreadsheet is quite reliable in its calculations, this should almost always equal the total we started with in cell G16. If it does not, then we know that some formula or routine in the spreadsheet has been damaged. We need to get it corrected before we continue. Selected formulas used for this summary and distribution appear below:

```
B6:   -> ='Fig 9-20 item calc'!$H$40
D6:   -> ='Fig 9-20 item calc'!$H$36
G14:  -> =0.12*D12
G18:  -> =G16/G12
H6:   -> =G6*G$18
I6:   -> =H6/B6
```

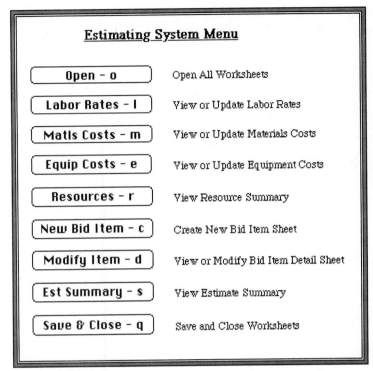

FIGURE 9-22
Control menu.

The overall coordination of the simplified bid system is handled by the menu in Fig. 9-22. Once the user *double-clicks* on its icon in a folder on the Macintosh, the menu and its backup macro sheet take over, display the menu shown in the figure, and lead the user through the various spreadsheets. Menu options can be selected either by clicking the buttons in the left column or by using equivalent keystroke combinations of the form "option-command-⟨letter⟩," where ⟨letter⟩ is the corresponding letter shown on the button. Once the user moves to a selected sheet, he or she can add resources to the labor and equipment cost tables or add items directly to the resource summary table. New blank worksheets can be created for each bid item from the summary sheet, and the summary sheet can be examined and modified. On each sheet that is called up, there is a button labeled **Return to Menu.** After the user has examined or modified the worksheet, clicking this button brings the menu in Fig. 9-22 back to the foreground of the screen.

The macros that support these menu choices are shown in Fig. 9-23. Like those for the Lotus 1-2-3 example in Fig. 9-14, they basically reproduce the commands that a user would give using the mouse and keyboard if performing the functions manually. But with a macro, just a single click on one of the buttons in Fig. 9-22 causes the recorded series of actions to take place. For example, selecting the **Open** button invokes the macro named **OpenEstSheets** in cells A1..A9.

	A	B	C
1	OpenEstSheets (o)		EstSummary (s)
2	=DIRECTORY("MacPaulson:Book:		=ACTIVATE("Fig 9-21 summary")
3	=OPEN("Fig 9-17 labor")		=SELECT("R18C1")
4	=OPEN("Fig 9-18 equip")		=RETURN()
5	=OPEN("Fig 9-19 Res sum")		
6	=OPEN("Fig 9-20 item calc")		BackToMenu (b)
7	=OPEN("Fig 9-21 summary")		=ACTIVATE("Fig 9-22 menu")
8	=ACTIVATE("Fig 9-22 menu")		=SELECT("R26C2")
9	=RETURN()		=RETURN()
10			
11	LaborRates (l)		SaveandClose (q)
12	=ACTIVATE("Fig 9-17 labor")		=ACTIVATE("Fig 9-21 summary")
13	=SELECT("R16C1")		=SAVE()
14	=RETURN()		=FILE.CLOSE()
15			=ACTIVATE("Fig 9-20 item calc")
16	MatlCosts (m)		=SAVE()
17	=RETURN()		=FILE.CLOSE()
18			=ACTIVATE("Fig 9-19 Res sum")
19	EquipCosts (e)		=SAVE()
20	=ACTIVATE("Fig 9-18 equip")		=FILE.CLOSE()
21	=SELECT("R18C1")		=ACTIVATE("Fig 9-18 equip")
22	=RETURN()		=SAVE()
23			=FILE.CLOSE()
24	ResSumry (r)		=ACTIVATE("Fig 9-17 labor")
25	=ACTIVATE("Fig 9-19 Res sum")		=SAVE()
26	=SELECT("R31C2")		=FILE.CLOSE()
27	=RETURN()		=RETURN()
28			
29	CreateItem (c)		
30	=RETURN()		
31			
32	ItemDetail (d)		
33	=ACTIVATE("Fig 9-20 item calc")		
34	=SELECT("R41C2")		
35	=RETURN()		

FIGURE 9-23
Macros for control menu.

Others activate selected windows so that the user can work on the selected worksheet. The macro in cells C11..C27 automatically saves and closes the worksheets at the end of a user session. Much more automation is possible with Excel macros, including internal manipulations like those shown for Lotus 1-2-3 in Fig. 9-14, but there is enough here to give an idea of the possibilities. Such macros enable a developer to use a spreadsheet environment to create standalone applications that require little actual spreadsheet expertise from their intended users.

Even after expanding upon the greatly reduced amount of data and other simplifying abstractions used in this example case, the fully developed system would still be a fairly straightforward application of spreadsheet technology. Users in construction have developed much more sophisticated systems in a wide variety of applications, including financial models for corporate performance and strategic planning, equipment cost systems, analysis and design computations for temporary structures, operations analysis and design, more complex estimating systems, and many others. But even the simple applications can be enormous time savers, and by cutting down on routine tedium they can make jobs much more interesting.

SUMMARY

Since their comparatively recent invention, microcomputer-based spreadsheet tools have become mainstays of business and engineering offices. In both capabilities and applications, they have evolved quickly from straightforward two-dimensional worksheets to include a wide variety of powerful enhancements.

The basic paradigm for an electronic spreadsheet is the familiar row-and-column-based worksheet format used by accountants, scientists, estimators, engineers, and many other professionals to organize, develop, and present their computations and procedures. However, electronic spreadsheets can make flexible use of references among cells and ranges on worksheets to replicate and substantially automate many of the tedious, complex, repetitive, and error-prone logical and arithmetic tasks that the user would otherwise have to perform. The computer screen can be maneuvered as a window roaming around a potentially vast worksheet surface, making very large models not only feasible but also convenient. With powerful functions accessible with a mouse click or keystroke, spreadsheets provide modeling and analytical capabilities that exceed the most advanced scientific calculators. Having built a model, the user can easily use "what if?" techniques to explore its limits, try alternative cases, and generally get a much better understanding of a problem. Spreadsheets also provide easy and attractive input and reporting capabilities. Perhaps most important, spreadsheet software is highly interactive and uses easily learned menu- and graphics-based techniques that enable almost any moderately educated person to put it to productive use with just a little bit of study or training.

Although even the simplest and most inexpensive spreadsheet packages have excellent capabilities that were almost inaccessible to the average computer user two decades ago, even more powerful extensions have been added and continue to evolve in this highly competitive marketplace. Those described in this chapter include further enhancements to their capacity, commands, formatting, and functions; protection and security; online, context-sensitive help information; flexible report design tools for screen displays and printed output; preparation and presentation of attractive and informative business graphics reports based on data contained in the spreadsheet; internal database definition and processing, as well as tools to access databases on remote computers; modeling concepts that extend to three dimensions or build links among multiple spreadsheets; programmable macro commands that can not only further automate routine tasks but also enable developers to employ spreadsheets as the means to create custom end-user applications that can be put to work productively by people who do not even understand the underlying spreadsheet technology itself. Such capabilities have also led to the development and sale of templates based on spreadsheet software that are predesigned to suit specific applications. Spreadsheet programs have also become one of several modules—together with word processors, file systems, communications, and graphics programs—that are combined in integrated package software wherein applications can be built that draw upon several modules concurrently, and where changes in one module automatically are updated wherever the results appear in other modules. Even standalone spreadsheet packages are now being designed to integrate easily with other popular types of programs.

As wonderfully versatile as they are, spreadsheets do have limitations in terms of their appropriate and inappropriate applications. In large measure, these limitations are defined by the boundaries where other types of programs can do the job better or more easily. Databases, simulation programs, word processors, expert systems, and procedural languages can all provide better development environments for solutions to problems within their domains. Software packages designed specifically for common applications such as scheduling, computer-aided design, and estimating can also provide more elegant solutions to problems that would be slow, awkward, or even impractical to solve with spreadsheets. As spreadsheet programs evolve, more and more of their limitations are removed, but one should always stop to think about what would be the best tool to use for a given situation.

This chapter presented two major examples. The optimum load-time computation for a scraper, implemented in Lotus 1-2-3 in an MS-DOS/PC environment, provided a solution to a modestly complex construction production problem. It also illustrated (1) the use of a few more spreadsheet capabilities, such as LOOKUP functions; (2) building of a custom menu with automated procedures using macro commands; and (3) graphic output.

The estimating example, built with Microsoft Excel running on an Apple Macintosh, showed how several worksheets can be linked together to form a more comprehensive application. Not only individual cells can be linked from one sheet to another but also functions such as LOOKUP can search whole ranges of data in one sheet to find the items needed in another. This example also used the Macintosh's graphics environment to show some more varied formatting in the spreadsheet designs. Finally, it illustrated a quite different application of programmed macros, combined with graphical button objects, to implement a menu-driven system to manage and control the access to the various worksheets in the system, as well as to open, save, and close the worksheets themselves automatically.

This chapter has just given an introduction to the power of computer-based spreadsheet packages, clearly a subject where the best way to learn is to practice. The reader is strongly encouraged to try the development problems near the end of this chapter using a suitable spreadsheet package. Further study in the manuals and texts suggested at the end of the chapter will also reinforce this learning experience. The knowledge and skills, once acquired, should enhance the capabilities of almost any engineering or management professional.

REVIEW QUESTIONS

1. Personal microcomputers and spreadsheets have evolved over about the same time period. What makes personal microcomputers such a suitable environment for spreadsheet applications? What, in turn, did the development of spreadsheet software do to increase the acceptance and utilization of personal microcomputers in business and engineering?

2. Explain how the ability to establish references among the cells of a worksheet gives the electronic version some major advantages over paper worksheets supported by calculators.

3. Explain how spreadsheet software enables the finite surface area of a computer CRT screen to cover a worksheet considerably larger than its own dimensions.

4. Advanced business and scientific calculators have many if not most of the same functions that are included with spreadsheet software. What advantages are there to the use of such functions in spreadsheets rather than calculators?

5. Briefly describe the relationship between the formula that produces the result that appears in a worksheet cell and the formatted appearance of the result in the cell itself. If the formula itself does not appear in the cell, how does one edit and make changes to it?

6. Describe the major categories of commands that usually are made available in the menu for a typical spreadsheet. What are two major alternative methods that spreadsheet program designers use to display the items on the command menu?

7. Describe the advantages of using spreadsheet software instead of manual calculation methods in situations where sensitivity testing and "what-if?" analysis are beneficial.

8. State three different ways that advanced spreadsheet programs make protection and security features available to spreadsheet application developers.

9. Why is *context-sensitive* help particularly useful in spreadsheet programs?

10. How does the ability to define forms for output reports differ from the ability to elegantly format the information contained within the rows and columns of the spreadsheet itself?

11. What is the typical scope of the graphics capabilities contained in an advanced spreadsheet program? How does this scope differ from that of a drawing program?

12. How does the internal database capability of spreadsheet programs differ from that of standalone relational database software?

13. What types of operations can one perform on information contained in the database section of a spreadsheet application?

14. Describe two different methods by which advanced spreadsheet packages allow application developers to move beyond the two-dimensional constraints of the basic programs. Compare the relative advantages and disadvantages of the two methods.

15. Programmable macro commands can greatly extend the capabilities of spreadsheet applications. Describe how the use of macro commands differs from the normal interactive use of spreadsheet software. Indicate two or more distinctly different types of application enhancements that can be implemented with macros.

16. Briefly describe the nature of integrated package software that includes spreadsheet capabilities among several application modules. Describe the relative advantages and disadvantages of integrated software compared to using separate programs to implement each different application type.

17. How is a custom template different from a fully implemented spreadsheet application?

18. Briefly state the limitations of spreadsheet programs in each of the following application areas compared to other types of software and suggest a more suitable alternative:

   • Word processing and report preparation
   • A database consisting of several large, interrelated files

- Simulation of the behavior of a complex earthmoving operation consisting of several interdependent production cycles
- Capturing the judgment of an experienced geotechnical engineer to provide advice on shoring trenches and open excavations in differing soil conditions
- Developing a CPM schedule for a construction project
- Preparing engineering construction drawings for a concrete structure

19. Briefly review and summarize the key concepts and techniques for the application of spreadsheet software that were illustrated by the optimum scraper load-time example.

20. Briefly review and summarize the key concepts and techniques for the application of spreadsheet software that were illustrated by the estimating system example.

DEVELOPMENT PROBLEMS

1. Cycle-Time Calculation

Chapter 8 provided an example of a vehicle's rimpull and cycle-time calculations implemented in the Pascal language. Refer to the explanation, criteria, and results achieved there, then reimplement the application using a suitable spreadsheet package. You may also wish to implement the extensions suggested in application development problem 2 at the end of Chapter 8. In addition, you can gain some experience with the graphical capabilities of your spreadsheet by plotting (*a*) a graph of the rimpull hyperbola and (*b*) a bar chart showing the time devoted to each of the four components of the cycle time. Use macros to automate some of the data entry and control procedures to make this into a custom application for a non-spreadsheet-oriented user.

2. Extend Scraper Load-Growth Curve Example

Near the end of the scraper example, we suggested some limitations to overcome and extensions to make that would enhance the capabilities of that model. Starting with that model, design and implement a set of extensions that fits your interests or meets any requirements that you have been given.

3. Extend Estimating Example

Starting with the estimating example presented in this chapter, design and implement the following extensions:

- Design a blank template that can be used for detailed bid item calculations of the type shown in Fig. 9-20.
- The **New Bid Item** button in Fig. 9-22 presently has a dummy macro that simply returns to the main menu when clicked with the mouse. Develop a new macro routine that, when the New Bid Item button is selected, will automatically duplicate the template from the first step, save the duplicate under a new name to be input by the user, and leave the blank template on the screen ready for use.
- Develop a macro routine that automatically inserts a new line item in the Bid Item Summary Sheet from Fig. 9-21 to correspond with the new detail sheet created in the previous step.

- Make a submenu and supporting macros that, when called from the **Modify Item** menu button of Fig. 9-22, will bring up a submenu or dialog box that will enable the user to choose which of the existing detail sheets to use.

4. Crane Selection

Common field engineering problems in construction relate to selecting a crane to handle a given load at a certain radius or, for a given crane, determining its safe operating limits (maximum load at a given radius, or vice versa). The objectives of this assignment are to illustrate the use of a computer to solve a practical construction problem and to enable you to become familiar with using generic microcomputer software in developing a preprogrammed application for users without computer skills.

BACKGROUND. The standard U.S. Power Crane and Shovel Association (PCSA) notation for crane ratings is summarized in Fig. 9-24. The **standard rating** (e.g., 40 tons) is measured using a 50-foot boom for its *rated* load in U.S. *tons* (2000 lb) at a *12-foot radius* from the *center of rotation* of the crane, and it is also expressed in terms of *hundredweight* (100 lb) at a *40-foot radius*. These loads are *tipping loads* and do **not** allow for a *factor of safety*. The center of rotation from which the radius is measured does not necessarily

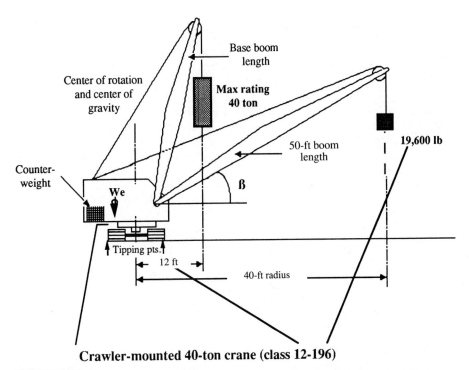

Crawler-mounted 40-ton crane (class 12-196)

FIGURE 9-24

Crane capacity rating and stability measurements (W_e = total equipment weight, β = boom angle). (Adapted from *Man the Builder*, Power Crane and Shovel Association (PCSA), Bureau of the Construction Industry Manufacturers Association (CIMA), *PCSA Technical Bulletin No. 1*), Milwaukee, Wisconsin, 1971, page 11, photo 2.)

coincide with the hinge point of the boom. The example shown in Fig. 9-24 is a class 12-196 crane, the meaning of which will best be understood by studying the figure itself.

If you make some simple moment calculations, you will notice that the live-load moment drops off as the radius increases from 12 feet to 40 feet (960,000 ft-lb drops to 784,000 ft-lb). The difference may in part be accounted for by the increasing moment from the weight of the boom, pulleys, cable, and hook as the radius increases. The effect will be greater for longer and heavier booms, jibs, and other special rigging.

In addition to this notation, the PCSA makes the following recommendations for *safety margins* in terms of the allowable safe loads as a percentage of the theoretical tipping load:

Crawler cranes: 75%

Truck-mounted (rubber-tired) cranes: 85%

There are many other details, such as the moment distance to the edge of the tracks, tires, or outriggers, the distance from the center of rotation to the hinge point of the boom, provisions for additional counterweight, and so forth, but for purposes of simplification in this assignment we will assume that all radius measurements and moment computations are taken from the center of rotation. Further assume that the center of gravity of the boom and rigging is at the midpoint of the boom.

SPREADSHEET REQUIREMENTS. Your spreadsheet should implement a well-designed application that will enable a user to answer some common crane rigging questions.

1. As a minimum, your spreadsheet should be able to do the following:

- Given a crane type (crawler or truck), load (in tons), radius (in feet), and lifting height (in feet), make some computations for capacity and geometric requirements and then go into a look-up table to select the minimum crane that will safely do the job. The table should contain the following machines:

Type	Rating, tons	Class*	Boom length, feet
Crawler	40	12-196	50
Crawler	75	12-340	100
Crawler	100	12-440	120
Crawler	150	12-600	180
Truck-mounted	50	12-250	60
Truck-mounted	120	12-520	150
Truck-mounted	200	12-800	200

*The second figure in the class notation given here already allows for the greater moment of larger booms where applicable.

- For a crane chosen from the preceding table and for a given specific load in U.S. tons, compute the maximum safe boom radius at which the crane can operate.
- For a crane chosen from the preceding table and for a given specific radius in feet, compute the maximum safe load that the crane can hoist.

- For a given crane, use the spreadsheet's graphics capability to plot an approximate curve of the maximum safe load (y axis) versus maximum radius (x axis) to form an upper bound within which the crane can safely operate. Use interpolation and extrapolation to allow for the boom moment, starting with the 12-foot and 40-foot ratings. Plot your points at 10-foot intervals, starting from 10 feet and extending out to a value large enough to include the full length of the boom in a horizontal position. The plot should be labeled and should be informative enough to be posted in the cab of the crane being considered.

2. The spreadsheet should have a structure that is intuitively easy to follow, with separate sections performing at least the following functions:

- Provide at least three different types of input error and range checking for the data from the user (e.g., specified radius within boom length, specified load within capacity of given crane, etc.). Prompt the user for corrections, where entry is erroneous or out of range, at the time of data item entry.
- Prepare and offer the option of printing on the CRT screen or on the printer a report summarizing the input data.
- Prepare and offer the option of printing on the CRT screen or on the printer an output report, containing at least the following information:

  - Title
  - The input data and answer requested for each of the types of inquiries given in part 1 above.

- Programmed macros should automatically drive the spreadsheet and guide the user through menu selections, data entry, and report selection, as appropriate to your design. In other words, your application should appear to the end user like a package program, with menu selections, automatic cursor movement to prompted data entry fields, protection of formula and label fields, and so on. The user should not have to know much if anything about how to use the spreadsheet program itself in order to use your application. Lotus 1-2-3 and Excel both have this type of programming capability, and it would be good experience to learn to use it.
- The program should include at least two examples of how a user can request optional help information pertinent to the context where such help is needed.

You may develop the program in the spreadsheet program of your choice on either the Apple Macintosh or the IBM PC or equivalents.

TEST CASES. In addition to input test data that you may wish to make up yourself, use the following test cases for your submission:

Case 1: Pick a truck-mounted crane to set a 20-ton load on top of an 80-foot building at a radius of 60 feet from the crane.

Case 2: Pick a crawler crane to set an 18-ton load on a bridge deck 50 feet high at a radius of 40 feet from the crane.

Case 3: Pick a crane to set a 40-ton load on a 10-foot-high pier located 60 feet out from the crane on the shore.

Case 4: For the 120-ton truck-mounted crane, what is the maximum radius at which it can safely handle a 30-ton load?

Case 5: For the 150-ton crawler crane, what is the maximum load that it can safely lift at a radius of 80 feet?

Case 6: Plot the diagram of maximum safe load versus radius for the 100-ton crawler crane.

SUGGESTIONS FOR FURTHER READING

A quick trip to any good bookstore reveals an abundance of books on major applications like spreadsheets. We offer a few suggestions for the three packages that are currently most popular, but we do so with the caveat that these selections will be the most rapidly outdated of those in any chapter in this book. They include a mixture of manuals and self-teaching books, but you might also examine their many competitors in the bookstore before choosing the ones that best suit your needs and interests.

Campbell, Mary. *Quattro Pro for Windows Handbook*. New York: Bantam Books, 1992.

Campbell, Mary. *1-2-3 Release 3.4*. New York: McGraw-Hill, 1993.

Cobb, Douglas, and Judy Mynhier, with Mark Dodge, Craig Stinson, and Chris Kinata. *Microsoft Excel 4 Companion,* Macintosh version. Redmond, Wash.: Microsoft Press, 1992.

Fielding, Justin. *The Lotus Guide to 1-2-3 for Windows*. New York: Brady, 1991.

Lotus 1-2-3 for DOS User's Guide, Release 2.4. Cambridge, Mass.: Lotus Development Corporation, 1992.

Microsoft Excel User's Guide, Version 4.0. Redmond, Wash.: Microsoft Press, 1992.

Quattro Pro for Windows User's Guide. Scotts Valley, Calif.: Borland International, 1992.

10

DEVELOPING APPLICATIONS WITH FILE AND DATABASE SOFTWARE

A man should keep his little brain attic stocked with all the furniture that he is likely to use, and the rest he can put away in the lumber room of his library, where he can get it if he wants it.

Sir Arthur Conan Doyle*

Construction professionals sometimes complain that they are required to build their projects out of paper. In effect, they are overwhelmed by the documentation and reporting requirements that seem to grow more demanding each year.

File-management and database systems will not reduce the requirements, but they can make coping with the resulting burden far easier. Applications of file-management and database systems can improve the management of materials and equipment, maintain records of correspondence and support papers that can later be retrieved to document a claim, assist in selecting workers and supervisors with the skills needed to make up effective work crews, and help in many other ways.

The history of file-management and database software for microcomputers has mostly been one of adapting existing concepts and systems from mainframe- and minicomputer-based software. Although there has not been a single conceptual breakthrough comparable to the one that launched spreadsheets on microcomputers, the personal computer environment certainly has stimulated developers to add capabilities that have made these systems more accessible to everyday users and has led to significant enhancements, particularly in graphical user interfaces.

*"Five Orange Pips" in *The Complete Works of Sherlock Holmes*, 1927.

This chapter introduces concepts and techniques of file and database systems commonly used in microcomputer applications. It builds upon the background concepts that were introduced in Chapter 4. Illustrative examples present (1) a simple file-management system for an application using a single file and (2) a relational database for an information system drawing on four interrelated application areas.

BASIC CONCEPTS OF FILE AND DATABASE SOFTWARE

Chapter 4 introduced fundamental concepts such as *data elements* being allocated *fields* and structured into *records* in a *file*, the data types that can be defined for fields, and the use of *keys* and *attributes* for searching, sorting and reporting. It also described *sequential* versus *direct-access* files. Finally, it used the term *database* to describe a more complex structure for information organization that can draw upon the equivalent of multiple files for very flexible sorting, searching, and reporting. Diagrammatic techniques for expressing different types of relationships among data types were also described (one-to-one, one-to-many, many-to-one, many-to-many). That chapter mentioned *network* and *hierarchical* as well as *relational* databases, but our focus here will be limited to single-file systems and relational databases, which are the two most commonly used with microcomputers.

File-Management Systems

Basic file-management systems are sometimes called "flat files" because they have a single two-dimensional structure of fields making up a record in one dimension, which we might think of as columns, and the number of records in another, which can be viewed as rows. Of course, one can use file-management software to create many independent files of this type, each with different layouts, but there may not be direct linkages among them. In Fig. 4-3 was a conceptual layout for a flat file. In geometry, this concept is similar to a two-dimensional spreadsheet, but there are major differences. First, the records in a file have the same structure from row to row, but, as we saw in Chapter 9, the design and type of contents in the individual rows can vary significantly with spreadsheets. Second, almost all file systems work from disk storage, holding only a fraction of the records in memory at any one time. In contrast, a spreadsheet normally reads the whole model and all of its data into RAM memory and accesses the disk mainly just to store updated copies of the model as it changes.

File system utilization consists of (1) design of the records, data-entry screens, output reports, and data management options; (2) data entry; (3) searching, selecting, and sorting information; and (4) preparing reports. There can also be options for exporting and importing data to and from other applications.

Examples of microcomputer-based file-management programs include Microsoft's File, Claris's FileMaker Pro, Software Publishing Corporation's Professional File, and Buttonware's PC-File. There are many others. Not one of these became so dominant as Lotus 1-2-3 did in the spreadsheet arena, and their market remains fairly competitive. Flat-file software is also commonly built into integrated applications such as Microsoft's Works, Spinnaker's PFS:WindowWorks, Lotus's Symphony, Ashton-Tate's Framework, Claris's ClarisWorks, and Symantec's GreatWorks.

One of the main virtues of file-management systems should be their simplicity for basic applications. However, the advantage of simplicity may be diminished to the extent that ongoing competition among developers adds features and options that are rarely needed and mainly add complexity, unless, perhaps, the vendor offers a dual level of capability. Normally, if the demands of an application are too complex for a simple file-management system, one should consider moving up to a database-management system instead of forcing the desired application into a more complex file-management system that may have outgrown its initial purpose.

Database-Management Systems

Relational database-management systems provide the capability to build integrated applications that draw upon the equivalent of multiple files. Individual files (usually called *tables* or *relations* in database systems) can focus on the primary information needs and responsibilities of certain individuals and groups. But relational databases also cut down on duplicate data entry of items that appear in two or more tables, since an item entered by an individual who is responsible for one area can then be referenced in the tables for other areas and be accessed by users in other groups. Relational database software also makes it relatively easy to make inquiries and prepare reports that draw upon multiple tables to extract desired information and put it into new and even custom formats that can facilitate analysis and understanding of relationships among different types of data.

In Fig. 4-7c is a conceptual outline of a relational database consisting of three different tables. Note that each table contains at least one key attribute in common with one or both of the others. These attributes establish the linkages between the tables that enable their integration for data entry and reporting. As an example of data entry, if the user responsible for the equipment roster table put in a new record or updated an existing one in that table, the code field would automatically be reflected in any inquiries or reports based on the maintenance table.

A later section in this chapter provides more detailed information about the way in which such a database can be set up and used in practice. The relational database example near the end of this chapter further illustrates how this type of software can help us in construction.

Examples of relational databases that have been popular on microcomputers include MicroRim's R:BASE, Borland International's Paradox and dBASE IV, and Microsoft-Fox Software's FoxPro, all of which were developed originally for IBM PCs and compatibles. Macintosh relational databases include Acius's 4th Dimension, Blyth Software's Omnis Seven, Provue's Panorama, and Microsoft-Fox

Software's FoxBASE+/Mac. Versions of the Oracle database system are also available on microcomputers and are compatible for integration with implementations of Oracle that run on mainframes and workstations. Although these various packages are similar in that they provide the means to implement relational database applications, they differ in their allowable specifications for data elements and table dimensions, menu functions, speed and flexibility for searching and sorting, report design and processing facilities, graphical output, computational capabilities, user interfaces and ease of use, programmability for building standalone applications, network integration, and in other ways. Some of these features will be discussed later.

WORKING WITH FILE-MANAGEMENT SOFTWARE

This section introduces some of the general capabilities provided for working with file-management software. It uses Claris's FileMaker Pro and a Macintosh computer to illustrate a few procedures used for implementing a simple file application.

There are four main stages to working with a file-management system: file design, data entry, file processing, and reporting. In file design, one normally calls upon some easy-to-use online routines to define the sizes and types of fields in the record structure, to design one or more data-entry forms for the user to employ when inputting or changing data, and to design the reports either to display on the screen or to print. There usually are built-in facilities for searching, sorting, selecting, and reporting information within the user-defined formats.

As an example, Table 10-1 contains headings and a few sample records for a file that lists power tools kept in a tool room. In addition to describing the tool and giving its inventory number, it shows its purchase date, cost, condition, present status, who checked it out and when, and when it is due back. The file also contains a picture of each tool, but this is not shown in this tabular layout.

Figure 10-1 is a picture of a sample screen taken while the developer of this application is describing the structure of a typical record. It uses FileMaker Pro's **standard layout**, which puts labels and data-entry fields in two parallel columns and has header and footer parts above and below the main body of the record. The drawing tools are illustrated in the panel along the left side of the figure, and FileMaker's implementation of the standard Macintosh pull-down menus are shown along the top. Other parts of this screen will be explained later.

Figure 10-2 is a picture of a screen taken while the developer is designing the data-entry and editing form that the tool-room clerk will see while looking for available tools and checking them out to specific workers. In this case the developer has shifted to FileMaker's **Browse** view and is using some test data to see how it will look to the users. This screen will also double as a report form in which the data are displayed if the user requests information about a single record. The small flip chart icon near the upper left corner enables the user to flip through the records one by one (by clicking the upper and lower pages to move backward

TABLE 10-1
Information for tool file

Inv. code	Tool type	Purchase date	Cost	Condition	Status	Who has it?	Date out	Date due
EW-02	Electric Welder	3/1/93	$1,650	Good	Available			
EW-04	Electric Welder	5/27/93	$1,714	New	Checked out	Jim Kelley	5/28/93	6/5/93
GR-07	Grinder	3/1/93	$100	Poor	In repair	Shop	5/7/93	6/13/93
GR-12	Grinder	4/1/93	$105	Good	Checked out	Sarah Michaels	5/20/93	6/8/93
GR-15	Grinder	4/22/93	$110	Good	Checked out	Rex Shea	5/12/93	6/25/93
GR-22	Grinder	5/28/93	$115	New	Available			
PB-05	Paving Breaker	1/20/93	$624	Fair	Available			
PB-08	Paving Breaker	4/21/93	$702	Good	Checked out	Joe Walsh	5/20/93	6/8/93
PD-10	Power Drill	12/23/92	$42	Poor	Broken			
PD-12	Power Drill	1/8/93	$47	Fair	Checked out	Mary Smith	5/18/93	6/8/93
PD-14	Power Drill	4/21/93	$52	Good	Checked out	Bob Nesbitt	5/20/93	6/11/93
PD-17	Power Drill	5/20/93	$55	New	Available			
PS-18	Power Saw	1/18/93	$72	Poor	In repair	Shop	5/24/93	6/10/93
PS-21	Power Saw	2/17/93	$60	Fair	Available			
PS-23	Power Saw	3/15/93	$65	Good	Checked out	Frank Jones	5/27/93	6/16/93
PS-27	Power Saw	5/13/93	$74	New	Checked out	Janet Thomas	5/17/93	6/3/93

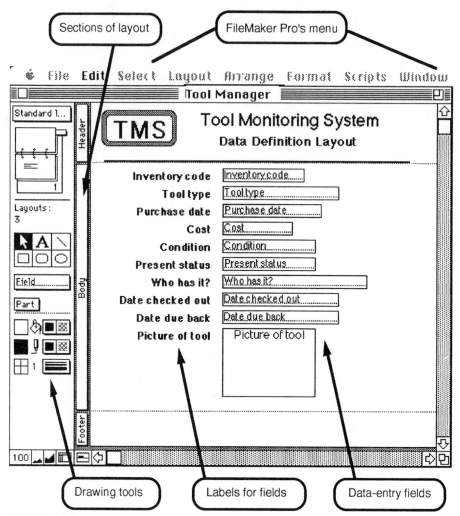

FIGURE 10-1
Standard layout of a typical record.

or forward); sliding the tab on the icon's right edge moves directly to various records in the file. (These operations can also be done with key combinations.)

Figure 10-3 is a picture of a screen taken when the developer is designing a tabular report for printing a set of records that may be requested later by a user. It has extra sections for a first-page title and a header for successive pages. There are also subsummary and summary fields, where the developer has placed, respectively, fields for subtotaling costs of sorted groups of tool types, and for the grand total cost for all the tools. Reports based on inquiries that use this report format are presented later in this chapter.

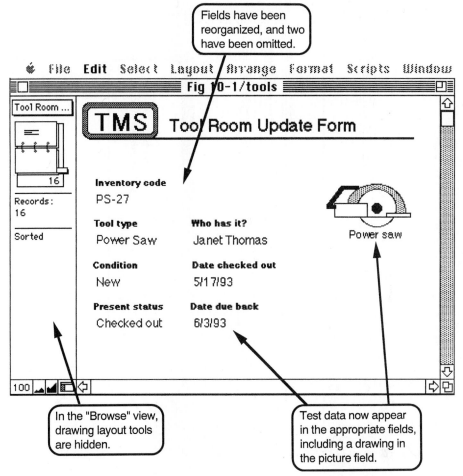

FIGURE 10-2
Data-entry screen design in progress.

In doing these design tasks, one can usually position and enter the text for titles, headers, and explanatory material simply by moving the cursor with a mouse or arrow keys to the place where an item should go and then typing in the text. Fields for data entry usually appear as rectangles (whose outlines may later be hidden) that can similarly be moved to the desired position. Once entered, the developer can still move these items about, just as one would do with objects in a drawing program. Most file-management software also allows a full array of fonts, sizes, and styles for text, plus different lines, boxes, and shading options that can be used for some very professional-looking report and form designs. Once the developer is satisfied, the file-application design can then be saved and also be protected (so that a user does not inadvertently spoil it).

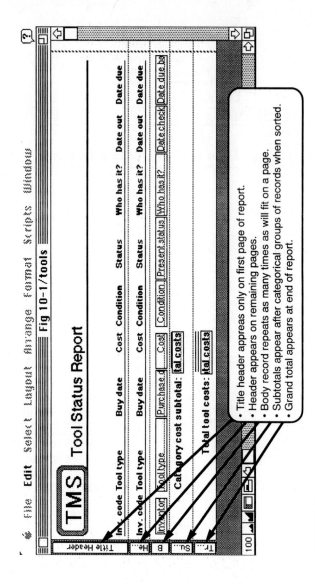

FIGURE 10-3
Report design in progress.

Once the application is complete, work using it is usually handled via keyboard commands or menu options, much like those described in Chapter 9 for spreadsheets. The options typically include the following:

- General file manipulation (start new file, open old file, save file, close file, etc.)
- Data entry (including the option to create new records, duplicate and modify an existing record, and retain like fields in order to expedite the typing of a new record)
- Editing (cut, copy, paste, insert, delete, etc.) both within record fields and for groups of records
- Searching or finding (browsing through the records in a file, locating specific information, etc.)
- Data manipulation (e.g., sorting records according to user-specified keys, selecting a record or records that match certain criteria, etc.)
- Reporting (display or print all records or those that meet specified criteria, in a sorted order or in categories requested by the user)
- Importing information from and exporting it to other application programs
- Macros (or *scripts*), as well as programmable buttons, to enable the developer to create menus and automate some functions, in a manner similar to that described in Figs. 9-22 and 9-23 for spreadsheets

With keyboard or mouse menu commands, the user is ready to take over from the developer and put the file-management application to work. Illustrations of such user actions will be given in context as part of the cases to be presented later in this chapter.

WORKING WITH DATABASE-MANAGEMENT SOFTWARE

Designing the records, individual files or tables, input screens, and output reports using modern relational database-management software is much like the process just described for file-management programs. But there is more of it—several different record layouts instead of just one, additional data-entry screens, more reports, extensive programmability, and so on. Furthermore, one must establish the key relationships between the tables in the database, and possibly predefine some macro routines for making everyday queries and extracting standard reports that draw from several different tables in the system.

In working with a database, the user has options similar to those listed for file systems. However, because there are multiple tables and several ways that one can combine data coming from several of them, there are much more powerful—and complex—search, selection, and reporting options. As defined in Microrim's R:BASE system, specific logical operations that typically can be performed either as the end result or as intermediate steps in responding to a query or reporting request include the following:

Append: This operation adds rows from a source table to a target table. Blank fields are inserted where necessary in the new records of the target table where such attributes are not included in the records of the source table.

Intersect: This operation forms a new table from two existing tables but only uses the rows that have matching attributes in common.

Join: Two existing tables are combined in this operation to form a new table, but the selection of new records is subject to user-defined logical relationships between the fields in specified columns.

Project: This operation forms a new table that is based on all or a subset of columns from a single existing table. One can specify further criteria to limit the selection of rows to be copied to the new table.

Subtract: This operation creates a new table from two existing tables but selects only the rows with specified attribute fields **not** in common between the two source tables.

Union: When two tables have at least one column in common, this operation combines all of the columns and all of the rows from two existing tables to form a new table. Duplicate entries are inserted in fields where necessary.

The second case example near the end of this chapter illustrates how several of these logical operations can be applied in practice.

Database Design

Relational database design involves setting up the data types and record layouts for one or more tables, establishing relationships among the separate tables, designing forms for data entry and inquiries, and designing reports that can be printed, viewed on the screen, or sent to a file. This section briefly shows how Borland's relational database program called Paradox appears when some of these design functions are underway.

The main menu for Paradox is shown in Fig. 10-4. The **View** mode allows the user to examine data in various tables, much like the **Browse** mode in FileMaker. **Ask** is the inquiry mode, and it enables the user to select information from single tables or to use the equivalent of the logical relationships just described to obtain combinations of information from multiple tables. **Report** and **Forms** access the design capabilities for these two types of entities. **Create** and **Modify** are where new tables and fields are defined or modified. New records are also entered under

Paradox main menu

Two tables shown in View mode

	View	Ask	Report	Create	Modify	Image	Forms	Tools	Scripts	Help	Exit

View a table.

SPECS	MODEL	DESCRIP	POWER	WEIGHT	LOAD
1	Cat 12G	Grader	101	13313	0
2	Cat 14G	Grader	149	18530	0
3	Cat 621E	Scraper	246	30479	21770
4	Cat 627E	Scraper	414	34670	21770
5	Cat 631E	Scraper	336	44000	34020
6	Cat 637E	Scraper	523	50843	34020
7	Cat 769C	Truck	336	30675	36300
8	Cat 773B	Truck	485	38660	52600
9	Cat 825C	Compactor	235	32400	0
10	Cat 966E	Loader	161	20324	5440
11	Cat 988B	Loader	280	42410	9600
12	Cat D10N	Tractor	522	79020	0
13	Cat D8N	Tractor	212	31383	0
14	Cat D9N	Tractor	276	42542	0

ROSTER	CODE	MODEL	BUYDATE	COST	JOB
1	CP-032	Cat 825C	8/04/93	315,000.00	CA-011
2	Gr-017	Cat 14G	11/08/91	212,000.00	CA-024
3	Gr-024	Cat 12G	5/21/92	155,000.00	CA-011
4	LD-021	Cat 966E	8/04/93	236,000.00	CA-024
5	LD-045	Cat 988B	7/08/92	389,000.00	CA-011

FIGURE 10-4
Paradox menu and **View** mode.

Modify. The **Image** mode allows the user to select alternative ways to display information, including graphical formats similar to those of spreadsheets. **Tools** provides a submenu of utility functions, such as directory changes; access to DOS; copying, renaming, or deleting tables, forms and reports; exporting and importing data to and from other applications; and so forth. With **Scripts** one plays or records a script. **Help** provides context-sensitive access to information pertinent to the task that is now being performed.

Figure 10-4 also contains two tables that have been presented on the screen in the **View** mode. The user can switch between these tables, open others, or use the arrow keys and others on the keyboard to move about within a particular table. The information contained in these particular tables is from the example of a relational database application to be discussed later in this chapter.

Figure 10-5 is the screen that appears after the developer selects **Create** from the main menu and enters the name of a new table (in this case, Costs). The field numbers automatically appear one by one in the left column, and the developer types in the name (e.g., HOURS) and specifies the corresponding data type (and field width in the case of the alphanumeric data type).

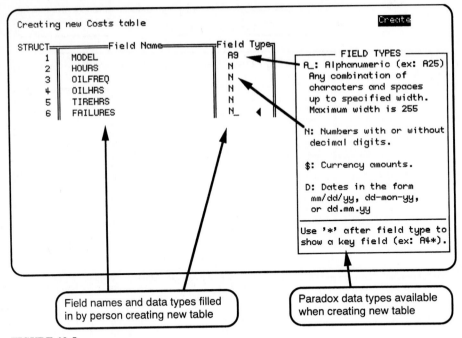

FIGURE 10-5
Paradox **Create** screen in use.

Figure 10-6 is a picture of a screen that appears when one is defining a form. It also shows the main menu of form-design options that are available when working in Paradox at this stage. This particular screen is a **Master Form** derived from a table that contributes the fields whose labels appear in the upper left quadrant. The shaded areas reserve space for fields drawn from three other tables that have been **Linked** to this one. Their field layouts are designed separately and do not appear in this master form until the form is actually used in the **View** mode or it is used for data entry. The appearance of this particular form when in use is shown in Fig. 10-14, which is part of the second application example later in this chapter.

Figure 10-7 is a picture of a report design screen as it is being used to create a report that draws fields from two different tables. At the instant this screen shot was taken, the developer was correcting the formula expression for the calculated field called Hours To Go. Note that the report has major sections similar to those used in FileMaker, including the overall report heading, a page heading and table column headings that appear on each page, and the layout of the actual data fields to be included in each record. Although the number, currency, and date types of data have default widths and formats when created, the developer has much more precise control over their layouts and appearance when designing a report.

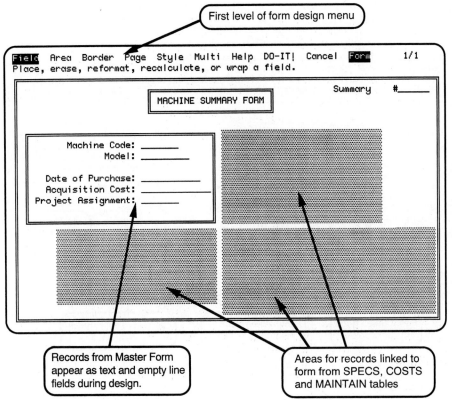

FIGURE 10-6
Paradox **Form** design screen in use.

In addition to this tabular type of report, Paradox also allows free-form report design so that one can create documents such as mailing labels, invoices, and so forth. As with FileMaker, it also allows calculations and summaries within groups of related records (e.g., for all the machines on a particular project) and comprehensive summaries for the whole report.

Figure 10-8 is an illustration of the tabular report that is printed using the design from Fig. 10-7. It can be accessed either while designing a report—a convenience for checking interim results—or from the **Report** option on the Paradox main menu. Although the design form itself was cluttered with the information and section lines needed to set up the report, the finished product can appear as neat and well organized as the developer can make it.

Not shown here is an example of the type of graphical reporting that is available in the Paradox system. Basically, Paradox has flexible capabilities to make a variety of bar charts, pie charts, line graphs, and other forms similar to

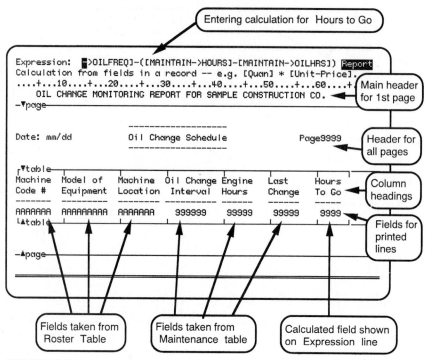

Entering calculation for Hours to Go

```
Expression: ■>OILFREQ]-([MAINTAIN->HOURS]-[MAINTAIN->OILHRS]) Report
Calculation from fields in a record -- e.g. [Quan] * [Unit-Price].
....+...10....+...20....+...30....+...40....+...50....+...60....+
    OIL CHANGE MONITORING REPORT FOR SAMPLE CONSTRUCTION CO.
─▼page
```
Main header for 1st page

```
Date: mm/dd          Oil Change Schedule              Page9999
```
Header for all pages

```
┌▼table┐
Machine  Model of   Machine  Oil Change Engine   Last    Hours
Code #   Equipment  Location  Interval  Hours    Change  To Go

AAAAAAA  AAAAAAAAA  AAAAAAA   999999    99999    99999   9999
└▲table┘

─▲page─
```
Column headings

Fields for printed lines

Fields taken from Roster Table

Fields taken from Maintenance table

Calculated field shown on Expression line

FIGURE 10-7
Paradox **Report** design screen in use.

those found in spreadsheets. Furthermore, we have not yet taken advantage of the excellent script programming capabilities available in Paradox, although they will be included in the application example later in this chapter. The next section gives an idea of the strong programming capability that is available.

Script Programming

Paradox offers three levels of programmable macros, which it calls scripts. The simplest is called **Instant Scripts**, which is a series of keystrokes recorded and temporarily retained in memory. This is similar to the macro recording function-ality that Chapter 9 described for Microsoft Excel. The user can automatically record some relatively complex actions, save the script, edit it, and combine it with others to substantially automate a database application. For example, if you are about to undertake a repetitive series of nearly identical operations—perhaps altering the same characters in a given field on a whole series of records—you could record the keystroke sequence as you change the field in the first record, then simply replay the script for all of the other records in the series. This level of scripting is easy and is useful to almost any type of Paradox user. However,

```
OIL CHANGE MONITORING REPORT FOR SAMPLE CONSTRUCTION CO.

                              --------------------
Date:   4/19                  Oil Change Schedule           Page   1
                              --------------------

Machine  Model of    Machine  Oil Change Engine   Last     Hours
Code #   Equipment   Location  Interval  Hours    Change   To Go
-------  ----------  --------  ---------- ------   ------   -----
CP-032   Cat 825C    CA-011       250     1950     1800      100
Gr-017   Cat 14G     CA-024
Gr-024   Cat 12G     CA-011
LD-021   Cat 966E    CA-024       200     2335     2200       65
LD-045   Cat 988B    CA-011       200     3270     3200      130
SC-014   Cat 637E    CA-011       300     4340     4230      190
SC-021   Cat 621E    CA-011       250     7200     7000       50
SC-022   Cat 621E    Yard         250     6280     6400      370
SC-032   Cat 631E    CA-026       300     7500     7230       30
SC-045   Cat 627E    CA-022       250      560      500      190
SC-046   Cat 627E    CA-024       250      780      600       70
TK-021   Cat 773B    CA-024       400     6200     6000      200
TK-033   Cat 773B    CA-011       400     2860     2700      240
TK-042   Cat 773B    Ca-022       400     1230      900       70
TK-064   Cat 769C    Yard         400     4500     4150       50
TK-076   Cat 769C    CA-011       400     2340     2000       60
TK-077   Cat 769C    CA-011       400     4620     2200     ****
```

FIGURE 10-8
Paradox **Report** output based on design in Fig. 10-7.

unless the user saves the script under a name other than the default name **Instant,** it will be overwritten the next time the user creates a different instant script.

At the next level up, Paradox offers a scripting language that it calls the Personal Programmer. With it, one can create menu commands, guide and process inputs from screen forms, automate complex queries, structure and produce reports, and so on. Yet, this is still a fairly easy-to-use development feature that people who do not have prior experience with procedural languages can still apply effectively. One can also create a start-up title screen and help screens. This level of programmability is also typical of several other relational database packages.

In Paradox, once you have defined a menu, created (or incorporated existing) commands, forms, and reports, and incorporated any predefined scripts, the Personal Programmer automatically generates the comprehensive scripts that control the application. Thus the Personal Programmer is like the program generators that were described briefly in Chapter 4. Furthermore, the Paradox Personal Programmer can automatically produce key components of the documentation needed to describe a particular application.

Paradox also offers a very powerful level of structured programmability that it calls the Paradox Application Language (PAL). At first glance, scripts written in this language look a lot like programs written in the Pascal, Modula-2, or Ada procedural languages. This fact is perhaps not so surprising given that Borland, the developer of Paradox, is also the developer of some of the most widely used development systems for Pascal and C programming. Indeed, PAL includes an editor and debugger similar to Borland's Turbo Pascal and Turbo C, it draws on an application-development "engine" consisting of C library functions and Pascal library functions and procedures, it provides links to *SQL* databases, and it offers commands, operators, and a library of functions comparable to full-scale procedural languages. In summary, Paradox combines the flexibility and ease of use of relational database systems with a full procedural language that is especially focused on database applications to make a very powerful and versatile development environment for information systems. Needless to say, at this level of programmability, it helps greatly if the developer has prior knowledge and experience with procedural language programming.

Paradox is suitable for users ranging from those with a modest level of computer literacy on up to professional application developers. It has thus become one of the most popular relational database packages on the market and is suitable for creating a wide range of construction applications.

ADVANCED FILE AND DATABASE CONCEPTS AND FEATURES

As with spreadsheets and other types of application-development software, there are numerous extensions and advanced concepts in file-management and database software that may either be incorporated into or be available as options for specific packages. Here we will briefly mention some of the common ones: enhancements to standard functions, protection and security, online help, arithmetic computations, graphical reporting, programmable macro commands, integrated software, and network data access.

Enhancements to Standard Functions

There is no clear boundary dividing standard from enhanced functions in either file-management or relational database software. Indeed, a relational database might be viewed as a major conceptual enhancement to a standard file system, and there can be a nearly continuous range of capabilities in between them. However, over the years certain areas within the basic capabilities have been enhanced without significant conceptual advances in the nature of the software. Examples of such enhancements, now commonly included even in low-cost packages, are facilitated sorting and searching (e.g., *query-by-example* templates), new editing capabilities, more options in the menus, enhanced report formatting (e.g., more text sizes, fonts, and styles; more flexible layout design), improved documentation and training materials, graphical data types, higher speed and capacity, and so on. Some of these enhancements might be considered by some people to be frills, but they will be useful for others. As with spreadsheets, the majority of users will be well served by the simpler and less expensive packages, and for them the quality, effectiveness, and ease with which the basic functionality is implemented is much more important than extra features.

Protection and Security

As with spreadsheets, one category of protection options can enable a developer to *protect the basic design* and functionality of a file or database application against inadvertent modification or damage by the user. Even in the most basic file-management systems, the menu options used to create or modify screen and report layouts, data structures, and the like should be separate from the options needed when using the systems for data entry, sorting, searching, reporting, and such. This *menu separation* itself provides some degree of protection. However, some packages allow the developer to hide development-oriented menu functions from the user and indeed to customize the menus to display only those options needed by the user for a given application. Other features might enable the developer to protect the layout of screens, reports, and record structures so that they can be modified only by a person who knows a *developer **password*** or has other means to gain access to the developer's tools. Finally, ***compilers*** are available for some popular database packages that totally separate the executable application code from the tools used to create it.

The other type of protection is *security for the data itself*. This is normally handled by *user passwords*. Almost all file and database packages provide the option for at least one level of password protection (i.e., a password is needed to allow the user to open the file or database in the first place), but more sophisticated schemes are available to provide *hierarchical password levels* to separate users into those who can have access only to look-up data and those who are authorized to add to, delete, or modify data in the system. Some protection schemes even allow one to specify what data fields and records certain users are allowed to see or modify. In relational database systems, the software might further enable the developer to specify different types of access (exclude, read only, modify, delete) for different tables within the system. This flexibility in refining password access enables a developer to set up procedures and responsibilities and to automatically monitor their use by different individuals or groups in a firm so that the information system will more closely match the human organization it serves.

Online Help

As with spreadsheet systems, online, context-sensitive help for learning and reference is available in the better file-management and database packages. It works much like that described in the corresponding section of Chapter 9. As well as providing single-keystroke access to online tutorial or reference material, some database packages provide a query-by-example facility. This guides the novice through the process of structuring complex queries or requesting reports requiring the user to describe the various logical relationships (e.g., information since a certain date) and specify the multiple steps of processing (e.g., unions, joins, etc.) to yield the desired information. In query-by-example mode, the system might provide screen forms containing questions, multiple choice menus or fill-in blanks to guide the user in structuring his or her request. An example will be given in the database case later in this chapter.

Arithmetic Computations

Although file and database systems usually do not try to approach the mathematical functionality of spreadsheets, it is handy at times to be able to perform some basic computations. Most commonly, one wishes to add up a column of numbers to put a total or subtotal at the bottom of a table or report. Another useful function is to count the total number of entries in a column, or just those that match a certain stated criteron (e.g., the number of invoices from vendor X). These two capabilities are commonly found in today's file and database software. Some also can add across rows and thus enable cross-footing to check row and column totals. Added calculator functionality commonly implemented in file and database software includes subtraction, multiplication, division, percentages, rounding, and so on. Beyond this we would indeed be encroaching on spreadsheets, and at this stage one might look either for integrated packages or for file or database software that can support direct linkages to spreadsheet or mathematical modeling software. Examples illustrating the use of basic mathematical capabilities have been provided in both FileMaker and Paradox illustrations in this chapter, though both of these packages do indeed offer a range of functions similar to those found in spreadsheets.

Graphics

Some file and database packages provide, as either standard or optional add-on capabilities, graphical reporting analogous to that found in spreadsheet software, that is, data plots in the form of line graphs, pie charts, bar graphs, and such. In this case the charts are based on numerical data contained in rows or columns within the file or database. Others allow pictorial information to be included as attributes within the records of a file, such as digitized portraits for a personnel file, or line drawings for a spare parts directory. In Fig. 10-2 is an example of a record from a file containing pictorial attributes.

Programmable Macro Commands

The more advanced packages, particularly for relational database applications, include powerful macro programming capabilities (some call them scripts) that enable a developer to predefine the steps followed as a result of various custom menu selections or keystroke commands that can be made available to users for convenient access. For example, rather than have the user figure out complex query and reporting procedures—even with the aid of query by example—the most likely queries and report needs might be anticipated by the developer and then be preprogrammed and placed in a menu of choices. Utility functions such as database backup, exporting of data to other applications, and disk space recovery can be preprogrammed for the user's convenience. Logical control capabilities (e.g., IF *condition* THEN *do this* ELSE *do that*) can allow contingent processing to take place without user intervention (e.g., automatically save the current file to disk IF more than 10 records have been added or modified in the course of a long session with the computer, and in any case before exiting the program).

Creation of macro routines may be done either by recording and editing an actual sequence of actions, or by typing a set of commands directly into a file of text. An example of such macros is included in the equipment database case near the end of this chapter.

Integrated Software

File-management modules are commonly included within integrated packages such as Microsoft's Works, Lotus's Symphony, and Claris's ClarisWorks. Some of their file modules are comparable to some of the good standalone file-management packages. As with other integrated capabilities, one must evaluate the capabilities of the functions one needs most against the capabilities of an equivalent collection of standalone packages.

At the time of this writing, no integrated package includes a full relational database capability. This is an area where, in order to achieve some integration of applications, one must evaluate the capabilities for linkages among the database package and others with which it is to be integrated to perform the desired applications.

Network Data Access

In many file and database applications, several people may need to access the file or database system simultaneously. An example is the equipment database described later in this chapter, where the equipment, estimating, project management, and accounting departments, as well as several individuals within them, might be working with the same files or tables of data. Figure 10-9 is a schematic

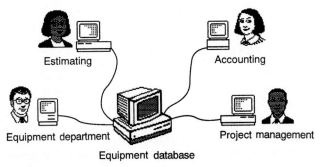

FIGURE 10-9
Distributed database access.

illustration of such a network distribution. For situations like this, some of the more advanced file- and database-management systems provide for *network data access*, where several people using different microcomputers interconnected by a network can be adding to, accessing, and modifying data at the same time—or almost at the same time.

Software with these capabilities must be designed carefully so that simultaneous or conflicting actions (e.g., somebody is deleting an employee while somebody else is moving him or her to a different job) do not cause errors and inconsistencies in the data. At a gross level, the system might provide file or *table locking*; in this case, while one user is entering or modifying data in a file, nobody else can have access to that file until the first user is done. In a frequently used file, this level of consistency protection would be very inconvenient. More commonly, one has record locking, and some systems even work at the level of individual data fields while modifications by one user are in progress.

Other concerns relate to efficiency of processing and related priorities. For example, two or more users might go in almost simultaneously with complex report requests or queries that involve lots of sorting, searching, and logical operations on multiple tables in a database; and each user's request might require that operations be handled quite differently. Unless copies of the relevant information are first downloaded to each user's workstation, it would be difficult to process the requests simultaneously, and one of the users might end up with a long wait before processing of his or her request even begins. Multiple indexing of tables can help overcome this problem.

It is also important to know whether the database is centralized (e.g., in a common network server) or decentralized (e.g., the various tables of a relational database are kept in subservers or even individual workstations in the departments most closely identified with them). It is also possible that some parts of the database might reside with computers in remote locations, necessitating telecommunications technology for integration.

It is beyond the scope of this section to give more than this brief taste of the issues involved in selecting software for network-based file and database processing. Different software packages address the issues (if at all) in different ways. This is an area where construction users should seek expert help in designing applications and in selecting software and hardware to implement them.

APPLICATIONS AND LIMITATIONS

File-management systems should certainly be considered for applications involving the entry, storage, retrieval, and reporting of data that can be configured into large numbers of similar records. Numerous construction administrative record-keeping tasks fall into this category: lists of tools, drawing logs, correspondence records, mailing lists, employee rosters, and so forth. If different but interrelated types of records are involved, then relational databases or other types of database software should be considered. Materials management systems, personnel systems, equipment information systems, and many other possibilities fall into this category. An equipment application is described in the example case later in this chapter, and a materials management database is included and described as a project for the reader to implement as a learning exercise.

File- and database-management systems can also be implemented for applications involving limited amounts of computation. It is assumed here, of course, that the selected software package has the necessary mathematical capabilities built in. Given this, one can then readily imagine several parts of an estimating system and even the standard modules of an accounting system (general ledger, accounts payable, accounts receivable, and perhaps even a payroll system) being implemented in today's more powerful database-management packages. These two applications, of course, are areas where even better custom software already exists, but they are still good illustrations of what might otherwise be appropriate.

On the other hand, most file- and database-management systems are limited in arithmetic capabilities beyond those found on a basic calculator. Applications needing complex scientific or engineering calculations usually are best done elsewhere, perhaps with a spreadsheet, mathematical modeling package, or procedural language. Similarly, the spreadsheet environment is far more suitable for interactive "What-if?" analysis wherein one makes several changes to input parameters and wants to see the result quickly.

At the other end of the spectrum, where the data file is small and simple, implementing a data management system within a spreadsheet or word processor might be better if one or both of these is already familiar to the user, particularly if the file is closely integrated with another application (e.g., a mailing list for a word-processing application or a reference look-up table for a spreadsheet). Word processing and spreadsheet packages commonly include limited file-management capabilities of this type within them, and they can be quite useful in the right context.

These limitations and alternatives notwithstanding, some level of file- or database-management software capability should be within the computer literacy of today's technical and administrative knowledge workers. Many of the packages are even simpler to master than spreadsheets and the more capable word processors, and they are very useful for many routine as well as complex chores. The examples that follow give a better idea of their advantages.

EXAMPLE: TOOL MONITORING SYSTEM IMPLEMENTED IN FILEMAKER PRO ON A MACINTOSH

This section follows up on the tool monitoring system introduced earlier in this chapter. Using the record layout, data-entry screen, and report designs that were described, it will illustrate how a user can modify, search, sort, and select records to answer inquiries and prepare reports for various purposes.

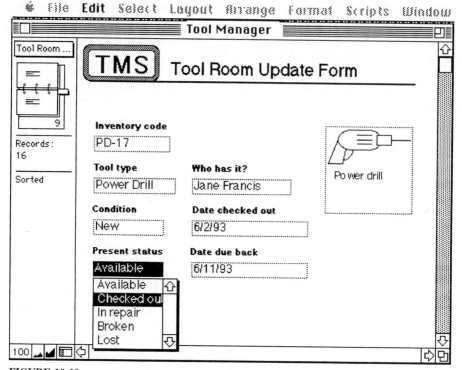

FIGURE 10-10
Data entry in progress for checking out a power drill.

The screens that were shown in Figs. 10-1 and 10-2 can facilitate data entry by enabling the user to select or create new records and to tab through the successive fields to enter or modify the data. In Fig. 10-10 we see a case where the tool-room clerk is partway through checking out power drill PD-17 to Jane Francis, having looked first to see that the tool's status was **Available.**

The flip chart icon in the upper left corner of Fig. 10-10 enables the clerk to page through the records in this file sequentially when in the **Browse** mode. However, if there is a large number of records, this could get very tedious and time-consuming. Thus file and database programs offer several ways to form queries that take the user directly to the desired information. For example, let us assume that on June 2, 1993, the tool-room clerk would like to find out who has the power saws checked out. The clerk could choose the **Find screen** from the **Select** menu, producing the form in Fig. 10-11, then formulate the query by typing the operator " = " followed by Power saw in the **Tool type** field, next type the operator "≥" followed by that day's date in the **Date due back** field, and click the **Find** button in the left panel to locate all people who still have power saws checked out. In this case, it would locate three records (refer to Table 10-1).

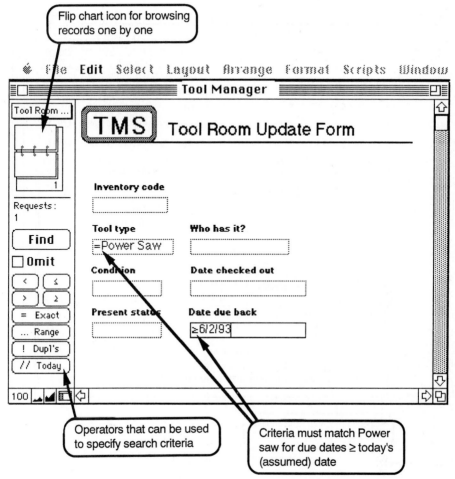

FIGURE 10-11
Using **Find** capability to search for records matching specified criteria.

Sorting is accomplished in a similar manner. For example, assume that we wish to sort the records by **tool type**, and within that do a secondary sort by **status**. Switching to the report design from Fig. 10-3, we could choose **Sort** from the **Select** menu, and specify our criteria as shown in Fig. 10-12. In the figure, the primary sort criterion has already been moved to the **Sort Order** box, and the user is in the process of selecting the secondary sort criterion. The user has also requested that both criteria be sorted in **ascending order**.

We can now proceed from this sort to show a typical report that is produced by the Tool Monitoring System. It appears in Fig. 10-13. Note the subtotals and

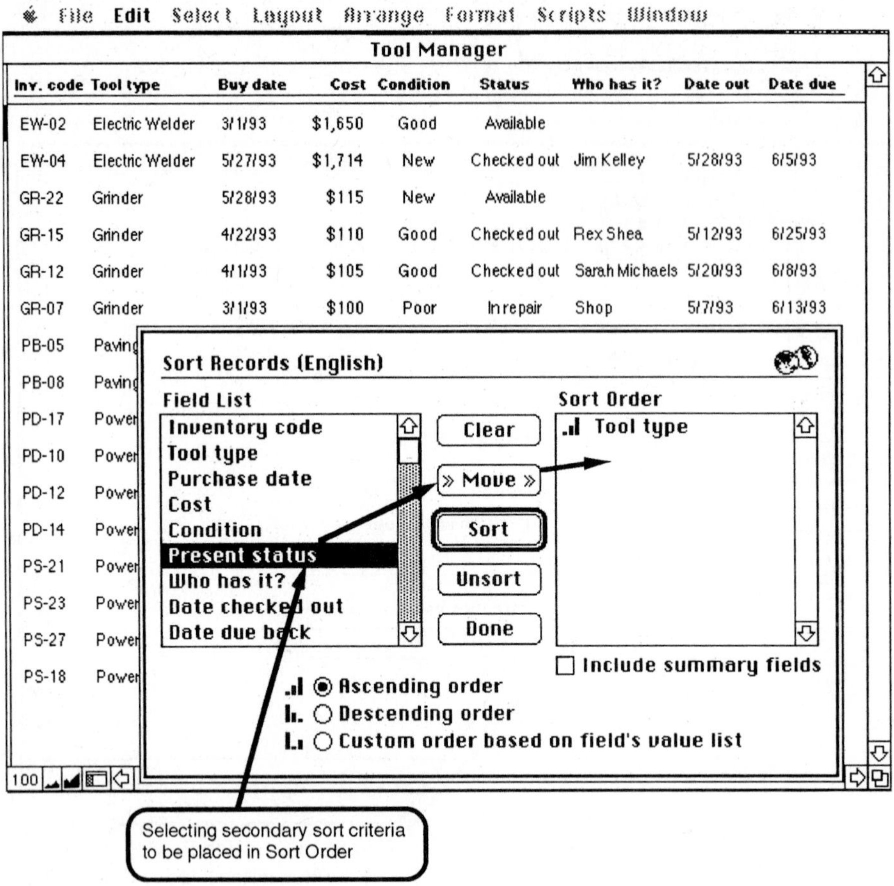

FIGURE 10-12
Specifying sort criteria for a file.

the grand total that are produced in the subsummary and summary fields that were specified in Fig. 10-3.

This overview of a top quality file-management system has been brief. Like most of the better products in this category, FileMaker Pro goes well beyond simple flat-file management and processing. We have seen that it includes graphics attributes and some very flexible graphic design tools. It also has the ability to perform look-ups from one file to another, though this falls well short of the relational database power that will be illustrated shortly. But for many if not most routine information-processing needs in a construction office, file-management systems offer plenty of features and capacity. They are a useful addition to a construction professional's bag of computer tools.

TMS Tool Status Report

Inv. code	Tool type	Buy date	Cost	Condition	Status	Who has it?	Date out	Date due
EW-02	Electric Welder	3/1/93	$1,650	Good	Available			
EW-04	Electric Welder	5/27/93	$1,714	New	Checked out	Jim Kelley	5/28/93	6/5/93
	Category cost subtotal:		$3,364					
GR-22	Grinder	5/28/93	$115	New	Available			
GR-15	Grinder	4/22/93	$110	Good	Checked out	Rex Shea	5/12/93	6/25/93
GR-12	Grinder	4/1/93	$105	Good	Checked out	Sarah Michaels	5/20/93	6/8/93
GR-07	Grinder	3/1/93	$100	Poor	In repair	Shop	5/7/93	6/13/93
	Category cost subtotal:		$430					
PB-05	Paving Breaker	1/20/93	$624	Fair	Available			
PB-08	Paving Breaker	4/21/93	$702	Good	Checked out	Joe Walsh	5/20/93	6/8/93
	Category cost subtotal:		$1,326					
PD-17	Power Drill	5/20/93	$55	New	Checked out	Jane Francis	6/2/93	6/11/93
PD-10	Power Drill	12/23/92	$42	Poor	Broken			
PD-12	Power Drill	1/8/93	$47	Fair	Checked out	Mary Smith	5/18/93	6/8/93
PD-14	Power Drill	4/21/93	$52	Good	Checked out	Bob Nesbitt	5/20/93	6/11/93
	Category cost subtotal:		$196					
PS-21	Power Saw	2/17/93	$60	Fair	Available			
PS-23	Power Saw	3/15/93	$65	Good	Checked out	Frank Jones	5/27/93	6/16/93
PS-27	Power Saw	5/13/93	$74	New	Checked out	Janet Thomas	5/17/93	6/3/93
PS-18	Power Saw	1/18/93	$72	Poor	In repair	Shop	5/24/93	6/10/93
	Category cost subtotal:		$271					
	Total tool costs:		$5,587					

FIGURE 10-13
Status report from Tool Monitoring System.

EXAMPLE: EQUIPMENT INFORMATION SYSTEM IMPLEMENTED IN PARADOX ON A PC

This example deals with an equipment information system consisting of four different components used by three different departments in a construction company—estimating, accounting, and equipment—as well as managers working on project sites. This section first provides a description of the problem, presents a general outline of a design for its solution, lists some example data, and then illustrates various ways in which the resulting database can be used in practice. The software used here is Borland International's Paradox, but the example could readily be converted to several other packages on a variety of computer systems.

Description

This simplified integrated relational database for a contractor's equipment information processing consists of four tables (files), each containing several attributes (columns) and a few typical records (rows). The four tables include the following:

1. Technical specifications for several types of machines
2. Inventory listing of each specific item of equipment owned
3. Ownership and operating costs by type of machine
4. Records and schedules of machine maintenance

These are typical of the kinds of information used by different parts of a contractor's organization that deal with construction equipment in some way.

In order to focus on the principles and procedures involved, we will keep the content, level of detail, and size of our database smaller than it would be in reality. Nevertheless, the example should help readers understand how to go about implementing similar systems on a practical scale. A similar example provided as a reader project at the end of this chapter will reinforce this knowledge. It focuses on materials management.

Objectives

In Table 10-2 are listed the basic structure and components of the four main tables in this database. The detailed design of how to organize, store, format, and retrieve this information will be conveyed via the Paradox implementation. General objectives are as follows:

1. Enable the user to add or delete records or modify selected fields within records, within any of the four tables, subject to certain password restrictions (e.g., read but cannot modify)
2. Provide easily understood prompting and CRT forms to guide the user through data entry, deletion, or modification
3. Provide validity checks on the range and type of data items input, with appropriate error messages and recovery procedures in case of user error
4. Provide preprogrammed routines to facilitate inquiries into the database and prepare CRT and printer reports drawing upon records and selected attributes from any one or more of the four tables

TABLE 10-2
Specifications and components of the equipment relational database
Database name: EQUIP

a: SPECS (standard specifications of machine types owned by the company)

Attributes:

#	Name	Type	Description	Example
1	MODEL	Alpha	Manufacturer's model number	CAT 637E
2	DESCRIP	Alpha	Describes type of machine	Scraper
3	POWER	Number	Combined engine power in KW	523
4	WEIGHT	Number	Empty weight in KG	50,843
5	LOAD	Number	Load capacity in KG	34,020
6	VOLUME	Number	Volume capacity in cubic meters	23.7
7	SPEED	Number	Maximum speed in KM/hour	48

b: ROSTER (inventory of each machine owned by the company, including acquisition date, cost, and location)

1	CODE	Alpha	Alphanumeric identification code for each machine owned	TK-064
2	MODEL	Alpha	Manufacturer's model number	CAT 769C
3	BUYDATE	Date	Date of machine purchase (y/m/d)	92/07/06
4	COST	Currency	Initial purchase cost	$100,000
5	JOB	Alpha	Job code where machine is located	CA-026

c: COSTS (estimating cost data for equipment types owned)

1	MODEL	Alpha	Manufacturer's model number	CAT 988B
2	OWNCOST	Currency	Hourly ownership cost (average)	$90.00
3	OPCOST	Currency	Hourly operating cost (average)	$80.00
4	HIMOD	Number	Cost multiplier for severe conditions	1.20
5	LOMOD	Number	Cost multiplier for easy conditions	0.85

d: MAINTAIN (information for equipment maintenance)

1	CODE	Alpha	Alphanumeric identification code for each machine owned	SC-024
2	HOURS	Number	Current engine hours	3124
3	OILFREQ	Number	Frequency of need for oil change (hours)	300
4	OILHRS	Number	Engine hours at last oil change	2894
5	TIREHRS	Number	Engine hours when new tires last installed	2100
6	FAILURES	Number	Cumulative number of in-service breakdowns	12

Sample information for each of the four main files used in this application is contained in Table 10-3, including some data designed to check for extremes and errors.

Description of the Application

The application that has been built to handle the database structure from Table 10-2 and the type of data from Table 10-3 consists of form and report designs

TABLE 10-3
Test data for the equipment database

a: SPECS

Model	Descrip	Power	Weight	Load	Volume	Speed
Cat 12G	Grader	101	13313	0	0	39
Cat 14G	Grader	149	18530	0	0	43
Cat 621E	Scraper	246	30479	21770	15	51
Cat 627E	Scraper	414	34670	21770	15	55
Cat 631E	Scraper	336	44000	34020	24	48
Cat 637E	Scraper	523	50843	34020	24	48
Cat 769C	Truck	336	30675	36300	24	69
Cat 773B	Truck	485	38660	52600	34	61
Cat 825C	Compactor	235	32400	0	0	15
Cat 966E	Loader	161	20324	5440	4	44
Cat 988B	Loader	280	42410	9600	6	41
Cat D10N	Tractor	522	79020	0	0	12
Cat D8N	Tractor	212	31383	0	0	11
Cat D9N	Tractor	276	42542	0	0	12

b: ROSTER

Code	Model	Buydate	Cost	Job
CP-032	Cat 825C	08/04/93	$315,000.00	CA-011
GR-017	Cat 14G	11/08/91	$212,000.00	CA-024
GR-024	Cat 12G	05/21/92	$155,000.00	CA-011
LD-021	Cat 966E	08/04/93	$236,000.00	CA-024
LD-045	Cat 988B	07/08/92	$389,000.00	CA-011
SC-014	Cat 637E	08/14/92	$520,000.00	CA-011
SC-021	Cat 621E	04/06/90	$285,000.00	CA-011
SC-022	Cat 621E	04/06/90	$285,000.00	Yard
SC-032	Cat 631E	03/22/90	$450,000.00	CA-026
SC-045	Cat 627E	09/14/94	$392,000.00	CA-022
SC-046	Cat 627E	09/14/94	$392,000.00	CA-024
TK-021	Cat 773B	01/08/91	$390,000.00	CA-024
TK-033	Cat 773B	03/22/93	$415,000.00	CA-011
TK-042	Cat 773B	05/04/94	$442,000.00	CA-022
TK-064	Cat 769C	11/27/92	$280,000.00	Yard
TK-076	Cat 769C	07/23/93	$320,000.00	CA-011
TK-077	Cat 769C	07/23/93	$320,000.00	CA-011

such as those that were illustrated in Figs. 10-6 to 10-8 plus menus and scripts to facilitate its utilization. This section gives a brief overview of these components.

Even without getting into the programmed version of the system, it is possible to use Paradox directly to browse, make queries, and obtain reports drawing on one or more of the tables. In Fig. 10-14 is a summary form that pulls together information pertaining to one particular machine from all four tables into one convenient screen. It also shows the top level of the standard Paradox menu hierarchy that can be used either for application design or utilization.

TABLE 10-3
(*continued*)

c: COSTS

Model	Owncost	Opcost	Himod	Lomod
Cat 12G	$32.00	$27.00	1.20000	0.85000
Cat 14G	$41.00	$38.00	1.20000	0.85000
Cat 621E	$62.00	$56.00	1.30000	0.80000
Cat 627E	$72.00	$64.00	1.30000	0.80000
Cat 631E	$92.00	$87.00	1.30000	0.80000
Cat 637E	$114.00	$95.00	1.30000	0.80000
Cat 769C	$54.00	$36.00	1.15000	0.90000
Cat 773B	$62.00	$57.00	1.15000	0.90000
Cat 825C	$56.00	$51.00	1.20000	0.85000
Cat 966E	$55.00	$52.00	1.35000	0.75000
Cat 988B	$78.00	$63.00	1.35000	0.75000
Cat D10N	$116.00	$105.00	1.20000	0.85000
Cat D8N	$77.00	$72.00	1.20000	0.85000
Cat D9N	$95.00	$90.00	1.20000	0.85000

d: MAINTAIN

Code	Hours	Oilfreq	Oilhrs	Tirehrs	Failures
CP-032	1950	250	1800	0	26
GR-017	6040	350	5740	4310	43
GR-024	4340	350	4020	3700	24
LD-021	2335	200	2200	1800	5
LD-045	3270	200	3200	1950	21
SC-014	4340	300	4230	4120	35
SC-021	7200	250	7000	6000	37
SC-022	6280	250	6400	5800	43
SC-032	7500	300	7230	6300	65
SC-045	560	250	500	0	2
SC-046	780	250	600	0	4
TK-021	6200	400	6000	4860	32
TK-033	2860	400	2700	2300	8
TK-042	1230	400	900	0	3
TK-064	4500	400	4150	2400	17
TK-076	2340	400	2000	2300	12
TK-077	4620	400	2200	2100	22

Figure 10-15 is an illustration of the overall structure of the equipment information system. This composite print of six screens shows the custom menus in action. Like the main Paradox menu, custom menu choices display horizontally at the top of the screen, with a second line to explain the action of the currently highlighted choice. The user moves from choice to choice with the left and right arrow keys—pressing the Enter key for the desired choice—or the user can simply type the first letter of the item desired. When there is a submenu of choices, it appears on the top line in place of the main menu.

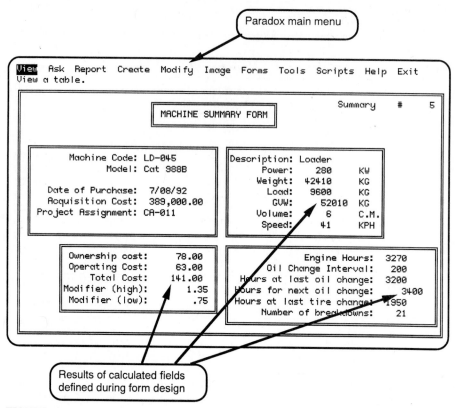

FIGURE 10-14
Using the composite summary form to review machine information.

In this custom application, the menus that appear to the user are quite different from the Paradox main menu, which is hidden once this application starts. The custom menu focuses on queries, reports, adding records, and editing data. Two additional items call up a help screen and leave the application. Under each of the first four menu items are second-level menus that direct the action to a specific table or, in some cases, produce forms or reports that combine information from two or more tables.

The menus and other actions in this application are controlled by programmed scripts that were generated automatically by the Paradox Personal Programmer after the author created the application. This program produced 47 different script files, with some 70 KB of code, all on the basis of comparatively simple menu, form, and report design activities carried out by the developer during a few hours of working directly on the screen under the structured guidance of the Personal Programmer. In Fig. 10-16 is listed some of the code from the opening script. It is but one of many pages. The developer did not have to write even one line of this code, which gives some idea of the power of such code generators.

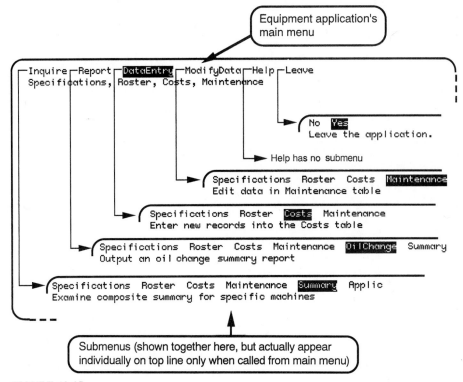

FIGURE 10-15
Equipment main menu and submenus as displayed on screen.

The script even documented its own code (see the lines preceded by semicolons) so that a knowledgeable human programmer could better read and understand the code to make special modifications or extensions later using the full version of the Paradox Application Language. Even such an advanced programmer could find it advantageous to start with the Personal Programmer and let it do most of the tedious work.

Application of the Database

This section provides two examples to show how one might go about using the database once it has been implemented.

In Fig. 10-17 we see the query-by-example selection procedure one would use to obtain a CRT screen report combining the following attributes from three of the four tables, with these conditions:

- Machines with power greater than 250 kW, purchased before January 1, 1993, that need an oil change in the next 100 engine hours.

```
; Equip

Echo Off
Clear
Reset
Cursor Off

; put up the greeting screen
@ 2, 0
Play "Equipg"

; ask for the password to the application; this password determines
;   the access to the tables in the application allowed for the
;   current user of the application.
@ 0, 0
Style Attribute SysColor(0)
?? fill(" ",160)
@ 1, 0
?? "Enter password for the application; [Esc] to cancel; [Enter] for
         no password."
@ 0, 0
?? "Password: "
Cursor Normal
zzzcolor = int(SysColor(0) / 16)
Style Attribute ((zzzcolor * 16) + zzzcolor)
Accept "a50" To pword
Style
EscEnter = not retval
Cursor Off

if (EscEnter) then
  Message "Cancelling the application"
  Sleep 2000
  Clear
  return
endif

if (pword <> "") then
  Password pword
endif


........

; Start the application

ReadLib "Equip1" Equip1Menu
Equip1Menu()
Release Procs Equip1Menu

Clearall
if (pword <> "") then
  UnPassword pword
endif
```

FIGURE 10-16
Personal Programmer script.

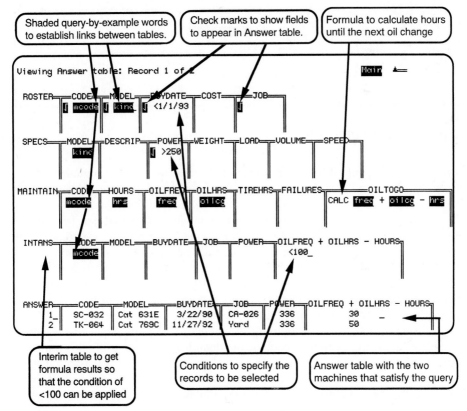

FIGURE 10-17
Query producing a CRT screen report.

- The report shows, for each machine that fits the criteria, its equipment identification code (e.g., SC-032), model (e.g., Cat 631E), purchase date, job location code, engine power, and the hours remaining until its next oil change.

Note the intermediate step (INTANS) used in obtaining the final result.

Figure 10-17 is an illustration of the concept of query by example. It was produced by working with Paradox directly, rather than under the custom menu-driven application that is being described here. The shaded example words in the key fields (i.e., **mcode** and **kind**) are just placeholders that link together the relevant key attributes from the different tables. Conditions (such as < 1/1/93 for the purchase dates to be selected, > 250 kW for the engine power, and < 100 for the remaining hours until the next oil change) are entered right in the related fields on the appropriate tables. Attributes to be included in the answer report are simply indicated by using a function key to display check marks in the desired fields.

Whereas the result displayed at the bottom of the screen in Fig. 10-17 is rather cluttered by all the other elements that led to its production, in Fig. 10-18

```
 4/26/92                        Standard Report                 Page    1

 CODE       MODEL        BUYDATE     JOB        POWER    OILFREQ+OILHRS-HOURS
 -------    ---------    --------    -------    ------   --------------------
 SC-032     Cat 631E     3/22/90     CA-026      336     30
 TK-064     Cat 769C    11/27/92     Yard        336     50
```

FIGURE 10-18
Results of query directed to a printer.

we see that the same information can be redirected as a neatly organized report to be displayed on the screen, sent to a printer, or stored in a file (where it might then be copied to a document being prepared on a word processor, for example).

In Fig. 10-19 is shown one of six predefined reports that can be obtained directly by going to the **Report** item on the custom menu and selecting one of the choices that are offered (in this case, **OilChange**). The report draws upon the **Roster** and **Maintenance** tables. The right column is computed by the same formula shown in the right column of Fig. 10-18. Routine documents such as invoices, cost reports, materials expediting reports, and so forth have a standard format that can be included in other applications in a similar manner.

In Fig. 10-20 we see another of the predefined reports requesting a listing, sorted in order of ascending equipment identification codes, of all equipment purchased since January 1, 1992. The latter choice of date is a variable that the user enters in response to a prompt just before the report is completed. It draws upon the **Roster** and **Costs** tables. The report includes columns with the following

```
 Date:    4/26            OIL CHANGE FORECAST        Page:   1

 Machine                       Engine   Change    Last     Hours to
  Code      Model     Project  Hours    Interval  Changed  Nxt Change
 -------   ---------  -------  ------   --------  -------  ----------
 CP-032    Cat 825C   CA-011    1950      250      1800        100
 LD-021    Cat 966E   CA-024    2335      200      2200         65
 LD-045    Cat 988B   CA-011    3270      200      3200        130
 SC-014    Cat 637E   CA-011    4340      300      4230        190
 SC-021    Cat 621E   CA-011    7200      250      7000         50
 SC-022    Cat 621E   Yard      6280      250      6400        370
 SC-032    Cat 631E   CA-026    7500      300      7230         30
 SC-045    Cat 627E   CA-022     560      250       500        190
 SC-046    Cat 627E   CA-024     780      250       600         70
 TK-021    Cat 773B   CA-024    6200      400      6000        200
 TK-033    Cat 773B   CA-011    2860      400      2700        240
 TK-042    Cat 773B   Ca-022    1230      400       900         70
 TK-064    Cat 769C   Yard      4500      400      4150         50
 TK-076    Cat 769C   CA-011    2340      400      2000         60
 TK-077    Cat 769C   CA-011    4620      400      2200      -2020
```

FIGURE 10-19
Output generated from a predefined report on a custom menu; note similarity to Fig. 10-18.

```
Date:   4/26      MACHINE COSTS IN SEVERE CONDITIONS     Page: 1

Machine                 Purchase  Ownership  Operating  Total O&O
  Code    Model           Date      Cost       Cost       Cost
-------   ---------     --------  ---------  ---------  ---------
CP-032    Cat 825C      8/04/93     67.20      61.20     128.40
Gr-024    Cat 12G       5/21/92     38.40      32.40      70.80
LD-021    Cat 966E      8/04/93     74.25      70.20     144.45
LD-045    Cat 988B      7/08/92    105.30      85.05     190.35
SC-014    Cat 637E      8/14/92    148.20     123.50     271.70
SC-045    Cat 627E      9/14/94     93.60      83.20     176.80
SC-046    Cat 627E      9/14/94     93.60      83.20     176.80
TK-033    Cat 773B      3/22/93     71.30      65.55     136.85
TK-042    Cat 773B      5/04/94     71.30      65.55     136.85
TK-064    Cat 769C     11/27/92     62.10      41.40     103.50
TK-076    Cat 769C      7/23/93     62.10      41.40     103.50
TK-077    Cat 769C      7/23/93     62.10      41.40     103.50
```

FIGURE 10-20
Printer report of machine costs in severe conditions.

attributes: equipment identification code, model number, purchase date, hourly ownership cost in severe conditions, hourly operating cost in severe conditions, and total ownership and operating cost in severe conditions. The three right-most columns are computed by multiplying the standard costs by the HIMOD modification factor for severe conditions (see Table 10-3c).

Finally, in Fig. 10-21 is a typical data-entry screen accessed under the **DataEntry** menu choice. The **ModifyData** choice uses the same screen when editing previously defined records. This particular form is used for adding records to or editing data in the SPECS table, where a Cat 777B truck is being added in this case. Forms for the roster, costs, and maintenance tables are similar.

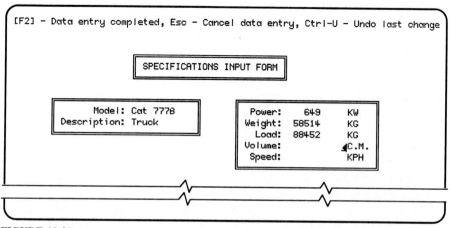

FIGURE 10-21
Data entry and editing form for specs.

This brief overview of the equipment information system has shown only a few of its reporting and querying capabilities. It has several more reports and query forms, and convenient data entry and editing forms as well. Th: xample, in turn, is a relatively simple one relative to the very extensive application-development capabilities that are found in relational database packages such as Paradox, dBASE, R:BASE, FoxBASE, and others. It is well beyond the scope of this chapter to go into detail for such powerful systems. Several books have been written on each of the major packages, and we list a few suggestions at the end of this chapter.

The best way to learn to use a relational database is to get access to a good software package, work through its tutorial, and try to implement a specific application. The development problem at the end of this chapter provides an example for this purpose.

SUMMARY

This chapter picked up where Chapter 4 left off, and provided specific information and examples about using file- and database-management systems in construction. Although most construction engineers and managers are not so familiar with these packages as they are with spreadsheets, file and database programs offer similar power and flexibility in handling a wide variety of practical construction applications.

Both file-management and database systems build from data elements that are placed in fields. The elements have standard data types, such as alphanumeric, numbers, currency, or dates. The fields are combined into records. Groups of records with the same format and data types become files. Simple file-management systems deal with information in one file at a time. Database-management systems go beyond this to allow relationships to be established among the various keys and attributes of multiple files, which they call tables or relations. This arrangement permits queries and reports to draw data from multiple files in varying combinations to meet the information needs of many different people and applications. They also allow logical operations to manage the tables, including append, intersect, join, project, subtract, and union.

Package software for file and database application development enables developers and users to create the structure of records, format screen forms, and output reports; to enter and modify data into the files and tables; to search, select, and sort information; and to prepare reports for printing or display. With microcomputers, they can store and process large volumes of information and keep it conveniently accessible to meet daily information needs and help solve problems.

Advanced capabilities found in both file and database packages include improved formatting, editing, and so forth. Most have protection and security both for the basic design of a custom application and for the data it contains. Hierarchies of passwords focused on different tables and access methods in a system can enable an application to closely reflect the organization and policies of the people it serves. Online, context-sensitive help and other aids can make the software easier to use. Some packages allow developers to create help files tailored to the users of custom applications. The more advanced packages also provide callable functions for mathematical, financial, logical, and other operations. Graphical capabilities can be offered to enhance the design of screens and reports and also as a data type for pictorial fields stored in the files themselves. Programmable macros, or scripts, make it possible to develop complete custom applications, and the power of these scripts can rival that of procedural languages in the more advanced systems. File-management capability is also found in integrated software but usually database-management is not. Finally, network data access can enable users at widely scattered computer systems to reach information that might either be stored centrally or be distributed across multiple computers in different locations.

This chapter presented two examples to illustrate the application of file and database systems. The first was a tool management system implemented on a Macintosh using Claris's FileMaker Pro. It could be used in the tool room on a construction site. The second was an equipment information system consisting of four different tables to serve varying needs in an organization, both in the office and on project sites. It was implemented in Borland's Paradox on an MS DOS-based PC.

As with spreadsheets, we have only skimmed the surface of the important subject of file and database systems. They have many valuable applications in construction and should be part of the working knowledge of today's construction professionals.

REVIEW QUESTIONS

1. How does the historical origin of file and database software for microcomputer applications differ from that of spreadsheets? In what ways has their evolution been similar?

2. Describe how the various data elements and structures in a file-management system are similar to those of a database system.

3. Describe the major ways in which file-management and database systems differ.

4. Some file-management systems have a LOOKUP capability, whereby a field from one file can make reference to a data field in another file for the purpose of either automatically inserting data from the other file into the one making the call or of making comparative checks. Is this the same idea as a relational database? Briefly explain your answer.

5. What information goes into the design of a record layout?

6. Briefly describe some of the functions that should be provided in the design of a data-entry form to be used on a CRT screen.

7. Explain the typical sections that can be included in designing a report layout for a file or database application.

8. Briefly describe how computer graphics can enhance the design of screen forms and tabular reports.

9. Describe how the spreadsheet-like business graphics techniques can effectively present the type of information contained in file-management and database-management systems.

10. In processing information from two different tables that are related by a common attribute, briefly state what can be accomplished with each of the following types of database operations, and what the result will be:

- Append
- Intersect
- Join
- Project
- Subtract
- Union

11. List some of the common data types used in file and database management.

12. Describe three different levels of script programming that can be used in database application development.

13. Describe the basic function of an automatic script code generator. In what ways can such a capability help a developer in creating a custom application?

14. Describe two different ways in which protection and security can be applied in a custom file-management application.

15. How can hierarchical passwords help adapt a distributed network database to the needs of a particular construction organization?

16. What is the advantage of being able to implement online, context-sensitive help in a custom database application?

17. At first it might seem excessive that advanced mathematical and financial functions would be included in a file-management or database package. Propose and briefly describe one engineering and one financial application that is suitable for the multiple-record, information-oriented nature of these packages and that could make particularly good use of such capabilities.

18. Consider an integrated package that includes both spreadsheet and file-management components. Describe an application that could appropriately integrate and best use each of these capabilities.

19. What special considerations must be given to files or databases intended for concurrent use by several individuals in a networked data access mode?

20. List a set of guidelines that would help you decide when an application is most suited for spreadsheet software, and when it is most suited for file- or database-management software.

21. Review and summarize the key concepts and techniques for the application of file-management software that were illustrated by the tool management example.

22. Review and summarize the key concepts and techniques for the application of database-management software that were illustrated by the equipment information system example.

DEVELOPMENT PROBLEM

This problem deals with a contractor's materials management system. In it you will first develop an integrated relational database consisting of four tables, each containing several attributes and a few typical records. The four tables will include (1) a file of vendor information, (2) a product file listing commonly needed construction materials, (3) a procurement tracking file, and (4) a purchase order file. These are typical of some of the information used by different parts of a contractor's organization that deal in some way with construction materials. After developing your database, you will apply it for a few typical query and reporting functions.

This problem is designed in terms of the capabilities of relational database software packages available for personal computers. Some of the benefit of doing the problem may also be gained if you only have simple file-management software available, but you will not be able to do the queries and reports drawing from multiple tables.

Design and Development

The basic structure and components of the four main tables in this database will be given in Tables 10-4 and 10-5 at the end of this problem, but the detailed design of how to organize, store, format, and retrieve this information is up to you. However, as a minimum it should be possible for the user to do the following:

1. Add or delete records or modify selected attributes (columns) within records within any of the four files, subject to password restrictions (e.g., read but cannot modify) that you feel are appropriate to impose
2. Guide the user through data entry, deletion, or modification by including easily understood prompting and/or CRT data-entry forms
3. Automatically check the validity and range of two or three example types of data input items, with appropriate error messages and recovery procedures in case of input errors
4. Make inquiries into the database and compose CRT and printer reports drawing upon records and selected attributes from any two or more of the four files

Application of Your Database

Once you have your database working, you should show that you know how to use it. As a minimum, do each of the following:

1. Show (by printing the equivalent command or macro file) the procedure you would use to obtain a **CRT screen report** combining the following attributes from the four files, with the conditions shown below. Print a copy (using the screen-print option on your computer) of the result.

   • A list showing all *projects* that have issued purchase orders for *framing lumber* since *September 1, 1995*. The report should include, for each project shown and in the order given, its project code, material description, quantity, unit of measure, unit price, total price, vendor name, purchase order number, and date of purchase order. The primary sort of the records should be by project number, and within that the ascending purchase order date should be the secondary sort.

2. Show (by printing the equivalent command or macro file) the procedure you would use to request a listing, *sorted* in order of ascending purchase *order dates*, of all *purchase orders* issued on *project number SU-94* to *Sunnyvale Lumber* and *Pine Cone Lumber* companies *since January 1, 1995*. The listing should include, in the order shown, columns with the following attributes:

   • P.O. number, P.O. date, vendor name, item description, total cost of P.O.

   Generate the hard-copy **printer report** resulting from this inquiry. At the bottom of the report, show the *total amounts* of the purchase orders issued to each of these two vendors since the given date.

TABLE 10-4
Specifications and components of the materials relational database
Database name: MATERIAL

---

a: VENDORS

Attributes:

#	Name	Type	Description	Example
1	VCODE	Alpha	Vendor code	PCL
2	VNAME	Alpha	Vendor name	Pine Cone Lumber
3	VADDR	Alpha	Vendor address	Sunnyvale, CA
4	VPHONE	Alpha	Vendor phone number	408-321-1234

---

b: PRODUCT

1	UCICODE	Number	UCI code of material product	06110
2	DESCRIP	Alpha	Description of product	Framing lumber
3	UNIT	Alpha	Unit of measure	MBM

---

c: TRACK (procurement tracking information)

1	PROJECT	Alpha	Project code	SU-94
2	PO	Number	Purchase order number	346721
3	RDATE	Date	Requisition date	1-May-95
4	PODATE	Date	Purchase order date	15-May-95
5	SHIP	Date	Shipping date	20-Jun-95
6	DELIVER	Date	Date of delivery	27-Jun-95
7	INSTALL	Date	Date of installation	10-July-95

---

d: PURCHASE

1	PO	Number	Purchase order number	346721
2	PROJECT	Alpha	Project code	SU-94
3	PODATE	Date	Purchase order date	15-May-95
4	UCICODE	Number	UCI code of material product	06110
5	QTY	Number	Quantity of material	22
6	VCODE	Alpha	Vendor code	PCL
7	UP	Currency	Unit price	$285

TABLE 10-5
Test data for the materials database

a: VENDORS (Listing of data in the VENDORS file)

VCODE	VNAME	VADDR	VPHONE
BBL	Bruce Bauer Lumber	Palo Alto, CA	415-948-1089
CSS	Calif. Shingle & Shake	Belmont, CA	415-592-8565
CW	Cabinet World	San Carlos, CA	415-592-8020
ELC	Economy Lumber	Oakland, CA	415-111-1111
GLC	Galaxy Lighting	Cupertino, CA	408-123-4567
H&J	Hubbard & Johnson	Palo Alto, CA	415-327-8580
KLM	K.L. Martin Plumbing	Menlo Park, CA	415-111-2222
MLC	Minton's Lumber Co.	MountainView, CA	415-968-9201
PSC	Pella Sales Co.	Redwood City, CA	415-366-2868
PCL	Pine Cone Lumber	Sunnyvale, CA	408-736-5491
RLS	Rayberg Lumber & Supply	Menlo Park, CA	415-323-4148
SLC	Statewide Lighting Co.	Sunnyvale, CA	408-765-4321
SVL	Sunnyvale Lumber Co.	Sunnyvale, CA	415-777-7777

b: PRODUCT (Listing of data for PRODUCT table)

UCICODE	DESCRIP	UNIT
3100	Concrete formwork	Sq.Ft.
3200	Concrete reinforcement	Tons-US
3300	Concrete, cast-in-place	Cu.Yd.
3400	Concrete, precast	Each
4200	Unit masonry	-0-
5100	Structural metals	Tons-US
6110	Framing lumber	MBM
6200	Finish lumber	Ft.
7200	Insulation	Sq.Ft.
7300	Shingles & shakes	Sq.-100
8100	Metal doors	Each
8200	Wood doors	Each
8500	Metal windows	Each
8600	Wood & plastic windows	Each
9100	Lath&plaster	Sq.Ft.
9250	Gypsum wallboard	Sq.Ft.
10000	Specialties	-0-
11000	Equipment for buildings	Each
12000	Furnishings	Each
13000	Special construction	-0-
14000	Conveying systems	-0-
14200	Elevators	Each
15000	Mechanical	-0-
15060	Pipe and pipe fittings	Lin.Ft.
16000	Electrical	-0-
16130	Outlet boxes	Each
12000	Sitework	-0-
12200	Earthwork	Cu.Yd.
1000	General Requirements	-0-

TABLE 10-5
(continued)

c: TRACK (Listing of data for TRACK table)

PROJECT	PO	RDATE	PODATE	SHIP	DELIVER	INSTALL
SU-94	940025	03/01/94	03/15/94	04/20/94	04/28/94	05/14/94
SU-94	940028	04/11/94	04/25/94	05/12/94	05/22/94	06/01/94
SU-94	940104	09/11/94	09/27/94	10/10/94	10/18/94	11/09/94
SU-94	940212	12/02/94	12/21/94	01/13/95	01/21/95	02/02/95
SU-94	950012	01/10/95	01/21/95	02/13/95	02/21/95	03/08/95
SU-94	950087	06/14/95	06/28/95	07/21/95	08/04/95	08/15/95
SU-94	950122	09/16/95	09/25/95	10/21/95	10/10/95	10/25/95
SU-94	950213	10/30/95	11/10/95	11/27/95	12/07/95	12/20/95
SU-94	950245	11/12/95	11/20/95	12/17/95	12/27/95	01/12/96
SU-94	960006	01/12/96	01/22/96	-0-	-0-	-0-
SU-94	960017	01/22/96	02/03/96	-0-	-0-	-0-
PA-95	950036	03/12/95	03/21/95	04/22/95	04/29/95	05/12/95
PA-95	950100	08/15/95	08/27/95	09/21/95	09/30/95	10/10/95
PA-95	950115	09/06/95	09/10/95	10/11/95	10/28/95	11/10/95
PA-95	950304	12/08/95	12/20/95	01/21/96	-0-	-0-
PA-95	960004	01/08/96	01/25/96	-0-	-0-	-0-
PA-95	960034	01/28/96	02/05/96	-0-	-0-	-0-
MP-95	950187	10/12/95	10/22/95	11/15/95	11/23/95	12/07/95
MP-95	950288	11/22/95	12/12/95	01/15/96	01/23/96	02/07/96
MP-95	960048	01/25/96	02/02/96	-0-	-0-	-0-

d: PURCHASE (Listing of data in PURCHASE table)

PO	PROJECT	PODATE	UCICODE	QTY	VCODE	UP
940025	SU-94	03/15/94	3100	360	SVL	$0.40
940028	SU-94	04/25/94	15060	220	KLM	$1.25
940104	SU-94	09/27/94	16000	54	H&J	$3.21
940212	SU-94	12/21/94	6110	4	PCL	$275.00
950012	SU-94	01/21/95	16000	36	GLC	$50.00
950087	SU-94	06/28/95	9250	1450	SVL	$0.20
950122	SU-94	09/25/95	7300	10	CSS	$76.50
950213	SU-94	11/10/95	8200	16	MLC	$45.00
950245	SU-94	11/20/95	6110	6	PCL	$176.00
960006	SU-94	01/22/96	8600	24	PSC	$210.00
960017	SU-94	02/03/96	6200	870	SVL	$0.45
950036	PA-95	03/21/95	3100	340	BBL	$0.55
950100	PA-95	08/27/95	4200	4400	H&J	$0.42
950115	PA-95	09/10/95	6110	7	PCL	$176.00
950288	MP-95	12/12/95	6110	5	PCL	$187.00
960048	MP-95	02/02/96	6200	800	BBL	$0.45
950304	PA-95	12/20/95	7300	16	CSS	$76.80
960004	PA-95	01/25/96	6200	800	SVL	$0.39
960034	PA-95	02/05/96	16130	48	H&J	$1.25
950187	MP-95	10/22/95	7200	3600	ELC	$0.10

SUGGESTIONS FOR FURTHER READING

Apart from the reference manuals for the two main programs for the examples in this chapter, the references here point to sources to pursue for an introduction and an in-depth treatment of the field of database management. As with spreadsheets, any good bookstore will offer numerous books dealing with specific database package software, but they become outdated too rapidly to be mentioned here. The books by Date and Ullman provide a more lasting background that transcends specific packages.

Date, C. J. *Database: A Primer*. Menlo Park, Calif.: Addison-Wesley, 1983. A brief introduction to database concepts by a leading authority in the field.

Date, C. J. *An Introduction to Database Systems*, 5th ed. Menlo Park, Calif.: Addison-Wesley, 1990. One of the most popular college texts for database courses.

FileMaker Pro User's Guide. Claris Corporation, 5201 Patrick Henry Drive, Santa Clara, Calif. 95052-8168, 1992.

Paradox User's Guide. Borland International, 1800 Green Hills Road, Scotts Valley, Calif. 95067-0001, 1990.

Ullman, Jeffrey D. *Principles of Database and Knowledge-Base Systems*. Rockville, Md.: Computer Science Press, Vol. 1, 1988, Vol. 2, 1989. Written by an excellent author who is widely recognized as a leading authority.

CHAPTER
11

DEVELOPING APPLICATIONS WITH AI-BASED EXPERT SYSTEM SOFTWARE

Our knowledge can only be finite, while our ignorance must necessarily be infinite.

Karl Popper*

Among the most persistently intriguing yet controversial branches of computer science is what has come to be known, perhaps too pretentiously, as artificial intelligence, or AI. Whereas mainstream computer applications have, for the most part, confined themselves to straightforward numerical computations, organizing and sorting information, and programmed data acquisition and control, AI boldly has sought to replicate certain functions involving distinct, nonnumerical reasoning processes that have, until recently, been viewed as innately human. Most of the successful applications of computing technology have been in areas where humans are weak: accurate and high-speed calculations, perfect retention and retrieval of vast amounts of information, and so forth. Early AI applications would scarcely have done credit to a child's abilities, let alone those of a working professional. Yet AI researchers have persisted and made substantial progress. The implementation of practical applications have rightfully focused management attention on its potential. AI technology has particularly strong implications for complex and experience-based fields like construction.

Artificial intelligence, in practice, does not attempt to replicate human thinking in a comprehensive way—at least, not yet. Walters and Nielsen (1988) define

---

Conjectures and Refutations, 1962.

artificial intelligence (AI) as "an interdisciplinary subfield of computer science that typically involves symbolic processing to approximate human behavior." Most AI research and development today falls within the following application areas:

- Natural language processing—understanding and producing written language (including automated translation between languages)
- Speech processing—recognition and synthesis of spoken language
- Machine vision and pattern recognition—capture, analysis, and understanding of visual (and other) images and patterns
- Robotics—mechanical devices capable of planned and controlled motion
- Planning—software that can recognize constraints and goals and develop a plan for meeting them
- General problem solving—programs that can reason symbolically to solve problems, usually based on formal logic theories
- Machine learning—machines that can accumulate and modify their knowledge based on new examples and experience
- Game playing—programs that play checkers, chess, and so forth
- Expert (or knowledge-based) systems—programs that capture and apply human expertise in specific areas for planning, design, and problem solving

Parsaye and Chignell (1988) provide a concise and readable history of AI and the interrelationships among these application areas as they evolved. Most of them, including game playing, have notable practical applications that increasingly are becoming a part of our everyday activities: speech synthesizers answering telephones, robots in factories, machine vision to sort mail and expedite supermarket checkout lines, and the like. We will return to some possibilities for construction in Chapter 16. This chapter, however, focuses on expert systems, also called knowledge-based systems, and shows how they can be applied in construction.

BASIC CONCEPTS

Dym and Levitt (1991, p. 11) define a *knowledge-based (expert) system* (KBES) as "a computer program that performs a task normally done by an expert or consultant and which, in so doing, uses captured, heuristic knowledge." Although *knowledge-based systems* is probably a more accurate term for this type of program, we will use the simpler and more common term *expert system* in this chapter. The word *heuristic*, in this context, is a rule of thumb, or experience-based guideline, that, although perhaps not subject to rigorous proof, has worked well in most cases where the expert has used it over the years.

Expert systems generally are appropriate for problems that have a narrowly and clearly defined, accessible knowledge base in a given specialty, and that are neither too trivial nor too hard for cognizant human experts. This section will define more explicitly the types of knowledge that go into an expert system, explain a typical *architecture* for this type of software, and discuss their common problem-solving strategies.

Types of Knowledge

While researchers strive to broaden the scope of expert systems applications, those that have been successful to date have embodied a relatively narrow, problem-specific body of knowledge, called the *domain* of the expert system. The knowledge incorporated in expert systems can include *deep knowledge,* which is the type of basic principles, facts, axioms, and laws of nature that one learns in school, and *surface knowledge,* which is the accumulation of experience and rules of thumb that an expert has gained from both successes and failures over many years of working within a problem domain. An expert system distills and organizes deep knowledge into an efficient and optimized form called *compiled knowledge*. It goes much further, however, by also attempting to capture the surface knowledge of one or more experts by means such as heuristic *rules*. One common type of expert system is thus called a *rule-based system*.

Expert systems also differ from more conventional programming techniques, such as those addressed in Chapter 8, by emphasizing *declarative knowledge* over *procedural knowledge*. As we saw in Chapter 8, procedural programs depend largely on predefined algorithms to accomplish their goals, with emphasis on *how* and in *what order* to solve a problem. They also clearly separate program algorithms from the data on which they operate. In contrast, *declarative programs* concentrate more on *what* is known about a problem and require much less explicit declaration of how to solve it or in what order to execute the rules. Expert systems more cleanly separate the knowledge about a problem domain (the *what*) from the inference and control procedures (the *how* and *when*) for manipulating that knowledge.

The source of knowledge is usually an expert in the problem domain, who may or may not know much about computers or expert system software. The expert may be involved personally through interviews, or expertise may be tapped indirectly via books, papers, or other sources. A *knowledge engineer* is a person who understands the expert system tool and can interface between the knowledge source and the computer to acquire and express the relevant expertise in a manner that is suitable to build the expert system application. These days, however, some expert system software in itself is becoming easy enough to use that the domain expert can work with it directly in many cases.

Knowledge Representation

Different kinds of expert systems can use several different methods to represent knowledge. Common representations include simple facts, logical expressions, rules, semantic networks, and frames.

Facts are attribute-value pairs or object-attribute-value (OAV) triplets that provide information to an expert system. For example, π approximately equals

3.1416; Jeff's birth year was 1972; the number of engines for a 767 aircraft is two. The previous examples are static facts, but there can also be transient facts (e.g., Laura's age is 18). A specific value (e.g., 45 degrees) for a general attribute (e.g., angle) is known as a *binding*.

In some expert systems, doubt or uncertainty about facts can be expressed as a *certainty factor* (CF)—also called a *confidence factor* (CNF)—in the range of -1.0 to $+1.0$, 0 percent to 100 percent, and so forth. For example, the cause of failure is "fatigue," with CF = 70 percent. Certainty factors are not the same as probabilities, which are a statistical result of a number of similar instances (e.g., 20 percent of the cylinder tests were below specification), but rather are an expression of one's degree of belief that something is true. In a sense, CFs quantify qualitative judgments so a computer can deal with them.

Logical expressions represent knowledge as assertions in logic—usually first-order predicate logic—in the formal language of mathematical logic. An example using the predicate calculus syntax of the Prolog language is as follows:

```
Supports (Column_C2, Beam_B4)
/* Comment: Column_C2 Supports Beam_B4 */
```

Logic is a powerful and flexible way of expressing knowledge that is relatively precise and well ordered, but has limitations when dealing with uncertainty.

Rules take the general form of IF <premise> then <conclusion>. The premise can assume the form of attribute-value or object-attribute-value representations. For example

```
IF the soil is hard
THEN use the bulldozer's rippers before excavating
```

The premise and the conclusion can also be compounded with logical AND, OR, and NOT operators, similar to those used in procedural languages. A rule is proved (AI folks say it *fires*) when its premise matches known or given facts in the expert system, and the result of this match then becomes part of the growing solution to a problem. Rules have advantages of simplicity and consistency in representing knowledge, but they have major limitations in expressing general relationships between different elements of knowledge.

Semantic networks represent objects or pieces of information as *nodes* and express relationships between them as *links*. In Fig. 11-1 is shown a small semantic network for information about Scraper 42 and its relationship to its machine classification and work environment. This is a powerful and flexible way of representing knowledge.

A concept that derives from such networks is **inheritance**, wherein properties of a subclass of a given class of **objects** are inherited by those under it. For example, Scraper 42 has the specific mechanical properties of a Cat 621E scraper (size, engine power, etc.), which itself inherits the general properties of scrapers (bowl, blade, bottom loading, etc.).

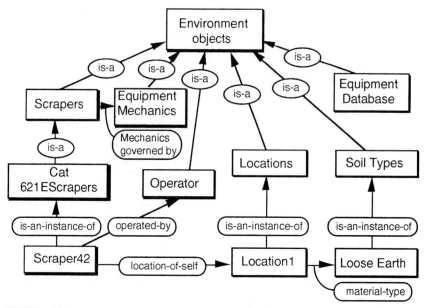

FIGURE 11-1
Semantic network. (From Boyd C. Paulson, Jr., Thomas Froese, and Lai-Heng Chua,
"Simulating the Knowledge Environment for Autonomous Construction Robot Agents,"
*Proceedings of the 6th International Symposium on Automation and Robotics in
Construction*. Burlingame, Calif., June 6–8, 1989, pp. 475–482.)

Frames generalize from the nodes in semantic networks to provide a structured representation of the properties and relationships of an object. They provide
a compact, modular, and hierarchical way of organizing knowledge in an expert
system that aligns well with the way many humans organize their own thinking.
Frames define objects via lists of attribute-value pairs. The attributes are called
slots, which in turn can declaratively encode values of an object's properties or
point to other frames, rules, or procedures that can *instantiate* (i.e., evaluate and
then assign) the values for the slots. Slots can also initially be given default values
to use until better knowledge becomes available.

Frames are usually organized into hierarchies, where more general expressions of objects (e.g., Scrapers in Fig. 11-1) are the parents of more specific
ones (e.g., Cat 621E). Such hierarchies can also provide for *multiple inheritance,*
where a lower frame can inherit attributes and values from two or more parents.

The typical syntax of a frame includes the frame name, its parent name, and
the attribute-value list. In Fig. 11-2 is a frame representation for an end-dump truck.

Architecture of Expert Systems

Although it is difficult to generalize in a field that is evolving so rapidly and that
uses so many different and creative approaches to solving problems involving expertise, some aspects of system architecture are common to many expert systems.

Name:	Cat 769C

Parent:	End-dump truck

Slots	Values
Power (kw)	336
Weight, empty (kg)	30,675
Gross vehicle weight (kg)	67,586
Max load (cu. meters)	23.5
Number of tires	6
Tire size	21.00-35, 32 PR (E-3)

FIGURE 11-2
End-dump truck represented as a frame.

This section will use an approach published by Maher (1987) in Chapter 1 of the book she edited for the American Society of Civil Engineers. She summarized her definition of the architecture in a figure similar to Fig. 11-3 (which also includes some alternative terminology added by Dym and Levitt [1991]).

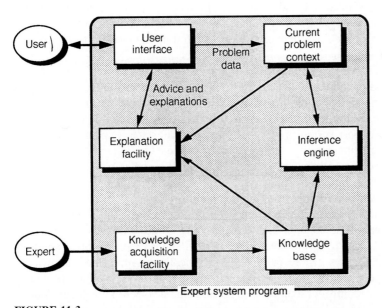

FIGURE 11-3
Architecture of a typical expert system. (Adapted from Maher [1987] and Dym and Levitt [1991]. Reproduced by permission of ASCE and McGraw-Hill.)

As a minimum, the ***user interface*** provides the menus, forms, or graphical means for the user to enter problem information, receive the solution in an intelligible form, and probe the reasoning behind it. Together with the explanation facility, it should provide a high level of *transparency* to enable the user to interact with and understand what is going on in the expert system.

The ***context*** is working memory that holds specific information about the current problem being addressed by the expert system and its user. It may start with specific input information that the user supplies about the parameters, then grow to accommodate the intermediate reasoning the expert system applies to deal with that problem, and finally receive the solution.

The ***inference engine*** controls the reasoning process that examines the data from the current problem context and consults the knowledge base to produce a solution to the problem and provide a supporting explanation. It can use one or more of the problem-solving strategies presented in the next section.

An *explanation facility* enables the user to probe the reasoning process that the inference engine uses to produce its solution. The user should be able to inquire about the process at intermediate stages as well as after the final solution becomes available. Like a respectable human expert, an expert system should do a good job at explaining itself. Means of explanation that may be implemented in various expert systems include frame hierarchies, rule traces, automatic explanation via near-natural language, rule graphs, and knowledge-based interactive graphics. See Chapter 6 of Dym and Levitt ([1991]) for descriptions of these techniques.

The *knowledge acquisition facility* enables an expert or a knowledge engineer to enter domain-relevant expertise into the knowledge base, or to modify and extend that information at a later date. It could also serve as an interface to other computer systems, such as databases or specialized application software such as a CAD system.

For a given domain of expertise, the ***knowledge base*** stores facts, usually as declarative knowledge, and heuristics about the problem domain, commonly as rules. The knowledge base should be subject to easy updating and expansion. At this point we should note that a different knowledge base can be substituted to generate a new expert system using the same user interface, explanation facility, knowledge acquisition facility, and inference engine. The latter combination of components is often called an expert system ***shell*** since it can hold different knowledge bases.

A significant variation on this architecture is the *blackboard model*. Blackboard systems decompose the knowledge base into multiple knowledge sources and use as a context the concept of posting messages between knowledge sources on a blackboard, which also maintains the current status of the problem solution. The knowledge sources can respond opportunistically to apply their knowledge when they see pertinent messages or changes in the state of the solution posted on the blackboard. A controller evaluates suggested changes from knowledge sources that have responded and selects the ones to fire on each cycle based on its own control rules.

Problem-Solving Approaches and Strategies

In problem solving we try to proceed from an initial set of facts and conditions to find a solution. Amarel* described a number of problem-solving tasks that fall under two broad approaches: *derivation* and *formation*. Derivation (also called *classification*) includes tasks such as diagnosis, interpretation, and monitoring, whereas formation (or *synthesis*) includes planning and design. Derivation approaches start from known problem conditions and deduce the best solution from a set of predefined solutions. Formation propagates from known information about the problem to create a solution at a higher level. Most expert systems use one approach or the other and are typed accordingly, but some are hybrids.

The strategy for problem solving controls how the expert system moves from its initial information about the problem to find a solution. The strategies are also called *inference mechanisms*. Maher (1987) classifies derivation control strategies as **forward chaining, backward chaining,** and *mixed chaining. Problem reduction, plan-generate-test,* and *agenda control* are strategies under the formation approach. Each is briefly described below:

- Forward chaining—involves bottom-up, data-driven reasoning. It moves from given data and applies rules to make inferences that can lead to a recommendation. This method is most appropriate when the amount of input to describe known facts is fairly small, yet there are many hypothetical solutions.
- Backward chaining—involves top-down, goal-driven reasoning. It proceeds from an assumed goal state or hypothesis back to known data or facts that can support or defeat the hypothesis.
- Mixed chaining—as its name implies, combines forward and backward chaining. It can concurrently reason forward from the data and backward from the goals. Assignment of initial probabilities to the goals guides the order of the search.
- Problem reduction—decomposes the problem into subproblems, solves these, and aggregates their solutions into a total solution.
- Plan-generate-test—uses an ordered approach to generate possible solutions and test them one at a time until one is found that meets the goal specifications of the problem. The planning element helps prune off inconsistent solution paths before they are needlessly explored in detail.
- Agenda control—assigns priorities to tasks in an agenda, with justifications and a method for each, and performs them top-down.

Another strategy for problem solving is *match,* including *simple* and *hierarchical match,* and *means-ends analysis.* A related search issue is defining criteria

*S. Amarel, "Basic Themes and Problems in Current AI Research," in V. B. Ceilsielske (ed.), *Proceedings of the Fourth Annual AIM Workshop,* Rutgers University Press, Rutgers, N.J., June 1978, pp. 28–46.

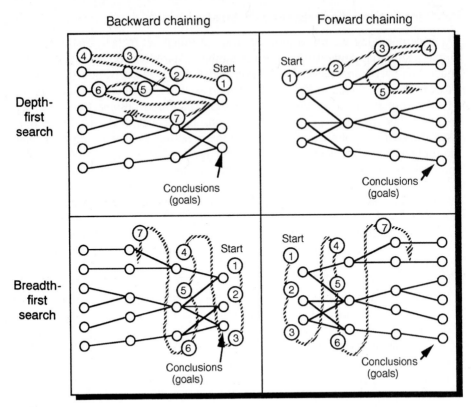

FIGURE 11-4
Depth-first versus breadth-first search strategies. (Adapted from Harmon and King, *Expert Systems: Artificial Intelligence in Business,* Copyright © 1985, p. 57. Reprinted by permission of John Wiley & Sons, Inc.)

for stopping, such as *satisficing* (stopping when a solution is acceptable) and *optimizing* (finding the best of the available solutions).

Search strategies over a search tree consisting of constraints, heuristics, and possible solutions also include *depth first* versus *breadth first*. Fig. 11-4 is an illustration of the difference in the context of forward-chaining and backward-chaining control strategies. The circled numbers and patterned lines indicate the order of the search.

WORKING WITH EXPERT SYSTEM SOFTWARE

Owing to the breadth of developments and technologies across the whole field of artificial intelligence, the range and variety of expert system software are far broader than we have seen thus far with spreadsheets and relational databases. Thus this section examines expert system software on two levels. First, we provide a general overview of the approach to developing expert system applications. Second, we briefly describe the spectrum of tools that is available for developing expert systems. We defer the specific workings of expert system software to the two example cases later in this chapter.

Developing Expert System Applications

The process of developing expert system applications generally parallels the over-all application development cycle described in Part III, Chapters 5 to 7. However, since it uses some specialized terminology and at the detailed development and implementation levels it has some unique characteristics, a brief review is in order here. Several books provide detailed explanations and practical guidance for this process, including those by Dym and Levitt (1991, Chapter 12), Harmon, Maus, and Morrissey (1988, Chapter 10), Parsaye and Chignell (1988, Chapter 8), and, extensively, Walters and Nielsen (1988).

In Fig. 11-5 is listed the typical sequence of steps in developing a moderately complex expert system application. As with the development of any complex system, the process is iterative rather than neatly sequential as shown on this simplified diagram.

The *feasibility analysis* is not unlike that described in Chapter 5. We define the scope of the knowledge domain to be captured and the task(s) to be performed,

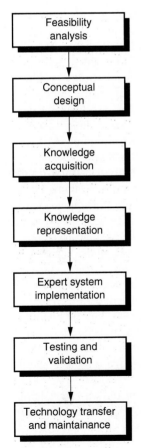

FIGURE 11-5

Expert system development cycle. (Adapted from Parsaye and Chignell, *Expert Systems for Experts,* Copyright © 1988, p. 296. Reprinted by permission of John Wiley & Sons, Inc.)

study existing systems and procedures, consider costs and benefits, and evaluate alternative application development methodologies. Should the conclusions point toward an expert system, we must further consider the source of the expertise that is to be encoded, the receptiveness of potential users to this type of application, and the suitability of the tasks selected for implementation in available expert system software. Most expert system developers agree that a key to success is selecting an appropriate and manageable task or set of tasks, usually within a narrow and well-defined scope. The need for expertise should be evident (as opposed to a fast number-cruncher or flexible database searcher, for example), and there should be an acknowledged source of that expertise.

The *conceptual design* formally specifies the tasks to be performed. It defines the scope and detail of the knowledge that will be needed to perform those tasks and proposes the strategy for acquiring that knowledge (use books or case studies, interview human experts, establish links to a database, etc.). This stage will also indicate the type of expert system tool that might be most suitable (e.g., simple rule-based, frame-based, hybrid, etc.). The development team should be identified, including knowledge source(s), the knowledge engineer, and others, as applicable. A schedule and a budget should be established to guide development. Sometimes creating a small prototype implementation will confirm the suitability of the proposed tool for the intended purpose.

Knowledge acquisition is the key step in building a high-quality expert system. The system can only be as good as the knowledge it contains. To begin, we face the problem of recognizing knowledge that is applicable to the problem, whether in published or human form. Even willing and cooperative human experts are often unsure about just what thought processes they use to reach their conclusions. It is insufficient for them just to say that they based their reasoning on experience or intuition or judgment. Knowledge engineers are trained to probe deeper into thought processes by asking for examples and analogies, posing hypothetical scenarios, watching experts work through specific cases—interrupting frequently to ask for details of how the analysis and solution is taking place. Interview notes, problems, flow charts, and other means help document what is learned about the domain expertise for the application. It takes great patience on both sides to achieve a successful transfer of a significant knowledge base.

In seeking the appropriate forms of *knowledge representation,* we refer to the many possibilities discussed earlier in this chapter (facts, logic, rules, certainty factors, semantic networks, frames, etc.) and try to model the acquired knowledge in one or more of these forms. Small- and medium-size knowledge bases might be represented entirely with facts and rules within a single context, but the hierarchy achievable with networks and frames helps structure larger knowledge bases in a form that is clearer and more manageable, including partitioning knowledge into separate contexts. Both the domain expert and the knowledge engineer should continue to collaborate in this step.

Expert system implementation is the process of taking the knowledge representation and encoding it into a computer application using a suitable expert system development program. The next section outlines some of the alternatives.

In addition to building the knowledge base itself via the knowledge acquisition facility, the developer may need to define specific input and output screens for the user interface and customize some parts of the inference engine and explanation facility. The examples presented later in this chapter provide more information about this stage.

The *testing and validation* stage is similar to that in other forms of software development. We basically want to be sure that the ultimate users of the system will be able to reach the same sorts of conclusions that the human expert would if faced with similar problems. We can do this with sample problems representing the full spectrum of situations the users are likely to encounter. The domain expert should also be available to review and confirm the validity of answers provided by the system. In contrast to the black box nature of typical application programs developed with procedural languages, many expert system development tools facilitate testing via their ability to make the ongoing reasoning behind their solutions transparent. The user can probe every step in the process, and—in the case of a bug in the system—the domain expert can help identify exactly where the process went astray.

Finally, in the stage called *technology transfer,* we put the system to work helping the users for whom it was designed. Additional computer hardware may be required for this purpose, and training should be available for any system of more than minor complexity. Developers must follow up periodically to see that the system is working as it was designed to work, and to solve unanticipated problems the users may have encountered. In *maintenance* of the expert system, we can modify and expand it over time—perhaps add new rules or change the structure of the knowledge base—and otherwise adapt the expert system to the changing work environment so that it grows with the increasing sophistication of the users. With such a broad range of rapidly evolving development software, long-term portability is also a major issue, and maintenance may evolve into total conversion from one system to another.

As a closing thought, we need to reconsider the nature of human expertise. Unlike a procedural program, where, for a given set of inputs, a predefined algorithm should always yield precisely the same answer, expert systems may produce different answers from each other and even from the human experts whose knowledge they encoded. Faced with any problem of reasonable complexity and where there is a range of possible solutions, even experts can disagree. The human who must act on the basis of a recommendation still has to use his or her own judgment and common sense.

Expert System Software

Expert systems have been developed with software ranging from FORTRAN to specialized development environments that cost tens of thousands of dollars. They run on microcomputers, workstations, minicomputers, and even mainframes. Needless to say, some development tools are more suitable than others for implementing a given expert system application design. This section describes the spectrum of software that might be used and then focuses on those tools that fit the objectives of this book.

Fig. 11-6, while perhaps oversimplified, is a representation of the spectrum of software that could be used for creating expert systems. The top of the diagram describes general categories, and the lower part mentions some typical programs within each category. The double-headed arrow connotes some progression in suitability (from left to right) and difficulty (from right to left). Although not specifically shown on the diagram, the third and fourth categories near the middle generally offer the most power and flexibility for creating expert systems, but at the expense of ease of use and, often, higher cost of software acquisition. The right side of the spectrum—the simpler and lower-cost rule-based tools—is where most construction expert systems are being developed, but we also owe attention to what the figure calls AI languages, such as LISP and Prolog.

At the left side of the diagram, conventional procedural languages can be and are being used to create expert systems. Pascal and C have been the most frequently used within this category, but to build anything but a small expert system this way takes a great deal of work. It would be like building a large house without using any power tools. However, many of the expert system development tools on the right half of the diagram have themselves been written in C or Pascal in order to achieve maximum speed, particularly compared to those written in symbolic AI languages such as Prolog and LISP. Thus these procedural languages definitely have an important if usually indirect role in modern expert system software.

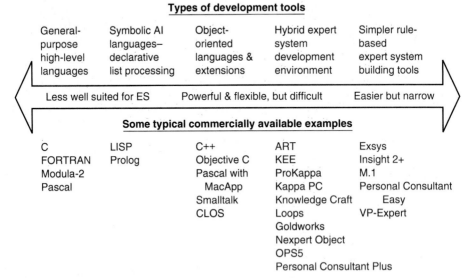

Types of development tools

General-purpose high-level languages	Symbolic AI languages– declarative list processing	Object-oriented languages & extensions	Hybrid expert system development environment	Simpler rule-based expert system building tools
Less well suited for ES		Powerful & flexible, but difficult		Easier but narrow

Some typical commercially available examples

C	LISP	C++	ART	Exsys
FORTRAN	Prolog	Objective C	KEE	Insight 2+
Modula-2		Pascal with	ProKappa	M.1
Pascal		MacApp	Kappa PC	Personal Consultant
		Smalltalk	Knowledge Craft	Easy
		CLOS	Loops	VP-Expert
			Goldworks	
			Nexpert Object	
			OPS5	
			Personal Consultant Plus	

FIGURE 11-6
Spectrum of expert system development software.

We have taken some liberties in combining LISP and Prolog under the general category of symbolic AI languages, but we did this because both are mainly identified with this branch of computer science. In practice, most texts classify LISP within or near conventional languages, and its list processing basis is really quite different from Prolog's declarative and logic-based approach. But it differs markedly from procedural languages (1) in computing primarily with symbolic expressions rather than numbers, (2) in representing both programs and data as linked list structures, and (3) in using recursion (a function that calls itself) extensively. Since LISP programs can be treated as data by other LISP programs, they can run and modify themselves or other programs while they run, a powerful attribute that AI programmers exploit to advantage. But this very flexibility resulted in many dialects of LISP, with many incompatibilities. In the 1980s, however, a version called *Common LISP* emerged as a standard and has been widely adopted. Most early versions of LISP ran interpretively, which is understandable given the ability of programs to modify themselves on the fly, but this fact also meant that application programs implemented in LISP ran slowly when scaled up to real-world problems. Converting successful applications to C or another procedural language thus became common, and the first symbolically coded version was then viewed as a prototype. Recently, however, LISP compilers have become available that translate completed applications directly into efficient machine-readable code that can be distributed for run-time applications.

Prolog (**Pro**gramming in **log**ic) is a flexible language that can implement a wide variety of applications. It originated in France, and standards for its most common version, Edinburgh Prolog, were developed in Great Britain and published by Clocksin and Mellish (1987). Prolog is widely used in Europe, but it was the adoption by the Japanese government of Prolog as the primary system and development language for its large-scale Fifth Generation Computing Project of the 1980s that made this language particularly well known. Why is this? Methodologies for logical reasoning in a formal way have evolved since Aristotle, and Prolog comes as close as any language to implementing these methods on a computer. It uses a simplified version of formal logic called predicate calculus and mostly uses a declarative style of programming. That is, instead of coding detailed procedures to accomplish a goal, you simply write a set of logical statements or axioms that state facts, relationships, and rules about objects. Prolog itself figures out the means to solve problems related to that information. The statements demand the precise syntactic form of Horn clauses (illustrated later) and take some getting used to, but they are not difficult to master. Prolog lends itself well to implementing expert systems that use backward chaining as their inference mechanism, but it can also implement other methods. Running a program is deceptively simple: You pose questions about the objects and relations contained in the application, and Prolog usually gives a YES or NO answer or else provides you the information that matches an unknown variable that you might have included in a question. We illustrate both the coding and some typical interactions in an example later in this chapter. Like LISP, most Prolog implementations run interpretively, and speed suffers when applications grow. Similarly, many developers consider it better for rapid prototyping of expert system applications but revert to other means for large-scale implementations.

The next step along the spectrum is object-oriented languages (of which Smalltalk is the best known) and powerful extensions to procedural languages that, in effect, turn them into object-oriented languages. They are playing an increasingly important role in AI programming and have advantages in structuring more complex expert systems, such as in implementing frame-based expert systems. C++ has emerged as the most widely used, but Pascal with the MacApp extensions has been fundamental to the development of much of the early Macintosh system and application software. Objective C became particularly well known when NeXT selected it as the primary system and development environment for their computers. Object-oriented languages have three key characteristics that distinguish them from conventional languages. First, they *encapsulate* both procedures and data into self-contained modules called *objects*. Second, objects communicate with each other mainly by sending *messages*. The availability of a growing library of tested and debugged objects enhances the ability to reuse code from program to program and thus can save development effort. Third, object-oriented programs exploit the properties of inheritance from parent objects to subordinate child objects, which further enhance program structure and efficiency.

The term *hybrid expert system development environments* in Fig. 11-6 is something of a catchall and encompasses a broad spectrum of AI expert system software and concepts. The term "hybrid" correctly implies that these environments usually include more than one inference or control mechanism, such as rules; frames and objects; forward, backward, and mixed chaining; and so on. Programs like Inference Corporation's ART, IntelliCorp's KEE and ProKappa and The Carnegie Group's Knowledge Craft are especially powerful and usually expensive development environments that define the high end of the scale. They typically run on UNIX workstations. Xerox's Loops and Verac's OPS5 are close to their LISP heritage and can require considerable programming expertise. Gold Hill's Goldworks, Neuron Data's Nexpert Object, IntelliCorp's Kappa PC and Texas Instruments' Personal Consultant Plus bring some of the power and flexibility of such environments to microcomputers, but still remain too complex for the average user. Suffice it to say, such systems are beyond the scope of this book, but they do show trends that will continue to work their way into the types of software that the application-oriented construction professional will be more likely to use.

The final category of Fig. 11-6, simpler rule-based expert system building tools, is oriented toward users who, although having little knowledge about artificial intelligence and only the basic concepts of expert systems, would like to build their own applications. For the most part these are nonstructured (one-level) rule-based tools that, for example, cannot build the type of hierarchical systems that use frames. They typically offer only one search strategy—usually backward chaining—and often have a predefined user interface. Happily, some of the best ones cost only a few hundred dollars and can run on personal computers. More important, with these tools we can build significant and useful expert systems for construction with a technically and economically reasonable amount of effort. Good examples include the Construction Industry Institute's *Modex* system for judging the suitability of construction operations for modularization (Modex, 1992) and the expert system by Amirkhanian and Baker (1992) for equipment selection for earthmoving operations. Both of these used WordTech's VP-Expert program, which we also employ for the construction safety example later in this chapter. Programs of this type can join spreadsheets, databases, and scheduling software as everyday tools for computer-savvy construction professionals.

Selection

Harmon, Maus and Morrissey (1988) concisely summarized some guidelines for expert system development tool selection in a diagram similar to Fig. 11-7. The vertical axis correlates somewhat with the spectrum in Fig. 11-6. The horizontal axis focuses on the size of the problem domain but could also include complexity of the problem domain or its degree of conceptual difficulty. For example, the body of knowledge in a domain may be of comparatively modest size, but it may not lend itself well to the flat rule-based structure of the simpler tools. In Fig. 11-7 we also note a boundary between do-it-yourself applications that the domain expert might implement directly and those that are so large or complex that a knowledge engineer should become involved at the interface between the domain

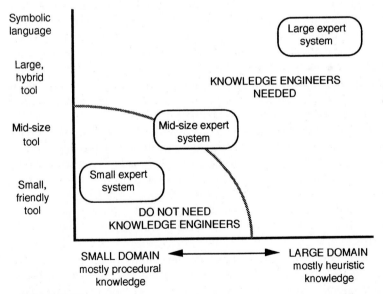

FIGURE 11-7
Expert system software selection. (Adapted from Harmon, Maus, and Morrissey, *Expert Systems Tools and Applications,* Copyright © 1988, p. 162. Reprinted by permission of John Wiley & Sons, Inc.)

expert and the tool. The diagram is simplistic, but it illustrates some key issues in selecting the right development tool for a particular application.

ADVANCED EXPERT SYSTEM CONCEPTS AND FEATURES

Because AI and expert systems remain a hotbed of research and development activity, attempting to classify advanced concepts and features—as we have done in earlier chapters—is a potential quagmire. *Advanced* is a relative concept, subject to one's point of view. We will thus confine ourselves to a brief mention of some topics that we have either not mentioned or just alluded to in earlier sections.

The hybrid category from the previous section—powerful systems that incorporate multiple inference and control techniques and that may also offer flexible user-interface-building capabilities—clearly are advanced relative to the simpler rule-based packages.

Advanced programs should also include flexible linkages to databases and perhaps to other types of application software. For example, the user of an expert system for construction planning might like to interrogate a design represented in a three-dimensional CAD system. Similarly, the languages LISP and Prolog, like some procedural languages, have versions with object-oriented extensions that are well suited to building complex expert-system applications and integrating them with databases, computational subsystems, and others.

Some programs have useful graphical attributes, starting with the ability to show tree-like traces of their reasoning, on up to animated dials and pictures that provide online feedback as to what is going on while the program runs. We are beginning to see the integration of expert systems with 3-D CAD systems. As a result, both the expert system and the human user can look at the objects that are the subject of their reasoning and problem-solving dialog, and the programs can even implement solutions by showing the recommended changes directly on the graphical representation of a structure, for example, "Place your scaffolding in this configuration to minimize costs and maximize productivity."

AI researchers are developing expert systems software that offer much stronger paradigms for reasoning—better means of qualitative reasoning, built-in techniques for reasoning about space and time, and abilities to represent beliefs, to learn, and to accumulate knowledge based on the cases encountered in ongoing applications. Some also go well beyond certainty factors for dealing with uncertainty and draw from a growing body of research called *fuzzy set theory*. This theory allows much more flexibility in representing the subjective and intuitive aspects of expertise within rules.

Expert systems are now moving into real-time monitoring and control applications, initially in aerospace and medical fields, but also in manufacturing, process plant operations, and perhaps soon in construction field operations. Such applications, particularly where human life is at risk, place the ultimate confidence in what only recently was an obscure academic research field whose boldest moves were in playing games.

APPLICATIONS AND LIMITATIONS

Most of the practical expert systems are found today in advisory or problem-solving situations. In cases where you would turn to a human expert, you might consider an expert system. In construction there are numerous such applications.

Estimating—generally considered more an art than a science, and certainly an area where an accumulation of experience is revered—is a prime candidate. What methods should we use? What productivity can we expect from a given crew and work configuration?

The engineering and design process could also benefit enormously from capturing human construction expertise about the cost and schedule implications of alternative design choices and from making this expertise available at the very moment when the designer is faced with those decisions. Perhaps such expert systems will become extensions of CAD systems. Research is already headed in this direction. Design Power's Design++ is already tightly integrated—both ways—with AutoCAD and can be called by the CAD system through an application program interface (API).

A number of expert systems have already been developed to address legal and contractual issues in construction. A major target is the evaluation and adjudication of claims. Think of all of the expert witnesses who seem to make a full-time job of going from one courtroom to another! Certainly the demand for such expertise exists.

Expert system developers have also addressed project planning, scheduling, and control. How long will an activity take? What will be the impact if it slips into the rainy season, or the winter months? Where should we focus expediting efforts?

A particularly important and promising area where expert systems could help out is occupational safety and health. Every year we seem to hear of more and more dangerous chemicals that pose hazards to the workplace, yet we construction professionals are poorly trained in the chemistry and human biology that would tell us what to do when faced with new problems. We can also use expert advice on safety aspects when designing a difficult construction operation, such as a multicrane heavy lift, an efficient and economical forming system, and so on. Expert systems could capture and broadly disseminate the knowledge of scientists and experienced consultants in these areas.

This list could go on indefinitely. Construction is very much a business of experience, expertise, and good judgment. It is no wonder that so many construction researchers quickly adopted expert system software and began to develop expert systems applications in their various areas of knowledge and interest. Some of these have been quite practical and have been well accepted by industry users.

But there are limitations. Top construction managers are broad generalists who can successfully investigate and solve a wide variety of problems in a single day. Most of the successful expert systems to date are narrowly focused specialists. Even if one had a large number of such systems available in a job-site toolbox, it seems that it would take a human expert just to know which tool to bring to bear on which situation. And in a fast-moving construction job, there may not be time to formulate problems in the rigorous manner desired by some expert system applications. All that we learned about the analysis and design of applications in Chapters 5 and 6 certainly applies well here. There needs to be a careful matching of the expert system tool and its resulting application to the needs of the situation faced by the construction professional in the field.

COMPACTION EXAMPLE IMPLEMENTED IN PROLOG ON A MACINTOSH*

Prolog has a number of special capabilities for rapidly building and implementing expert system applications (Bratko, 1990). Apart from the AI-related capabilities and international acceptance mentioned earlier in this chapter, an important aspect is portability across a wide variety of computer platforms—assuming that the developer stays close to the Edinburgh standard.

Quintus Prolog, developed in England by Logic Programming Associates, is an enhanced version of the Edinburgh standard Prolog that has object-oriented extensions. The Macintosh version of this language—MacProlog (Quintus, 1990)—also includes modern capabilities for building interfaces with graphical and windows-oriented features that make it easier for the user to work with a well-designed application. An alternative on the Macintosh, with similar interface-building capabilities, is Advanced A.I. Systems' AAIS Full Control Prolog (1991). In this section, however, we will stick close to the standard aspects of the language so that readers can try out the code using still other versions of Prolog, such as Borland's popular Turbo Prolog (Smith, 1988) and Arity's Arity Prolog that run on standard DOS PC computers.

Description of Example Problem: Compactor Selection

This example deals with the selection of the appropriate compaction equipment to use in specific soil conditions. Construction soil compaction involves the interaction of engineering specifications, materials properties, standard test methods, equipment types and procedures, and environmental conditions. The purposes can be to increase strength, reduce permeability, and reduce settlement of soil embankments and backfills.

Soil compaction is one of many construction operations that is as much art as science, involving qualitative judgment as well as quantitative analysis. It lends itself well to the type of experience-based knowledge that can be captured in an

---

*The Prolog program for this example was first developed by Dr. Hossam El-Bibany, Assistant Professor of Architectural Engineering at The Pennsylvania State University, and then modified by the author.

expert system. Three major components of the problem include the following:

- Material characteristics (gradation, cohesiveness, abrasiveness, etc.)
- Moisture content
- Compactive effort

Material characteristics are commonly represented in the Unified System of classifying soil characteristics shown in Table 11-1. For a given soil type, there is an optimum moisture content, above or below which it is more difficult to compact the soil. A variety of different tools and machines can deliver compactive effort, but for best results the optimum machine should be selected for a given set of soil conditions. The relative compaction density and moisture content can be measured by Standard and Modified (for high loads) Proctor (AASHTO) tests.

Compaction machines have been designed to provide varying degrees of the following types of compactive effort (from *Caterpillar Performance Handbook*, 19th ed., Peoria, Ill., October, 1988, p. 517):

- Static weight (or pressure)
- Impact (or sharp blows)

TABLE 11-1
Unified System of soil classification

Symbol	Name
Coarse-grained soils (less than 50% pass No. 200 sieve)	
GW	Well-graded gravel
SW	Well-graded sand
GP	Poorly graded gravel
SP	Poorly graded sand
GM	Silty gravel
SM	Silty sand
GC	Clayey gravel
SC	Clayey sand
Fine-grained soils (50% or more pass No. 200 sieve)	
ML	Low plasticity silt
CL	Low plasticity clay
OL	Low plasticity organic
MH	High plasticity silt
CH	High plasticity clay
OH	High plasticity organic
PT	Peat or organic

Source: S. W. Nunnally, *Managing Construction Equipment,* ©1977, pp. 12, 134. Adapted by permission of Prentice Hall, Englewood Cliffs, N.J.

- Kneading action (or manipulation)
- Vibration (or shaking)

General categories of such machines include the following:

- Tamping-foot rollers
- Grid or mesh rollers
- Vibratory compactors
- Smooth steel drums
- Pneumatic rollers
- Segmented pad roller

Most of these can be either self-propelled or towed. Empirical tables and charts, such as those in Table 11-2 and Fig. 11-8, respectively, can help field construction people select the best machine for a given set of soil conditions, or at least find the best compromise among machines that might be available. The volume of

TABLE 11-2
Soil compaction guide

Soil type*	Compaction equipment† Recommended	Suitable	Maximum dry density (pcf) Modified AASHO
GW	VR, VP	PH, SW, SP, GR, CT	125–140
GP	VR, VP	PH, SW, SP, GR, CT	110–140
GM	VR, PH, SP	VP, SW, GR, CT	115–145
GC	PH, SP	SW, VR, VP, TF, GR, CT	130–145
SW	VR, VP	PH, SW, SP, GR, CT	110–130
SP	VR, VP	PH, SW, SP, GR, CT	105–135
SM	VR, PH, SP	VP, SW, GR, CT	100–135
SC	PH, SP	SW, VR, VP, TF, GR, CT	100–135
ML	PH, SP	TF, SW, VR, VP, GR, CT	90–130
CL	PH, SP	TF, SW, VR, GR, CT	90–130
OL	PH, SP	TF, SW, VR, GR, CT	90–105
MH	PH, SP	TF, SW, VR, GR, CT	80–105
CH	TF, PH, SP	VR, GR, SW	90–115
OH	TF, PH, SP	VR, GR, SW	80–110
PT	Compaction not practical		

Source: S. W. Nunnally, *Managing Construction Equipment*, ©1977, pp. 12, 134. Adapted by permission of Prentice Hall, Englewood Cliffs, N.J.

*See Table 11-1 for key to symbols.

†Symbols

CT = Crawler tractor 10–30t	SW = Smooth Wheel 3–15t
GR = Grid roller 5–15t	TF = Tamping foot 5–30t
PH = Pneumatic roller 10–50t	VP = Vibrating plate < 1t
SP = Segmented pad 5–30t	VR = Vibrating roller 3–25t

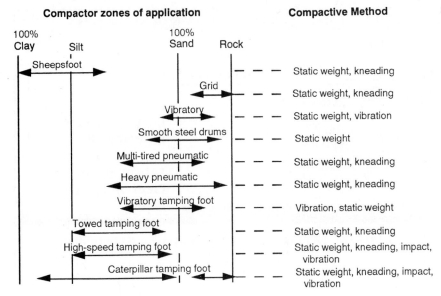

FIGURE 11-8

Compactor zones of application versus compactive method. (From *Caterpillar Performance Handbook*, 19th ed., Caterpillar Inc., Peoria, Illinois, 1988, p. 518. Reproduced by permission of Caterpillar Inc.)

material to be compacted is not reflected here, nor are the costs and productivity of alternative kinds of machines.

For purposes of this example, we found that the codes **SP** and **SW** were used for both soil types and compaction machines. Although human intelligence can easily keep each in its appropriate context when reading Table 11-2, we will make life easier for the "artificial intelligence" of this example by changing the code for the Smooth Wheel roller (SW) to **FW** (for Flat Wheel) and the code for the Segmented Pad roller (SP) to **PD** (for PaD).

Implementation in Prolog

Although we will use Quintus MacProlog for this example, our program will be simple enough and fairly standard so that it could be reimplemented easily in other Prolog systems. MacProlog works from the usual array of Macintosh pull-down menus. In Fig. 11-9 we see the main menu bar with the **Eval** pull-down menu displayed. Choosing **Query** from this menu pops up the dialog box shown in the center of the figure. Also note the background window containing a portion of our code called **1. Simple Compactors**, and the echo of our queries and the responses in the **Σ Output Window** at the bottom of the figure. This is a typical programming and execution configuration while you are working with MacProlog. A graphical display window is also available.

The primary means for entering code is simply a text editor, so we will not give details of that process here. Fig. 11-10 is the full listing of the facts stated

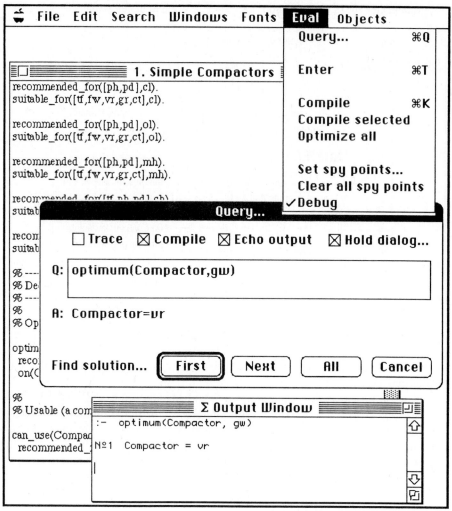

FIGURE 11-9
Running a program under MacProlog.

in the **database** section of the code, and Fig. 11-11 is the **rules** section. Note how closely the database replicates the facts contained in our "expert knowledge source," that is, Table 11-2. Apart from the dry density information, the database contains the same information expressed as Prolog statements. The first line after the comments in Fig. 11-10 (comments are those lines beginning with %) provides the program with a list of codes that define soil types used in this application. (That line is not actually used in the part of the program that we concentrate on here but is used in another section of code that we have left out so as to simplify the example.) The next pair of clauses and the similar fact clause pairs that follow

```
%  -------------------------------------
%  PROLOG PROGRAM FOR COMPACTOR SELECTION
%  -------------------------------------


%  -------------------------------------
%  Database (Facts)
%  -------------------------------------

soils([gw,gp,gm,gc,sw,sp,sm,sc,ml,cl,ol,mh,ch,oh,pt]).

recommended_for([vr,vp],gw).
suitable_for([ph,fw,pd,gr,ct],gw).

recommended_for([vr,vp],gp).
suitable_for([ph,fw,pd,gr,ct],gp).

recommended_for([vr,ph,pd],gm).
suitable_for([vp,fw,gr,ct],gm).

recommended_for([ph,pd],gc).
suitable_for([fw,vr,vp,tf,gr,ct],gc).

recommended_for([vr,vp],sw).
suitable_for([ph,fw,pd,gr,ct],sw).

recommended_for([vr,vp],sp).
suitable_for([ph,fw,pd,gr,ct],sp).

recommended_for([vr,ph,pd],sm).
suitable_for([vp,fw,gr,ct],sm).

recommended_for([ph,pd],sc).
suitable_for([fw,vr,vp,tf,gr,ct],sc).

recommended_for([ph,pd],ml).
suitable_for([tf,fw,vr,vp,gr,ct],ml).

recommended_for([ph,pd],cl).
suitable_for([tf,fw,vr,gr,ct],cl).

recommended_for([ph,pd],ol).
suitable_for([tf,fw,vr,gr,ct],ol).

recommended_for([ph,pd],mh).
suitable_for([tf,fw,vr,gr,ct],mh).

recommended_for([tf,ph,pd],ch).
suitable_for([vr,gr,fw],ch).

recommended_for([tf,ph,pd],oh).
suitable_for([vr,gr,fw],oh).
```

FIGURE 11-10

Database of facts for the compactor selection example.

```
%  ---------------------------------------
%  Rules
%  ---------------------------------------
%
%  Optimum (a compactor type may be used only if it is recommended)

optimum(Compactor, Soil):-
            recommended_for(RecommendCompactors,Soil),
            on(Compactor,RecommendCompactors).

%
%  Usable (a compactor type may be used if it is recommended or suitable)

can_use(Compactor, Soil):-
            recommended_for(RecommendCompactors,Soil),
            on(Compactor,RecommendCompactors).

can_use(Compactor, Soil):-
            suitable_for(RecommendCompactors,Soil),
            on(Compactor,RecommendCompactors).
```

FIGURE 11-11
Declarative part of the compactor selection example.

are *predicate clauses* that relate *arguments* (e.g., **vr** and **gw**) via the *predicate* (e.g., **recommended_for**). The first pair can be interpreted as

> "Compactor types **vr** and **vp** are recommended when you have soil type **gw**, but compactor types **ph**, **fw**, **pd**, **gr**, and **ct** could also be suitable if you have soil type **gw**."

The Prolog syntax in a predicate clause such as **recommended_for(vr,gw)** is called *prefix* notation—that is, the predicate precedes its arguments. Our English translation used *infix* notation—that is, the predicate is between the arguments—which is typical of this spoken language. Other languages—both computer and human—use *postfix* notation.

A further implication of our pair structure is that compactors listed on the first line of each pair are preferable to those on the second, but you might use one of those on the second if it is all you have available. But as we shall see, the list of facts alone is insufficient to enable Prolog to understand that implication. In any case, expressing straightforward factual knowledge is rather easy once you understand the basic syntax of these clauses.

Even at this stage—with just the list of facts—we could begin asking questions in this application. For example, if we typed

```
recommended_for([vr,vp],gw)?
```

where the question mark turns one of our fact clauses into a question, Prolog would answer YES. But if—thinking that it ought to match one compactor even more easily than both—we typed

```
recommended_for(vp,gw)?
```

Prolog would answer NO because we have not exactly echoed the syntax of the fact. Similarly, if we typed

```
suitable_for([vr,vp],gw)?
```

Prolog would also answer NO because it finds no matching **suitable_for** fact in the database. It does not understand the implication that if a vibrating roller (vr or vp) is **recommended_for** a well-graded gravel (gw) soil—an even tougher standard—it should also be **suitable_for** that soil. Thus we need a more powerful way of reasoning in this program, and Prolog complies by allowing us to create *rules* such as those shown in Fig. 11-11.

The rules section uses only three rules, in the syntax built up from first-order-logic *Horn clauses*, to enable us to tap into the knowledge in the database. In general, the goal clause to the left of the :- sign—called the *head* of the rule—is true only if each of the clauses to the right of the :- sign—which together constitute the *body* of the rule—is true. The first rule says that if a given **Compactor** matches an item in a list represented by the variable **RecommendCompactors**, that is,

```
on(Compactor,RecommendCompactors)
```

where **RecommendCompactors** is a variable that also correlates by position to an item or list of items in one or more of the **recommended_for** clauses in the database, and **Soil** is a positional variable that correlates to one of the given soil types, then the compactor is the **optimum** choice for the **Soil** that is associated with the **RecommendCompactors** item or list. The **on** keyword at the beginning of the last clause in each rule enables Prolog to be satisfied upon matching just one element of a list, thus overcoming one of the earlier errors we encountered when querying the rules directly. Although that explanation may be a bit hard to understand, it should become more clear with a couple of examples.

To start, let us give the program a soil type and ask it to come back with an optimum compactor for that soil. We begin by selecting **Query** from the **Eval** menu, which brings up the aforementioned **Query** dialog box. We enter

```
optimum(Compactor,gw)
```

in the **Q** (for question) section of the dialog box and click the **First** button. This would again produce the result **vr** in the **A** (for answer) part of the dialog box, as it did for the first query back in Fig. 11-9. If there are more solutions, pressing the **Next** button would show the next item (**vp** in this case). Exhausting the available compactors in the database that satisfy the **optimum** clause produces the response

```
No more solutions
```

as shown in Fig. 11-12. As an alternative to clicking **First** and then **Next**, we could have clicked the **All** button to find and display all of the solutions at once.

```
┌──────────────────────────────────────────────────┐
│                    Query...                        │
│  ☐ Trace   ☒ Compile   ☒ Echo output   ☒ Hold dialog... │
│                                                    │
│  Q: ┌──────────────────────────────────────────┐  │
│     │ optimum(Compactor,gw)                    │  │
│     │                                          │  │
│     └──────────────────────────────────────────┘  │
│                                                    │
│  A: No more solutions                              │
│                                                    │
│  Find solution...  ( First )  ( Next )  ( All )  ( Cancel ) │
└──────────────────────────────────────────────────┘
```

FIGURE 11-12
Query dialog box for finding the optimum compactor for an input soil type.

Either approach gets us to the state shown in the **Σ Output Window** in Fig. 11-13.

Before proceeding, note that a pair of rules is used to define the **can_use** part of the declarative section shown in Fig. 11-11. The first is about the same as the **optimum** statement; it recognizes that optimum choices should also be regarded as suitable. The second statement adds the information from Table 11-2 about compactors that might be suitable, though not optimum, for a given soil type. This overcomes the second of our earlier dilemmas when the query

```
suitable_for([vr,vp],gw)?
```

gave NO as the wrong answer. We will illustrate how to make such a query shortly.

Once we have set up our database and rules, we can use any of the variables and constants as inputs to find their counterparts as outputs in our questions

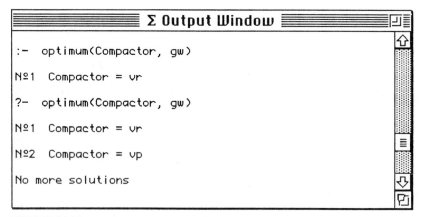

```
╔══════════════════ Σ Output Window ═══════════════╗
║                                                   ⬆ │
║ :-  optimum(Compactor, gw)                          │
║                                                     │
║ №1  Compactor = vr                                  │
║                                                     │
║ ?-  optimum(Compactor, gw)                          │
║                                                     │
║ №1  Compactor = vr                                  │
║                                                     │
║ №2  Compactor = vp                                ≡ │
║                                                     │
║ No more solutions                                 ⬇ │
╚═══════════════════════════════════════════════════╝
```

FIGURE 11-13
Σ Output Window for recommended compactors.

without further programming. Unlike procedural languages, the declarative logic programming style of Prolog enables us to do that. In other languages, we would have had to write separate procedures for each type of query. Let us illustrate this capability of our Prolog program. If we wish to input a particular compactor to see what soil types it could best be used for, we can type a query such as

```
can_use(vp,S)
```

Here we are asking the types of soils that can be handled well by a vibrating plate (vp) compactor. We click the **All** button and find that there are nine such soils. (There would only have been four if we had requested **optimum** instead of **can_use**.) The dialog box and recommended soils for this case appear in Fig. 11-14.

Note that in the question part of this dialog box we did not spell out the word "Soil" as the variable to match but simply shortened it to "S" instead. The variables are position sensitive, and it does not matter what characters you type

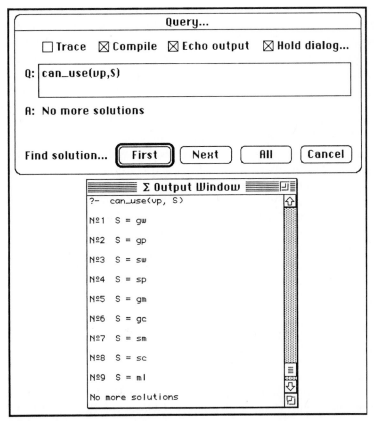

FIGURE 11-14
Soil types that can be handled by a given compactor.

in a given position. But note that the output answers are equally cryptic, for example,

№ 1 S = gw

so it might be better to be more descriptive with the inputs to improve the outputs that are echoed back to us.

Also note the order in which the answers appear in Fig. 11-14. This order reflects the depth-first, backward-chaining processing of a standard Prolog program. In this case, it first searched through all of the recommended_for rules, producing the first four answers, and then came back and searched the suitable_for rules to produce answer numbers 5 to 9.

In this example we have given only a brief idea of how Prolog works and have said even less about the MacProlog development environment. Both the Prolog language and MacProlog are powerful development tools. In preparing this chapter, we extended the example to calculate compactor production and also implemented a full Macintosh-style menu and graphical interface, but space does not permit us to continue here.

The reader with interests in a powerful AI development environment is encouraged to investigate the effectiveness of Prolog programming. For another type of construction application, study the paper by Raynar and Smith (1993) titled, "Intelligent Positioning of Mobile Cranes for Steel Erection," which describes a Prolog program that offers guidance to help steel erectors minimize the number of times the crane has to be relocated while putting up a structure. A paper by El-Bibany, Chua, and Paulson (1990), which builds upon earlier work by Chan and Paulson (1987), describes an object-oriented Prolog-based system for tracking and propagating the effects of constraints that evolve as decisions take place among the various parties involved in the design and construction of a project. Thus it can flag subsequent decisions that may cause conflicts with the existing design or construction plans (e.g., a designer might try to specify materials that will cause an overrun in the project budget, or a component that is too heavy for available lifting equipment), so that they can be resolved before they go on to become problems further down the line (e.g., crews stopped while managers and engineers scurry to reroute a pipe that would otherwise cut through a beam). There are many other useful possibilities for Prolog applications in design and construction.

SAFETY EXAMPLE IMPLEMENTED
IN VP-EXPERT ON A PC

WordTech's VP-Expert is an inexpensive but effective rule-based expert system development tool with many features typically found in much larger systems. It runs easily within the memory limits of minimal DOS-based PCs yet offers a friendly, colorful, user-oriented, menu-driven interface that, if anything, makes it easier to use than spreadsheet programs. Coded in C, it is fast and produces lively interactive applications. Although dominantly backward chaining, it allows limited forward chaining via commands in rule conclusions and further supports some measure of hierarchy in its rule sets. It provides a convenient induction mechanism to set up a rule structure based on tabular input from an appropriately structured text file or database file. It is based on attribute-value pairs but also allows subscripted attributes that in turn allow efficient processing of successive records in databases or spreadsheets. The developer can embed explanatory text with any rule that can be called up when a user selects **Why?** or **How?** from the VP-Expert menu. Fully implemented confidence factors deal with uncertainty. Users can trace the logic leading to recommendations and solutions, and it is easy to change selected inputs to run alternative "What-if?" scenarios. VP-Expert extends its power by providing interfaces to relational databases using dBASE format, to spreadsheets using Lotus format, and to text files. Further enhancements include graphical display of tree structures and interactive sliders and dials for inputs and outputs of an executing expert system.

Our expert system application deals with construction safety and specifically is designed to enable a contractor to perform a self-evaluation of how its record will appear to today's increasingly safety-conscious owners in a competitive contractor evaluation and selection process. Research underlying the organizational and behavioral approach to construction safety was conducted by Clarkson Oglesby, Raymond Levitt, Nancy Samelson, Jimmie Hinze, and others at Stanford University in the 1970s. Levitt and Samelson summarized the results of numerous research studies in a book entitled *Construction Safety Management.** They found a strong correlation between the safety performance of construction companies and the attitudes, behavior, and policies of managers and supervisors from president through project managers and superintendents to forepersons. Happily, they also found, under supervisors with excellent safety records, that safety performance was strongly and positively correlated with productivity, lower costs, and higher quality of the work. The findings and recommendations of this research have been adopted and widely implemented by member firms of such prestigious organizations as the Business Roundtable and the Construction Industry Institute. Contractors that have fully committed themselves to the policies and procedures that go with these programs have experienced accident rates that are well below industry averages and have enjoyed related benefits in employee morale and productivity as well as major cost savings.

In the 1980s, Levitt built two expert systems that captured the methods and recommendations of the Stanford safety research and made them available to owners and contractors interested in improving the safety performance on their

---

*Raymond E. Levitt and Nancy M. Samelson, *Construction Safety Management,* 2d ed., John Wiley & Sons, New York, 1993.

projects.† Our example here basically follows the outline of SafeQual but has been simplified to fit the constraints of a textbook illustration.

Scope of Application

The attributes contained in this expert system, and some of the features within the main categories, appear below:

- Experience modification rating (EMR)—this is a modifier that insurance companies apply to standard industry book rates to determine the rate for a particular firm. The EMR is determined by a number of factors, but most significant is the contractor's actual accident history in recent years. Typical EMRs run from less than 0.5 for excellent safety records to over 1.5 or even 2.0 for poor firms. Since standard book rates can themselves run 10 percent of payroll, firms with high EMRs are at a serious competitive disadvantage and often do not last long.

- OSHA incidence rate (OIR)—rate is the frequency of recordable accidents, as defined by the federal Occupational Safety and Health Administration, per 200,000 worker hours (derived based on 100 workers at 40 hours per week times 50 work weeks per year). Typical construction industry OIRs are in the range of 5 to 9, but some outstanding firms have consistently reduced them below 1 whereas others are criminally high.

- Safety training—components of a good in-house training program include formal orientation procedures for new workers; weekly toolbox meetings that effectively stress safety; printed materials regularly given to workers; task-specific training for new or potentially dangerous construction procedures; first-aid training for workers and supervisors; and professional safety training for all supervisory personnel.

- Tracking of supervisors' safety records—safe companies now quantify and track safety records by individual forepersons, superintendents, and managers, with detailed breakdowns by type and frequency of accidents or hours worked without accidents. Senior managers give high priority to monitoring and reviewing these records and use them in their employee evaluation, promotion, and compensation procedures.

- Professional safety staff—medium-size and larger firms now maintain professional safety staffs that report at higher levels in the project and company organizational hierarchies. Better firms have higher ratios of such people, they are professionally trained and have solid experience, and they are compensated accordingly.

---

†Raymond E. Levitt, "Howsafe: A Microcomputer-Based Expert System to Evaluate the Safety of a Construction Firm," Spring Convention, American Society of Civil Engineers, Seattle, Wash., 1986. Raymond E. Levitt, *SAFEQUAL: A Knowledge-Based System for Prequalification or Selection of Construction Contractors Based on Expected Safety Performance*, Building Knowledge Systems, Inc., Stanford, Calif. 1989.

- Operational policies—numerous operational policies bring safety to the jobsite. Good supervisors explicitly design safety into field construction procedures. They provide the appropriate personal protection gear and safety equipment and enforce its use even by firing noncompliers. They have frequent self-inspections by knowledgeable personnel. And they have meaningful awards and recognition events for workers, crews, and supervisors with excellent records.

Obviously, many of these factors are qualitative and judgmental in nature. Many do not lend themselves well to traditional computer programming environments. Let us see how they can be incorporated into an expert system application.

Building the VP-Expert Safety Application

VP-Expert is mainly a menu-driven program, particularly when running an existing application. But building a new application takes place in an embedded text editor that works much like those found in standalone DOS applications. That is, it relies extensively on the function keys and other special keys found on a typical PC keyboard and uses a text-based rather than graphical user interface.

In the front layer of Fig. 11-15 we see the main menu screen that the developer encounters after moving past the introductory title screens of VP-Expert. The main menu is the second line from the bottom of the screen, and below it is the submenu for **Consult**, the item to which the cursor currently points. One selects options by typing the indicated numbers, by typing the first letter of a menu item, or by using the cursor keys to position a highlight over the item and pressing Enter. The main functions are mostly self-explanatory, beginning with online **Help**. **Induce** accesses a structured text or database file that automatically creates a set of rules that can form the basis for further application development. **Edit** accesses the text editor. **Consult** moves to the menu shown on the bottom line, to be explained in a moment. **Tree** displays the result of a trace as a structured tree, with a choice of either a text or a graphical format. **FileName** and **Path** provide the typical DOS utility functions of loading a particular knowledge base file or of first changing the DOS path to access files in a different directory. **Quit** on this top-level menu leaves VP-Expert.

Moving down one level, we are in the **Consult** menu, which appears on the next-to-bottom line of the upper layer in Fig. 11-15 and with **Go** selected, in the second layer at the bottom of the figure. **Help**, which appears in all submenus, reaches online help but in a context-sensitive manner specific to the currently active menu. **Go** begins a consultation session using an existing expert system. After you have run a consultation, **WhatIf** enables you to change initial inputs and try out alternative scenarios. **Variable** permits the user to probe the final values of different attributes, whereas **Rule** helps you select and display any rule in the knowledge base. **Set** enables a user to record the logical flow of a consultation, speed up or slow down the display, or alter the display window configuration. **Edit** accesses the aforementioned text editor from this level too, whereas **Quit** in this case returns to the main menu. Details of how to work VP-Expert's main menu

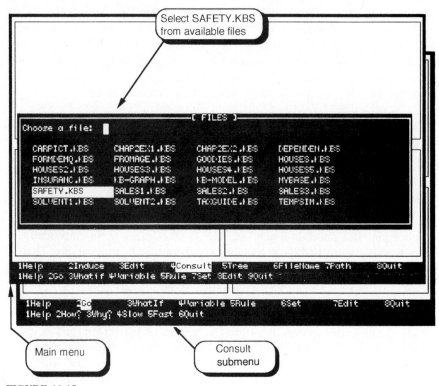

FIGURE 11-15
VP-Expert screens and menus.

and its hierarchy of submenus are contained in the WordTech *VP-Expert* manual (1989), in Wang and Mueller's *Illustrated VP-Expert* (1989), and in Dologite (1993).

In Fig. 11-16 we see the default display configuration that appears when one selects **Go** and gets into an expert system consultation. The screen divides into three active windows in addition to the menu bar. At the top is the main **Consultation window** where user input and output takes place. In this case it shows the first two questions posed by the safety expert system, plus the user's answers—direct numerical responses here. Once an application is complete, the developer can simplify its appearance by making this the sole window occupying the whole screen. On the lower left is the **Rules window**, which shows the rules and results scrolling by in the sequence in which they come up during a consultation. In effect, it provides a view into VP-Expert's inference engine while it is running. Unless the trace option is turned on, the various steps only appear momentarily as the system moves along. On the lower right is the **Values window**, which notes the attribute values in their intermediate and final stages during a consultation session. It provides another detailed look at what is going on inside, but again is a fleeting glimpse unless the trace is recording the action in a separate file.

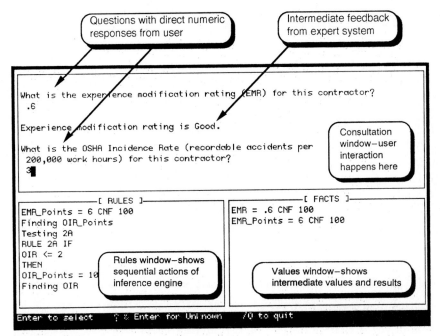

FIGURE 11-16
VP-Expert screen windows.

While we maneuver around VP-Expert using the menus, we build an expert system in a text editor. We can do this directly either by using VP-Expert's internal editor or by accessing from VP-Expert a standalone editor that might be more familiar. Either produces a text file, and if it has the suffix **.KBS**, VP-Expert can find it. As mentioned, VP-Expert can also induce an initial set of rules from tabular information in a text or database file, but we will not go into that feature in this example. All of these methods lead to the same place: a knowledge base for a given application.

Rather than go into the details of text editing, we will now skip directly to the example expert system application thus produced. Representative parts of the safety expert system are presented in Figs. 11-17 to 11-19. In Fig. 11-17 we see the **ACTIONS** section, which produces the introductory text, sets the two main goals for the consultation (develop a composite numerical safety rating to assign to **Safeval** and find which of two alternative recommendations to put in **Saferec**), and then displays the results.

Some of the 16 rules contained in the knowledge base appear in Fig. 11-18. Rules 1A to 1D determine which range the user's specified Experience Modification Rating (**EMR**) falls within, then assigns a corresponding portion of the evaluation points and comments on the given EMR. Rules 2A to 2D, not shown, work similarly to assign points for the OSHA Incidence Rate (**OIR**). Rule 3 assigns points directly in proportion to the number of training methods employed by the contractor. It also checks to be sure that the response is within the given

```
! ---- EXAMPLE EXPERT SYSTEM FOR SAFETY PERFORMANCE EVALUATION ----

ACTIONS

    DISPLAY "                    *** CONTRACTOR SAFETY EVALUATION SYSTEM ***

This expert system enables a contractor to perform a self-evaluation
of how its management policies and performance record will measure up
when compared to others regarding construction safety.

    Press any key to begin the consultation . . . ~"

    CLS

! State overall goals for the consultation:

    FIND Safeval

    EMR&OIR = (EMR_Points + OIR_Points)

    FIND Saferec

! Clear screen and then display the overall rating and recommendation:

    CLS
    DISPLAY "
The contractor's overall safety rating is {#Safeval}.

The recommended level of acceptability is: {#Saferec}.

        Press any key to end consultation . . .~";
```

FIGURE 11-17
VP-Expert ACTIONS section for safety expert system.

range. Rules 4 and 6, not shown, are similar but multiply the user's input number by 2 to get the points to assign. Rules 5A to 5C examine the user's selection from an **ASK** with three **CHOICES** options, and assign the corresponding points to the choice that matches. Rule 7 first reconfirms that all input responses were in the appropriate range then adds up all the points to compute **Safeval**. Rule 8, which uses the results of the **EMR&OIR** sum in the **ACTIONS** section, employs four criteria to recommend whether the contractor "has good safety" or "must improve."

Figure 11-19 is the user input section. It uses **ASK** statements to describe the requested information and tell the user how to enter it. The first two take direct numeric inputs for the attributes EMR and OIR. The third, fourth, and sixth questions show a menu of criteria and ask the user for a numeric response indicating how many of them apply to this contractor. The fifth **ASK** question includes a **CHOICES** statement, which enables the user to select from three defined options to assign to the attribute **Staff**.

This example is relatively simple, and only uses a few of the many possible capabilities that are found in an expert system building tool such as VP-Expert. It

```
RULE 1A
      IF EMR <= 0.5 THEN EMR_Points = 10
          DISPLAY "Experience modification rating is Excellent.";
RULE 1B
      IF EMR > 0.5 AND EMR <= 0.8 THEN EMR_Points = 6
          DISPLAY "Experience modification rating is Good.";
RULE 1C
      IF EMR > 0.8 AND EMR <= 1.1 THEN EMR_Points = 3
          DISPLAY "Experience modification rating is Average.";
RULE 1D
      IF EMR > 1.1 AND EMR <= 1.3 THEN EMR_Points = 1
          DISPLAY "Experience modification rating is Poor."
          ELSE EMR_Points = 0
          DISPLAY "Experience modification rating is Unacceptable.";
      . . . . . . . .

RULE 3
      IF Training >= 0 and Training <= 6 THEN T_Points = (Training)
        Else DISPLAY "Training methods response must be between 0 and 6."
              T_Points = -1;
      . . . . . . . .

RULE 5A
      IF Staff = None THEN S_Points = 0;
RULE 5B
      IF Staff = Part_time THEN S_Points = 2;
RULE 5C
      IF Staff = Full_time THEN S_Points = 4
          ELSE S_Points = -1;
      . . . . . . . .

RULE 7
      IF EMR_Points > -1 AND OIR_Points > -1 AND T_Points > -1
        AND R_Points > -1 AND S_Points > -1 AND P_Points > -1
          THEN Safeval = (EMR_Points + OIR_Points + T_Points
                          + R_Points + S_Points + P_Points)
          ELSE Safeval = -99
          Display "*** Error in one or more inputs.  Correct and retry.";

RULE 8
      IF EMR&OIR>8 AND Safeval>25 AND EMR_Points>=3 AND OIR_Points>=3
          THEN Saferec = Contractor_has_good_safety
          ELSE Saferec = Contractor_must_improve
      BECAUSE "To be recommended, the contractor should be at least at the
industry average in BOTH the EMR and OIR. The sum of these two is taken
to be sure that at least one of them is above average. Considering all
of the inputs, the overall safety rating should at least exceed 25,
which requires either an exceptional EMR + OIR, or at least average
contributions from most other categories.";
```

FIGURE 11-18
VP-Expert RULES section for safety expert system.

does not use graphics, access a database or spreadsheet, use custom windows, or exploit other internal commands and functions. Nevertheless, it does outline the basic structure of a rule-based expert system. The following section examines a typical consultation.

```
ASK EMR : "
What is the experience modification rating (EMR) for this contractor?";

ASK OIR : "
What is the OSHA Incidence Rate (recordable accidents per
 200,000 work hours) for this contractor?";

ASK Training : "
How many of the following training methods are normally used
 by this contractor?
   A -- Project safety orientation sessions for new workers
   B -- Weekly toolbox meetings that effectively emphasize safety
   C -- Printed safety materials regularly given to workers
   D -- Task specific training for new or dangerous procedures
   E -- First-aid training for workers and supervisors
   F -- Professional safety training for all supervisory personnel";

ASK Records : "
How many of the following record tracking or evaluation methods are
 normally used by this contractor?
   A -- Accident records broken down in detail by supervisor
   B -- Senior managers give high priority to monitoring these records
   C -- Senior managers use these records for employee evaluation,
           promotion and compensation procedures";

ASK Staff : "
Using the cursor keys and by pressing the ENTER key, select the option
below that best describes the use of professional safety staff in this
company.";

CHOICES Staff : None, Part_time, Full_time;

ASK Policies : "
How many of the following field operations policies are normally used
 by this contractor on most jobsites?
   A -- Safety is explicitly considered and designed into work methods
   B -- Jobsites provide personal protection gear and safety equipment
   C -- Supervisors rigorously enforce safety equipment and rules
   D -- They have frequent self-inspections by knowledgeable personnel
   E -- They have meaningful awards and recognition procedures for
           workers, crews and supervisors with excellent safety records.";
```

FIGURE 11-19
VP-Expert input section for safety expert system.

Example Consultation

Upon starting the safety expert system, the user first sees a brief title and introductory screen, then VP-Expert's inference engine goes to work. Rules and interim results pass by in the Rules window until finally VP-Expert needs some input from the user. We saw the first two questions of a typical interaction in Fig. 11-16. After each user input, VP-Expert works for a while, displaying attribute values in the Values window as they are determined. In Fig. 11-20 we see the status of the system at the moment after EMR and OIR inputs have been inputted and produced values of **EMR_Points** and **OIR_Points** of 6 each. The **CNF** (confidence factor) of 100 for each value results because the user did not specify a lesser value with

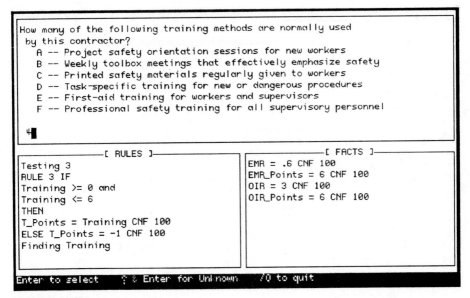

How many of the following training methods are normally used
  by this contractor?
   A -- Project safety orientation sessions for new workers
   B -- Weekly toolbox meetings that effectively emphasize safety
   C -- Printed safety materials regularly given to workers
   D -- Task-specific training for new or dangerous procedures
   E -- First-aid training for workers and supervisors
   F -- Professional safety training for all supervisory personnel

4

```
----------[ RULES ]---------
Testing 3
RULE 3 IF
Training >= 0 and
Training <= 6
THEN
T_Points = Training CNF 100
ELSE T_Points = -1 CNF 100
Finding Training
```

```
---------[ FACTS ]---------
EMR = .6 CNF 100
EMR_Points = 6 CNF 100
OIR = 3 CNF 100
OIR_Points = 6 CNF 100
```

Enter to select   ? & Enter for Unknown   /Q to quit

FIGURE 11-20
VP-Expert Safety expert system paused at a consultation input.

the inputs, and internal rules did not reduce them either. At the time Fig. 11-20 was snapped from the screen, the safety expert system had paused to ask the user the next question to find the number of methods to assign (internally) to the attribute **Training**.

In Fig. 11-21 we see the screen when the question with the **CHOICES** method is being used for input. By now, additional facts have appeared in the Values window, and the user is in the process of choosing **Part_time** as the input for the variable **Staff**. In this case, the user can select only one of the options, but VP-Expert also allows **Plural** variables that in turn can take multiple inputs from such a CHOICES menu.

After the user has successfully made all of the inputs, the safety expert system concludes with the screen shown in Fig. 11-22. In this case, the total of points assigned to Safeval is 28. Given that this exceeds the minimum of 25 and that the results also satisfy the other criteria in Rule 8, the recommendation for Saferec is favorable.

Having completed the initial consultation, the user can use the **WhatIf** option from the **Consult** menu to select and modify any of the variables used in the consultation—including internal ones as well as input values—and rerun the consultation. Since all of the other inputs stay the same and need not be reentered, the user can sit back and watch the action in the Rules and Values windows while the inference engine reasons away and comes to its revised conclusions. Figure 11-23 is a composite of two screens. In the lower one the user has selected WhatIf and is in the process of selecting EMR to input a new value. Having changed the EMR value from 0.6 to 1.2, we see in the upper screen of the figure

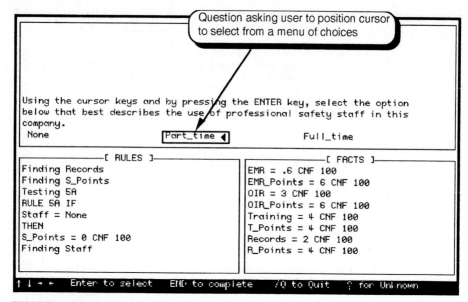

FIGURE 11-21
Selecting an input from a VP-Expert Safety CHOICES menu.

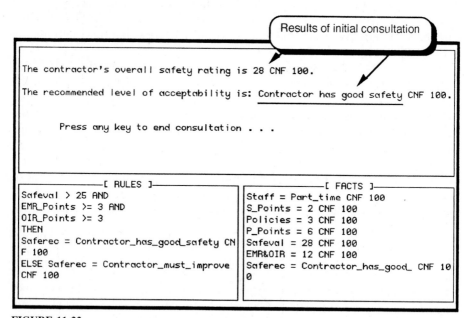

FIGURE 11-22
Results of the initial VP-Expert Safety expert system consultation.

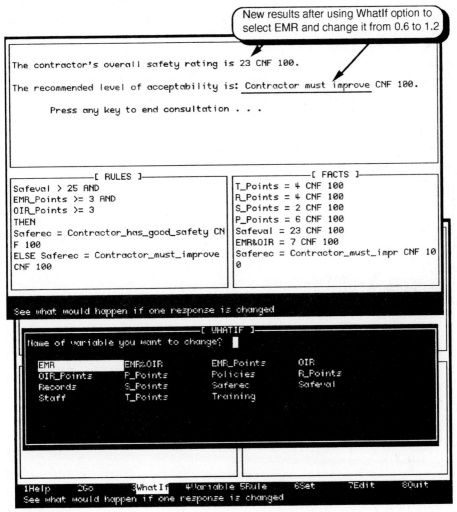

FIGURE 11-23
VP-Expert Safety expert system during and after **WhatIf** input.

that the new safety rating from the computation of **Safeval** drops to 23, and the recommendation drops to "Contractor must improve."

With a high-speed computer, the actions and interim results appearing in the Rules and Values windows fly by too quickly to be easily understood. VP-Expert overcomes this feature in several ways, all from Consult menu. First, one can pause to use the **4Variable** and **5Rule** menu options to inspect individual variable values and rules midway through a consultation. Second, one can use the option **Slow** under the **Set** menu option literally to slow down the rate at which VP-Expert works its way along. Third, one can select **Trace** under the **Set** option to record a complete history of a consultation in a **Trace file** (suffix **.TRC** appended to the name of the KBS application name) and then print and peruse this text file at leisure later on.

Figure 11-24 is a trace file for the interaction that we have been following. Notice how it begins with the action **FIND Safeval** (refer to the ACTIONS section of Fig. 11-17). It looks through the Rules, and the first place that it finds Safeval on the left side of an assignment clause is in Rule 7. Thus the next step in the trace is **Testing 7**. If you look at that assignment clause in Rule 7, you will note that it is the sum of the six variables that make up the quantitative part

```
Testing SAFETY.kbs
(= yes CNF 0 )
!    Safeval
!    !    Testing 7
!    !    !    EMR_Points
!    !    !    !    Testing 1A
!    !    !    !    !    EMR
!    !    !    !    !    !    (= .6 CNF 100)
!    !    !    !    Testing 1B
!    !    !    !    (= 6 CNF 100 )
!    !    !    OIR_Points
!    !    !    !    Testing 2A
!    !    !    !    !    OIR
!    !    !    !    !    !    (= 3 CNF 100)
!    !    !    !    Testing 2B
!    !    !    !    (= 6 CNF 100 )
!    !    !    T_Points
!    !    !    !    Testing 3
!    !    !    !    !    Training
!    !    !    !    !    !    (= 4 CNF 100)
!    !    !    !    (= Training CNF 100 )
!    !    !    R_Points
!    !    !    !    Testing 4
!    !    !    !    !    Records
!    !    !    !    !    !    (= 2 CNF 100)
!    !    !    !    (= (Records * 2) CNF 100 )
!    !    !    S_Points
!    !    !    !    Testing 5A
!    !    !    !    !    Staff
!    !    !    !    !    !    (= Part_time CNF 100 )
!    !    !    !    Testing 5B
!    !    !    !    (= 2 CNF 100 )
!    !    !    P_Points
!    !    !    !    Testing 6
!    !    !    !    !    Policies
!    !    !    !    !    !    (= 3 CNF 100)
!    !    !    !    (= (Policies * 2) CNF 100 )
!    !    (= (EMR_Points + OIR_Points + T_Points
                          + R_Points + S_Points + P_Points) CNF 100 )
(= (EMR_Points + OIR_Points) CNF 100 )
!    Saferec
!    !    Testing 8
!    !    (= Contractor_has_good_safety CNF 100 )
```

FIGURE 11-24
VP-Expert Safety expert system trace of a consultation.

of the evaluation. Therefore, at the fourth level over in the trace, the inference engine works its way through rules 1A to 6, requesting inputs as necessary and assigning the corresponding point values to each component. It then comes out to compute the total for Safeval. The next step is to compute EMR&OIR, and this then enables Rule 8 to come up with the qualitative recommendation assigned to **Saferec**.

Even at this level of simplicity, there is considerable value to such an expert system. Although it would be possible to implement a similar system using a procedural language or some other type of computer software, it is worth noting that, using a tool designed specifically for such an application (VP-Expert in this case), it took a beginner less than six hours to produce the example shown here, and he encountered few obstacles in the process. Expert systems do have their place in the toolbox of today's construction professional.

SUMMARY

Artificial intelligence (AI) is among the newer and more advanced fields of computer science, and it comes the closest yet to replicating the thinking processes of human beings. Although it is not nearly so widely known in construction as the established methods that focus on computations and data processing, it has strong potential in this experience- and judgment-oriented industry. General applications of AI range from language and speech processing through machine vision and robotics to general planning and problem solving and are now extending to machine-based learning. But the AI application area that has been most successful thus far is in replicating narrow areas of human expertise through software called knowledge-based expert systems, more commonly shortened to expert systems.

AI researchers and application developers who build expert systems have taken a fresh look at how computers can assist human activities and inevitably have introduced new concepts and terminology. They try to define more explicitly the types of knowledge that constitute the knowledge domain of an expert system, and generally classify this knowledge into deep and surface knowledge. They pay special attention to the heuristic thinking used by human experts—rules of thumb and the like—that is based on years of experience with what works and what doesn't. Expert systems differ from conventional computer programs by emphasizing declarative knowledge over procedural knowledge.

Developers of expert systems use a variety of techniques to represent the knowledge contained in an expert system. Common computer-based methods include facts, certainty factors, logical expressions, rules, semantic networks, and frames. Some of these conveniently organize and connect elements of knowledge with hierarchical representations, with links, or with properties called inheritance.

The architecture of typical expert systems is summarized in Fig. 11-3. Components include a user interface, the context, the inference engine, an explanation facility, a knowledge acquisition facility, and the knowledge base. An inference engine can employ several different problem-solving strategies. Two broad categories of strategy are derivation (or classification) and formation (or synthesis).

Common control strategies used in expert systems include forward chaining, backward chaining, or mixtures of these two. Within them are the depth-first versus breadth-first search options. Other control strategies are problem reduction, plan-generate-test, and agenda control.

Designing and building expert system applications is similar to the general approach to software development described in Chapters 5 to 7, but this process has some special emphases. For example, a professional knowledge engineer can facilitate the knowledge acquisition phase for moderate to complex expert systems. Developers also have available a whole new range of methods for knowledge representation.

There exists a wide variety of software tools for building expert systems, and they are evolving rapidly. Although it is feasible to implement expert system applications in conventional procedural languages such as Pascal and C, even these have been enhanced with object-oriented extensions to make them more effective in this area. Symbolic AI languages such as LISP and Prolog are even better suited to the task, although their common implementations as interpreters rather than compilers can cause speed problems as applications scale up in size and complexity. The software market also contains a range of specialized and powerful hybrid expert system development environments, though some can be very expensive and require professional staff who know the specific tools. For getting started with expert systems applications in construction, the best options at present are within a large number of low-cost rule-based expert system building tools. Their simpler architectures limit their ability to represent knowledge and implement problem-solving strategies, and most have a predefined user interface, but they still can address a variety of practical construction problems that would otherwise require calling in a human expert, and they can do so economically and effectively.

Although it is somewhat arbitrary to separate basic from advanced features and capabilities in a rapidly evolving field such as this, some useful enhancements include linkages to other software, such as databases and analytical programs; graphical capabilities in the user interface; stronger methods for dealing with uncertainty and replicating qualitative reasoning; and adaptation to real-time monitoring and control systems.

It is still too early to tell where in the construction industry AI and expert systems will have their most valuable applications. Obvious candidates include estimating, getting constructibility into design, legal and contractual issues (including evaluation and resolution of claims), project planning and control, and occupational safety and health. All of these applications rely extensively on human expertise, and some of this can be captured in a computer-based expert system. Given the range of interesting construction applications, it is not surprising that developers are already at work creating useful applications, and some have already made it to the field. But we must also be aware of the limitations. Construction managers, estimators, and other professionals are often broad generalists, capable of understanding and solving a wide variety of technical, economic, and human problems in a single day. Expert systems thus far have been specialized in their expertise and they quickly hit their limits when the scope of a problem expands beyond their defined boundaries. Most work better as assistants to human professionals, and these systems are far from replacing such people.

The first example in this chapter showed how an expert system implemented in Prolog can provide useful advice to a field engineer or superintendent regarding the selection of the appropriate type of compactor for a given soil type. It highlighted some advantages of logic programming over procedural language programming.

The second example employed a powerful and flexible yet inexpensive and easy-to-use rule-based expert system called VP-Expert to address construction safety. It focused on evaluating the managerial and organizational aspects of safety and, if developed further, it could assist a contractor in self-evaluation and improvement.

REVIEW QUESTIONS

1. What general characteristics of construction engineering and management planning and problem solving give them strong potential for expert system applications?
2. Give one example from construction that would be classified as deep knowledge, and give a second example that would be classified as surface knowledge.
3. Which of the following would be classified as heuristics?

   - Don't place concrete if rain seems imminent.
   - The volume of a rectangular footing equals length × width × height.
   - Well-trained workers are more productive.
   - If another truck is waiting, load the present one a bit light.

4. Briefly describe the role of a knowledge engineer in the development of an expert system for an application such as estimating concrete-placing productivity.
5. Which of the following are methods of knowledge representation in an expert system?

   - Certainty factor
   - Explanation facility
   - Logical rule
   - Frame
   - Multiple inheritance
   - Inference engine

6. Which of the following best describes the function of the inference engine in an expert system?

   - Stores the knowledge that infers the solution to a particular problem
   - Controls the reasoning process that leads to a solution
   - Provides the user with an interface so the user can see the inference process
   - Explains what is going on while the system works on a problem

7. Give an example of how a backward-chaining control strategy would work in trying to solve a scheduling problem where a critical milestone has to be reached by a certain date.

8. Briefly describe how the knowledge representation stage of expert system application development differs from the system design stage of conventional software application development.

9. What should be the domain expert's role, if any, in the testing and validation stage of developing an expert system application?

10. Which of the following types of development software are suitable for developing expert system applications?

   - Hybrid expert system development environment
   - Pascal
   - Simulation program
   - Object-oriented language
   - Prolog
   - Database package
   - VP-Expert
   - Spreadsheet package

11. State two of the main ways in which an object-oriented language such as Smalltalk differs from a procedural programming language such as FORTRAN.

12. Which expert system development language comes closest to the formal logical reasoning of predicate calculus using a declarative style of programming?

13. This chapter briefly described a number of advanced concepts and features that could enhance an expert system. State which of the advanced features would be most useful for the following applications:

   - Automated control of an excavator
   - Dealing with uncertainty in deciding a markup in the bidding process
   - Using an expert system as a user interface for an estimating cost database
   - Providing constructibility input to a structural designer

14. Briefly compare and contrast the nature and limitations of computer-based expert system expertise relative to that of an experienced construction professional.

DEVELOPMENT PROBLEMS

For hands-on experience with an expert system development tool to which you may have access, you might first consider the following:

1. Reimplement this chapter's compactor selection example in an environment other than Prolog.

2. Reimplement the safety example in a system other than VP-Expert.

3. You could then develop new expert systems for other applications of similar scale, such as the following: **How to handle the influence of weather on production operations.**

Build an expert system that will use information from the evening weather report about tomorrow's forecast to help a superintendent decide whether and how to undertake the operations that have been scheduled for the following day. From the weather report you might consider the following:

- Temperature
- Precipitation
- Cloud cover
- Humidity
- Wind velocity
- Degree of certainty

To keep the problem manageable, narrow your focus to a small range of distinctly different types of field operations, each of which varies in its sensitivity to the weather, such as these:

- Foundation excavation
- Forming for concrete
- Wood framing
- Roofing
- Exterior painting
- Landscaping
- Placing concrete outdoors
- Steel erection
- Masonry
- Interior finishes

Your expert system might also consider the overall context of the project, such as schedule pressure, size of crew(s) affected, the stage of construction (relative to the overall schedule, degree of enclosure), cost of delays, and so forth, but the main focus should be on providing recommendations on whether or not to work on a particular type of activity under given meteorological circumstances.

The program could begin by asking the user for specific parameters from the actual weather forecast and for the types of operations on which the superintendent would like to work if weather were not a factor. It could then apply a series of rules inferring the likely impacts of different weather conditions on the various types of operations. Obvious examples are, don't paint in the rain, don't place concrete when it's freezing or raining hard, don't erect steel in a strong wind, and the like. More subtle advice could deal with trade-offs in productivity versus temperature, and such.

If time permits, you should try your hand at knowledge engineering by contacting professional superintendents to try to learn how they think through decisions of this type. What numbers do they use in their rules of thumb? How do they decide when to call and tell a crew to stay home?

As an enhancement, your expert system could go on to provide recommendations regarding methods to mitigate the impacts of weather, such as enclosures, heaters, and so forth. What would be the cost and productivity trade-offs?

SUGGESTIONS FOR FURTHER READING

AI and expert systems are rapidly developing fields, so one must read rather widely to keep track of what is going on. Although there are too many references here to annotate individually, most have been cited in the text of this chapter so that you can see their relevance in context. However, for an introduction, overview, and some added depth in the field, particularly for engineering and construction readers, we recommend Dym and Levitt (1991), Harmon, Maus and Morrisey (1988), Maher (1987), Parsaye and Chignell (1988), Walters and Nielsen (1988), and Winston (1992).

AAIS Full Control Prolog Reference Manual. Mountain View, Calif.: Advanced A.I. Systems, 1991.

AI Magazine. American Association for Artificial Intelligence (AAAI), 445 Burgess Drive, Menlo Park, Calif. 94025.

Amirkhanian, Serji N., and Nancy J. Baker. "Expert System for Equipment Selection for Earthmoving Operations." *Journal of Construction Engineering and Management,* ASCE, Vol. 118, No. 2, June 1992, pp. 318–331. Describes a 930-rule expert system written in VP-Expert.

Bratko, Ivan. *Prolog Programming for Artificial Intelligence,* 2d ed. Reading, Mass.: Addison-Wesley, 1990.

Chan, Weng-Tat, and Boyd C. Paulson, Jr. "Exploratory Design Using Constraints." *Journal of Artificial Intelligence in Engineering Design and Management,* Vol. 1, No. 1, December 1987, pp. 59–71.

Clocksin, W. F., and C. S. Mellish. *Programming in Prolog,* 3d ed. New York: Springer-Verlag, 1987.

Dologite, D. G. *Developing Knowledge-Based Systems Using VP-Expert.* New York: Macmillan, 1993 (includes disk).

Dym, Clive L., and Raymond E. Levitt. *Knowledge-Based Systems in Engineering.* New York: McGraw-Hill, 1991.

El-Bibany, Hossam, Lai-Heng Chua, and Boyd C. Paulson, Jr. "Coordination Between Project Participants through Constraint Management." *Proceedings of the 7th International Symposium on Automation and Robotics in Construction,* Bristol, England, June 5–7, 1990, pp. 505–512.

Harmon, Paul, and David King. *Expert Systems: Artificial Intelligence in Business.* New York: John Wiley & Sons, 1985.

Harmon, Paul, Rex Maus, and William Morrissey. *Expert Systems Tools and Applications.* New York: John Wiley & Sons, 1988.

Harmon, Paul, and Brian Sawyer. *Creating Expert Systems for Business and Industry.* New York: John Wiley & Sons, 1990.

IEEE Expert. IEEE Computer Society, 10662 Los Vaqueros Circle, Los Alamitos, Calif. 90720-2578.

Levitt, Raymond E. "Expert Systems in Construction: State of the Art." In Mary Lou Maher, ed. *Expert Systems for Civil Engineers: Technology and Application.* New York: American Society of Civil Engineers, 1987, pp. 85–112.

Maher, Mary Lou, ed. *Expert Systems for Civil Engineers: Technology and Application.* New York: American Society of Civil Engineers, 1987.

Modex: Modularization Decision Support Software—User's Guide. Special Publication 29-2, Construction Industry Institute, Austin, Tex. May 1992.

Parsaye, Karman, and Mark Chignell. *Expert Systems for Experts.* New York: John Wiley & Sons, 1988.

Quintus MacProlog Reference Manual. Mountain View, Calif: Quintus Computer Systems, 1990.

Raynar, Karl A., and Gary R. Smith. "Intelligent Positioning of Mobile Cranes for Steel Erection." *Microcomputers in Civil Engineering,* Vol. 8, No. 1, 1993, pp. 67–74.

Smith, Peter. *Expert System Development in Prolog and Turbo Prolog.* New York: Halsted Press, 1988.

Turbo Prolog Reference Manual. Scotts Valley, Calif.: Borland International, 1988.

VP-Expert: Rule-Based Expert System Development Tool. Orinda, Calif.: WordTech Systems, 1989.

Walters, John R., and Norman R. Nielsen. *Crafting Knowledge-Based Systems: Expert Systems Made Realistic.* New York: John Wiley & Sons, 1988.

Wang, Wally, and John Mueller. *Illustrated VP-Expert.* Plano,Tex.: Wordware Publishing, 1989.

Winston, Patrick H. *Artificial Intelligence,* 3d ed. Reading, Mass.: Addison-Wesley, 1992.

PART
V

APPLICATION
PACKAGES

Now, here, you see, it takes all the running you can do, to keep in the same place. If you want to get somewhere else, you must run at least twice as fast as that.

Lewis Carroll*

C hapter 5, which described the feasibility study and analysis for a new construction application, expressed strong reservations about undertaking custom software development efforts if good commercial package solutions in the application area already exist. Chapter 7 provided guidance for evaluation and procurement of these packages. But what kinds of solutions are available that apply well in construction? The next four chapters introduce key application areas and present two or three representative examples in each.

There is no attempt here to present a comparative survey of current software in each field. Such a survey would be out of date before this book could even be published. Lewis Carroll's Alice would appreciate that problem. For current information on rapidly changing products, the reader must refer to current

---

Through the Looking Glass, Chapter 2, 1872.

trade magazines and journals, attend trade shows and conferences, stay in touch with active user groups, and consult software directories. Numerous weekly and monthly magazines—some general, some focused on particular types of computers and software—often contain information relevant to construction applications. The journals of some contractor associations and professional groups also publish surveys and articles about computer applications. There are trade shows and conferences focused specifically on computer applications in architecture, engineering, and construction. Directories are available both in printed form and as online computer databases and bulletin boards.

The selections of the applications for illustration in these chapters are not intended to imply endorsements. Some are indeed widely used and may be predominant in their market, whereas others illustrate a contrasting approach. Once this book has been in print for a while, the versions shown here will be quite out of date. It is the concepts and principles behind the applications that are important for our purposes. Before going out to acquire software in any of these categories, the reader should follow the procedures that were outlined in Chapters 5 and 7 and apply them to the market that exists when you read this book, not when it was written.

Only some of the most common and widely used types of commercial applications are represented here: estimating in Chapter 12, project planning and scheduling in Chapter 13, accounting and cost control in Chapter 14, and operations simulation in Chapter 15. Some entrepreneurs offer products in almost all of the many application areas surveyed in Chapter 2, but most of these companies are small and specialized. Space available in a book like this precludes covering more than a sampling, hence we have focused on the big ones. Significant omissions among the latter still include *computer-aided design (CAD)* and *computer-aided engineering (CAE).* Our plan is to include these later in a parallel volume that will be aimed more at the engineering design professions.

CHAPTER
12

ESTIMATING

When we mean to build,
We first survey the plot, then draw the model;
And when we see the figure of the house,
Then we must rate the cost of the erection;
Which if we find outweighs ability,
What do we then but draw anew the model
In fewer offices, or at last desist
To build at all?

William Shakespeare*

There are numerous methods for preparing cost estimates for construction projects. Estimates range in scope and detail from educated guesses to detailed contractor bids. The latter are based on a relatively complete set of plans and specifications; and a thorough, accurate, and detailed estimate is much broader in concept than merely determining costs. The estimator must practically build the project on paper, assessing quantities not only of the contract materials reflected in the drawings but also of the temporary materials, such as formwork for concrete, and temporary plant. Such estimates may require the estimator to hypothesize alternative methods for different components of the project, determine the labor, equipment, and materials required by each method, evaluate the productivity and costs, and select those methods that, taken together, will complete the project on schedule and at the lowest overall cost. Computers can assist the estimator every step of the way.

Although various computer-based estimating techniques can be used from project concept through design and into construction, we will focus on detailed

Parts of this chapter have been adapted and condensed from Chapter 11 of *Professional Construction Management*, 3d ed., by Donald S. Barrie and Boyd C. Paulson Jr., McGraw-Hill, Inc., New York, NY, 1992. Reprinted by permission of McGraw-Hill, Inc. Details of estimating methods and procedures are contained in that book.
*Henry IV, Part 2, I:3

estimates prepared from completed plans and specifications. This chapter explores how computers can assist in preparing the cost estimates used by construction managers, general contractors, and subcontractors. The stages at which they best apply include the following:

1. Project location and tracking
2. Collection of cost and productivity data
3. Quantity take-off
4. Establishment of work methods and productivity rates
5. Estimation of direct and indirect costs
6. Compilation, analysis, and bidding

We briefly examine several computer applications that are now at work in construction estimating. Most help take the drudgery out of estimating, improve accuracy, and leave estimators with more time to explore alternatives that can reduce construction costs. Estimating software currently available commercially falls into three categories: the first focuses on data collection (steps 1 and 2); the second on quantity take-off (step 3); and the third addresses cost-related topics covered in steps 5 and 6. However, given such a wide range of applications, we can provide details about only a few and briefly mention or illustrate others. Simulation software, spreadsheet packages, and engineering design and analysis programs can assist in step 4, but they will not be emphasized in this chapter.

We begin by fitting estimating into an overall integrated computer information system for construction. We then review fundamental components and procedures of the estimating process and briefly illustrate some applications that can help at various stages. Next, we examine a few advanced techniques implemented in some packages, then outline key applications and limitations of today's construction estimating software. We close with two somewhat more detailed examples; the first focuses on quantity take-off for a site-grading project, and the second on cost estimating for a building project.

ESTIMATING AS PART OF AN INTEGRATED SYSTEM

This is the first of four chapters dealing with construction applications, and most of these applications are being increasingly integrated in various ways. We thus digress for a moment to look at estimating in its overall integrated context.

For construction contractors, estimating normally occurs at the transition between project design and project construction. For this reason, it draws input from many sources and provides its output to various project control systems.

Several components of project planning and control systems are shown in Fig. 12-1, as well as their relationship to estimating. Several of the links in the figure now exist in electronic form. CAD systems can provide a bill of materials and quantities to estimating; in-house and commercial cost and productivity databases also connect directly to popular computer-based estimating systems; both manufac-

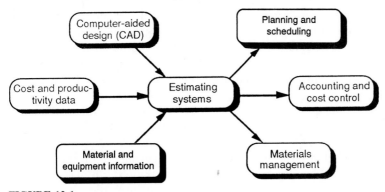

FIGURE 12-1
The place of estimating in integrated computer systems.

turers and information services publish electronic versions of traditional materials specification catalogs and equipment productivity reference books. Outputs from the estimate can include a budget and work breakdown structure upon which to build job-costing and cost-engineering systems; activity time durations, costs, and resource loadings to expedite the preparation of computer-based schedules; and detailed information for materials procurement and tracking systems.

As we go along in the application chapters, we will point out some opportunities for integration and mention specific packages that implement such links. Keep Fig. 12-1 in mind, especially as you read through Chapters 12, 13, and 14. Chapter 16 will explore even more extensive and advanced integration concepts.

BASIC CONCEPTS OF ESTIMATING SOFTWARE

The contractor's estimate is the foundation for a successful project. It must be low enough to obtain the work, yet high enough to yield a profit. To begin, the contractor must obtain information about projects available for bidding. Based on plans and specifications, detailed bid estimates first require a careful tabulation of all the quantities for a project or portion of a project; this is called a *quantity take-off*. Almost all successful contractors now estimate production components with separate evaluations for work hours, labor, materials, equipment usage, and subcontractors. These quantities and resources are then multiplied by unit costs, and the resulting sum represents the estimated *direct cost* of the facility. The addition of *indirect costs*, home office overhead, profit, escalation, and contingency results in the total estimated project cost. We explain each of the main steps in more detail as follows.

Types of Detailed Estimates

Detailed estimates fall into two broad categories: (1) *unit-price*, and (2) *fixed-price*. There are several estimating packages that focus on each category. Unit-price estimates usually apply to heavy construction jobs such as dams, tunnels, and highways. Here the unit prices are set constant, while quantities vary within limits inherent in the nature of the work. Quantities may overrun or underrun owing to a number of potential causes, such as additional foundation excavation to solid rock, poor ground conditions, excessive water in tunnels, or other factors usually associated with the anticipated accuracy of geological and geophysical interpretation.

Fixed-price estimates include those for *lump-sum* and *guaranteed-maximum-price* contracts. On lump-sum projects, the estimate can be developed, in the absence of changed conditions, on the basis of a complete set of plans and specifications. On negotiated fast-track projects featuring a guaranteed maximum price, many, but not all, of the detailed drawings are usually complete; firm subcontracts and material prices may have been obtained for a sizable portion of the work; prices for major equipment are known, and some of the work may have been completed; an added contingency allows for the remaining unknowns.

Project Location and Tracking

Before you can estimate a project, it is necessary to find out about it. McGraw-Hill's *Dodge Reports* has long been a primary bidders' information service. On a targeted subscription basis, they send out frequent reports of new projects in various parts of the country. These reports track projects through the stages of planning, design, bidding, and into construction, give important dates, and tell contractors and suppliers how to contact the owner, designer, and contractor (once bid). The service also helps owners and managers get the word out about their projects to improve bidding competition. Contractors can limit the information for which they pay to subscribe by specifying criteria regarding type of project, region, and size.

Recently, this service has gone online in the form of *Dodge DataLine*[2]. One accesses it using Dodge's communication software, available for both Apple Macintoshes and IBM PC compatibles. With this program, an estimator can access Dodge's centralized database via communications utilities such as Tymnet and TeleNet (often accessible from a local call) and then explore the same data that are the source of the printed *Dodge Reports*. One can limit searches by even more criteria than the mailed version—up to 17 in all.

The system is menu-driven and easy to use. You begin with a main menu and proceed from there to specify search criteria on a screen such as that shown in Fig. 12-2*a*. You can then move to subordinate screen menus that allow you to be very specific about project type (we chose commercial or industrial warehouse or medical), locations (we chose Hawaii or California), size ranges (we said over $1,000,000), and stage of planning, design or bidding (we focused on those from planning on up to bidding). This narrowed search produces a summary list of projects that fit the criteria, and from this one can call up the details of specific projects on the list, such as that shown in Fig. 12-2*b*. You can modify the criteria as you go along to further narrow the search. For example, in this case, after seeing an initial list where this project was one among many, we decided to focus

```
                    Dodge DataLine2(c) Search Menu

DATE RANGE: 1/5/94 TO 1/6/94

Type  search category letter(s) and press <ENTER>.  Example:  PVA.
The database will be searched for  Dodge Reports issued within the
DATE RANGE  above.  To change the DATE RANGE, include C (Change Date
Range) as your first search category.  Pressing <ENTER>, without
specifying search categories will retrieve all reports issued
in your subscription territory since the date of your last access.

                         SEARCH CATEGORIES


        C--Change Date Range      A--Action Stage
        G--Geography              B--Bid Date
        P--Project Type           O--Ownership
        V--Valuation              I--Trades/Materials/Equipment
        S--Square Footage         Z--Search Filters
        H--Story Height           T--TEXT SEARCH


   F1Help                 F2TrackingLists           F3AutoSearch
   F7Logoff               F8PowerSearch             F10MainMenu
```
```
 Esc for Command?, Home for Status  ||    Capture Off    ||  On: 00:02:50
```

(*a*)

```
    ==============================================================
         DODGE REPORT 18  of 35 (c) 1993, by McGraw-Hill Inc.
    ==============================================================
    Dodge # 92- 439413-3      400228                    Planning
    Date: 12/20/93  Last: 09/03/93  First: 07/15/92
    Project Codes: 190, 205, 070
    Estimate: $90,000,000 - $100,000,000
    ----------------------
    Hospital/Outpatient Treatment Ctr/Parking Structure
    Honolulu, HI (Honolulu) Tripler Army Medical Center 96819
    Design development underway-working drawings to begin second or
    third quarter of 1994-bidding and construction not expected
    until early 1995
    Method of Contracting: GC to be Competitively Bid
    Owner: Dept of Veterans Affairs--Contracting Officer, 300
        Ala Moana, P O Box 50188, Honolulu, HI 96850 (808-541-1442)
    Architect: Architects Hawaii Ltd, Walter Muraoka, 1001
        Bishop  Ste 300, Honolulu, HI 96813 (808-523-9636)
    Structural Information: Buildings: 4 / Stories Above Grade: 4
    Additional Features: convert existing nursing wing for VA use -
    add outpatient bldg - construct 60-bed nursing home - 97 bed
    acute in-patient facility - parking structure for 700 cars

 <ENTER>=More U=PgUp N=NxtRept P=PrevRept O=Sort #=View# L=List S=SelDnld T#=Trak#
 TX=NegTrack   F1Help F5Download  F7Logoff F8SearchMenu F9ModifySearch  F10MainMenu
```

(*b*)

FIGURE 12-2
Dodge DataLine[2] screens: (*a*) menu from which to select search categories and boundaries; and (*b*)
detailed report on a specific project. (From *Dodge DataLine Demo Disk*, F.W. Dodge/
McGraw-Hill, New York 1991.)

only on competitively bid projects, and this project was then the eighteenth of 35 on a narrowed list.

This type of online information will become increasingly available as the network utilities and information services continue to expand in the 1990s and into the next century. Quick access to such information will improve the knowledge and competitiveness of the firms that adopt the new technologies to obtain it.

Collecting Specifications and Cost and Productivity Data

Information that is useful in estimating is becoming widely available electronically. The famous *Sweet's Catalog*, published by the Sweets Group of McGraw-Hill, consists of over a dozen annually updated volumes that collect information from most major materials and equipment suppliers and organize it according to the Masterformat classification system, by the Construction Specifications Institute (CSI). It is now available in a *CD-ROM* format called SweetSource that can be viewed on PCs running Microsoft's Windows system program. Not only the text of specifications but also illustrations, photographs, and CAD detail drawing files for a wide range of building products are online and can be copied and incorporated into shop drawings and submittals. It is indexed—and thus can be searched— by CSI Masterformat code, manufacturers, trade names, key words, and product areas.

For information on construction equipment, Dataquest's Machinery Information Division has teamed up with the Grantlun estimating software company to produce Dataquest Equipment Cost Estimator (DECE). This program produces detailed cost estimates for all components of machine usage on a project. Figure 12-3a is a screen from DECE that lists categories from its large equipment database and shows that a subset of machines has been selected for a fleet called SAMPLE. Detailed assumptions, components, and results of cost calculations for a specific machine are shown in Fig. 12-3b.

Although DECE focuses on the equipment cost component of estimating, it also provides a wealth of online information and considerable flexibility for an estimator to modify the parameters and assemble numerous separate fleets for different kinds of operations within a construction project. But the system does not address productivity. This aspect of estimating is picked up when these costs are combined with a more comprehensive estimating program such as the Hard Dollar Estimating Office System created by the same developer, Grantlun.

Caterpillar Inc. produces and markets a variety of equipment-related programs for maintenance management, production and cost estimating, and financial calculations. One Caterpillar program, called Fleet Production and Cost Analysis (FPC), can produce results similar to those shown in Fig. 12-3, or at least do so for earthmoving fleets, and includes production computations (e.g., cost per cubic meter). Chapter 15 will illustrate another of Cat's programs, VEHSIM, which makes a detailed comparative analysis of different machines running over one or more haul roads whose characteristics can be specified in great detail.

```
                        F L E E T
                F l e e t   C a r d      DataBase Date: 6-19-1993
================================================================================
 Fleet ID: SAMPLE          Description: Demonstration Fleet v1.0
================================================================================
     Equipment Catagory        | Count|      Equipment Catagory        | Count
 ------------------------------ | -----| ------------------------------ | -----
     Air Tools and Equipment    |   2  |  Marine                        |   0
     Crushing and Conveying     |   0  |  Pile Driving                  |   2
     Asphalt &  Bituminous      |   6  |  Pumping                       |   0
     Compaction                 |  21  |  Road Maintenance              |   0
     Concrete                   |   0  |  Shop Tools                    |   0
     Drilling                   |   0  |  Trailers                      |   0
     Tractors & Earthmoving     |   0  |  Trucks                        |   0
     Excavating                 |   1  |  Tunneling                     |   0
     Motors & Generators        |   0  |  Miscellaneous                 |   0
     Hoists & Derricks          |   1  |  Custom                        |   0
     Lifting                    |   1  |  Last Rev:06/27/93     Total   |  32
================================================================================


================================================================================
 Add      Copy      Delete      Help     Modify      Note     Search     Exit
 Look-up and page through the equipment cards
```

(*a*)

```
                         E S T I M A T I N G
    DECE History Estimate
 ==============================================================================
 | Fleet ID: SAMPLE          Description: Demonstration Fleet v1.0            |
 | Model: BG-220B            Mfr: BARBER-GREENE        Eq No: DECE32473       |
 | Config: Standard Screed Width:8'0' SV-80 |Wheel Drive:4                    |
 | Power:  Diesel           Yr Manf: 1989               Alt Eq No:            |
 |---------------------------------------------------------------------------|
 |  CWT  |Cubes|   Base   | Econom|  Use | OVERHEAD/Hr |      OWNERSHIP/Hr    |
 |       |     |   Price  | Life  |  Hrs |             | Deprectn       CFC  |
 |-------|-----|----------|-------|------|-------------|---------------------|
 D 250.3 |  38 |  147,610 | 6,600 |  820 |    0.00     |  18.29  |    9.28   |
 H       |     |        0 |     0 |    0 |    0.00     |   0.00  |    0.00   |
 E       |     |  152,000 | 7,800 |  820 |    0.27     |  16.21  |    9.36   |
 |---------------------------------------------------------------------------|
 |          OPERATING/Hr                  |  T O T A L  C O S T  >   Short    |
 | Labor   Parts  Sply  Tires  Fuel  Lube | MONTH   WEEK    DAY      HOUR     |
 |----------------------------------------|----------------------------------|
 D 15.11 | 24.95 | 4.99 | 0.81 | 3.75 | 0.39 |13,652.32|3,822.64|955.65 |143.35 |
 H  0.00 |  0.00 | 0.00 | 0.00 | 0.00 | 0.00 |    0.00 |   0.00 |  0.00 |  0.00 |
 E 19.63 | 25.69 | 5.14 | 0.96 | 4.33 | 0.45 |14,442.55|4,043.91|1,010.9|151.64 |
 |==========================================================================|
 | Calc-Method    Est-Forms    Help    Notes    Summary-Sheets    Exit       |
 | Access Estimating Forms to adjust machine Details and Basis Values        |
```

(*b*)

FIGURE 12-3
Example screens from Dataquest DECE system: (*a*) DECE fleet summary screen; and (*b*) DECE cost estimating details for a paver. (From *Dataquest Equipment Cost Estimator (DECE)*, San Jose, Calif. Dataquest, 1991. Phone: 800-669-3282.)

Caterpillar's programs provide, in effect, an interactive computer version of its technical publications, such as the *Performance Handbook*, long a staple of the contractor's bookshelf. The computer versions can be much more convenient, and they automatically perform in minutes calculations that would take hours or days to do with traditional estimating methods.

Perhaps of broadest interest is the availability in electronic form of the R. S. Means Company's construction cost databases. Means is one of the largest and most widely used general information sources for construction cost and productivity data and has long published an extensive variety of books, manuals, and forms. They not only address general building and heavy construction but also have books and cost manuals focused on the needs of specialty contractors, facility managers, and architects.

Their electronic database is available in various forms. The most accessible is the MeansData for Lotus package, which can connect directly to spreadsheet-based estimating systems such as the one illustrated in Chapter 9. Indeed, it even comes with a preformatted estimating form should you not wish to create your own. Tens of thousands of cost items can be searched by CSI codes or key words. There are separate files for building, residential, and heavy construction costs, and for specialty areas like concrete, electrical, mechanical, plumbing, lighting, interiors, and site landscaping. Some calculations are built in. Regional cost factors can be used to modify cost data. Users can also add their own data to the database. The data can be accessed directly from—and selected items can be transferred automatically to—Lotus 1-2-3 spreadsheets.

The type of regional cost modifiers available from MeansData is illustrated in Fig. 12-4a. The **T** in the tag column indicates that we have picked cost modifiers for Los Angeles. Fig. 12-4b is a part of the item list that appears when we are selecting temporary fencing for our site. At the time of this writing, there are over 60,000 such items in MeansData, each with detailed cost components.

In Fig. 12-5 is the MeansData spreadsheet with its calculations for two items picked for this demonstration, plus the estimate summary with indirects, contingency, profit, and so forth added on. Note that each item can be broken down into costs for labor, materials and equipment to install the computed quantities; and the total worker-hours (manhours in Fig. 12-5) also appear. Any of these numbers can be overridden by the estimator. Since this is basically a Lotus 1-2-3 spreadsheet, its data can be exported and integrated with other applications just like any other Lotus data.

Major commercial estimating system developers, such as Timberline Software (Precision Estimating), G2 Inc. (G2 Estimator) and Management Computer Controls (MC^2), have also plugged the Means database directly into their systems, which makes a very flexible and powerful combination of estimating software with an extensive cost database.

Before using published information—be it a cost index, unit price, or whatever—for estimating and controlling construction work, one must thoroughly understand exactly what the information does and does not include. For example, consider a unit price for labor for laying brick. Which of the various cost

```
Criteria STATE=add_row
File CCI                    Index CCI_ST

Record   tag  CITY            STATE  MATL_01  INST_01  MATL_02  INST_02
   10         ANAHEIM         CA     101.0    120.9    104.6    112.6
   11         BAKERSFIELD     CA      99.1    110.1     97.0    111.3
   12         FRESNO          CA      99.6    112.9     94.7    120.2
   13     T   LOS ANGELES     CA      99.3    123.7     97.8    116.0
   14         OXNARD          CA      99.4    118.9    101.9    104.6
   15         RIVERSIDE       CA      99.6    119.7     98.7    111.6
   16         SACRAMENTO      CA      99.8    115.1     86.4    105.1
   17         SAN DIEGO       CA     101.7    115.0     95.3    108.3
   18         SAN FRANCISCO   CA     103.3    143.7    102.1    116.4
   19         SANTA BARBARA   CA     105.3    119.4    125.2    112.3
   20         STOCKTON        CA      99.2    114.8    121.0    113.5
   21         VALLEJO         CA      99.4    129.0    106.9    114.9
```

(*a*)

```
File DATA                   Index DATA

Record   tag  LINE_NO       DESC1
   246         0153040100   TEMPORARY FENCING ;CHAIN LINK, 6' HIGH
   247         0153040200   TEMPORARY FENCING CHAIN LINK 6' HIGH TO 500',
   248         0153040250   TEMPORARY FENCING CHAIN LINK 6' HIGH OVER
   249         0153040350   TEMPORARY FENCING PLYWOOD PAINTED 2"X4" FRAME
   250     T   0153040400   TEMPORARY FENCING PLYWOOD PAINTED 4"X4" FRAME
   251         0153040500   TEMPORARY FENCING WIRE MESH ON 4"X4" POSTS,
   252         0153040550   TEMPORARY FENCING WIRE MESH ON 4"X4" POSTS,
   253         0153060100   WINTER PROT, PLASTIC ON WOOD FRAME TO CLOSE
   254         0153060200   WINTER PROT, TARP OVER SCAFFOLD, 8 USE, NOT
   255         0153060300   WINTER PROT, PREFAB FIBERGLASS PANEL, STEEL
   256         0154800010   WATCHMAN MONTHLY BASIS UNIFORMED MINIMUM
   257         0154800100   WATCHMAN MONTHLY BASIS UNIFORMED MAXIMUM
```

(*b*)

FIGURE 12-4
Examples of database information from MeansData for Lotus: (*a*) cost modifiers with Los Angeles "tagged" and (*b*) section from the index with record 250 "tagged." (From *MeansData for Lotus*, Kingston, Mass.: R.S. Means Co., 1991.)

components does it include? Is it just the base wage, or does it add fringes, insurance, and wage premiums? What does it assume for productivity? Does it consider regional differences, learning-curve variations with quantity, environmental effects, work schedule, or management approach?

When tempered with a contractor's own experience and judgment of the company's operations, such published sources can be a valuable source of supplementary information. Ultimately, however, the contractor's personal experiences—as documented accurately, consistently, and in a well-organized and readily accessible manner—should be the primary source for estimating and controlling construction work.

```
A14: [W16] '0153040400                                           MENU
File  Search  Estimate  Print  Global  User  Quit
Select input file
          A           B         C         D         E         F        G
1                          *** MeansData Estimate ***
2     ==========================================================================
3     Estimate:      93-12            Date:
4     Description:   Building 4
5     Project:       MegaCorp         Bid Date: 3/1/93
6     Location:      Orangetown       Job #:    9312-B
7     Sq. footage:   120,000          City indx:LOS ANGELES, CA
8     ==========================================================================
9     Line #         Description
10    --------------------------------------------------------------------------
11                   Manhours    Matl    Labor   Equipment   Sub     Total
12    ==========================================================================
13
14    0153040400     TEMPORARY FENCING PLYWOOD PAINTED 4"X4" FRAME
15                   8' HIGH                          100.00 L.F.
16    Unit Costs        0.22      6.45     5.83       0.00    0.00     12.28
17    Total Costs         22       645      583          0       0      1228
18
19    01 GENL REQUIREM     22       645      583          0       0      1228
20
21    0205541200     SITE DEMOLITION,WALL,BRICK,SOLID
22                                             123.00 C.F.
23    Unit Costs        0.07      0.00     1.64       1.21    0.00      2.84
24    Total Costs          9        $0     $201       $148      $0      $350


      -----------------------------------------------------------------------

35    --------------------------------------------------------------------------
36                   Manhours    Matl    Labor   Equipment   Sub     Total
37    ==========================================================================
38    01 GENL REQUIREM     22      $645     $583         $0      $0     $1228
39    02 SITEWORK           9        $0     $201       $148      $0      $350
40
41    TOTAL                31      $645     $783       $148      $0     $1578
42
43    SALES TAX         5.00%      $32
44    MATERIAL MARK-UP 10.00%      $65
45    LABOR MARK-UP    57.00%               $447
46    EQUIPMENT MARK-U 15.00%                           $22
47    SUB MARK-UP       0.00%                                   $0
48
49    TOTAL BEFORE CONTINGENCY    $742    $1,231       $171      $0    $2,143
50    CONTINGENCY       5.00%                                          $107
51    BOND              1.20%                                           $26
52    PROFIT           12.50%                                          $268
53
54    JOB TOTAL                                                      $2,544
```

FIGURE 12-5
Estimating sheet from MeansData for Lotus. (From *MeansData for Lotus*, Kingston, Mass.: R.S. Means Co., 1991.)

Quantity Take-off

Quantity take-off is the detailed enumeration of all materials and components that go into a construction project, including temporary items such as scaffolding and formwork. In the case of a unit-price project, some but not all of the quantities will be specified in the contract as the basis for payment. The costs of those not so specified must be lumped or prorated into those that are. In a fixed-price project the breakdown does not affect the mechanism for payment except perhaps as a measure for progress payments, but accuracy is even more important because there are no adjustments for variations in quantity.

Computer programs can assist in the quantity take-off by helping estimators to measure, count, compute, and tabulate quantities, lengths, areas, volumes, and so forth of objects found in the plans and specifications. One way would be to read drawings conventionally and type the data into a spreadsheet or custom take-off program. But an approach that is particularly convenient is to connect a large digitizing tablet to the computer, lay plans on the tablet, and then use a highly accurate cursor/keypad to count and measure the items directly on the drawing. A typical configuration appears in Fig. 12-6. In a sense, this is the reverse of CAD, given that the drawings may have been produced by the designer's computer in the first place. The estimator in part is putting the geometric information back into his or her computer. It seems obvious that one might bypass this printing and reentry and simply pass the electronic drawing files directly to the estimating program. That is exactly the type of integration implied in Fig. 12-1, and it is becoming more and more common. But some designers are reluctant to pass on their electronic drawings for liability reasons, and others want to charge steep prices for this service, so the need remains for digitizers and other aids to quantity take-off

FIGURE 12-6
Using a digitizer for quantity take-off.

Methods, Resources, and Productivity

The most challenging aspects of estimating are figuring out the methods to do a piece of work; identifying and quantifying the resources of workers, equipment, and materials to do the work in an acceptable amount of time and quality; and determining their productivity. Costs can then follow as a matter of course. This phase of estimating will continue to be dominated by experienced human beings. Nevertheless, computers can provide assistance along the way.

Historical records of past methods are often fragmented, and much of the related knowledge and experience resides in people's heads. But the indexing, classification, storage, and retrieval capabilities of computers can help us get at this information in a more systematic manner. For example, which project managers and superintendents have experience with certain kinds of bridge construction? What past projects implemented new ways of placing concrete? Where can we find their archives? There is also a trend toward using expert system software to document and preserve some of the knowledge and experience of top estimators so that they can be shared with younger employees, even after the experts retire.

The new CD-ROM and multimedia technologies offer considerable potential for archiving actual video footage and firsthand explanations of innovative methods, cost-effective approaches to constructibility, and unique designs. Tying them into conventional information systems and computer networks could indeed yield a useful knowledge store. But as of this writing, this technique has yet to be exploited significantly in construction.

Resource estimates largely depend on the methods that have been chosen. For a given method, we again rely primarily on experience and judgment to figure out the appropriate numbers of each type of craft and machine, but spreadsheets and other tools can help us in machine and crew balancing computations and in production estimates.

To some extent, the historical cost files of computer estimating systems provide worker-hour factors per unit of various kinds of construction work. But these also must be tempered by experience and judgment as related to the specific needs and characteristics of a given project. Quantitative modeling and analytical computer tools, such as the simulation programs described in Chapter 15, can also help here.

Estimating Construction Labor Costs

One of the most difficult topics in preparing a detailed estimate is labor, so it needs extra attention. The basic approach is to divide labor costs into two main components and develop them separately: (1) costs in money terms and (2) productivity. The first includes costs associated with wages, fringe benefits, payroll-based insurance and taxes, and wage premiums. Though calculations of the money components can be complex, most of the parameters can at least be readily and accurately quantified. The second factor is productivity—the amount of work that a worker or crew can accomplish in a defined period of time.

Estimating the *money component* of labor costs is difficult in construction. Reasons for this situation include the scope and variety of the work involved, the craft structure of labor unions, and the regional and local autonomy of labor and employer collective bargaining units. There are literally thousands of different wage rates, fringe benefits, insurance rates, and work rules, and there are exceptions to almost all of them. Superimposed upon these are federal, state, and local laws, taxes, and special programs.

Even if given identical money elements for labor, contractors vary widely in the way they analyze, combine, and distribute them for estimating and control purposes. Many now prefer to put all elements into a direct hourly rate. Others still prefer to split off various elements into the indirect or overhead category. Some estimate by craft and some by crew. Some combine scheduled overtime with straight time to produce an "average" hourly rate, and others keep them separate. There is not one correct method, but several, and there are generally good reasons why a particular contractor uses a specific method in a given type of work. Therefore, rather than force a single method, estimating software should be flexible in handling the major money elements, including basic wages, fringe benefits, payroll insurance and taxes, and wage premiums.

In Fig. 12-7*a* we see how the HCSS Estimating System from Heavy Construction Systems Specialists, Inc. handles craft data entry. It does not provide a detailed breakdown of the tax and fringe burden at this stage. Payroll insurance burdens are handled in the individual work items. Even within this given craft, the workers' compensation rates will vary, depending on the kind of work they are doing. Part of the HCSS crew composition report, which details the makeup of crews and equipment assigned to different tasks appears in Fig. 12-7*b*. These crews can be referenced from individual bid-item calculations.

Of the two main labor cost factors, *productivity* is far harder to determine than the money component. Whereas wages and other money components may stay essentially constant over the duration of an operation, productivity can fluctuate wildly. To estimate and control productivity, one needs not only accurate, consistent, and up-to-date records but also a great deal of experience and judgment. Many of the factors influencing labor productivity are highly qualitative in nature, and judgment is needed to develop the type of quantitative information that is required. They include the effect of location and regional variations, learning curve effects, work schedule (overtime and multishift), work rules, weather and other environmental effects, experience of the workers employed, morale on the job site, safety, and motivation. Computer programs can store some historical information related to these topics, but they serve best by providing a standard and convenient framework wherein the estimator can express his or her own experience and judgment.

Hourly wages and fringe benefits often change in construction. Different sections of the country have vastly different hourly rates and fringe benefits. Applying unit costs for labor from one section of the country to another is impractical, and comparing today's costs with those of a few years ago is equally difficult. Keeping productivity records in worker-hours, however, neutralizes the money component.

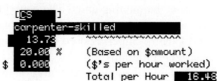

```
                           SETUP LABOR RATES


            Labor Code       [CS    ]
            Description      carpenter-skilled
            Rate per MH        13.73      ~~~~~~~~~~~~~~~~~~~
            Taxes              20.00 %    (Based on $amount)
            Fringes        $   0.000      ($'s per hour worked)
                                          Total per Hour    16.48


 ESC=EXIT F1=HELP F4=COPY/MORE F6=DELETE F8=PrintCodes F9=CLEAR F10=SAVE
 [TAB=FIND   PgUp=PREV   PgDn=NEXT]   UpArrow=PrevField   UpArrow to Unit
```

(a)

```
HEAVY CONSTRUCTION SYSTEMS SPECIALISTS                   02/09/93  17:38
DEMO   HEAVYBID/Q ESTIMATING SYSTEM DEMO                          Page   1

                          CREW COMPOSITIONS

************************************************************************
          Crew AAC1    -   conc. paving crew
                                                         Base    Burdened
Labor Code         Labor Description       Members       Rate      Rate
   LF          labor-foreman                 1.00       12.25     14.70
   LS          labor-skilled                 7.00       11.65     13.98
   OP          operator-paver                2.00       13.61     16.33
         10.00 MH/CH  $145.2200 /CH   $14.5220 /MH    ** BURDENS INCLUDED **

Equip Code       Equipment Description  Pieces    Total/Hr
   8AC1    air compressor 250 cfm         1.00      18.40
   8CP1    conc. paver - 25'              1.00      25.60
     Equipment Only:    $44.0000 /CH    $4.4000 /MH
 Labor + Equipment:   $189.2200 /CH   $18.9220 /MH


************************************************************************
          Crew AAC2    -   conc. place crew
                                                         Base    Burdened
Labor Code         Labor Description       Members       Rate      Rate
   LF          labor-foreman                 1.00       12.25     14.70
   LS          labor-skilled                 7.00       11.65     13.98
   OP          operator-paver                1.00       13.61     16.33
          9.00 MH/CH  $128.8900 /CH   $14.3211 /MH    ** BURDENS INCLUDED **

Equip Code       Equipment Description  Pieces    Total/Hr
   8AC1    air compressor 250 cfm         1.00      18.40
     Equipment Only:    $18.4000 /CH    $2.0444 /MH
 Labor + Equipment:   $147.2900 /CH   $16.3655 /MH

************************************************************************
```

(b)

FIGURE 12-7
HCSS input screen and output report: (*a*) HCSS craft data-entry screen; and (*b*) HCSS crew
composition report. (From *HCSS Demo Disk*, Houston Tex.: HCSS, 1991.)

For example, a contractor may determine that unit productivity for foundation forms is anticipated to be 0.80 unit hours per square meter (m^2). If labor costs are \$15.00/h including fringes, the unit cost for labor is \$12.00/m^2. If labor costs are \$20.00/h, the unit cost becomes \$16.00/m^2. If a contractor records worker-hours, performance can be reviewed in all projects and actual productivity can be compared with estimated productivity in a straightforward manner. Estimating software should support these capabilities.

Materials, Equipment and Subcontractors

In the early stages of estimate preparation, an estimator usually inserts *plug prices* for items that he or she expects to contract out to suppliers and subcontractors. Sometimes a detailed estimate is prepared as the basis for subsequent *make or buy* decisions, such as whether to fabricate or rent forms, or whether to perform directly or to subcontract excavation. Initially, sources such as the aforementioned MeansData and DataLine[2] can provide general price figures to use here, and these may be sufficient for smaller items in a bid. But more analysis and detail are needed for the higher-cost items in the estimate.

An important capability of estimating software for this phase is to be able to help the estimator subdivide parts of the work into packages that fit the craft structure and practices of potential suppliers and subcontractors and to track and analyze their quotes when they come in. The more clearly the work packages can be defined, the easier it will be to analyze the quotes. But even then it will not be a matter of just selecting lowest prices from those that come in. Suppliers and subcontractors often qualify their bids with exclusions, conditions, and combination options; and we need to be able to analyze the cost and schedule impacts of these qualifications quickly.

Indirect Costs

Contractors vary widely in how they define and handle the indirect and overhead costs in their estimates, and various estimating software packages vary accordingly. Some costs, such as the tax and insurance burdens that go with payroll, can readily be identified with the direct work in the field. Even so, some contractors prefer to total them up and keep them as a separate overhead item in their bid. Other costs, such as a major overhaul on a piece of equipment, are more difficult to tie to specific work items, but many estimators nevertheless use a variety of techniques to prorate some of these intermittent costs to any direct work with which they logically can be associated. Still other items, such as home office expenses, are treated as overhead by most contractors when it comes to recovering these expenses on the revenue-generating work the company performs, but even here some firms charge services such as accounting and purchasing directly to the projects for which they are performed.

Good estimating packages offer flexibility to match a given contractor's preferences for dealing with overhead and indirect costs, rather than force contractors

```
HEAVY  CONSTRUCTION  SYSTEMS  SPECIALISTS                                    Page    1
DEMO    HEAVYBID/Q  ESTIMATING  SYSTEM  DEMO                         02/09/93   19:30
                                INDIRECT COST DETAIL REPORT


Item No.     Description    Quantity           Unit          Constr    Equip-
  Detail                              Unit     Cost   Labor  Matl&Exp  Ment      Total

90000        INDIRECT COSTS      1.00  LS
    3900     small tools     303,000.00  LAB$   0.030          9,090             9,090

90000        INDIRECT COSTS      1.00  LS
    3932     Builder risk   1575000.00  JOB$   0.012         18,900            18,900
    8PK1     pickup    8 MO    1,408.00  HR     4.000                   5,632    5,632
    QQ02     superintendent      8.00  MO      4,500  43,200                   43,200
                                                      43,200  18,900   5,632   67,732

***********************************************************************************

      $76,822.00           *** REPORT TOTALS ***         43,200   27,990   5,632   76,822
```

FIGURE 12-8
HCSS indirect cost report. (From *HCSS Demo Disk*, Houston, Tex.: HCSS, 1991.)

to change the way they estimate and bid work. An edited and condensed version of the indirect cost report from the HCSS program appears in Fig. 12-8. Note that small tools are computed as a fraction (0.030) of direct labor costs, and builder's risk insurance is a fraction (0.012) of total direct costs, rather than as unit costs per se.

Compilation, Analysis, and Bidding

The bid preperation process usually becomes very hectic in the final hours as subcontractors and suppliers telephone and fax in their bids and prices for various parts of the work. Here a computer can really pay off in its speed and accuracy. As the numbers come in, they must be subjected to careful and yet rapid analysis and checking, and the selected ones must be included and reflected in cost computations.

Once the estimator has computed the direct and indirect costs, management must give consideration to matters such as the risks and contingencies associated with the work, the competition that will be faced in bidding, and how much markup to put on the bid. Figure 12-9 is a markup worksheet from the HCSS program. Note that the higher uncertainty associated with labor productivity causes the contractor to put a higher markup on that portion of direct costs than, say, on subcontracts, where the risks are largely assigned to someone else. The composite markup in this case is 10.81 percent.

In a unit-price bid, the estimator must take the indirects, markup, and other costs not included in the direct-cost bid items and prorate these costs to the unit prices of the direct bid items, which are the basis for payment. A calculation of this type was included in the spreadsheet example at the end of Chapter 9 and will not be repeated here. Unit-price bid-oriented programs like HCSS and Hard Dollar also allow the contractor to use a technique called *unbalancing* to shift costs from

```
                    BID  SUMMARY  PROCESS

    Indirect/Addon  Proration  Costs     L,T,X   X
      (Labor,Total,X=TotalLessSub)

     Cat.    Markup %    Total Costs    Markup Amt.
     Labor     20.00    x    345,921   =     69,184
     PermMatl  10.00    x    263,684   =     26,368
     ConstrExp 10.00    x     74,565   =      7,456
     Equip     10.00    x    278,137   =     27,813
     Sub        5.00    x    458,887   =     22,944
                       =============================
     Total     10.81    x  1,421,196   =    153,767

     Addons       Description       Total
        1.      _____         0
        2.      _____         0
                        Total Markup:       153,767
                        Balanced Bid: $1,574,963
    - - - - - - - - - - - - - - - - - - - - - - - - - - -
    F1=HELP.  F10=SAVEtheScreen&COMPUTE.  ESC=EXIT
```

FIGURE 12-9
HCSS screen for entering markup calculations. (From *HCSS Demo Disk*,
Houston, Tex.: HCSS, 1991.)

items that are likely to underrun to those that may overrun, either to shelter risks or
to exploit errors in the quantities. Most programs also provide numerous analysis
reports for the labor, material, equipment, and subcontract costs, and summary
reports to provide a better overall feeling for the bid. Figure 12-10 is a typical
summary commonly called a recap report.

Some owners, particularly in the public sector, also complicate bidding matters by specifying goals for subcontracting to various types of disadvantaged
business enterprises (DBEs—small businesses [SBE], firms owned by minorities
[MBE], women [WBE] or disabled people). Some estimating software is very
helpful in enabling a contractor to classify likely bid components and the status
of bidders in advance and then to come up rapidly with the bid combination that
meets the requirements of the owner at the lowest feasible cost. In Fig. 12-11 we
see how HCSS splits out the DBE section in its subcontractor summary. Grantlun's
Hard Dollar estimating program is considered to be particularly strong in this type
of capability. Some bidding software even prints the supplier and subcontractor
listings that must be submitted at bid time to certain owners.

Finally, the availability of portable computers connected to cellular telephones makes it possible to take the last-minute bid preparation right to the door of
the bid room. This helps to cut down the likelihood of errors in that high-pressure
process and can improve the chances of success in this competitive business.

```
HEAVY CONSTRUCTION SYSTEMS SPECIALISTS              02/09/93  19:07
DEMO    HEAVYBID/Q ESTIMATING SYSTEM DEMO

                        ESTIMATE RECAP

                    DIRECT      INDIRECT        TOTAL    % OF TOTAL
Labor            302,721.38    43,200.00    345,921.38      24.3%
PermMatl         263,684.62                 263,684.62      18.6%
ConstrMatl        46,575.64    27,990.00     74,565.64       5.2%
Equipment        272,505.42     5,632.00    278,137.42      19.6%
Subcontract      458,887.50                 458,887.50      32.3%
                 ----------                 ----------
Total Costs    1,344,374.56    76,822.00  1,421,196.56     100.0%

% of Total           94.59%        5.41%      100.00%

Markup                                                       0.0%
                                          ============  (% of costs)
COST + MARKUP  ---------------------->    $1,421,196.56

BALANCED BID TOTAL
BID TOTAL                                 $1,576,102.95
```

FIGURE 12-10
HCSS recap report. (From *HCSS Demo Disk*, Houston, Tex.: HCSS, 1991.)

```
*H. C. S. S.*                                          02/09/93
                     SUBCONTRACTOR USE REPORT          Page   1


Bid    Activity  Detail              Quan   Unit    Rate    %     Total
item

   5      00     4                   1.00   SET  85,440.00 100.0  85,440.00
   6      00     4                   1.00   LS   32,500.00 100.0  32,500.00
                 -------------------------------------------------------
 Subtotals:      4    Subcontract                          100.0 117,940.00

   2      01     41               2,065.00  LOAD    83.50  100.0 172,427.50
   3      03     41               1,544.00  LOAD    83.50  100.0 128,924.00
   4      01     41               4,168.00  CWT      9.50  100.0  39,596.00
                 -------------------------------------------------------
 Subtotals:      41   Subcontract - DBE                    100.0 340,947.50

                                                           ============
                 REPORT TOTAL ====================>            458,887.50
```

FIGURE 12-11
HCSS subcontractor use report. (From *HCSS Demo Disk*, Houston, Tex.: HCSS, 1991.)

ADVANCED ESTIMATING SOFTWARE CONCEPTS AND FEATURES

In general, estimating programs are not large-volume software products like databases and spreadsheets. Most come from smaller developers who tend to focus their products on certain sectors of the construction industry, most notably (1) general building, (2) heavy and highway, and (3) specialty contractors. Thus it is not clear in this market what constitutes an advanced feature and what is just a different focus. Nevertheless, we will give some indication of things to look for.

Added Features and Functions

As with any type of software, developers of estimating programs can enhance the programs' basic functions. We could have more flexibility to define human and equipment resources and combine them into crews. Some programs allow tremendous amounts of detail in subdividing costs and work item categories. Most provide a large variety of detailed and summary reports. Heavy-construction-oriented programs have a lot of flexibility in prorating indirects, contingency and profit to the bid items, including the power to unbalance by using something other than a straight linear distribution factor. More and more programs are helping contractors evaluate suppliers and subcontractors not only on price but also in various categories that meet the social goals of owners. A wide range of features and capabilities helps distinguish the advanced programs from the more ordinary ones.

User-Friendliness and Online Help

Although most estimating programs lack many of the advanced user interfaces and custom macro programming capabilities of mass-market databases and spreadsheets, the better ones do provide clear menu structures that minimize the need to memorize keystroke commands, and many of them provide very good online help documentation. The Hyper·Estimator and Hyper·Remodeler programs from Turtle Creek Software, although limited mainly to residential and remodeling construction, provide a very strong graphical user interface that makes them exceptionally simple to use within their intended application areas. They are among the few Macintosh-based estimating products.

Report Form Definition

Some estimating programs allow custom report writing. Flexibility in report layout is very helpful. For example, in making the transition to a new system, an office might employ this capability to create forms and reports with a design similar to those already familiar to the company's estimators.

Integration with Cost Databases

We have already mentioned the ability to integrate with the R. S. Means published cost data products. Some programs can integrate with other similar products, such as those published by the mechanical (MCAA) and electrical (NECA) contractors associations, Richardson Engineering Services (particularly strong in industrial construction), and the National Construction Estimator. Although this type of database integration is becoming increasingly common in this competitive marketplace, it still could be called an advanced capability. Note, however, that access to these databases can add hundreds or thousands of dollars to the price of single-workstation systems.

Range Estimating

A few systems offer a powerful capability for dealing with uncertainty called *range estimating*. Basically, the estimator comes up with not only the most probable cost for key bid items but also a minimum-to-maximum range, expressed either in absolute terms or as a plus-or-minus percentage. The system can then compute probabilities of the overall estimate by taking simulated runs through the calculations hundreds of times, using a weighted random function (e.g., beta distribution) to pick off numbers within each range. The results give not a single bid price, but a range of prices plotted or tabulated as a probability distribution function. This can give the estimator a very good feeling for the risks and uncertainty associated with the estimate and hence help assess the amount of contingency that ought to be included in the markup. Alternatively, the estimator can devote more time to the items with the greatest uncertainty to obtain more information, look at other methods, or perform more calculations to reduce these unknowns. The disadvantage of all of this is that it can take more time to develop the ranges for each bid item, but this time can be reduced by selectively estimating uncertain elements in more detail so as to focus attention mainly on the items that need it. This selectivity actually reduces the effort that is customarily and unwittingly wasted on estimating low-risk items.

Interfaces to Other Software

At the outset of this chapter, we mentioned the possibility of integration between CAD programs and estimating programs. Whereas a growing number of commercial estimating systems do indeed tap into programs such as AutoCAD, this new capability is clearly in the advanced category. The major barriers here are not technical but institutional, since most designers work in different organizations from most construction estimators. But within design-construct organizations, this capability can be a real competitive advantage, and traditional design firms and construction firms will have to learn to work more closely together to exploit such capabilities if they wish to keep their share of the construction market.

It is already becoming much more common to interface from the estimating program to project management and administrative software for cost control, scheduling, and procurement. The technical feasibility is well established, and the institutional barriers are minimized because these functions can all happen within the contractor's own organization. But again, this capability is implemented in varying degrees, starting with programs that can only dump out files in database, spreadsheet or ASCII text format. More advanced programs will be providing the types of direct linkages that enable a change in one program, say the estimate, to be reflected immediately in the others.

Artificial Intelligence and Expert Systems

In Chapter 11 we pointed out that estimating is one of the most promising application areas for artificial intelligence and expert systems. Although little if any of this application is found in current commercial products, a number of university, government, and industry research efforts are underway to bring it about. It would be surprising if, sometime in the 1990s, the results of these efforts do not start finding their way into advanced estimating software products.

APPLICATIONS AND LIMITATIONS

Computer-based estimating programs are dealing increasingly well with the data collection, computational, and clerical aspects of estimating. They archive and retrieve large volumes of resource, cost, and productivity information, perform calculations quickly and accurately, and present results in an organized, neat, and consistent manner. All of these virtues are of tremendous help in the high-pressure environment in which most construction estimators often find themselves.

But the systems do not replace the knowledge, judgment, and experience of a veteran estimator, and it will be some time before new computer technologies such as artificial intelligence and expert systems make significant contributions in this area. This aspect of estimating remains more of an art than a science. Thus it is very important, in selecting estimating software, to pay close attention to the human interface so that the software adapts well to and supports the estimators in their work and becomes almost a natural extension of their intellectual activities. Unfortunately, the human interface in several of the commercially available estimating programs considerably lags that found in many of the large-volume mass-market programs, such as spreadsheets and word processors.

Before being too critical of estimating software developers, we should recognize that they are not working in a high-volume market. Construction estimating is a specialized business application, and good programs focus well on the client needs. What they lack in fancy interfaces, many make up in a wide range of features and the flexibility to adapt to the numerous approaches different firms take to estimating and in the huge variety of methods and components found in construction projects.

The more specialized nature of construction estimating tends to make the price per user of estimating software packages quite high—usually in the range of thousands of dollars rather than a few hundred. In turn, this high price seems to compel some vendors to use copy protection schemes to protect themselves against users who might steal even though their inconvenience penalizes honest users. These devices, for example, hardware keys or plugs that connect to a computer I/O port, have long since been shunned in the mass-market software. Some also limit the use of their products to one-year intervals, requiring expensive upgrades to continue. To some extent these price and protection practices may inhibit the more rapid adoption of computer-based estimating. But do not be overly concerned about prices as such. An expensive package can still deliver sufficient gains in estimating productivity and quality to pay for itself many times over in the first year it is used. But a buyer should be careful about evaluating the alternatives before acquiring a system, because the benefit/cost ratios could vary significantly from one product to another.

As a closing note in this section, we should comment briefly on why, of all the chapters in the tool and application parts of this book, we devote relatively little attention to Macintosh-based software in Chapter 12. The fact is that the estimating software market is indeed quite lopsided in favor of DOS PC-based software. There are a few Macintosh products, such as Turtle Creek Software's MacNail Estimating and Hyper·Estimator, but even those are most suitable for residential and remodeling jobs or perhaps some small commercial building work. Also, most of the vendors of Macintosh estimating software are indeed very small firms, some even home-based. Similarly, at the time of this writing, few estimating products take advantage of the object-oriented environment of Microsoft Windows, perhaps because of the difficulty of programming for that type of graphical user interface compared with doing so for traditional text-oriented tools with which most estimating programs have been developed. Whether this situation will change very quickly for such low-volume, specialty software is uncertain.

SITE-GRADING QUANTITY TAKE-OFF USING AGTEK'S EDGE SYSTEM

This first example deals with site grading for a small subdivision layout. The purpose is to illustrate computer-assisted quantity take-off with the Sitework Engineering module that is part of AGTEK's EDGE system, which is used by numerous excavating and grading contractors. This system employs a digitizing tablet like that shown in Fig. 12-6, though it will be difficult to convey that aspect here. Instead, we will show some of the screen images and calculation results that develop while the estimator works on the digitizing tablet with a cursor/keypad.

Basically, the system works by evaluating the differences between the natural ground surface, as reflected in the contour plan, and the designed finished grade. With the two surfaces thus defined, the computer uses the average-end-area calculation method to compute cut-and-fill volumes, and it can print out a

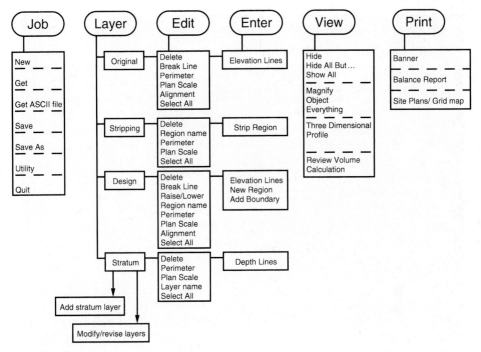

FIGURE 12-12
Menu structure of Sitework Engineering.

multicolored grid plan that tells the surveyors exactly how to stake the building site for excavation and grading.

In Fig. 12-12 is a summary of the structure and options of the pull-down menus included in the Sitework Engineering program. Here one can choose options to create or save a file for the **Job** (similar to a File menu in other programs); select **Layer** to enter data for the original surface, finish grade, stripping, and so on; use **Edit** to modify previously entered data; go to **Enter** to input the perimeter, elevation lines, and so forth; use **View** to see and modify screen reports; and select **Print** to output various graphical and tabular reports. This fairly simple and convenient interface enables estimators to keep their attention focused on the drawings.

The estimator begins by typing in general information about the site (e.g., depth of topsoil to strip), locating the 0-0 origin for the X-Y coordinates used in the system, establishing the boundaries, and entering the scale for a drawing that has been taped to the digitizer tablet. He or she then traces natural ground contours with the cursor, periodically recording points along the way by pressing the Enter button on the keypad. This produces a contour image on the screen, as well as an audit list of points, as shown in Fig. 12-13 for the contour line at elevation 45 ft. The system interpolates in between the points and produces smooth

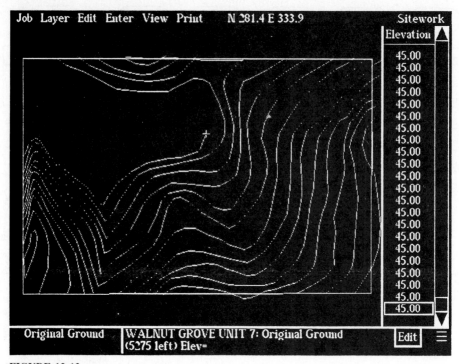

FIGURE 12-13
Data entry for natural contours.

contour lines. Similarly, the estimator hits key points on the grading plan to digitize that surface.

Once the natural ground surface has been defined, the estimator can request a screen or printer report with a three-dimensional perspective view, such as that in Fig. 12-14. As shown in the subordinate menu at the bottom of the screen, one can change the viewing **Angle** up or down, **Rotate** around the surface, **Zoom** in or out for varying amounts of detail, and **Magnify** the vertical scale the better to show the relief on mildly contoured surfaces. Although this type of figure is visually interesting and informative in its own right, it is particularly useful for detecting errors in input. For example, points specified at incorrect elevations will stand out like big spikes upward or downward, and one can quickly correct them.

In Fig. 12-15 the estimator is in the midst of entering the corners to define the boundaries at the grade level of one of the building pads. The audit trace at the right of the screen shows that the current one is at elevation 47 ft, and all or parts of the data for three others also appear in that list. Note at the top of the screen the coordinates of the point, which are relative north and east of the origin for the data entry, also shown in Fig. 12-13.

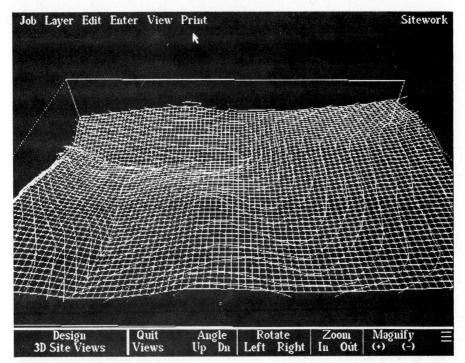

FIGURE 12-14
3-D perspective view of natural contours.

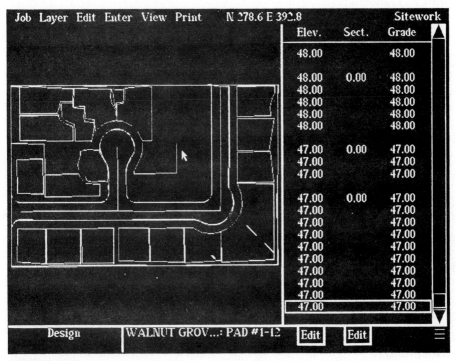

FIGURE 12-15
Data entry for design grade.

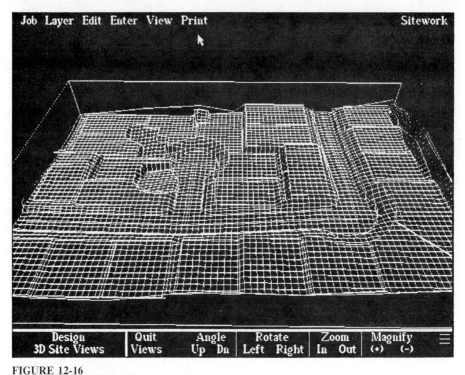

FIGURE 12-16
3-D perspective view of design surface.

Figure 12-16 is a three-dimensional view of the site-grading design. At the bottom of the screen is the same type of control submenu as the natural contours view in Fig. 12-14. Note how clearly it displays the roads, building pads, and side slopes. The irregular areas on the left half show the program's ability to work around preserved trees and other natural conditions.

Figure 12-17 is a brief summary report that appears on the screen after you have finished entering both the natural contour surface and the design surface. It

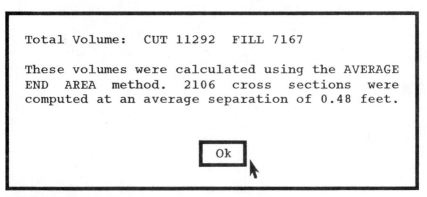

FIGURE 12-17
Summary report for quantities of cut and fill.

gives both cut and fill quantities and the number of cross sections that it used in the calculation. This is a much more accurate result than one could obtain with a planimeter or other traditional manual technique, but of course it still can be no more accurate than the original field survey and the resulting drawing.

Once the information is in the computer, it can then produce both on-screen reports such as that in Fig. 12-18 and printed reports such as those indicated in the pull-down menu in the figure. A particularly useful printed report uses a red/blue/black ribbon to print a grid plan showing cut sections in red, fill in blue, and cut/fill depths in black print at each grid intersection. A surveyor can use this printed report in the field to set out the grade stakes for the excavation and grading operation. With a multicolored ribbon, you can print reports that appear much like the screens in Fig. 12-14, 12-16, and 12-18, but in colors, plus cross sections not shown here.

Figure 12-19 is a tabular report that summarizes the surface area and the volumes of the cut portions and fill portions of the site excavation and grading plan. In Fig. 12-20 the compaction factor is added (in this case, 1.15 cy of cut material will be needed to place 1.00 cy in the fill). Taking the difference in the resulting volumes of cut and fill gives the net amount of material to export or import. The balance change shows how much volume would be gained or lost per 0.1 ft of raising or lowering the design grade. Using such information, one

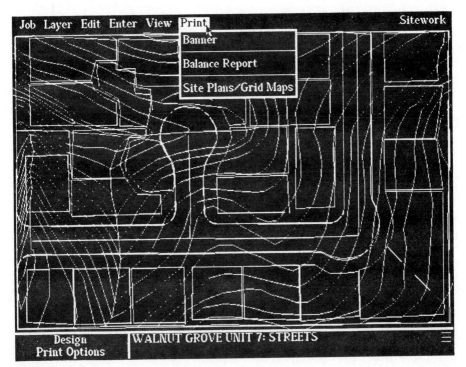

FIGURE 12-18
Composite natural/design plan with screen menu request to print.

```
VOLUMES   REPORT
JOB NAME : WALNUT GROVE UNIT 7
PRINTED : 04/08/1993 10:34:57 AM

STRIPPING REGION        DEPTH      AREA      VOLUME
-----------------       -----      ----      ------
SITE STRIPPING           0.20     259821      1925
```

| | | | | | AREA, SF | | | VOLUMES, CY (AFTER STRIPPING) | |
REGION	LAYER	MATERIAL	SECT	TOTAL	CUT	FILL		CUT	FILL
PAD #1-12	NAT'L GROUND	SUITABLE	0.00	109601	22863	86738		999	6471
PAD #13-23	NAT'L GROUND	SUITABLE	0.00	99677	82369	17308		6776	585
SUB TOTAL:	NAT'L GROUND	SUITABLE		209278	105232	104046		7775	7056
STREETS	NAT'L GROUND	SUITABLE	1.50	42890	37759	5131		3517	111
JOB TOTAL	NAT'L GROUND	SUITABLE		252168	142991	109177		11292	7167

```
                              SITE AREA:  260803
                           UNSPECIFIED:     8635
```

These volumes were calculated using the AVERAGE END AREA method.
2106 cross sections were computed at an average separation of 0.48 feet.

FIGURE 12-19
Report of volumes developed by Sitework Engineering.

BALANCE REPORT
JOB NAME : WALNUT GROVE UNIT 7
PRINTED : 04/08/1993 10:34:56 AM

REGION	LAYER	MATERIAL	VOLUMES, CY (AFTER STRIPPING)		COMPACTION		VOLUMES CY		EXPORT (IMPORT)		BALANCE CHANGE
			CUT	FILL	CUT	FILL	CUT	FILL	UNSUIT	SUIT	PER .1 FT
PAD #1-12	NAT'L GROUND	SUITABLE	999	6471	1.00	1.15	999	7441	0	(6442)	467
PAD #13-23	NAT'L GROUND	SUITABLE	6776	585	1.00	1.15	6776	673	0	6103	425
SUB TOTAL:	NAT'L GROUND	SUITABLE	7775	7056			7775	8114	0	(338)	
STREETS	NAT'L GROUND	SUITABLE	3517	111	1.00	1.15	3517	128	0	3389	183
JOB TOTAL			11292	7167			11292	8242	0	3050	1075

FIGURE 12-20
Report of cut-and-fill balance developed by Sitework Engineering.

can propose alternatives to the design grade to better balance the cut and fill and to reduce off-site hauling or import of material if it is appropriate. In this case, raising the design grade 0.3 feet at 1075 cy per 0.1 foot would just about take care of the 3050 cy that would otherwise have to be hauled off-site. Indeed, being able to compute and analyze such information rapidly has been a real competitive advantage to firms using this type of materials take-off system. Owners will often be impressed with money-saving suggestions and may award a contract for that reason.

This example has illustrated just one of the modules from AGTEK's system. They also have software products for taking off underground utility work, asphalt and concrete paving materials, and a template program suited to producing sections and mass diagrams for linear earthwork projects such as highways and canals. Other vendors, such as Timberline, make digitizer-based take-off products oriented more toward building construction. One thing they all have in common is a tremendous potential to save time and improve the accuracy of the quantity take-off process. See page 400 for contact information.

WAREHOUSE ESTIMATE USING TIMBERLINE'S PRECISION ESTIMATING SYSTEM

This example illustrates the most widely used general and building construction estimating system—Timberline Precision Estimating (PE). PE is actually a series of products that particularly emphasize the kind of component-based unit-price estimating that is commonly used in residential and building construction. For our purposes we will use the low-end product from Timberline's four-step product range, PE Light. Like other PE products, PE Light is a DOS-based program that is controlled from text menus and function keys. It has four main areas of functionality: (1) database review and editing, (2) quantity take-off and cost estimating, (3) report generation, and (4) utility options such as setting system parameters for storage, display, and printing hardware. PE Light comes with three small but expandable standard databases (home builder, general contractor, and metric general contractor) that are organized approximately according to the 16-section CSI Masterformat, so it is possible to begin making an estimate immediately. It also includes a small example building estimate one can use for a quick and easy-to-follow tutorial exercise. We will return to a more detailed description of PE Light in a moment.

For this example we draw information from the Easyway's Mountaintown Warehouse Project, which is described in detail in Chapter 4 and Appendix A of McGraw-Hill's *Professional Construction Management*, 3d ed., by Donald S. Barrie and Boyd C. Paulson, Jr. It is to be a dry-storage warehouse of some 151,600 ft^2, with pallet racks and some flow racks for individual items. Its principal components follow:

- Yard paving of heavy duty concrete
- Floor slab that is loading dock high

- Special hardened topping on floor slab
- Double-tee precast concrete walls
- Structural-steel framing with metal roof deck
- Sprinklered throughout
- Built-up roof
- Required utilities (water, fire, storm and sanitary sewers, natural gas, electrical)
- Natural-gas unit heaters
- Mercury vapor lighting
- Rail spur for incoming items
- Usual appurtenances

Figure 12-21 is a plan view of the warehouse. For more detailed information about the project itself, please refer to the aforementioned text. To simplify this illustration, we have consolidated, changed, and omitted some of the work items shown in the Barrie and Paulson book. For example, we have grouped several

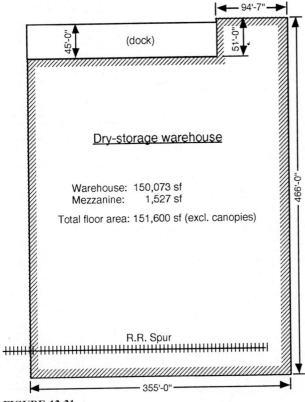

FIGURE 12-21
Plan view of Mountaintown dry-storage warehouse.

specialty items, such as mechanical and electrical, and assumed that we will just subcontract them. We have also changed some production and cost figures to be able more readily to use the Timberline database and to illustrate some of its additional capabilities (such as computing material waste).

The structure of the menus in PE Light appears in Fig. 12-22. Making selections beneath the levels shown here takes one to the various editing and data-entry screens, some of which we illustrate later. The MAIN MENU has three working options, plus END. Its principal subordinate menus (EDIT DATABASE and PRODUCE ESTIMATE) appear beneath the main menu. The UTILITY OPTIONS menu is also accessed from the main menu and appears to its right. Its choices are fairly self-explanatory. Subordinate menus under EDIT DATABASE and PRODUCE ESTIMATE, also shown to the right, access database editing options and group two different categories of report-printing choices.

From the EDIT DATABASE menu, we see that one can store and edit several different types of information that apply from one estimate to another.

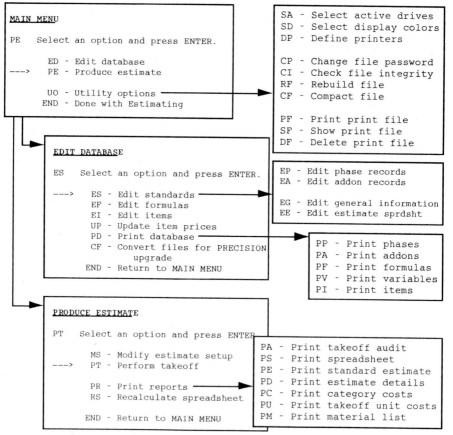

FIGURE 12-22
Menu structure of PE Light.

These include *standards* for our company information and for the structure of our cost database; a wide variety of *formulas* (areas, volumes, conversions, etc.) for various types of quantity take-off calculations (e.g., lumber quantities based on the length and height of walls); definitions of work *items*, their units of measure, formulas, resource productivity, and unit costs; and the unit *prices* for each item in the subcategories of labor, materials, and other. This menu also controls printing for all or parts of the database and is used for converting the PE Light database files to the more complex structure required by more advanced programs in the Timberline Precision Estimating product line.

The format used for the estimate for a particular project is a *spreadsheet* (i.e., an estimator's spreadsheet—similar to but not to be confused with the generic spreadsheet programs described in Chapter 9). We access various parts and capabilities within the spreadsheet via menus and function keys to set up the project, select work items, compute quantities, and determine the add-ons such as overhead and profit. Given work items and their quantities, PE Light draws item descriptions, unit costs, and production rates directly from the database wherever possible, but the estimator can override any of the stored information to customize the estimate for a particular project. Understanding the interaction between the database and the estimating spreadsheet is the key to working successfully with the Timberline Precision Estimation series. We will now examine these components in more detail.

Database Operations

Within the EDIT DATABASE menu, *Standards* basically defines the structure of the categories and work items within the database. It is also the place where we can enter general information about our company to use in report headings, cover pages, and the like. We can further specify the type of add-ons that we commonly apply to the direct costs of an estimate as well as their default percentages.

Choosing the suboption **Edit phase records** enables us to add, review, or modify the number, description, and units of measure for items in the database and to specify their level in the hierarchy. In the case of the standard general contractor database included with PE Light, the CSI Masterformat structure includes the following main categories, which PE Light calls *Group Phases* or *Divisions*:

```
 1.000   GEN CONDITIONS
 2.000   SITEWORK
 2.200   DEMOLITION
 3.000   CONCRETE
 4.000   MASONRY
 5.000   STEEL
 6.000   FRAMING
 7.000   WATERPRF & INSUL
 8.000   DOORS & WINDOWS
 9.000   FINISHES
10.000   SPECL CONDITIONS
```

```
11.000   EQUIPMENT
12.000   FURNISHINGS
13.000   SPECIAL CONST
14.000   CONVEYING SYSTEM
15.000   MECHANICAL
16.000   ELECTRICAL
```

Within these, the three decimal digits to the right of the 16 main Group Phase categories enable us to define numerous subordinate *Phases*. Many of these have been predefined for us in the standard database, particularly in the sections more likely to be handled by a general contractor's own work forces. A sampling from the concrete section is as follows:

```
3.101    Footing Concrete
3.105    Footing Forms
3.108    Footing Steps
3.109    Keyway
3.115    Footing Rebar
3.116    Set Grade Pins
3.121    Wall Concrete
3.125    Wall Forms
3.126    Pilaster Forms
```

Within these phases, PE Light further allows us to define one or more work *Items* and to assign them alphanumeric code of up to four characters. The following are a few of the items under 3.121–Wall Concrete. The code has been used here to convey the type of concrete (e.g., cg40 means concrete with gravel, 4000 psi).

```
cg40     Wall Conc 4000 psi,1/2" Gravel
cg45     Wall Conc 4500 psi,1/2" Gravel
cg50     Wall Conc 5000 psi,1/2" Gravel
cg60     Wall Conc 6000 psi,1/2" Gravel
cg70     Wall Conc 7000 psi,1/2" Gravel
cp30     Wall Conc 3000 psi,Peagravel
cp35     Wall Conc 3500 psi,Peagravel
cp37     Wall Conc 3750 psi,Peagravel
```

Although this list seems fairly detailed, most of the specialty subcontractor areas (mechanical, electrical, etc.) in this standard PE Light general contractor database have very few subcategories predefined, and even the general contractor topics are highly variable in amount of detail. But the database is fully editable, so we can add as much detail as we wish. It just takes time. The higher levels of Timberline's PE products can tap into the rich capabilities of the Means or Richardson databases, which can contain over a hundred thousand items.

```
┌─────────────────────────────────────────────────────────────────────┐
│ Estimating Item Maintenance                                 3-28-93  │
│                                                                       │
│ Phase      3.121  Wall Concrete                                       │
│   Item      c 20    Desc  Wall Conc 2000 psi                          │
│                                                                       │
│ Takeoff unit  cuyd                        Waste    5%                 │
│      Formula  Vol CY - Conc Wall          Categories  LM              │
│ ────────────────────────────────────────────────────────────────     │
│                  Labor                  Material                      │
│                 ─────────              ─────────                      │
│ Apply waste?              NO                      YES                 │
│ Order unit              hour                     cuyd                 │
│   convert       .75000 hour/cuyd                                      │
│   round            N                       N                         │
│                                                                       │
│ Price                  15.000                   60.000                │
│   date chgnd           3-28-93                  3-28-93               │
│                                                                       │
│ ────────────────────────────────────────────────────────────────     │
│ F1 Done          F3 Previous item    F5 Renumber item    F7          │
│ F2 Delete item   F4 Next item        F6                  F8          │
│ Help available (F10)                                                  │
└─────────────────────────────────────────────────────────────────────┘
```

FIGURE 12-23
Database item data entry and editing screens of PE Light.

To illustrate some of the types of screens we encounter in the database section, Fig. 12-23 is an entry/editing screen for Phase 3.121, work Item c 20–Wall Concrete. This item (c 20) one is measured in cubic yards (cuyd), allows 5 percent waste, and has labor and material components in its costs. Labor is converted at 0.75 work hours per cubic yard. In Fig. 12-24 we see the section of the item price database containing the labor wage and material unit price for this item. In the formula part of the database, the formula referred to as "Vol CY–Conc Wall" appears as follows:

```
┌─────────────────────────────────────────────────────────────────────┐
│                                                                       │
│   Formula name   Vol CY - Conc Wall                                   │
│                                                                       │
│        Length ' * Wall Height ' *( Width " /12)/27                    │
└─────────────────────────────────────────────────────────────────────┘
```

Any items added to or edited in the database are available in any project estimate spreadsheet that subsequently draws on that database. The information subsequently can be modified in the spreadsheet, but those changes do not reflect back into the database. (But one does have the option to access and change some parameters in the database while working within the spreadsheet.) Also, once data have been copied into an estimate, they do not change automatically if changes are subsequently made to the corresponding items in the data base. You must

```
╔══════════════════════════════════════════════════════════════════════════╗
║ Estimating           Phase  3.121 Wall Concrete      Last price change  3-28-93 ║
║                                                                            ║
║ UPDATE PRICES                                                              ║
║                                                                            ║
║   Item Description                    Labor          Material          Other ║
║   r114 Ftng Reb #11, Grd 40        15.000/hr       420.000/tn              ║
║   r116 Ftng Reb #11, Grd 60        15.000/hr       440.000/tn              ║
║      3.116 Set Grade Pins                                                  ║
║      10 Set Grade Pins Ftng        15.000/hr          .200/lf              ║
║      3.121 Wall Concrete                                                   ║
║   c 20 Wall Conc 2000 psi  ████████15.000/hr        60.000/cy              ║
║   c 25 Wall Conc 2500 psi          15.000/hr        60.500/cy              ║
║   c 30 Wall Conc 3000 psi          15.000/hr        62.000/cy              ║
║   c 35 Wall Conc 3500 psi          15.000/hr        63.000/cy              ║
║   c 37 Wall Conc 3750 psi          15.000/hr        64.000/cy              ║
║   c 40 Wall Conc 4000 psi          15.000/hr        65.000/cy              ║
║   c 45 Wall Conc 4500 psi          15.000/hr        66.000/cy              ║
║   c 50 Wall Conc 5000 psi          15.000/hr        68.000/cy              ║
║   c 60 Wall Conc 6000 psi          15.000/hr        70.000/cy              ║
║   c 70 Wall Conc 7000 psi          15.000/hr        72.000/cy              ║
║                                                                            ║
║ F1 Done              F3 Previous page    F5              F7 Goto phase/item ║
║ F2                   F4 Next page        F6              F8                 ║
║ Help available                                                             ║
╚══════════════════════════════════════════════════════════════════════════╝
```

FIGURE 12-24
Database price data entry and editing screens of PE Light.

either manually make the corresponding updates in the spreadsheet or delete the affected items and reselect and recopy their source items from the database.

Notice at the bottom of the Fig. 12-23 screen the function-key menu that enables us to delete (F2) or renumber (F5) the item currently displayed, move to adjacent items (F3 and F4), display a help screen (F10, available on most parts of the system), or exit this section (F1). Similar choices appear at the bottom of Fig. 12-24.

Project Estimating

The main options under PRODUCE ESTIMATE are to enter or modify the basic setup, perform the take-off, recalculate the spreadsheet, and print reports. **Modify estimate setup** enables us to define report headers specific to a particular estimate; enter general information about the project, client and architect; and modify the format and content of the spreadsheet form used for take-off and estimating. Figure 12-25 is an example screen from the Mountaintown Warehouse Project. A second screen in this section (reached by pressing function key F4 shown in the function-key menu at the bottom of this screen) enables the user to modify the standard spreadsheet column selections.

In **Perform takeoff** we mainly focus on selecting work items from the database that are pertinent to the project we are estimating and on computing their quantities. In doing so we can draw on formulas stored in the database and also use a convenient online calculator for one-time calculations. Formula definition includes a standard set of arithmetic operators (+, −, *, /, etc.) and a basic set

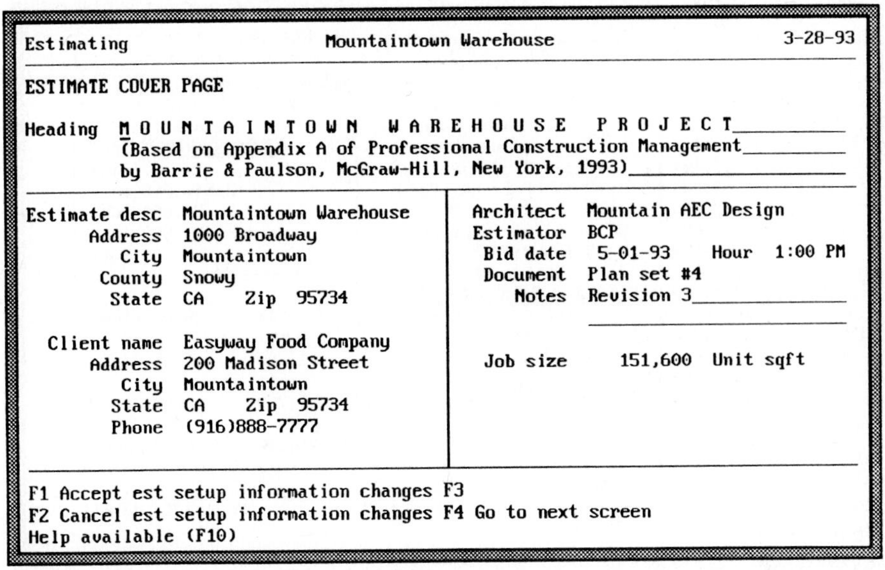

```
╔════════════════════════════════════════════════════════════════════════╗
║ Estimating              Mountaintown Warehouse              3-28-93      ║
║                                                                          ║
║ ESTIMATE COVER PAGE                                                      ║
║                                                                          ║
║ Heading   M O U N T A I N T O W N    W A R E H O U S E    P R O J E C T_ ║
║           (Based on Appendix A of Professional Construction Management__ ║
║           by Barrie & Paulson, McGraw-Hill, New York, 1993)_____  ║
║                                                                          ║
║ Estimate desc  Mountaintown Warehouse │ Architect  Mountain AEC Design   ║
║      Address   1000 Broadway          │ Estimator  BCP                   ║
║         City   Mountaintown           │ Bid date   5-01-93   Hour 1:00 PM║
║       County   Snowy                  │ Document   Plan set #4           ║
║        State   CA    Zip  95734       │   Notes    Revision 3_____  ║
║                                       │                                  ║
║  Client name   Easyway Food Company   │                                  ║
║      Address   200 Madison Street     │ Job size      151,600  Unit sqft ║
║         City   Mountaintown           │                                  ║
║        State   CA    Zip  95734       │                                  ║
║        Phone   (916)888-7777          │                                  ║
║                                                                          ║
║  ─────────────────────────────────────────────────────────────────────  ║
║ F1 Accept est setup information changes F3                               ║
║ F2 Cancel est setup information changes F4 Go to next screen             ║
║ Help available (F10)                                                     ║
╚════════════════════════════════════════════════════════════════════════╝
```

FIGURE 12-25
General project information data-entry and editing screen.

of functions (ABS, AVG, COS, IF, MAX, MIN, POWER, ROUND, SIN, SQRT, SUM, TAN, TRUNC). We also select the add-ons (home office and supervisory overhead, insurance, contingency, bonds, profit, etc.) from those that may have been predefined in the database or create new ones as needed. Alternatively, we can start the whole project from a *template* that we may have predefined for types of projects we do repeatedly, and modify it to suit the items and quantities in the actual project at hand.

Figure 12-26 is a take-off screen with two superimposed layers of pop-up windows. From the spreadsheet, we selected **Takeoff**, and from that we used **F10-Help** to access from the database the list of phases shown on the top layer. We can simply press Return at this level to select a highlighted item and place it on the take-off form, which is the next layer down. On the take-off layer we only need to fill in the quantity (on the right side of the second layer) and PE Light can then go to the database again to get the labor and materials prices. The prices are partially visible on the bottom layer. If the quantity calculation is available in a predefined formula, one can also use function key F4 to call up a formula window (not shown here) when working at the take-off level, and fill in the various parameters that go into the calculation. Alternatively, one can press F3 to bring up a simple calculator window to perform a calculation for a unique item.

In Fig. 12-27 we see the underlying spreadsheet screen with the **F6 Detail** option on. (F6 turns to **F6 Detail off** when this window appears, so you can close the detail window by pressing F6 again.) This shows not only the take-off quantity and the unit and total labor and material costs for each item but also,

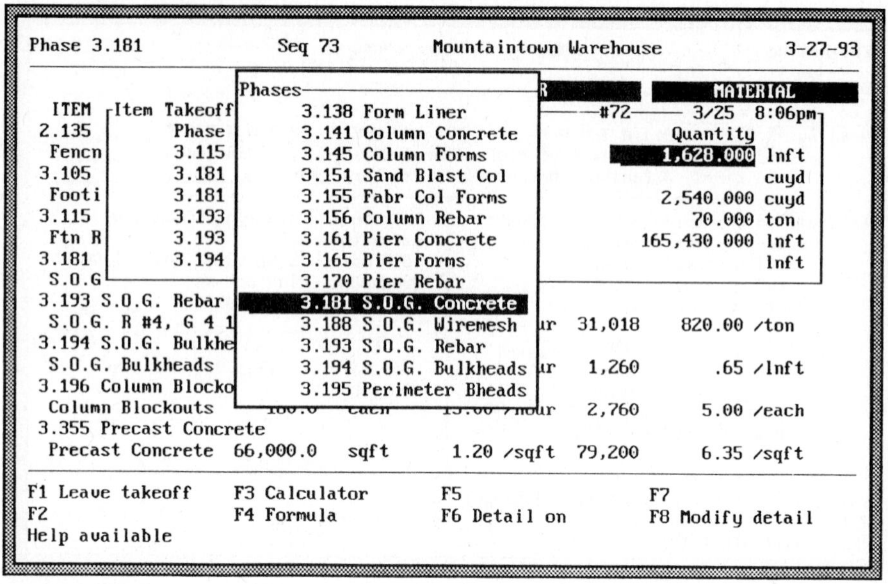

```
┌─────────────────────────────────────────────────────────────────────┐
│ Phase 3.181          Seq 73          Mountaintown Warehouse   3-27-93 │
│                                                                       │
│                  ┌Phases────────────────────┐R ──────  MATERIAL       │
│    ITEM  ┌Item Takeoff│ 3.138 Form Liner     │  ──#72──── 3/25  8:06pm─│
│   2.135  │    Phase   │ 3.141 Column Concrete │        Quantity         │
│   Fencn  │   3.115    │ 3.145 Column Forms    │    1,628.000  lnft      │
│   3.105  │   3.181    │ 3.151 Sand Blast Col  │               cuyd      │
│   Footi  │   3.181    │ 3.155 Fabr Col Forms  │   2,540.000   cuyd      │
│   3.115  │   3.193    │ 3.156 Column Rebar    │      70.000   ton       │
│   Ftn R  │   3.193    │ 3.161 Pier Concrete   │ 165,430.000   lnft      │
│   3.181  │   3.194    │ 3.165 Pier Forms      │               lnft      │
│   S.O.G  └────────────│ 3.170 Pier Rebar      │                         │
│   3.193 S.O.G. Rebar  │ 3.181 S.O.G. Concrete │                         │
│   S.O.G. R #4, G 4 1  │ 3.188 S.O.G. Wiremesh │r  31,018   820.00 /ton  │
│   3.194 S.O.G. Bulkhe │ 3.193 S.O.G. Rebar    │                         │
│   S.O.G. Bulkheads    │ 3.194 S.O.G. Bulkheads│r   1,260      .65 /lnft │
│   3.196 Column Blocko │ 3.195 Perimeter Bheads│                         │
│   Column Blockouts    └───────────────────────┘r   2,760     5.00 /each │
│   3.355 Precast Concrete                                               │
│   Precast Concrete  66,000.0   sqft    1.20 /sqft  79,200   6.35 /sqft │
│                                                                       │
│ F1 Leave takeoff    F3 Calculator    F5              F7               │
│ F2                  F4 Formula       F6 Detail on    F8 Modify detail │
│ Help available                                                        │
└─────────────────────────────────────────────────────────────────────┘
```

FIGURE 12-26
Spreadsheet with Take-off and Phases windows overlaid.

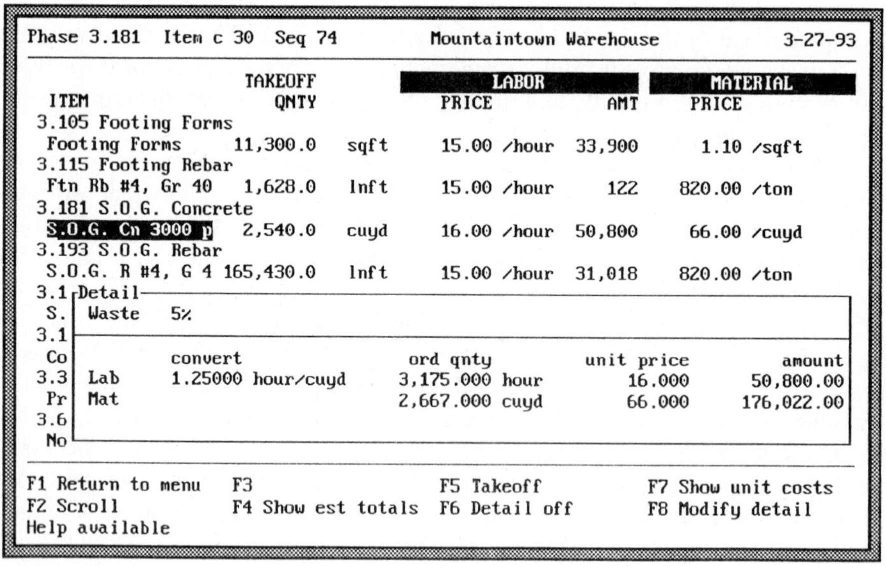

```
┌─────────────────────────────────────────────────────────────────────┐
│ Phase 3.181  Item c 30  Seq 74      Mountaintown Warehouse    3-27-93 │
│                    TAKEOFF         LABOR           MATERIAL            │
│    ITEM            QNTY       PRICE         AMT    PRICE               │
│ 3.105 Footing Forms                                                   │
│  Footing Forms    11,300.0   sqft   15.00 /hour  33,900   1.10 /sqft  │
│ 3.115 Footing Rebar                                                   │
│  Ftn Rb #4, Gr 40  1,628.0   lnft   15.00 /hour     122  820.00 /ton  │
│ 3.181 S.O.G. Concrete                                                 │
│  S.O.G. Cn 3000 p  2,540.0   cuyd   16.00 /hour  50,800   66.00 /cuyd │
│ 3.193 S.O.G. Rebar                                                    │
│  S.O.G. R #4, G 4 165,430.0  lnft   15.00 /hour  31,018  820.00 /ton  │
│ 3.1 ┌Detail────────────────────────────────────────────────────────┐ │
│ S.  │ Waste   5%                                                     │ │
│ 3.1 ├────────────────────────────────────────────────────────────── │ │
│ Co  │        convert          ord qnty        unit price     amount  │ │
│ 3.3 │ Lab   1.25000 hour/cuyd  3,175.000 hour     16.000   50,800.00 │ │
│ Pr  │ Mat                      2,667.000 cuyd     66.000  176,022.00 │ │
│ 3.6 │                                                                │ │
│ No  └────────────────────────────────────────────────────────────── │ │
│                                                                       │
│ F1 Return to menu   F3              F5 Takeoff     F7 Show unit costs │
│ F2 Scroll           F4 Show est totals F6 Detail off F8 Modify detail│
│ Help available                                                        │
└─────────────────────────────────────────────────────────────────────┘
```

FIGURE 12-27
Spreadsheet with detail window turned on.

in the pop-up detail window in the bottom half of the screen, supporting details for the currently highlighted item, such as the allowance for material waste, the labor production rate, and the quantities of labor and material consumed. Just as with generic spreadsheets, one can move the cursor around the spreadsheet and position it on quantities, unit prices, or total prices. At any of these one can make changes and see the result on the screen. By moving the cursor to the right border, we can scroll over to the unit and total prices under the presently hidden columns Other and Total. Moving the cursor up and down at the top and bottom of the spreadsheet scrolls the screen to reveal other Phases and items in the spreadsheet. By pressing the spacebar, we can also pop up a small GoTo window wherein we can type a specific phase number and item code to jump directly to that part of the spreadsheet. With a little experience, one can thus navigate fairly quickly around the various parts of the estimate.

Note that the PRODUCE ESTIMATE menu in Fig. 12-22 does not have a specific menu option for "estimate costs." The reason is that PE Light pulls unit production and/or cost figures for the various required labor, material, and other resources directly from the database and performs the extensions automatically. If the database is up-to-date and 100 percent applicable to the project at hand, this may be sufficient. However, once the numbers have been pulled in, we have complete flexibility to override any unit prices, production rates, waste percentages, or other parameters to tune the estimate precisely to the needs of the project. Once we have reached an appropriate interim or completion stage, we can then select **Recalculate spreadsheet** followed by **Print reports** to choose from a variety of reports available to print. The latter can be in various levels of detail, from a single total cost, through the Group Phase summary, to Phase and even work Item details, all with unit and total costs for labor, materials, other, and for the line-item sum of these. After reviewing reports and discussing them with others, we can go back in and modify any portion of the database or the project spreadsheet.

Pressing function key **F4 Show est totals** from the spreadsheet level (Fig. 12-27) takes us to the totals screen in Fig. 12-28, which also summarizes the add-ons that go into the total bid price. It includes subtotals for labor, material, and other (e.g., equipment, subcontracts, etc.) and, at the bottom, the total unit price per square foot (the overall parameter used as a gross measure in this particular project). The column of letters near the add-on percentages tells the basis for applying that add-on (L = labor, M = material, O = other, C = current total from previous line, and A = all categories). Any change in any quantity or unit price anywhere in the estimating spreadsheet will be reflected immediately in these totals and add-ons, making it easy to use the estimate for "What-if?" analyses as bid time draws near.

Reports

There are two main categories of PE Light reports: database and estimate. Their choices were listed in the two print submenus in Fig. 12-22. Any of these reports can be sent to the primary or alternate printer, to the screen, to a disk file, or

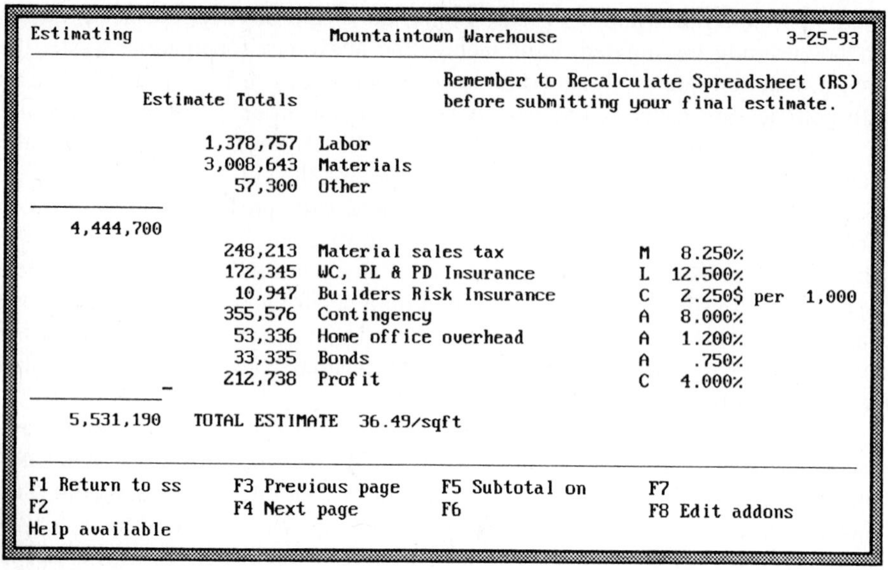

```
Estimating                        Mountaintown Warehouse                    3-25-93

                                       Remember to Recalculate Spreadsheet (RS)
            Estimate Totals            before submitting your final estimate.

              1,378,757  Labor
              3,008,643  Materials
                 57,300  Other

    4,444,700
                248,213  Material sales tax          M    8.250%
                172,345  WC, PL & PD Insurance        L   12.500%
                 10,947  Builders Risk Insurance      C    2.250$ per  1,000
                355,576  Contingency                  A    8.000%
                 53,336  Home office overhead         A    1.200%
                 33,335  Bonds                        A     .750%
             _  212,738  Profit                       C    4.000%

    5,531,190  TOTAL ESTIMATE  36.49/sqft


  F1 Return to ss        F3 Previous page    F5 Subtotal on       F7
  F2                     F4 Next page        F6                   F8 Edit addons
  Help available
```

FIGURE 12-28
Estimate of add-ons and totals.

to another device. Particularly in the estimate reports, available standard options provide a wide variety of formats, content, and levels of detail. One can also print limited ranges of information, such as the phases pertinent to a specific craft or subcontractor. Other options include sort orders, combining like items, various types of rounding, and inserted notations. But if these are still insufficient, Timberline also markets two versions of Timberline Report, a custom report-generating package that can draw upon any data fields stored in the estimate files and present them in a user-defined output format.

Database reports essentially just reproduce the information contained in the database in the categories phases, add-ons, formulas, variables, and items. We have mentioned all of these except variables, which are defined on the left side of formulas. Once defined, variables can be used in subsequently defined formulas, and the user can add new variables and formulas to those that come with Timberline's databases. Since much of the database information will be output in some of our Mountaintown Warehouse project reports, we will not devote space here to showing database reports as such.

The estimate reports for a particular project include the following:

- Take-off audit—shows the sequence in which items were taken off, including quantities, costs, items, date and time of take-off, and totals
- Spreadsheet—contains all columns from the spreadsheet and can be organized by phase and item category
- Standard estimate—shows all quantities, unit prices, and extensions in phase order

- Estimate details—includes standard content plus conversions, labor productivity, and material waste factors
- Category costs—shows each category; its total labor, material, other, and combined costs; and labor requirement
- Take-off unit costs—provides unit costs and quantities for all items and categories
- Material list report—shows item descriptions, take-off quantities (e.g., square feet of polyethylene), and order quantities (may be different from take-off quantity, e.g., rolls), unit and extended price, and grand totals within divisions

Printing even a subset of the many reports generated for the warehouse takes hundreds of pages, so we will only give a few examples here. Figure 12-29 is

```
Example Construction Co.              Estimating Standard Report          3-25-93     Page 5
                                       Mountaintown Warehouse             10:30 am

                                   <------LABOR----->   <------MATRL---->  OTHER   TOTAL
ITEM DESCRIPTION        TAKEOFF QTY  UNIT PRICE  AMOUNT  UNIT PRICE AMOUNT AMOUNT  AMOUNT

  6.000  FRAMING
------------------------
  6.110  Studs 2x4
2410 Studs - 2x4x10'       2.70 mb  15.000/hr    2,430   820.000/mb  2,325    -     4,755

  6.115  Studs 2x6
26 Studs 2x6               1.60 mb  15.000/hr    1,440   820.000/mb  1,378    -     2,818

  6.230  Plywood Sheathng
34 Plywood Shthing 3/4" 2,000.00 sf 15.000/hr      900     1.200/sf  2,520    -     3,420

  6.310  Roof Nailers
10 Roof Nailers           10.00 mb  15.000/hr   18,000 1,160.000/mb 12,296    -    30,296

  6.420  Sheetrock
30 Sheetrock 1/2"       3,200.00 sf 16.000/hr    1,280      .600/sf  2,016    -     3,296

  6.700  Cabinets-Shelves
sub Cabinets & Shelves    40.00 lf  20.000/lf      800    30.000/lf  1,200          2,000

                                                  ---------------------------------------
                                    FRAMING       24,850               21,734        46,584
                                                  ---------------------------------------


  7.000  WATERPRF & INSUL
------------------------
  7.100  Built Up Roofing
sub Built Up Roofing   165,000.00 sq   .360/sq   59,400     .440/sq 72,600          132,000

  7.206  Insulation
sub Insulation Sub     165,000.00 sf   .200/sf   33,000     .300/sf 49,500           82,500

  7.600  Flashing & Metal
10 Misc. flashings       4,600.00 lf  4.250/lf   19,550    5.000/lf 24,150    -      43,700

  7.705  Roof Hatch
10 Roof Hatch              49.00 ea  16.000/hr   10,584 1,000.000/ea 49,000   -      59,584

                                     ---------------------------------------------------
                                     WATERPRF & INSUL 122,534         195,250       317,784
                                     ---------------------------------------------------
```

FIGURE 12-29
Section from Estimating Standard Report—item detail level.

a section of an Estimating Standard Report at the detail level. Headings selected in this case include unit prices for labor and materials, but not for other and the item totals. Group Phase totals appear at the bottom of each section.

Figure 12-30 is the Estimating Category Costs report at the summary level for the whole project. This report also is available at the Phase and Item detail levels. Note that even at the summary level it indicates the total labor hours required for each category.

Figure 12-31 is a section of the Estimating Material List. If the take-off quantity is in the same units of measure as the order quantity, it does not appear.

```
Example Construction Co.      Estimating Category Costs       3-25-93     Page 1
                              Mountaintown Warehouse          10:38 am

  GROUP    DESCRIPTION             LABOR         MATRL        OTHER          TOTAL

  1.000    GEN CONDITIONS         137,325        67,455       14,800        219,580
                              Labor hrs: 5,600.00
  2.000    SITEWORK               173,975       468,468       42,500        684,943
                              Labor hrs: 10,000.964
  3.000    CONCRETE               356,560       706,045                   1,062,605
                              Labor hrs: 18,279.015
  4.000    MASONRY                 27,540        15,390                      42,930
  5.000    STEEL                  151,913       750,309                     902,222
                              Labor hrs: 7,543.50
  6.000    FRAMING                 24,850        21,734                      46,584
                              Labor hrs: 1,598.00
  7.000    WATERPRF & INSUL       122,534       195,250                     317,784
                              Labor hrs: 661.50
  8.000    DOORS & WINDOWS         16,584        74,300                      90,884
                              Labor hrs: 382.40
  9.000    FINISHES                85,796        36,968                     122,764
 10.000    SPECL CONDITIONS         2,396         8,430                      10,826
                              Labor hrs: 159.70
 11.000    EQUIPMENT                6,405        25,600                      32,005
                              Labor hrs: 427.00
 15.000    MECHANICAL             166,760       389,612                     556,372
 16.000    ELECTRICAL             106,120       250,140                     356,260

               ESTIMATE TOTALS

           1,378,757  Labor                 44,652.079 hrs
           3,009,702  Materials
              57,300  Other
        -------------
        4,445,759
                       248,300  Material sales tax        M      8.250%
                       172,345  WC, PL & PD Insurance      L     12.500%
                        10,949  Builders Risk Insurance    C      2.250$ per    1,000
                       355,661  Contingency                A      8.000%
                        53,349  Home office overhead       A      1.200%
                        33,343  Bonds                      A       .750%
                       212,788  Profit                     C      4.000%
        -------------
        5,532,494  TOTAL ESTIMATE   36.49/sqft
```

FIGURE 12-30
Estimating Category Costs report—summary level.

```
Example Construction Co.      Estimating Material List         3-25-93    Page 3
                              Mountaintown Warehouse           10:42 am

   DESCRIPTION              TAKEOFF QTY    ORDER QTY     UNIT PRICE        AMOUNT

SITEWORK
===============
Bulk Fill
----------------
   Bulk Fill                               26,250.0     cuyd     3.00     78,750.00

Asphalt Paving
----------------
   Asphalt Material Type H                     63.0     ton     30.00      1,890.00
   Crusher Run Base                            88.0     cuyd    12.60      1,108.80
                                                                         -----------
                                                                           2,998.80
Concrete Curb
----------------
   Plain Conc Curb 3000 psi  1,060.0 lnft      44.6     cuyd    76.00      3,389.60

Concrete Paving
----------------
   S.O.G. Conc 3000 psi                     4,042.5     cuyd    68.00    274,890.00

Storm Drainage
----------------
   Storm Drainage                           2,690.0     lnft    16.00     43,040.00

Catch Basins
----------------
   Catch Basins                                 6.0     each 1,400.00      8,400.00

Manholes
----------------
   Manholes                                     4.0     each 1,960.00      7,840.00

Railroad Work
----------------
   Railroad Sidings Sub                       320.0     lnft    22.00      7,040.00

Fencing
----------------
   Fencing Subcontractor                    4,680.0     lnft     9.00     42,120.00

                                                                         ===========
                                                        SITEWORK         468,468.40
```

FIGURE 12-31
Section from Estimating Material Report.

Thus the only item showing a different take-off quantity from its order quantity is "Plain Conc Curb," which is taken off in linear feet but converted to cubic yards to order the concrete material.

Limitations and Alternatives

Timberline's Precision Estimating Light is a useful program for small and medium-size residential and building projects. It handles the commodity take-off and unit pricing quite well for a project like the Mountaintown Warehouse. But it does have several limitations in capacity, application, and functionality, some of which are addressed by more advanced Timberline estimating products. Capacity limits are 500 line entries, only three resource categories (labor, material, and other, where "other" would have to double for both subcontracts and equipment), and 18 spreadsheet column headings. PE Light also does not allow a work-package take-off breakdown, tracking of subcontractor quotations, nonuniform distribution of markups to different phases, or multi-user operation; and it does not provide interfaces to CAD and scheduling programs, digitizer-based quantity take-off subsystems, the Means database, and so forth. All of these limits are overcome or greatly relaxed at the next step up to regular Precision Estimating. Another inconvenience at the PE Light level is not having separate tables for labor and equipment costs. If you get a new wage rate, you must go in and update every separate related work item in the database (there could be dozens or hundreds of them), or at least do so for the ones used in a given estimate. Equipment costs are not included separately at all. The Plus and Extended versions of Precision Estimating allow unlimited rate tables and the grouping of labor resources into defined crews that can then be linked to work packages or items. The Extended version further allows full coding of work-breakdown structure and reporting schemes that better tie the estimate to schedules; resource and cash requirements analysis; pricing of alternates to the standard bid package; cost escalation; and several other useful capabilities. Moving up to these capabilities, of course, costs more but can save time and effort in the long run.

Although Precision Estimating is fairly versatile, contractors working mainly in heavy construction may find it constraining. It is definitely designed more for a commodity take-off and pricing approach to estimating than for one that builds on methods and resources. The latter approach is much more common in heavy construction, and programs like the aforementioned Grantlun Hard Dollar Estimating System and the HCSS Estimating System have been designed with it in mind. Because of these differences in the estimating application area, it is all the more important to go through the procedures described in Chapters 5 and 6 to make the choice that best suits the needs of a given company.

SUMMARY

This chapter examined computer-based approaches that can support several stages of estimating costs of a construction project. They include project location and tracking, collecting cost and productivity data, quantity take-off, establishing work methods and productivity rates, estimating direct and indirect costs, and compilation, analysis, and bidding. We further showed how estimating is but part of an overall project planning and management environment, and thus estimating software can integrate with CAD systems, scheduling applications, accounting and cost control, and others.

We next set forth some basic concepts for estimating costs of projects. There are many legitimate approaches for estimating project costs, and each has its appropriate applications and limitations. Of importance in any method, however, is that (1) it correctly accounts for all the various cost and productivity factors in constructing a project, and (2) it is applied uniformly and consistently from one project to another. Detailed bid estimates are best prepared from completed plans and specifications. They are based upon actual quantity take-offs that are multiplied by unit prices developed by the estimator. Bid estimates typically include lump-sum or unit-price material and subcontract quotations.

Several kinds of estimating software have recently become available that can feed data into the main estimating system. These include project location and tracking programs such as McGraw-Hill's Dodge DataLine[2]; materials specification information such as SweetSource, an electronic version of McGraw-Hill's Sweets Catalog; equipment cost information such as the Dataquest Equipment Cost Estimator and Caterpillar's Fleet Production and Cost program; and estimating databases such as the R. S. Means Company's MeansData and Richardson's engineering databases, both of which can integrate with popular estimating programs.

Quantity take-off can be performed online within estimating programs using the various formulas and calculation facilities they provide. But using an electronic digitizing tablet with appropriate computer interface software can provide a real boost in accuracy and productivity.

Figuring out the methods to do the work, selecting the most appropriate labor and equipment resources, and determining their productivity are among the major challenges of estimating. This area is still one where the best asset is human experience and judgment. But computers can assist these valuable people by archiving records from past projects, by simulating field operations, and eventually by capturing and making available the knowledge of experts in various techniques and specialties.

Labor costs were divided into two main categories. The first dealt with strictly financial aspects. These included basic wages, fringe benefits, insurance, taxes, and wage premiums. Most of these are readily quantifiable and can be handled with ease by computer programs. The second introduced factors related to productivity, which require considerably more human judgment.

We concluded our overview of estimating by briefly examining cost analysis for materials, equipment, and subcontractors; the estimation and application of indirect costs; and final compilation, analysis, and bidding. The computer's speed and accuracy is particularly valuable as bid day nears, and incoming quotes from suppliers and subcontractors require changes almost to the moment of bid submission.

There is so much variety in estimating software and its utilization that it is difficult to say just what constitutes advanced features. But some helpful additions include software for custom report writing, and close integration with databases and other software such as CAD, scheduling, and cost control. Range estimating is a newer concept that allows for uncertainty in various parts of an estimate and provides guidance as to which parts of an estimate contain the highest risks. Although there has been little progress to date, estimating is also an attractive target for new developments involving artificial intelligence and expert systems.

In addition to the brief illustrations of specific programs in earlier sections, we provided two examples of estimating software at work. The first showed the general sequence of performing a quantity take-off using a digitizer and the Sitework Engineering program. The second was Timberline's Precision Estimating system, where we used the low-end PE Light version to prepare an estimate for the Mountaintown Warehouse Project. Although these examples pointed out many of the advantages of computer-based estimating, they also showed some of the differences and limitations to look out for when selecting an estimating package for a particular construction organization.

For all their power and convenience, estimating programs are far from ready to replace experienced human estimators. Estimating is often appropriately called more of an art than a science, and top estimators are some of the most valuable people in an organization. But computer-based estimating software can become a strong asset to these people, reinforcing their abilities by cutting through routine information searches and calculations, by providing a standard and consistent format for their work, and by giving them more time for the challenging experience- and judgment-based parts of the process.

REVIEW QUESTIONS

1. In an integrated applications environment, what types of information could most usefully be transferred from the following?

   - A CAD system to an estimating system
   - An estimating system to a job-cost accounting system
   - An estimating system to a project scheduling system

2. Briefly state how a building construction contractor could best take advantage of project location and tracking software such as Dodge DataLine.

3. What are the key advantages that an online computer version of a project location and tracking system has over a mailed paper version? What disadvantages might there be to this method?

4. Which of the following types of information can be obtained from commercially available construction information services? Give an example of each one that applies.

   - Construction materials
   - Equipment costs
   - Equipment fleet production information
   - Labor cost and productivity data

5. Briefly describe three ways in which a computer might assist an estimator in determining the methods, resources, and productivity for a heavy construction task such as building a cofferdam.

6. Assume that a construction cost information system contains some historical project labor cost data that are one to three years old for a variety of locations. What factors

should be taken into account before using such information to help in estimating current work?

7. In question 6, what would have been a more useful form in which to keep the historical information from field production operations?

8. State at least three ways in which computer-based estimating systems can assist a contractor in the frantic last-minute efforts to receive and analyze supplier and sub-contractor quotes and prepare a competitive and accurate bid submittal.

9. What is the most significant difference in the way an estimating system handles the indirect costs and markup on a unit-price bid as opposed to a lump-sum bid?

10. How can an estimating system help a contractor meet an owner's social goals in the employment of disadvantaged business enterprises?

11. Briefly describe the advantages of each of the following advanced features found in some estimating programs:

- Nonlinear distribution of indirect costs to unit-price bid items
- User-definable report forms
- Integration with published cost databases
- Range estimating

12. What is the major limitation that prevents more full automation of the estimating process?

13. Relative to mass market software applications such as spreadsheets and databases, what market factor primarily accounts for the higher costs of estimating software?

14. Briefly describe the procedure for quantity take-off for site grading using a digitizer-based system such as the Sitework module from AGTEK's EDGE system.

15. Briefly describe the principal approach taken by Timberline's Precision Estimating Light package to preparing a cost estimate for a building. What are some of the limitations of this low-end version of Timberline's product line?

APPLICATION PROBLEMS

The first step with any new estimating program should be to work through its tutorial material and try it with examples that may have been provided by the vendor. For hands-on experience with the program after that step, you might consider the following:

1. Reenter the warehouse example from this chapter and use it as a basis for further exploration. Look well beyond the limited number of estimating capabilities and reports that were included in this chapter. You will need to consult the Barrie and Paulson *Professional Construction Management* text (see following reading list) for more detail on this project.

2. If you have access to a digitizer, obtain a drawing with a limited amount of detail, and of a type that fits the kind of work for which the software was designed, and experiment with a quantity take-off of that project.

3. Prepare a tabular or narrative summary comparing human versus computer strengths and weaknesses in estimating. On this basis, define some important characteristics of the human-computer interface that will make the system easy and effective to use.

SUGGESTIONS FOR FURTHER READING
AND EXPLORATION

The following references focus mainly on tools mentioned in this chapter. Several of the vendors make the demo disks available at little or no cost, and they are a good means for further exploration and learning.

Building Construction Cost Data. R.S. Means Company, Inc., P.O. Box 800, Kingston, Mass. 02364. Published annually.

Barrie, Donald S., and Boyd C. Paulson, Jr. "Chapter 11: Estimating Project Costs," *Professional Construction Management*, 3d ed. New York: McGraw-Hill, 1992.

Dataquest Equipment Cost Estimator (DECE). Dataquest, 1290 Ridder Park Drive, San Jose, Calif. 95131-2398.

Dodge DataLine Demo Disk. F. W. Dodge / McGraw-Hill, 1221 Avenue of the Americas, New York 10020.

Fleet Production and Cost Analysis. Caterpillar, Inc., Peoria, Ill. 61602, 1991.

G2 Estimator. G2 Inc., 10656 Executive Drive, Boise, Idaho 83704, 1989.

Hard Dollar Estimating Office System. Grantlun Corporation, 2055 E. Centennial Circle, Tempe, Ariz. 85284, 1993.

HCSS Demo Disk. HCSS, 6222 Richmond, Suite 325, Houston, Tex. 77057, 1991.

Hyper·Estimator/Hyper·Remodeler. Turtle Creek Software, 651 Halsey Valley Road, Spencer, N.Y. 14883, 1989.

MC². Management Computer Controls, Inc., 2881 Directors Cove, Memphis, Tenn. 38131, 1991.

MeansData for Lotus. R. S. Means Company, Inc., P.O. Box 800, Kingston, Mass. 02364, 1991.

Neil, James. *Work Packaging for Project Control*, Publication No. 6-5. Austin, Tex.: Construction Industry Institute, November 1988.

Paydirt Earthwork Estimating Systems. Spectra-Physics Laserplane, Inc., 5475 Kellenburger Road, Dayton, Ohio 45424, 1992.

Precision Estimating. Beaverton, Ore. 97005: Timberline Software, 1991.

Reich, Mark, and Jud Youell. *Estimating and Timberline Precision*. Albany, N.Y.: Delmar Publishing, Inc., 1994.

Richardson Engineering Services, Inc., P.O. Box 9103, Mesa, Ariz. 85214, 1993.

Sitework Engineering System. AGTEK Inc., 396 Earhart Way, Livermore, Calif. 94550, 1990.

SweetSource Demo Disk. Sweet's Electronic Publishing, 99 Monroe Ave., NW, Suite 400, Grand Rapids, Mich. 49503-2651, 1992.

CHAPTER 13

PROJECT PLANNING AND SCHEDULING

Time present and time past
Are both perhaps present in time future
And time future constrained in time past.

T. S. Eliot*

Planning, scheduling, and control of the functions, operations, and resources of a project are among the most challenging tasks faced by construction managers. Normally, this responsibility involves coordinating design with construction to produce the necessary plans and specifications, to package them along recognized trade and subcontractor specialties, and to engage contractors who are best qualified to carry out their work efficiently and economically in conjunction with other contractors on the site. In the construction phase, the contractors normally provide the detailed planning, scheduling, and control needed to sequence operations properly and to allocate the resources involved efficiently.

This chapter deals with computer-based methods available to construction professionals for accomplishing these objectives and offers practical guidelines for their effective application. The ***critical path method (CPM)*** underpins the software presented in this chapter, so we begin with some background on that technique and show how it can be applied using a computer. Advanced concepts and features of CPM software are next described, including flexible logic, calendar dating, resource allocation and leveling, network-based cost control and cash

Parts of this chapter have been adapted and condensed from Chapter 12 of *Professional Construction Management*, 3d ed., by Donald S. Barrie and Boyd C. Paulson, Jr., McGraw-Hill, New York, 1992. Reprinted by permission of McGraw-Hill, Inc. Details of planning and scheduling methods and procedures are contained in that book and in specialized planning and scheduling books mentioned in the references at the end of this chapter.

*Four Quartets, 1943.

401

flows, time/cost trade-offs, graphical reporting, and others. We then present guidelines for applying CPM computer programs in project control. Finally, two examples illustrate the application of representative commercial programs: MacProject and Primavera Project Planner.

BASIC CONCEPTS OF PLANNING AND SCHEDULING

The critical path method is one of several analytical tools and graphical techniques for planning, scheduling, and controlling operations and resources. Others include bar charts, progress curves, matrix schedules, and linear balance charts. Bar charts and progress curves themselves are readily produced by CPM software and thus are common features of such programs. It is important to emphasize at the outset, however, that none of these is in and of itself the *plan* for the project. The complete plan, if it exists at all, exists only in the minds of the planners. All the tools mentioned here are merely abstract means to aid the planners in *organizing* and *documenting* their thinking and assumptions and in *communicating* that thinking to persons responsible for putting the plan into action.

Bar Charts

A *bar chart* graphically describes a project as consisting of a well-defined collection of tasks or activities whose completion marks its end. An *activity* is a task or closely related group of tasks whose performance contributes to overall project completion. A typical activity in a bar chart for a building project could be "Excavate foundation."

A bar chart is generally organized so that activity labels are listed in a column at the left side of the diagram or in the schedule itself on or to the left of the activity symbols. A horizontal time scale extends to the right, with a line corresponding to each activity in the list. A *bar* representing the progress of each activity is drawn between its corresponding scheduled start and finish times along its horizontal line. Figure 13-1 is a simple bar chart for a small concrete gravity-arch dam.

Bar charts have a number of advantages over other scheduling techniques. Their simple graphical form results in relatively easy general comprehension. This fact has led to their common acceptance as a good form of communication in industry, where virtually all levels of management understand them. Also, they are fairly generalized planning and scheduling tools, so they require less revision and updating than more sophisticated systems. This feature is especially helpful in the turbulent early stages of an engineering and construction project when frequent changes and revisions are a fact of life.

Used by themselves, bar charts have limitations. First, they become large and cumbersome as the number of line activities increases. Second, although the planner who prepared the bar chart no doubt considered the logical interconnections and constraints of the various activities in the project, this logic is not

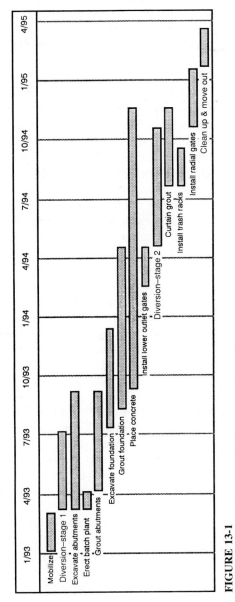

FIGURE 13-1
Example of a bar chart schedule for constructing a dam.

403

expressed in the diagram. Third, it is difficult to use a bar chart for forecasting the effects that changes in a particular activity will have on the overall schedule. It is thus limited as a control tool. However, when bar charts are integrated with CPM software, where they are readily produced as output reports based on the CPM logic, these limitations are substantially overcome, and the bar chart option adds its communication advantages to the CPM technique.

Progress Curves

Progress curves, also called *S-curves*, graphically plot cumulative progress on the vertical axis against time on the horizontal axis. Progress can be measured in terms of money expended, surveys of quantity of work in place, worker-hours expended, or any other measure that makes sense. Any of these can either be expressed in terms of actual units (dollars, cubic meters, etc.) or be normalized to a percentage of the estimated total.

The shape of a typical S-curve results from integrating progress per unit of time (day, week, month, etc.) to obtain cumulative progress. On most projects, expenditures of resources per unit time tend to start slowly, build up to a peak near the middle, then taper off at the end. This causes the slope of the cumulative curve to start low, increase in the middle, then flatten near the top. A simple example appears in Figs. 13-2a and 13-2b.

Once the project is underway, actual progress can be plotted and compared with that which was planned. It is then possible to make projections based on the slope of the actual progress curve. Figure 13-3 is an illustration of the basic concepts of planning, reporting, comparing, and projecting progress using S-curves. Actually, four curves appear in this figure, two of which use the critical path network concepts of early start and late start discussed later in this chapter. The scheduling of an activity closely correlates with the timing of its resource expenditures (money, labor, materials, etc.) and, of course, to its accomplishment. It follows that if all activities in a project are scheduled as early as possible (an early-start schedule), the progress will take place and be reported earlier. Conversely, if all activities are scheduled at their late starts, progress takes place and is reported later. The upper and lower curves in Fig. 13-3 are illustrations of these ideas. Curve *ES* is the cumulative early-start progress curve. Curve *LS* is the cumulative late-start progress curve. Note that the two curves start and end at the same point, but that at any other point the *LS* curve falls below or to the right of the *ES* curve. The planned and reported progress should fall between these two extremes, as shown by curves *P* and *R*, respectively, on Fig. 13-3.

When money is the resource measured on progress curves, *cash flows* may be shown graphically by plotting one progress curve for expenditures on the same graph with a second curve for income. A third curve representing the financing required or cash surplus at any time results from subtracting the expenditures ordinate from the income ordinate at each point in time. Figure 13-4 is a typical cash flow graph.

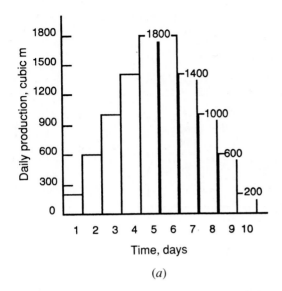

(a)

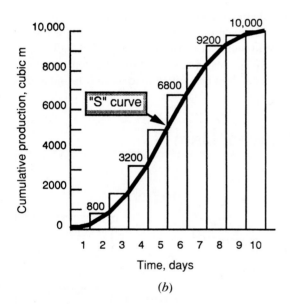

(b)

FIGURE 13-2
Development of a progress curve: (a) daily production,
and (b) progress curve, or S-curve.

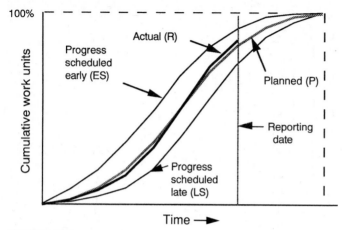

FIGURE 13-3
Early, late, and actual progress.

Concepts of Critical Path Networks

The critical path method enables planners and managers to analyze thoroughly the timing and sequential logic of all operations required to complete a project before committing time, money, equipment, labor, and materials for engineering and construction. Planning with CPM focuses expert attention on potential trouble spots and indicates where extra effort can best be applied to reduce costs and delays

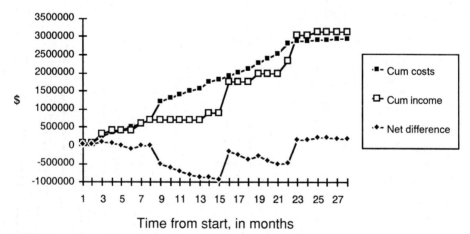

FIGURE 13-4
Cash flows for construction of dam in Fig. 13-1 (data exported from MacProject and plotted by Excel).

without wasting resources. In addition, computer-based CPM permits relatively easy revision of the schedule and facilitates the simulation and evaluation of the impact of changes. It thus becomes an excellent control tool during project execution.

Essential elements in all CPM networks are *activities*, their *durations*, and *logical interrelationships* among them. Given these, one can compute each activity's *early start, late start, early finish, late finish, total float*, and *free float*. These computations also yield the total expected *project duration*, and, of great importance, they focus attention upon the most *critical activities* and hence the *critical path* for the project. This powerful concept greatly aids management in setting its priorities for allocating resources to operations.

As mentioned for bar charts, an *activity* is a task or closely related group of tasks whose performance contributes to the completion of the overall project. An activity should be sufficiently well integrated that it can be rescheduled as a unit. One example of an activity could be "Construct column footing." Most activities, including this one, can be further subdivided into component activities. The degree of breakdown depends upon the size and type of project, its requirements, and the purpose of the schedule. In this case, component activities could include "excavate," "fabricate forms," "assemble rebar cage," "set forms," "fine grade," "set rebar cage," "place and finish concrete," and "strip forms." These activities consume both time and resources. In scheduling, activities might also represent administrative procedures, such as "client approval of plant layout"; show delays that require time but no resources, such as "Winter shutdown" and "Spring floods"; and allow for shipment and delivery of equipment and materials, such as "Order ready-mix."

In the *arrow diagram* form of the CPM, directed lines, or *arrows*, represent activities. Figure 13-5 is an arrow network for the concrete footing. The direction of each arrow indicates the direction of progress. Uniquely labeled nodes (1, 2, ..., 8) placed at the beginning and end of each arrow identify the activity symbolically. Activities are joined in a logical manner determined by the construction

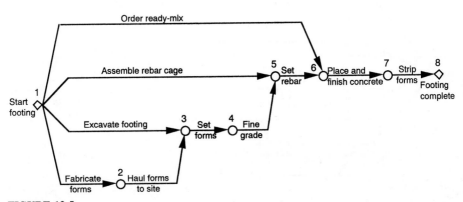

FIGURE 13-5
Arrow diagram for concrete footing construction.

sequence of the operation being performed, the methods used, the time allowed, and the resources available. A group of activities may be represented in a graphical form called a **network**. This network itself may be but a small *subnetwork* of the overall project network.

Precedence diagramming (sometimes called the *activity-on-node* or *circle-and-connecting-line* method) is an alternative way of representing a critical path network and has some advantages. It is the opposite of arrow diagramming in that the nodes represent activities and the connecting lines show the logical relationships among them. Precedence diagrams eliminate the need for dummy arrows to represent the logic correctly and are therefore much easier to construct and modify. Figure 13-6 is a precedence diagram equivalent to the arrow network in Fig. 13-5. Note that in this case a delay of seven days appears on the link before "Strip forms" to allow for cure time. Also, the rounded boxes are *events* to mark the beginning and end of this operation.

Logical properties of networks include *precedence, succession,* and *concurrence.* The start of a particular activity is permitted by the completion of all preceding activities or by the start of the project. In the example, both "Order ready-mix" and "Set rebar" are *predecessors* of "Place and finish concrete." Both predecessors must be completed before "Place and finish concrete" can start. Similarly, the completion of an activity either permits the start of a following activity or marks the end of the project. In the example, completion of "Fabricate forms" permits its direct *successor,* or *follower,* "Haul forms to site," to commence. Other activities may logically proceed simultaneously, in which case they are logically *concurrent.* In the example, "Set forms" and "Assemble rebar cage" are concurrent activities. Planners must logically think through these relationships to construct a network. The process of constructing a network, in turn, strongly aids in such logical thought processes.

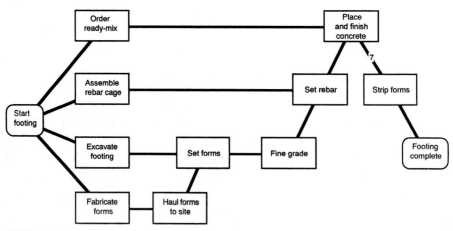

FIGURE 13-6
Precedence diagram for concrete footing construction.

WORKING WITH PLANNING AND SCHEDULING PROGRAMS

This section discusses how we define activities, prepare and organize networks, estimate activity durations, enter data into a computer program, perform schedule calculations, and generate reports. Most topics apply equally well regardless of which computer-based scheduling software is used.

Defining Activities

The level of detail to which a network's activities are subdivided depends upon a number of factors, including the schedule's purpose, the type and size of project, and the preferences of management. In scope, activities should be defined and detailed so as to correspond to work items subject to cost control in a work breakdown structure. In general, activity durations should be long enough so that management can take corrective action if schedule reports indicate that this is necessary. Where possible, an activity should be contained entirely within one supervisor's responsibility.

Preparing a Network

Although the technical details of network preparation are beyond the scope of this chapter (see Moder, Phillips, and Davis or other CPM texts), it is worth reviewing the general procedure to indicate where a good planner will include the needs of users. The steps listed here also show that the mechanics of the process are fairly simple; they apply to large as well as small networks.

1. Begin by learning about the project itself. Study the plans and specifications, visit the site, and seek input from key parties involved in the project.
2. Make a preliminary listing of some key activities, keeping in mind the guidelines for defining activities.
3. Put a key activity on the diagram—first, last, or in between—but make a start. This can be done on a paper draft or directly on a computer screen.
4. Ask yourself the following questions:

   • What must be completed immediately before this activity can begin?
   • What activities can follow once this activity is complete?

5. Put these new activities on the diagram.
6. Repeat steps 4 and 5 until you have a reasonably comprehensive diagram. Its organization at this stage will probably be chaotic, but tidiness can wait.
7. Reexamine the plans, specifications, and other sources of information to be sure that all parts of the project are covered and none is duplicated.
8. Check the network logic for errors such as possible recursive loops, and be sure that each activity has a defined start and finish point.

9. At this stage you have a rough diagram. Before preparing the final version, check it with those parties consulted in step 1 to confirm that it represents their thinking.

10. Repeat the earlier steps as needed to produce a satisfactory draft network. Only at this stage should one proceed to an elegant presentation.

Organizing the Network

The physical organization of activities on a network should proceed according to some rational pattern representing the needs and characteristics of the project. Possible criteria for organization, which may be used alone or in combination, include the following:

1. By responsibility: trades, type of work, supervisors, and so forth
2. By geographic area and facility within the project
3. By time scale
4. By activity cost code

Each type of organization mentioned here is made possible in the plotter-drawn networks of one or more computer programs.

Estimating Durations

Having defined and organized all activities into a logical CPM network, one must estimate their durations. An activity's *duration* is the expected amount of time, expressed in consistent time units, required to complete the activity from start to finish. The time units may be days, weeks, or even hours or minutes, just so that all activities use the same units.

Entering Activity Data and Network Logic into a Computer Program

With today's interactive scheduling software, data entry generally parallels initial rough drafting of the network and thus interacts with the steps described thus far. This section provides some brief illustrations of how activities and network logic are input and how the computer checks the input.

In Fig. 13-7 are two screen shots taken at the time that activity data were being entered into the Primavera Project Planner in arrow diagramming form (it also allows precedence diagramming). In this case, the scheduler enters the I and J nodes, description (title), and duration. Figure 13-7a is an entry form where one enters or edits data for individual activities. In Fig. 13-7b, one enters or edits data for multiple activities on one screen in tabular form.

In Fig. 13-8 we see the **Chart** window from MacProject Pro. In this case, the scheduler can place boxes representing activities directly on the screen and connect

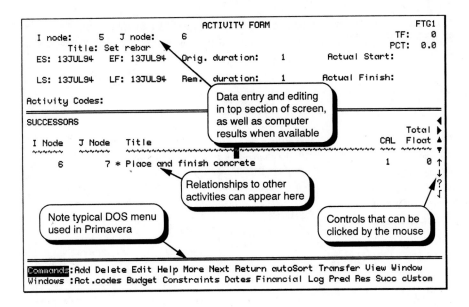

FIGURE 13-7
Entering activity data into Primavera: (*a*) using Form mode; and (*b*) using Table mode.

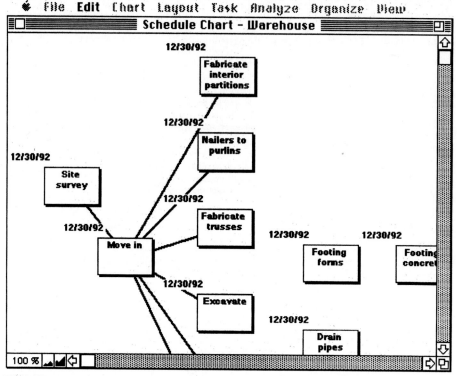

FIGURE 13-8
View of a network being constructed in MacProject.

them by moving the mouse-guided cursor from one box to another. Descriptions and other data can be entered via dialog boxes or into separate tabularly oriented windows called the **Outline** view, the **Project Table** (a table summarizing all of the input and computed results for each activity), or the **Task Timeline** (a bar chart view). Data entered in any of these windows automatically update in the others, making for very flexible data entry.

Both Primavera and MacProject allow variations of the forms-oriented methods represented in Fig. 13-7 and the direct diagram construction method of Fig. 13-8. Both programs have entry sections for resource and cost data and also analysis and reporting options. A few of these are illustrated later in this chapter, but for the most part they are beyond the scope of what we can include here.

One pitfall is the unintentional creation of *loops* in the network. For example, if in Fig. 13-5 the arrow for "Assemble rebar cage" were entered in the opposite direction, two cyclical logic loops would result. Loops such as these have no logical beginning or end; their presence indicates an error in a CPM network. In this case, the error is easy to spot, but some fairly lengthy loops can creep into more complex networks and be fairly difficult to detect by hand. Most CPM computer programs have routines to search out and trace any loops that may be present. Loops must be corrected before processing can continue.

CPM Time Calculations

So far, we have only discussed the activity definition, duration estimates, and logic elements in CPM networks. With these, the computer can perform schedule calculations. However, the user should understand the following procedure:

1. A *forward pass* calculation
2. The *backward pass* calculation
3. *Float* calculations
4. Determination of the *critical path(s)* in the network

In Table 13-1 the notation and mathematical formulas used in such calculations are defined.* Regardless of whether the logic is shown by arrow or precedence

TABLE 13-1
Summary of CPM notation and equations

Notation

$D(x)$	=	Estimate of duration for activity x	(a)
$ES(x)$	=	Earliest (expected) start time for activity x	(b)
$EF(x)$	=	Earliest (expected) finish time for activity x	(c)
$LS(x)$	=	Latest allowable start time for activity x	(d)
$LF(x)$	=	Latest allowable finish time for activity x	(e)
$TF(x)$	=	Total float for activity x	(f)
$FF(x)$	=	Free float for activity x	(g)
S	=	Project start time	(h)
T	=	Target project completion time	(i)

Equations for calculating the CPM parameters

Forward pass

$ES(x)$	=	S for beginning activities	(1)
	=	Max (EF(all predecessors of activity x))	(2)
$EF(x)$	=	$ES(x) + D(x)$	(3)

Backward pass

$LF(x)$	=	T for ending activities	(4)
	=	Min (LS(all followers of activity x))	(5)
$LS(x)$	=	$LF(x) - D(x)$	(6)

Floats

$TF(x)$	=	$LS(x) - ES(x)$	(7)
	=	$LF(x) - EF(x)$	(8)
$FF(x)$	=	Min (ES(all immediate followers of activity x)) $- EF(x)$	(9)

Critical path
A critical path is a continuous chain of activities with the minimum total float value. By summing activity durations, it is the longest duration path through the network. There may be more than one critical path in various parts of the network.

---

*The form of the notation used in this chapter is based upon that used by Jerome D. Wiest and Ferdinand K. Levy in *A Management Guide to PERT/CPM*, ©1969, p. 31. Adapted by permission of Prentice-Hall, Englewood Cliffs, N.J.

notation, the methods for network computations are similar. The notation and formulas in Table 13-1 provide all that is required for standard CPM network calculations. The following paragraphs summarize the computations performed by a computer-based CPM scheduling program.

The calculation procedure called the *forward pass* establishes the *earliest* expected start and finish times for each activity in the network, and also the total project duration. Definitions (a), (b), (c), and (h) of Table 13-1 apply to this calculation. It begins with S, the project start time or date. The set of rules that defines the procedure for the forward pass calculations is called an ***algorithm***. Only three rules are required: equations (1), (2), and (3) in Table 13-1.

The early finish of the last activity is the maximum early finish in the project and thus becomes the earliest completion date of the project. The total *project duration* is therefore calculated to be Max(EF) − S. This step completes the forward pass calculations.

The calculation procedure called the *backward pass* establishes the *latest* allowable start and finish times for each activity that will still permit the overall project to be completed without delaying beyond the scheduled completion date. The nomenclature used in the backward pass is given in definitions (a), (d), (e), and (i) of Table 13-1; and the rules expressed as equations (4), (5), and (6) describe the algorithm for the backward pass.

The project completion time T may become the early project completion time from the forward pass, or it may be the contractual completion date. If this backward pass started with the completion time from the forward pass, the minimum late start of all the activities should equal the start time S of the project. This result confirms the calculations and completes the backward pass.

The *total float** is the maximum amount of time that the activity can be delayed without extending the completion time of the overall project. However, such a delay may postpone the early start of one or more of its following activities. The total float for each activity may be calculated directly as the difference between the activity's late start and its early start, or as the difference between its late finish and its early finish. Refer to definition (f) and equations (7) and (8) in Table 13-1.

Free float is the maximum amount of time an activity can be delayed without delaying the early start of any of its followers. It is the minimum of the early starts of an activity's immediate followers minus the activity's early finish. This is expressed mathematically with definition (g) and equation (9) in Table 13-1.

All activities on the critical path have zero float—total or free—unless the target completion date T for the backward pass is not set equal to the earliest project finish date from the forward pass. Free float for any activity is less than or equal to its total float.

A *critical path* is a continuous chain of activities from the beginning to the end of a network with the minimum float value. By summing activity durations,

---

*Float is sometimes called *slack.*

we find that the critical path is the longest path through the network. There may be more than one critical path running in parallel in various parts of the network.

In Fig. 13-9 are the calculations for the precedence network from the earlier concrete footing example. Times are in working days from the start at day 1 rather than in calendar days. The legend describes the location of the ES, EF, LS, LF, Duration, and Slack (Total Float) for each activity. The program that made this chart, MacProject Pro, actually sets the EF and LF to one day less than the formulas of Table 13-1 would say (i.e., $EF = ES + Dur - 1$). It does this to show the finish date at the *end* of the activity's last day, not the beginning (i.e., start) of the next day.

CPM Schedule Reports

This section briefly describes some of the types of reports that typical computer-based scheduling software can produce. Specific illustrations appear in the example cases later in this chapter.

Almost all CPM packages produce *tabular reports*. One type simply summarizes the *input data*—activity labels, descriptions, durations, and labels of immediate followers or predecessors. Another gives *results* of the calculations, such as ES, LS, EF, LS, TF, and FF, and possibly adds asterisks, boldface type, or otherwise designates those on the critical path. Based on these tabular reports, numerous sorting, selection, and summarization options are possible. Activities can be sorted by labels or related codes and by the calculated ES, LS, EF, LF, and float times.

Selectivity and subreporting capabilities in CPM software produce subreports of activities in certain categories (e.g., for craft forepersons as defined by the labels or codes of their work type) or highlight critical activities or those whose start times will occur in the next month or so. Careful sorting and selection can help management effectively focus on important areas within the information that these systems can produce, and get the right information to the right people at the right level of detail and in time for decisions and corrective action.

Most programs also offer *graphical reports*. These can include bar charts, activity networks, and S-curves. More advanced versions also include various types of resource plots and cash flow diagrams. The quality and flexibility of graphic output vary considerably, as do the types of output devices supported. Some make good use of color when available. Sorting and selectivity options are similar to those of tabular reports.

ADVANCED PLANNING AND SCHEDULING CONCEPTS AND FEATURES

The preceding sections described the basic CPM scheduling features that should be available in any commercial program. These include data entry, network computations, and at least some reporting of the network data and the results of the computations. Most programs include far more than this.

There have been numerous useful extensions to the standard elements of activities, time, and logic in networks. The strict logic requirements of basic

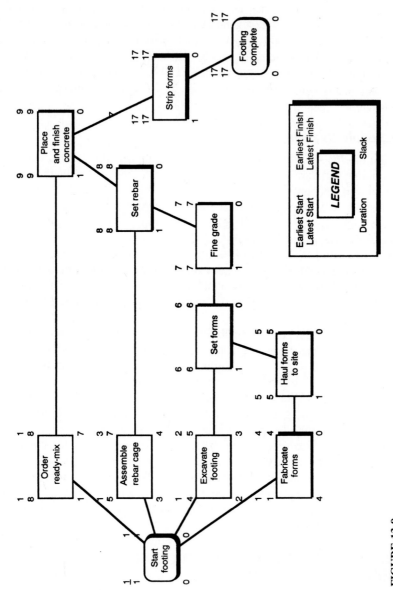

FIGURE 13-9

An example of CPM calculations.

CPM have been relaxed somewhat by the incorporation of techniques permitting activities to be overlapped or to have delays inserted between them. Actual calendar dates, including weekends, holidays, and shift schedules, can be reflected in most systems. In some cases, the methods of probability and statistics have been applied to both time and logic. By identifying resources of labor, materials, and equipment with activities, planners and managers can take advantage of powerful analytical techniques for resource allocation to ensure that the project can be completed within finite resource limits, and they can use resource leveling to improve the efficiency and continuity of resource utilization. Financial management enhancements include analytical procedures for time/cost trade-off analysis, network-based cash flows, and network-based cost control—and, of course, modern software features such as interactive tutorials, online help information, macros, and the ability to integrate with other types of software come with good programs.

These and other possibilities will be discussed in general terms next. It is beyond the scope of this chapter to explain them in detail, but most are described in textbooks such as those listed at the end of this chapter.

Additions within the Basic Functions

Enhancements to the basic functionality of CPM scheduling software are similar to those of other types of package programs. These include graphical user interfaces, easier data-entry editing techniques, variety in fonts and symbols, and so on. Particularly useful, however, is more flexibility in representing the logical relationships between activities. In addition to the standard linkage from the finish of one activity to the start of its follower(s), any or all of the following possibilities are common in CPM programs:

- Start-to-start (the succeeding activity can start at a specified *lag time* after its predecessor starts—e.g., allow backfill to follow two days after pipe laying begins)
- Start-to-finish (the finish of a succeeding activity is tied into a time at least a given lag time after the start of its predecessor—e.g., to meet a contract requirement)
- Finish-to-finish (one activity should finish at or a specified lag time after the finish of another activity—e.g., to coordinate a start-up procedure)
- Finish-to-start plus a lag time (a delay is specified after one activity finishes and before a following activity can start—e.g., to allow paint to dry or concrete to cure)

Each of these is illustrated graphically in Fig. 13-10.

Calendar-Based Computations

Most good commercial CPM-based scheduling software allows the use of actual calendar dates for computing the various scheduled times in a network. Indeed,

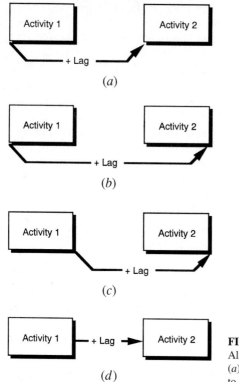

FIGURE 13-10
Alternative activity-to-activity logic relationships:
(*a*) start-to-start; (*b*) start-to-finish; (*c*) finish-to-finish; and (*d*) finish-to-start with lag factor.

they can allow multiple calendars to handle different craft or subcontractor work schedules and overtime and multiple-shift work. They can even specify exceptions by activity to standard calendars. The flexibility and variety of these options vary considerably, however. To provide a better understanding of what to look for, this section describes a method of constructing calendar-scaled CPM diagrams. The basic unit of time for this purpose is 1 day.

To convert the *working-day schedule* to a *calendar-dated schedule*, the start day is designated on the calendar. Succeeding parts of the working-day schedule are then compared to the calendar on a day by day basis. Extra days are inserted for nonworkdays, such as weekends and holidays. The overall procedure is quite straightforward and yields corresponding calendar dates for each workday in the schedule.

To illustrate the construction of a calendar-dated diagram, assume that the example project from the previous sections starts on Friday, July 1, 1994. Also assume a standard 40-hour, five-day workweek. The calendar for July 1994, taken from a setup screen in MacProject, appears in Fig. 13-11. A holiday occurs on the 4th. The 2nd, 3rd, 9th, 10th, 16th, 17th, 23rd, 24th, 30th, and 31st fall on weekends. The remaining days are available for work.

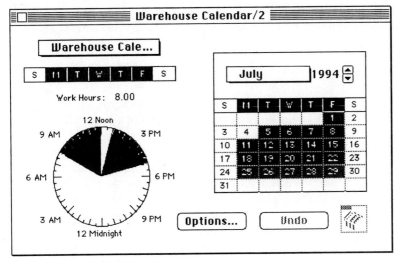

FIGURE 13-11
Calendar specification screen from MacProject Pro.

In Fig. 13-12 we see Primavera's calendar-dated bar chart corresponding to the schedule in Fig. 13-9. Note that allowing for the July 4 holiday and four two-day weekends along the way makes the completion date July 26 rather than 17 working days.

Probability and Statistics

A version of computer-based network scheduling systems developed in the 1950s was the *Program Evaluation and Review Technique (PERT)*. In addition to being event-oriented rather than activity-oriented, it provided for three time estimates

```
Day                 04      11      18      25      01      08
                    JUL     JUL     JUL     JUL     AUG     AUG
PRED   SUCC         94      94      94      94      94      94
~~~~~  ~~~~~  ~~~~~~~~~~~~~~~~~~~~~~~~~~~~~~~~~~~~~~~~~~~~~~~~~~~~~~
   1      6    ▪...++++..++▨ Order ready-mix  .        .
   1      3    ▪...▪+▨ Excavate footing        .        .
   1      5    ▪...+▨..▨ Assemble rebar cage   .        .
   1      2    ▪...▪ Fabricate forms .         .        .
   2      3    .    ▪ Haul forms to site       .        .
   3      4    .      ▪ Set forms     .        .        .
   4      5    .      .▪ Fine grade .          .        .
   5      6    Set rebar ▪    .         .        .        .
   6      7    .         .  ▪ Place and finish concrete
   7      8    .     Cure ▪..▪▪▪▪▪..▪         .        .
   8      9    .       . Strip forms.▪         .        .
```

FIGURE 13-12
An example of a calendar-dated CPM schedule.

for each activity (optimistic, most likely, and pessimistic), and used these as a three-point beta distribution by more heavily weighting the most likely estimate as follows:

$$\text{Weighted mean} = \frac{\text{optimistic} + 4 \times \text{most likely} + \text{pessimistic}}{6}$$

The schedule computation could then apply probability methods to compute the likelihood for various project durations rather than just accept a single deterministic estimate.

Some program developers went even further and allowed probabilistic branching down alternative contingent paths in a network (e.g., if activity A has X outcome, proceed next to activity B, otherwise proceed to C). Given these possibilities, still other developers added the concept of recursive simulation to network scheduling.

Needless to say, working with such systems is far more complex than just dealing with deterministic CPM calculations. Obtaining three estimates for every activity duration is more time-consuming and expensive than one. For this and other reasons, such features are not commonly found in commercial CPM programs, but one example is Primavera Systems' Monte Carlo program, and custom programs have been developed in industrial and construction companies that deal with risky projects like offshore oil platforms.

Resources

Resource management extensions are common in CPM scheduling software. These occur in three main categories: resource aggregation, resource allocation, and resource leveling. In each case the scheduler can specify one or more resource types (crafts, machines, etc.) to be available in the project and assign specific amounts of one or more of these to individual activities. Some programs also allow groups of resources to be collected together as crews and be assigned as a unit. One might also be allowed to vary the overall availability levels of individual resources over the life of a project—for example, starting at one level, building up to a peak, and then tapering off.

In the case of *resource aggregation*, the program computes the total number of resources of each type used by activities on any date and tabulates or plots their usage profile over the life of the project (time is the horizontal axis). A typical output report for the laborers appears in Fig. 13-13. The upper half is a bar chart report from a CPM network with activities labeled A to K. Activity durations appear in white, and their floats are in gray on each bar. The lower half is a profile of daily laborer requirements that results from laborers being assigned to each activity as shown in the second column. The black horizontal line at 10.00 on the vertical axis indicates that there is an availability limit of 10 laborers that is exceeded in the two areas shown without a black outline. In some programs, this is the only automatic resource computation capability available, and they leave it to the user to reschedule activities in periods where total usage exceeds availability or to level resource usage where the profile fluctuates excessively.

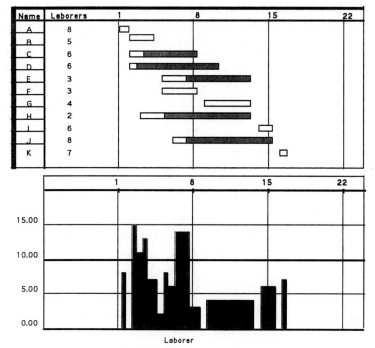

FIGURE 13-13
Example of resource aggregation.

Resource allocation routines observe the upper limits imposed on resource availability and automatically interrupt or reschedule activities as necessary to stay within these limits. If insufficient resources are available, these procedures will then increase the overall project duration, but they try to minimize this delay. They usually allow the user to choose among various criteria for rescheduling (e.g., move those with the most float), and some programs allow the user to be selective in specifying whether or not an activity can be interrupted. Some also have flexible limits on resource availability (e.g., 50 carpenters, but allow encroachments of up to 10 percent more on any given day if necessary to avoid delaying the overall project schedule).

Resource leveling algorithms generally ignore overall resource availability and usually do not have the option of delaying the project schedule. Instead, they concentrate on minimizing fluctuations in resource usage from one day or week to the next and thus reduce the corresponding inefficiencies of hiring and firing workers and mobilizing and demobilizing equipment.

Advanced computer programs can handle both of these types of constraints and then some. For example, Fig. 13-14 is a screen with which the user works when using a powerful interactive method of simultaneously leveling and allocating resources in MacProject. The top message advises of a specific conflict encountered on day 5 when processing the example from Fig. 13-13. The middle graphical

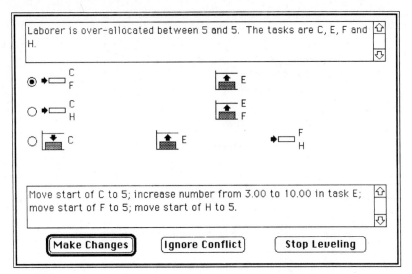

FIGURE 13-14
An example of resource allocation.

section offers a choice of methods for modifying the schedule, whereas the lower message tells exactly what will be done. MacProject also allows the user to turn on fully automatic leveling, but there is considerable benefit in using the interactive method to better understand the changes as they are made.

In Fig. 13-15 we see the results of one series of changes made to the schedule in Fig. 13-13 via the interactive method in MacProject. Note that the whole resource profile stays under the limit of 10 laborers, but to do so we extended the schedule three days. The final schedule, in Fig. 13-16, adds little bulldozer icons to the portions that had to be shoved back in time and also adds some other numerical information to a few of the activities and their connectors.

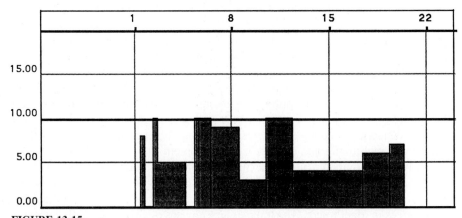

FIGURE 13-15
An example of a resource-constrained profile.

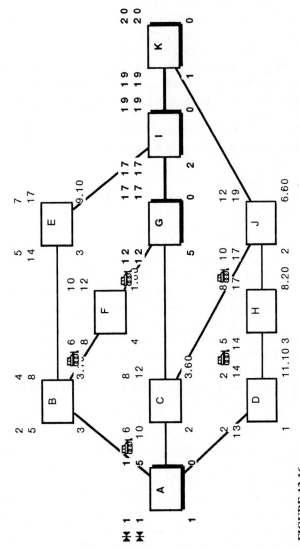

FIGURE 13-16
Modified CPM network with resource constraints.

423

The fractional times result from using resources at less than 100 percent of capacity on some activities in order to make time available for others.

Cost Control

Numerous programs have some facility to add costs to their scheduling systems. The simplest approach is to include an estimated cost as an attribute for each activity, and perhaps to allow an overall daily or weekly overhead or indirect-cost figure whose total will vary with the computed duration. Programs with resource management capabilities might allow unit costs to be entered for each resource and then compute the totals for each activity to which they are assigned. But the most comprehensive approach is an integrated cost and scheduling system, and these typically follow the pattern of a work breakdown structure (WBS). Most of these cost-based approaches also allow development of project cash flows, particularly if forecast income or payments can be associated with activities. Charts such as Fig. 13-4 can then be printed out.

Time/Cost Trade-offs

Earlier, we mentioned PERT's use of three time estimates. If we also have cost estimates associated with each of these three activity times, or even if we only have two cost and time estimates for selected activities, we can perform what are called *time/cost trade-offs*. The assumption is that to proceed at a faster rate than what is normal for an activity will cost more money, at least in terms of direct cost. This situation may arise because we have to work overtime or use more expensive equipment or materials. There may be some savings in indirect cost if the overall project duration shortens as a result. In getting ready for a time/cost trade-off analysis, we can compute the time/cost slopes for various activities, as shown in the upper part of Fig. 13-17. Here we have simple two-point lines for each critical activity in a schedule (those with no float shown as shaded extensions on the bar chart).

The idea of combining this concept with CPM is to focus initially on activities along the critical path to find activities that we might accelerate. There will be no savings in schedule if we expedite noncritical activities. For those on the critical path, start by accelerating the ones with the lowest cost/time slope—that is, those where time can be purchased at the lowest unit cost. If this increased cost per day is lower than the overhead costs per day, one could actually reduce overall project cost by spending money to accelerate the activities with minimum slope. This reduction could be particularly advantageous if the overall project duration would otherwise be in an area where the contractor would incur liquidated damages or penalties, or conversely where there were incentives for early completion. Activity B in Fig. 13-17 has the lowest slope, and thus would be the best place to start spending money to expedite the schedule. Then we would move to Activity G, and so on until the desired project duration was achieved. Figure 13-18 is an illustration of the composite time/cost trade-off curve. Eventually, as we come

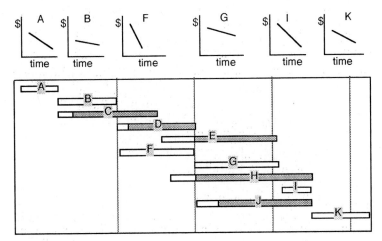

FIGURE 13-17
An example of activity time/cost curves for a CPM schedule.

to activities with higher and higher costs, we will reach a point of diminishing returns where direct costs of reducing time exceed the savings in indirect costs. Also, at some stage parallel paths may become critical, and two or more activities may have to be accelerated concurrently to reduce the overall schedule. Nevertheless, this approach is far better than just accelerating the project as a whole, and it optimizes the application of resources to buy time.

Since this seems like a valuable technique, you may be disappointed to learn that it is seldom found in commercial CPM scheduling programs, in part because it is a tremendous burden for users to estimate all of the extra time and cost points. Also, to look at time and costs without considering resource constraints simultaneously can be unrealistic and thus misleading with respect to the expected gains. However, an understanding of the principles behind this technique can still be applied quite effectively with existing CPM scheduling software. The results of the CPM computations first focus our attention on the activities that can make a difference, and with a few selective time/cost estimates we can then choose the best activities to accelerate.

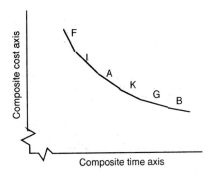

FIGURE 13-18
Composite time/cost trade-off curve for schedule compression.

Graphics

Since schedules based on CPM networks are very dominantly a graphical technique, the graphical user interface and reporting capabilities are very important. In the early decades of computer-based CPM, however, automated graphics were almost nonexistent, and reams of tabular output were a poor substitute. When graphical reports became available in the 1970s, they were found in programs costing tens of thousands of dollars, and their quality often was not very good. Fortunately, in the 1980s quality graphical output became available even in programs costing a few hundred dollars, and no program should be considered seriously unless it has strong capabilities in this area.

With a graphical user interface, the user can draw and manipulate network diagrams directly on the CRT screen. Graphical report options can include bar charts, networks and subnetworks in various formats and levels of detail, resource usage profiles, cash flow diagrams, and progress curves. They can be displayed on color CRT screens or be printed and plotted in color or black and white on almost any computer output device, ranging from impact and laser printers to large multi-pen plotters. Graphical output has thus become common on job sites and is valued for its communication capabilities. Several of the illustrations in this chapter exhibit these modern graphics capabilities.

Integration with Other Software

As one of many types of project management software, a CPM scheduling program must be able to integrate with other programs in many cases. Examples of important interfaces include those to estimating, CAD, cost engineering, and materials management.

ESTIMATING. The first versions of a project schedule often evolve in the estimating process. Estimating and scheduling share the same need for time estimates, and if costs and resources are added to schedules, they have these in common too. Therefore, in addition to adding some of these characteristics directly to their scheduling programs, some vendors have provided interfaces to estimating software. For example, Primavera Project Planner can link to Timberline's Precision Estimating to download cost and resource data to assign to network activities according to a common work breakdown.

CAD. In some ways, CAD models and CPM schedules are just two different representations of the same project. The objects being designed also need to be scheduled, and they may have a number of attributes in common. There have thus been numerous research efforts to establish links between the two. One successful product is Bechtel's Construction CAE, which will be described in Chapter 16. This is a very powerful planning, analysis, communication, and control tool, and one can hope that more features of such advanced programs will become available in the lower-cost CAD and scheduling software used on microcomputers.

COST ENGINEERING. As mentioned earlier, it is common to find cost-related features in scheduling programs, and some versions are sophisticated and well enough integrated that they can serve both scheduling and cost engineering for some projects by themselves. However, there can also be important reasons to have interfaces between independent scheduling and cost-engineering programs as well, and some developers have catered to this need.

MATERIALS MANAGEMENT. Since many activities involve materials that must first be procured over a period of time, it may seem logical to include procurement activities in a network diagram. However, since there can be a dozen or more steps from requisition through approvals, purchase order, delivery, and acceptance and, further, since there may be multiple procured items per activity, detailed procurement activities could soon overwhelm the primary work activities in the schedule. It is better to build links between the CPM project schedule and the materials-management system of the project. In some cases this materials-management capability is an optional package available from the CPM scheduling software developer. In other cases, one would look for the possibility of establishing links to independent materials-management software or to database or spreadsheet software in which such systems are often implemented.

General Software Capabilities

As with other types of computer software, one can find interactive tutorials, online help files, and other capabilities in some CPM scheduling packages. One should evaluate their quality, effectiveness, and value versus the needs and capabilities of the likely users and applications.

APPLICATIONS AND LIMITATIONS

Planning and scheduling packages should be evaluated with respect to their suitability for documenting the characteristics of the planned project, the knowledge and level of sophistication of those who are expected to use them, the desired scope and level of detail for scheduling, and the means available for updating and revision. One should not immediately discard simpler tools when more powerful ones seem appealing. Similarly, even the more powerful techniques can be mismatched to the physical and managerial characteristics of certain projects.

Package software varies considerably in ease of use, flexibility, and power. However, these need not be mutually exclusive. Some of the more flexible and capable systems are among the easiest to use. MacProject is a good example. Programs like Primavera have far more features and high capacities, but they are more difficult to learn. Although they can handle even very large projects, they tend to be more suitable where scheduling engineers are the main day-to-day users.

From day one onward, those responsible for the management and control of a project have more information than did the planners who prepared the schedule for the project's guidance. Each day brings new information about the project. If the network is indeed to be a viable control tool on the project, it must be subject to revision, change, and improvement. Scheduling software should readily support the need to update, change, and forecast the status of a project. In the following sections are comments on some specific issues related to float, collecting input data, updating, and project documentation.

Managing Float

Both free float and total float give management flexibility in scheduling a project's activities. Judicious rescheduling of activities within their float ranges can effectively even out resource usage and make a smoother, more efficient job. On the other hand, float is a valuable commodity that can be wasted if used indiscriminately. Total float, especially when it is shared through a series of activities in a chain, must be regarded as community property among those activities. If a superintendent or supervisor carelessly lets an early activity in the chain slip to its late start and finish times, thinking that there is plenty of leeway, the remaining activities in the chain will become critical. Considerable managerial flexibility is thus lost. There is no fixed answer to this problem; it is a matter of policy that must be resolved by management on each project. Scheduling software should support management in its efforts to allocate float to various activities in the schedule.

Obtaining Update Information

Sources of information for schedule control are in part the same as those for cost engineering. They include labor and equipment time sheets, field quantity reports, and various kinds of trend reports. Particularly useful, however, are preprinted forms requesting actual and estimated time and resource information associated with specific activities. One good type of information form, and one that minimizes the amount of writing requested, provides space for field supervisors to overwrite information on one of two duplicate copies of subreports sent to the field. Used in this way, the subreport becomes a "turnaround" document such as that shown in Fig. 13-19. When returned to the office, it serves as input for the next schedule update.

Updating Considerations

There are several procedures for updating network-based schedules. With low-cost computer processing, it is feasible to carry out complete activity-by-activity computations on a periodic or demand basis. In updating we wish to compare planned with actual progress and to calculate variances to highlight problem areas. Parameters of interest include the following:

1. Floats, especially in critical and near-critical activities
2. Changes in the critical path
3. Logic changes, including new and deleted activities

Textbook Example PRIMAVERA PROJECT PLANNER Warehouse Project

REPORT DATE 17JAN93 RUN NO.16 Example Warehouse Project START 1JAN93 FIN 11FEB93
TURNAROUND DOCUMENT FOR SCHEDULE DATA DATE 1JAN93 PAGE NO. 1

ACTIVITY ID	ORIG DUR	REM DUR	%	CODE	ACTIVITY DESCRIPTION	SCHEDULED START	SCHEDULED FINISH
1	1	1	0		Site survey	4JAN93	4JAN93
2	3	3	0		Move in	5JAN93	7JAN93
xxxx	------	xxxxxx	------	xxxxxx	------ xxxxxx ------	xxxxxx ------	xxxxx
27	2	2	0		Paint exterior walls	8FEB93	9FEB93
30	2	2	0		Clean up	10FEB93	11FEB93

FIGURE 13-19
Turnaround document printed by Primavera Project Planner.

4. Resource usage, especially to predict constraints and reduce idleness

5. Changes in durations

6. Activities completed, and percentage complete on those in progress

One often hears debates among scheduling people about the best frequency with which to update schedules: daily? weekly? monthly? The debates, however, miss the point. The important thing is not how often a schedule is updated, but how accurately the schedule reflects the actual conduct of the work.

Documentation for Changes, Claims, and Disputes

Changed conditions, change orders, delays, claims, and disputes occur in some measure on almost all projects of significant size. Thus, a body of law and accepted practices has evolved that helps to achieve just, equitable, and fair resolution of disputes. In contracts, these take the form of clauses for changed conditions, change orders, delays, contract time, liquidated damages, disputes, and claims. But the process only starts here.

The effects of changes can be subdivided into three main categories: (1) direct costs, (2) time extension, and (3) impact costs. Even the first two can be difficult to assess, and impact costs are almost certain to provoke disagreement. More and more emphasis is put on thorough documentation and analysis to short-circuit many such claims before they become a real issue. Critical path networking methods have become accepted as one of the most powerful tools for documenting and resolving the differences in these instances. One approach, sometimes called *factual networks*, has been used for nearly two decades for the documentation of projects.* There are numerous special techniques for constructing such networks, but most involve adding some special graphical symbols to represent different types of delays and their impacts on arrow or CPM networks.

WAREHOUSE PROJECT† IMPLEMENTED IN MACPROJECT ON A MACINTOSH

This example deals with a small warehouse project with a compressed duration that serves our purposes of illustration. It demonstrates a few of the many capabilities of MacProject Pro, a powerful but user-friendly application suitable for use by construction professionals.

---

*J. M. Antill and R. W. Woodhead, *Critical Path Methods in Construction Practice*, 2d ed., John Wiley & Sons, New York, 1970, Chapter 11.

†This is not the Mountaintown Warehouse from Chapters 12 and 14, but rather is one contrived by John Fondahl, Emeritus Professor of Civil Engineering at Stanford University, to illustrate resource allocation techniques.

In Fig. 13-20 we see the 30-activity schedule for this project in precedence network form. We can create activity boxes of almost any size and place them anywhere on the diagram, as long as the logic flows from left to right. In addition to activity descriptions within the boxes, we can select any of the parameters input to and computed by the program, and position them at any location around the boxes, as shown in the legend on Fig. 13-20. If we want more detail, the items can be stacked several layers deep.

This schedule only has one level of detail, but MacProject Pro, like most of the more powerful scheduling programs, permits the hierarchical creation and interrelating of multiple main projects and subprojects. This in turn allows us to do companywide resource allocation and financial management, as well as to prepare different parts of the schedules and selectively focus on different subcontractors and supervisors.

Figure 13-21 is the bar chart produced for this project. In this case only the planned bars are shown, but MacProject also allows us to show a second bar for each activity that will display actual progress as it occurs so that we can easily see how we are doing. Critical-path activities are black, noncritical activity durations are white, and float times are gray. Also note here that we have chosen a format option that lists the activity labels in the left column, as contrasted to the adjacent position selected in Fig. 13-1.

In Table 13-2 are listed the resources that we plan to use on this project, their daily costs, as well as the limits that we have set on each (4.00 for all of them in this case). Note that this table also alludes to MacProject's flexibility to let us use a separate calendar for each resource (e.g., some crafts may prohibit weekend work). "Accrual method" tells whether or not we are aggregating resources over multiple projects. "Level" indicates whether to allow MacProject to apply its resource allocation and leveling algorithms to the particular resource. MacProject allows us to set individual activity-leveling priorities in the range of 0 percent to 100 percent, where 0 percent blocks leveling of that particular activity. "Gather" tells whether we should collect resources from multiple projects for purposes of our analysis.

In Table 13-3 we see how resources have been allocated to the activities in this project. Activities with multiple resources appear on multiple lines. Those with no resources do not appear at all. Workdays is an aggregate measure of effort. For example, in the first activity 2 carpenters times 2 days equals 4 workdays. Note that MacProject allows us to set fractional levels of effort. Thus a piping designer might concurrently spend 50 percent of his or her time on one project or activity and 50 percent on another. This table also automatically computes the cost for each resource on each activity.

The initial resource requirement profiles for each of the three resources used in this project appear in Fig. 13-22. In this case the aggregations are based on the early-start schedule with no resource allocation or leveling applied. Thus we note significant overages for carpenters (pale areas exceeding the limit of 4), which will have to be rectified. The laborers and ironworkers are all right as is.

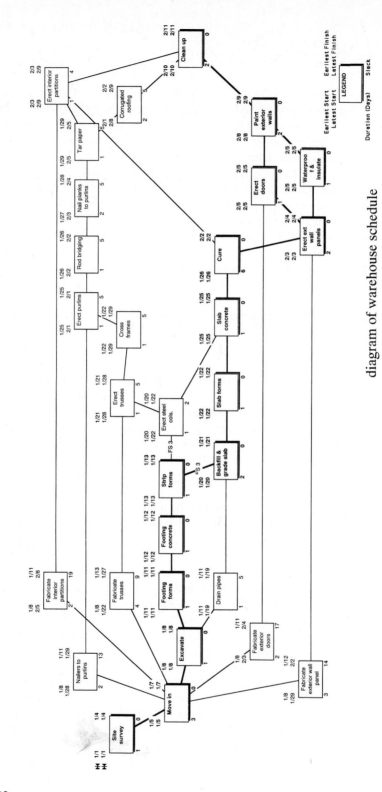

FIGURE 13-20

Network diagram of warehouse schedule. (Note: FS is the equivalent of FF in Table 13-1.)

diagram of warehouse schedule

432

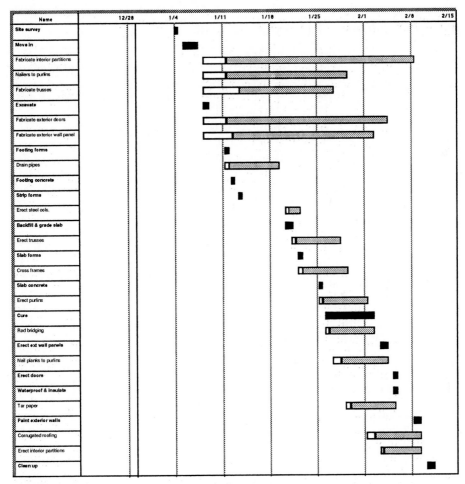

FIGURE 13-21
Bar chart for warehouse project.

TABLE 13-2
Resource availability for warehouse project

Resource	Cost/Day	# Available	Calendar Name	Accrual Method	Level?	Gather?
Carpenter	160.00	4.00	Warehouse Calendar/2	Multiple	True	True
Laborer	120.00	4.00	Warehouse Calendar/2	Multiple	True	True
Ironworker	200.00	4.00	Warehouse Calendar/2	Multiple	True	True

TABLE 13-3
Resource requirements for warehouse activities

Name	Resource	Work-Days	Number	% Effort	Duration	Allocation Cost
Nailers to purlins	Carpenter	4	2.00	100	2	640
Excavate	Laborer	4	4.00	100	1	480
Fabricate exterior doors	Carpenter	4	2.00	100	2	640
Fabricate int. partitions	Carpenter	8	4.00	100	2	1,280
Fabricate trusses	Carpenter	16	4.00	100	4	2,560
Footing forms	Carpenter	4	4.00	100	1	640
Footing forms	Laborer	2	2.00	100	1	240
Footing forms	Ironworker	2	2.00	100	1	400
Footing concrete	Carpenter	1	1.00	100	1	160
Footing concrete	Laborer	2	2.00	100	1	240
Strip forms	Carpenter	1	1.00	100	1	160
Strip forms	Laborer	1	1.00	100	1	120
Backfill & grade slab	Laborer	4	2.00	100	2	480
Erect steel cols.	Ironworker	4	4.00	100	1	800
Erect trusses	Ironworker	4	4.00	100	1	800
Slab forms	Carpenter	4	4.00	100	1	640
Slab forms	Ironworker	2	2.00	100	1	400
Cross frames	Carpenter	2	2.00	100	1	320
Cross frames	Ironworker	2	2.00	100	1	400
Slab concrete	Carpenter	1	1.00	100	1	160
Slab concrete	Laborer	2	2.00	100	1	240
Erect purlins	Ironworker	4	4.00	100	1	800
Rod bridging	Ironworker	4	4.00	100	1	800
Erect ext wall panels	Carpenter	8	4.00	100	2	1,280
Nail planks to purlins	Carpenter	4	2.00	100	2	640
Erect doors	Carpenter	2	2.00	100	1	320
Waterproof & insulate	Laborer	2	2.00	100	1	240
Tar paper	Laborer	2	2.00	100	1	240
Corrugated roofing	Carpenter	4	2.00	100	2	640
Erect interior partitions	Carpenter	4	4.00	100	1	640
Clean up	Laborer	8	4.00	100	2	960
Fabricate ext. wall panel	Carpenter	12	4.00	100	3	1,920

Similar resource profiles are shown in Fig. 13-23 after the scheduler and MacProject have worked interactively to revise the schedule to fit within resource constraints. The carpenters now stay within the limit of 4.0, but the schedule is extended. By interrupting some activities, a human scheduler can solve the warehouse project without extending the schedule at all, but on the first pass through MacProject we did not do this. One could try any number of alternative approaches, all the while gaining a better understanding of the project schedule and its constraints.

MacProject has numerous reporting options, several of which have already been shown in this chapter. Table 13-4 is a summary report that gives an overview of the activities, their status, important project dates, planned and actual project costs and income, resources, and relationships to other projects. Table 13-5 is a summary cash flow report. This could be exported to a spreadsheet program to produce a graphical report like that shown in Fig. 13-4.

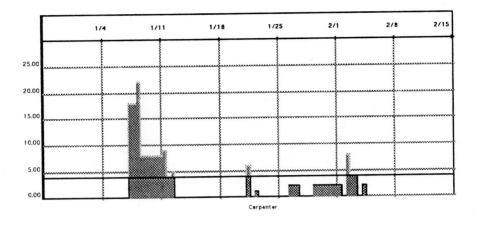

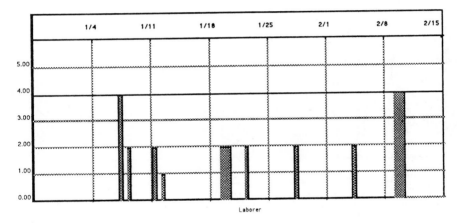

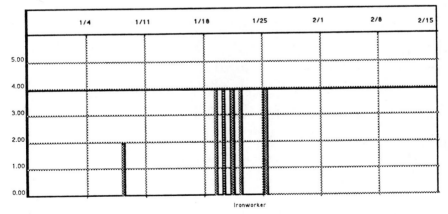

FIGURE 13-22
Early-start resource profiles for craftworkers on warehouse project.

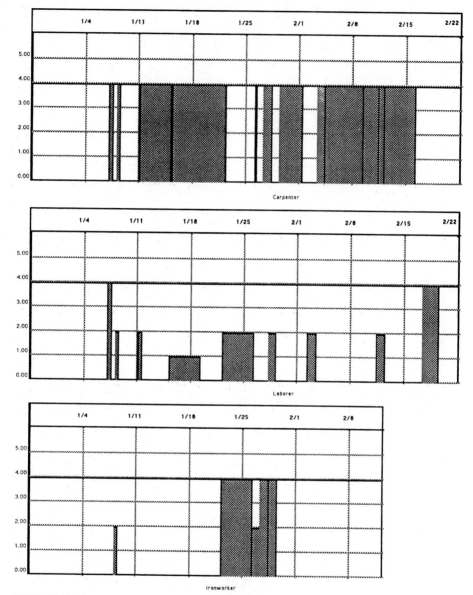

FIGURE 13-23
Leveled resource profiles for craftworkers on warehouse project.

TABLE 13-4
Overview report for warehouse dated 12/31

Warehouse
12/31
Project Overview

• Tasks and Activities	
Tasks	30
Milestones	0
Supertasks	0
Completed Tasks	0
Completed Milestones	0
Summary Activities	0
• Project Dates	
Project Start Date	1/1
Project Finish Date	2/11
Actual Start	1/1
Actual Finish	2/11
• Project Costs	
Total Planned Cost	44,630
Total Planned Income	48,010
Total Actual Cost	0
Total Actual Income	0
• Other	
Projects in Family Tree	0
Projects in Resource Scope	0
Project % Done	0
Number of Resources	3
Number of Calendars	1

TABLE 13-5
Weekly cash flow for warehouse project

Starting	Plan Costs	Plan Incomes	Actual Costs	Actual Incomes	Ending	Plan Cumulative	Actual Cumulative
12/28	300	0	0	0	1/4	-300	0
1/4	19,760	22,280	0	0	1/11	2,220	0
1/11	3,720	2,300	0	0	1/18	800	0
1/18	10,030	11,400	0	0	1/25	2,170	0
1/25	4,660	5,610	0	0	2/1	3,120	0
2/1	5,840	6,420	0	0	2/8	3,700	0
2/8	960	0	0	0	2/15	2,740	0

WAREHOUSE PROJECT IMPLEMENTED IN PRIMAVERA ON A PC

This second application uses the same example to provide some comparisons with MacProject when demonstrating features of Primavera Project Planner (P3), and to show some additional kinds of reports that are available in either program.

P3's interface is heavily menu- and form-oriented, using conventional text screens; it is not so graphical as MacProject, but it does have a subsystem named PENGUIN that permits some graphical input, interaction, and output. P3 has a capacity for large and complex projects (up to 100,000 activities) and has a wide variety of reporting options (over 100 standard reports plus the ability to create custom reports). These and other capabilities and features have made it something of a standard in the construction industry.

The main menu for P3 appears in Fig. 13-24. In it we see three options for data input, a section for defining activity classification codes, resources and cost accounts, CPM network and leveling calculations, and access to submenus for a variety of reports.

For the most part, one interacts with P3 via forms or tables, as shown in Fig. 13-7. One can enter and examine a large number of scheduling, resource, and cost parameters this way. Figure 13-25 is a form one can use to select and order the parameters to appear on a particular table.

P3 is particularly rich in its reporting options. Figure 13-26 is the overview menu of the report categories just within the tabular report section, and a similar

```
                                                                    WHSE
                        PROJECT DATA MENU
                        ~~~~~~~~~~~~~~~~~~

        Project data:       Calendar.....................1
                            Activity data................2
                               Forms.....................F
                               Tables....................T
                               Penguin...................P

        Dictionaries:       Activity codes...............3
                            Resource.....................4
                            Cost accounts................5
                            Custom data items............6

        Calculations:       Schedule/Level...............7
                            Global Change................8

        Reports/Graphics:   Execution....................9
                            Reports......................X
                            Graphics.....................G

                            Configuration options........C
                            Return to project listing....R
                            Exit.........................X
```

FIGURE 13-24
Main menu of Primavera Project Planner.

```
                              ACTIVITY TABLE                         WHSE

  Ref.No.: TD-01   Title: ACTIVITY DATA
 ═══════════════════════════════════════════════════════════════════════
  CONTENT                    Resource:              Cost Account:

  Data Item      Column #  | Data Item     Column # | Data Item    Column #
  ────────       ────────  | ─────────     ──────── | ─────────    ────────
                           |                        |
  Activity ID       1      | Suspend Date      0    | Qty this Period   0
  Activity Title    2      | Resume Date       0    | Qty to Complete   0   ◄
  Early Start       4      | Act. Code  1 - 1  0    | Qty at Completion 0   ►
  Early Finish      5      | Act. Items 1 - 1  0    | Quantity Variance 0   ▲
  Late Start        6      | Resource          0    | Budget Cost       0   ▼
  Late Finish       7      | Cost Account      0    | Cost to Date      0   ↑
  Actual Start      0      | Res. Designator   0    | Cost this Period  0   ↓
  Actual Finish     0      | Driving Res. Flag 0    | Cost to Complete  0   ?
  Total Float       8      | Units Per Day     0    | Cost at Completion 0  ▌
  Free Float        9      | Resource Lag      0    | Cost Variance     0
  Calendar ID       0      | Resource Duration 0    | Res. Items 1 - 1  0
  Orig. Duration    3      | Resource Percent  0    |
  Rem. Duration     0      | Budget Quantity   0    |
  Percent Complete  0      | Qty to Date       0    |
 ═══════════════════════════════════════════════════════════════════════
  Commands: Add Delete Edit Help More Next Return Transfer Window eXecute
  Windows : Content List
```

FIGURE 13-25
Selection form for data to appear on activity table.

```
                                                                    WHSE
                          Types of Reports
                          ~~~~~~~~~~~~~~~~~~
                 Schedule (tabular) reports........1

                 Bar chart reports...............2

                 Network logic diagrams..........3

                 Resource reports................4

                 Cost reports....................5

                 Export data files...............6

                 Import data files...............7

                 Activity matrix report..........8

                 Custom report writer............9

                 Return to Project Data Menu......R
                 Exit...........................X
```

FIGURE 13-26
Main menu for P3's tabular reports.

```
                            SCHEDULE REPORT                            WHSE

Ref.No.: SR-21    Title: SCHEDULE REPORT SUMMARY BY SUBPROJECT
═══════════════════════════════════════════════════════════════════════════

        Ref.No.                          Title
        ~~~~~~~       ~~~~~~~~~~~~~~~~~~~~~~~~~~~~~~~~~~~~~~~~~~~~~~~~~
        SR-01    CLASSIC SCHEDULE REPORT - SORT BY ES, TF
        SR-02    SCHEDULE REPORT - SORT BY TF, ES
        SR-03    SCHEDULE REPORT WITH ACTIVITY BUDGETS                    ◀
        SR-04    SCHEDULE REPORT SHOWING LOGS                             ▶
        SR-05    SCH REP - SKIP PAGES BY ACTIVITY ID CODE                 ▲
        SR-06    SCHED REP - PREDECES ORS AND SUCCESSORS                  ▼
        SR-07    SCHED REP - DETAILED PRECEDENCE ANALYSIS                 ↑
        SR-08    SCHEDULE SUMMARY REPORT WITH BUDGETS                     ↓
        SR-09    SCHEDULE REPORT SUMMARIZED WITH BUDGETS                  ?
        SR-10    TURNAROUND DOCUMENT FOR SCHEDULE                         ▯
        SR-11    SCHEDULE REPORT WITH RESOURCE USAGE
        SR-12    SCHED REP COMPARISON TO TARGET SCHEDULE
        SR-13    SCHED REP - LINE SKIP BY COMBINED CODES
        SR-14    SCHEDULE REPORT ORGANIZED BY DEPARTMENT

═══════════════════════════════════════════════════════════════════════════
Commands : Add Delete Edit Help Level More Next Return Transfer Window eXecute
Windows  : Content Format List Order Selection Target
```

FIGURE 13-27
Partial list of tabular schedule reports in P3.

range is available if one selects Graphics from the main menu of Fig. 13-24. Each menu item in Fig. 13-26 leads the user to further submenus containing anywhere from two to 21 separate reports. Figure 13-27 is the first screen of a menu listing 21 schedule-related reports. These options tend to be overwhelming at first, but once a project manager has found the reports that are desired by the owner, management, and staff on a particular project, it is fairly easy to set them up and use them on a regular basis.

Space only permits a small sampling of the many reports provided in this system. We will concentrate on tabular reports, although Primavera also provides a wide range of detailed and summary graphical reports for networks, bar charts, resources, and costs similar to those shown for MacProject. The reports included here have been edited and condensed to fit within the book's page margins, since most of Primavera's tabular reports use the traditional 132-column width of mainframe computer reports.

Figure 13-28 is Primavera's Classic Schedule Report for the warehouse project—"classic" in the sense that it is typical of the reports generated by mainframe computer CPM programs in the days of yore. It lists the activity labels, durations, descriptions, and the results of the schedule calculations. It is sorted by early start as the primary sort and, in the case of ties, by total float as the secondary sort.

Figure 13-29 is part of one of Primavera's printer-output bar charts. It is typical of the type of mainframe reports that were widely used before plotter-drawn graphics became economical enough for regular use.

```
----------------------------------------------------------------------------
Textbook Example              PRIMAVERA PROJECT PLANNER        Warehouse Project

REPORT DATE 17JAN93 RUN NO.12 Example Warehouse Project   START 1JAN93 FIN 11FEB93
CLASSIC SCHEDULE REPORT - SORT BY ES, TF              DATA DATE  1JAN93  PAGE NO. 1

--------- ----  ------------------------------  -------  -------  -------  ------- -----
ACTIVITY ORIG          ACTIVITY DESCRIPTION      EARLY    EARLY    LATE     LATE  TOTAL
   ID    DUR                                     START   FINISH   START   FINISH FLOAT
--------- ----  ------------------------------  -------  -------  -------  ------- -----
       1    1  Site survey                      4JAN93   4JAN93   4JAN93   4JAN93    0
       2    3  Move in                          5JAN93   7JAN93   5JAN93   7JAN93    0
       6    1  Excavate                         8JAN93   8JAN93   8JAN93   8JAN93    0
       5    4  Fabricate trusses                8JAN93  13JAN93  22JAN93  27JAN93    9
       4    2  Nailers to purlins               8JAN93  11JAN93  28JAN93  29JAN93   13
       8    3  Fabricate exterior wall panels   8JAN93  12JAN93  29JAN93   2FEB93   14
       7    2  Fabricate exterior doors         8JAN93  11JAN93   3FEB93   4FEB93   17
       3    2  Fabricate interior partitions    8JAN93  11JAN93   5FEB93   8FEB93   19
       9    1  Footing forms                   11JAN93  11JAN93  11JAN93  11JAN93    0
      10    1  Drain pipes                     11JAN93  11JAN93  19JAN93  19JAN93    5
      11    1  Footing concrete                12JAN93  12JAN93  12JAN93  12JAN93    0
      12    1  Strip forms                     13JAN93  13JAN93  13JAN93  13JAN93    0
      14    2  Backfill & grade slab           20JAN93  21JAN93  20JAN93  21JAN93    0
      13    1  Erect steel cols.               20JAN93  20JAN93  22JAN93  22JAN93    2
      15    1  Erect trusses                   21JAN93  21JAN93  28JAN93  28JAN93    5
      16    1  Slab forms                      22JAN93  22JAN93  22JAN93  22JAN93    0
      17    1  Cross frames                    22JAN93  22JAN93  29JAN93  29JAN93    5
      18    1  Slab concrete                   25JAN93  25JAN93  25JAN93  25JAN93    0
      19    1  Erect purlins                   25JAN93  25JAN93   1FEB93   1FEB93    5
      20    6  Cure                            26JAN93   2FEB93  26JAN93   2FEB93    0
      21    1  Rod bridging                    26JAN93  26JAN93   2FEB93   2FEB93    5
      23    2  Nail planks to purlins          27JAN93  28JAN93   3FEB93   4FEB93    5
      26    1  Tar paper                       29JAN93  29JAN93   5FEB93   5FEB93    5
      28    2  Corrugated roofing               1FEB93   2FEB93   8FEB93   9FEB93    5
      22    2  Erect exterior wall panels       3FEB93   4FEB93   3FEB93   4FEB93    0
      29    1  Erect interior partitions        3FEB93   3FEB93   9FEB93   9FEB93    4
      24    1  Erect doors                      5FEB93   5FEB93   5FEB93   5FEB93    0
      25    1  Waterproof & insulate            5FEB93   5FEB93   5FEB93   5FEB93    0
      27    2  Paint exterior walls             8FEB93   9FEB93   8FEB93   9FEB93    0
      30    2  Clean up                        10FEB93  11FEB93  10FEB93  11FEB93    0
```

FIGURE 13-28
Classic CPM schedule report.

Figure 13-30 is the resource profile for the carpenters before leveling. The "E" character used for the bars implies that this profile corresponds to the early-start schedule.

In Fig. 13-31 we see part of a schedule report that includes the resource usage for each activity as well as the early and late activity scheduled times. Primavera includes similar reports with cost allocations and has very flexible methods for breaking the resources and costs down by supervisor assignments, subprojects, and other ways. It can summarize all of these at various levels of aggregation. This flexibility is very important to achieving the objectives of selectivity and subreporting in order to get just the right information to the people who need it without overburdening them with irrelevant data.

Figure 13-32 is a partial list of the activity-by-activity actions taken when the scheduler asked Primavera to level the warehouse network. One can specify

```
--------------------------------------------------------------------------------
Textbook Example              PRIMAVERA PROJECT PLANNER            Warehouse Project

REPORT DATE 17JAN93 RUN NO.11  Example Warehouse Project START 1JAN93 FIN DATE 11FEB93
BAR CHART BY ES, TF                              DATA DATE 1JAN93        PAGE NO.1
                                                                 DAILY-TIME PER.   1
--------------------------------------------------------------------------------
        .........ACTIVITY DESCRIPTION.......        04     11     18     25     01     08
ACTIVITY ID  OD   RD  PCT CODES FLOAT SCHEDULE JAN    JAN    JAN    JAN    FEB    FEB
-----------  ---  ---- --- ----- ----- -------- 93     93     93     93     93     93
                                                --------------------------------------

Site survey                            EARLY   *E     .      .      .      .      .
         1   1    1   0           0             *.     .      .      .      .      .
Move in                                EARLY   *.EEE  .      .      .      .      .
         2   3    3   0           0             *.     .      .      .      .      .
Excavate                               EARLY   *.     E      .      .      .      .
         6   1    1   0           0             *.     .      .      .      .      .
Nailers to purlins                     EARLY   *.     E..E   .      .      .      .
         4   2    2   0          13             *.     .      .      .      .      .
Fabricate exterior doors               EARLY   *.     E..E   .      .      .      .
         7   2    2   0          17             *.     .      .      .      .      .
Fabricate interior partitions          EARLY   *.     E..E   .      .      .      .
         3   2    2   0          19             *.     .      .      .      .      .
Fabricate exterior wall panels         EARLY   *.     E..EE  .      .      .      .
         8   3    3   0          14             *.     .      .      .      .      .
Fabricate trusses                      EARLY   *.     E..EEE .      .      .      .
         5   4    4   0           9             *.     .      .      .      .      .
Footing forms                          EARLY   *.     E      .      .      .      .
         9   1    1   0           0             *.     .      .      .      .      .


  XXXXXX  ------  XXXXXX  ------  XXXXXX  ------  XXXXXX  ------  XXXXXX  ------

Nail planks to purlins                 EARLY   *.     .      .      . EE    .      .
        23   2    2   0           5             *.     .      .      .      .      .
Tar paper                              EARLY   *.     .      .      . E     .      .
        26   1    1   0           5             *.     .      .      .      .      .
Corrugated roofing                     EARLY   *.     .      .      .     EE       .
        28   2    2   0           5             *.     .      .      .      .      .
Erect interior partitions              EARLY   *.     .      .      .     . E      .
        29   1    1   0           4             *.     .      .      .      .      .
Erect exterior wall panels             EARLY   *.     .      .      .     . EE     .
        22   2    2   0           0             *.     .      .      .      .      .
Erect doors                            EARLY   *.     .      .      .     . E      .
        24   1    1   0           0             *.     .      .      .      .      .
Waterproof & insulate                  EARLY   *.     .      .      .     . E      .
        25   1    1   0           0             *.     .      .      .      .      .
Paint exterior walls                   EARLY   *.     .      .      .      .     EE
        27   2    2   0           0             *.     .      .      .      .      .
Clean up                               EARLY   *.     .      .      .      . EE
        30   2    2   0           0             *.     .      .      .      .      .
```

FIGURE 13-29
Bar chart report (printer version).

in advance numerous alternative scheduling heuristics, such as whether to give priority by early start, minimum total float, and the like; whether activities can be interrupted once underway; and whether to work from the beginning or end of the network. But once calculations get underway, they proceed to completion in a batch mode rather than in the interactive mode that we saw on Fig. 13-14 for MacProject. Indeed, the interactive mode would be tedious and impractical for

```
- - - - - - - - - - - - - - - - - - - - - - - - - - - - - - - - - - - - - - - - - - - - - - - - - - - - - -
  Textbook Example                PRIMAVERA PROJECT PLANNER            Warehouse Project

  REPORT DATE 17JAN93 RUN NO.10   RESOURCE PROFILE START DATE 1JAN93 FIN DATE 11FEB93
  RESOURCE PROFILE - DAILY, MASK WEEKENDS                  DATA DATE 1JAN93  PAGE NO.   1

- - - - - - - - - - - - - - - - - - - - - - - - - - - - - - - - - - - - - - - - - - - - - - - - - - - - - -
  RESOURCE C      -Carpenter                                          TIMESCALE -DAILY
        20..*...............................................................................
          .*.    E     .     .     .     .     .     .     .     .     .     .     .     .
  M       .*.    E     .     .     .     .     .     .     .     .     .     .     .     .
  D       .*.    E     .     .     .     .     .     .     .     .     .     .     .     .
          .*.    E     .     .     .     .     .     .     .     .     .     .     .     .
  P       15..*....EE.............................................................................
  E       .*.    EE    .     .     .     .     .     .     .     .     .     .     .     .
  R       .*.    EE    .     .     .     .     .     .     .     .     .     .     .     .
          .*.    EE    .     .     .     .     .     .     .     .     .     .     .     .
  D       .*.    EE    .     .     .     .     .     .     .     .     .     .     .     .
  A       10..*....EE.............................................................................
  Y       .*.    EE    .     .     .     .     .     .     .     .     .     .     .     .
          .*.    EEE   .     .     .     .     .     .     .     .     .     .     .     .
          .*.    EEE   .     .  . E  .     .     .     .     .     .     .     .     .
          .*.    EEE   .     .  . E  .     .     .     .     .     .     .     .     .
        5.=====EEE=======E====  . E====================================================
          .*.    EEEE  .   E. .  . E  .     .     .     .     .     .     .     .     .
          .-----EEEE------E-------EE---------------------------------------------------
          .*.    EEEE  .   E. .  . EE .     .     .     .     .     .     .     .     .
          .*.    EEEE  .   E. EE EEEEE.     .     .     .     .     .     .     .     .
        0..*....EEEE......EE. EE.EEEEE...........................................................
          .   04    11    18    25    01    08    15    22    01    08    15    22    29    05
              JAN   JAN   JAN   JAN   FEB   FEB   FEB   FEB   MAR   MAR   MAR   MAR   MAR   APR
              93    93    93    93    93    93    93    93    93    93    93    93    93    93
```

FIGURE 13-30
Resource profile report (printer version).

the large projects on which Primavera thrives, but we do lose some feel for the impact of resource constraints by turning the calculations over to the "black box."

Finally, in Fig. 13-33 we see the classic CPM schedule as it appears after resource leveling. The result is different from that derived with MacProject, in part because we specified different criteria and in part because we allowed a secondary limit of six units (versus the primary limit of four) for each resource (both shown in Fig. 13-30). Compare specific activity schedule times with those in Fig. 13-28 to see how the leveling algorithm took advantage of some of the float that was available in the non-resource-constrained schedule.

In this chapter, we have only been able to give a small glimpse of the power and flexibility of MacProject and Primavera. Both are excellent tools for project management, and mastering this type of capability is essential for today's construction professionals. These programs can deal with schedule, resource, cost, procurement, and contractual aspects of all projects—from small and simple to large and complex. The capabilities of such programs are improving all the time, and any project can put them to work to enhance planning and control. Effectively applied, they can help overcome many of the current shortcomings of project management.

```
-----------------------------------------------------------------------------
Textbook Example                PRIMAVERA PROJECT PLANNER           Warehouse Project

REPORT DATE 17JAN93 RUN NO. 17  Example Warehouse Project  START 1JAN93  FIN DATE 11FEB93
SCHEDULE REPORT WITH RESOURCE USAGE                        DATA DATE 1JAN93         PAGE NO. 1

-------- ---- - ---------- ------------------------- ------- ------- ------- ------- -----
ACTIVITY ORIG                 ACTIVITY DESCRIPTION    EARLY   EARLY    LATE    LATE  TOTAL
    ID   DUR       CODE                               START  FINISH   START  FINISH FLOAT
-------- ---- - ---------- ------------------------- ------- ------- ------- ------- -----
      6   1                        Excavate           8JAN93  8JAN93  8JAN93  8JAN93    0
              L Laborer
                4.00 MD  /DAY       4

      5   4                     Fabricate trusses      8JAN93 13JAN93 22JAN93 27JAN93    9
              C Carpenter
                4.00 MD  /DAY      16

      4   2                   Nailers to purlins       8JAN93 11JAN93 28JAN93 29JAN93   13
              C Carpenter
                2.00 MD  /DAY       4

      8   3                 Fabricate exterior wall    8JAN93 12JAN93 29JAN93  2FEB93   14
              C Carpenter
                4.00 MD  /DAY      12

      7   2                Fabricate exterior doors    8JAN93 11JAN93  3FEB93  4FEB93   17
              C Carpenter
                2.00 MD  /DAY       4

      3   2               Fabricate interior part.     8JAN93 11JAN93  5FEB93  8FEB93   19
              C Carpenter
                4.00 MD  /DAY       8

      9   1                     Footing forms         11JAN93 11JAN93 11JAN93 11JAN93    0
              C Carpenter
                4.00 MD  /DAY       4
              L Laborer
                2.00 MD  /DAY       2
              I Ironworker
                2.00 MD  /DAY       2

XXXXXX ------ XXXXXX ------ XXXXXX ------ XXXXXX ------ XXXXXX ------ XXXXXX
-----------------------------------------------------------------------------
```

FIGURE 13-31
Activity resource usage combined with schedule report.

SUMMARY

This chapter introduced several important methods and concepts for planning, scheduling, and controlling the operations and resources of engineering and construction projects. Basic scheduling tools commonly implemented in computer programs include bar charts, progress curves, and critical path networks. Depending on the nature of the project and the needs and capabilities of its staff, each of these has advantages and disadvantages. Regardless of focus, any analytical tool for planning and control should be evaluated with respect to two main criteria. First, how well does it document the thinking of the planner? Second, how well does it communicate the planner's thinking to the people charged with the execution of the project? No matter how accurate and meticulously prepared, a plan is of little value if its language is foreign to its users.

Textbook Example

Primavera Project Planner

Warehouse Project

Activity Id	Resource	Daily Usage	Rem Dur	Early Start				Delayed by Pred				Delayed by Res				Early Leveled	
				Tf	Date	Norm	Max	Tf	Date	Norm	Max	Tf	Date	Norm	Max	Start	Finish
5 Fabricate trusses	C	4.00	4.00	9	08JAN93	4.00	6.00					5	14JAN93	4.00	6.00	14JAN93	20JAN93
15 Erect trusses	I	4.00	1	5	21JAN93	4.00	6.00									21JAN93	21JAN93
17 Cross frames	C	2.00	1	5	22JAN93	0.00	2.00					4	25JAN93	3.00	5.00	25JAN93	25JAN93
	I	2.00			22JAN93	2.00	4.00						25JAN93	4.00	6.00	25JAN93	25JAN93
4 Nailers to purlins	C	2.00	2.00	13	08JAN93	4.00	6.00					11	12JAN93	3.00	5.00	12JAN93	13JAN93
19 Erect purlins	I	4.00	1	5	25JAN93	2.00	4.00	4	26JAN93	4.00	6.00					26JAN93	26JAN93
21 Rod bridging	I	4.00	1	5	26JAN93	0.00	2.00	4	27JAN93	4.00	6.00					27JAN93	27JAN93
23 Nail planks to pur	C	2.00	2	5	27JAN93	4.00	6.00	4	28JAN93	4.00	6.00					28JAN93	29JAN93
26 Tar paper	L	2.00	1	5	29JAN93	4.00	6.00	4	01FEB93	4.00	6.00					01FEB93	01FEB93
28 Corrugated roofing	C	2.00	2	5	01FEB93	4.00	6.00	4	02FEB93	4.00	6.00					02FEB93	03FEB93
8 Fabricate exterior	C	4.00	3	14	08JAN93	4.00	6.00					3	26JAN93	4.00	6.00	26JAN93	28JAN93

FIGURE 13-32

Impact of leveling within resource constraints.

```
-------------------------------------------------------------------------------
Textbook Example              PRIMAVERA PROJECT PLANNER          Warehouse Project

REPORT DATE 17JAN93 RUN NO.19 Example Warehouse Project  START 1JAN93   FIN 11FEB93
CLASSIC SCHEDULE REPORT-SORT BY ES, TF                    DATA DATE 1JAN93 PAGE NO. 1


--------  ----  -------------------------------  -------  -------  -------  -------  -----
ACTIVITY  ORIG        ACTIVITY DESCRIPTION        LEVELED  LEVELED   LATE     LATE    TOTAL
   ID     DUR                                        ES       EF     START   FINISH  FLOAT
--------  ----  -------------------------------  -------  -------  -------  -------  -----
       1     1  Site survey                      4JAN93   4JAN93   4JAN93   4JAN93      0
       2     3  Move in                          5JAN93   7JAN93   5JAN93   7JAN93      0
       6     1  Excavate                         8JAN93   8JAN93   8JAN93   8JAN93      0
       9     1  Footing forms                    11JAN93  11JAN93  11JAN93  11JAN93     0
      10     1  Drain pipes                      11JAN93  11JAN93  19JAN93  19JAN93     5
      11     1  Footing concrete                 12JAN93  12JAN93  12JAN93  12JAN93     0
       4     2  Nailers to purlins               12JAN93  13JAN93  28JAN93  29JAN93    11
      12     1  Strip forms                      13JAN93  13JAN93  13JAN93  13JAN93     0
       5     4  Fabricate trusses                14JAN93  20JAN93  22JAN93  27JAN93     5
      14     2  Backfill & grade slab            20JAN93  21JAN93  20JAN93  21JAN93     0
      13     1  Erect steel cols.                20JAN93  20JAN93  22JAN93  22JAN93     2
      15     1  Erect trusses                    21JAN93  21JAN93  28JAN93  28JAN93     5
      16     1  Slab forms                       22JAN93  22JAN93  22JAN93  22JAN93     0
      18     1  Slab concrete                    25JAN93  25JAN93  25JAN93  25JAN93     0
      17     1  Cross frames                     25JAN93  25JAN93  29JAN93  29JAN93     4
      20     6  Cure                             26JAN93   2FEB93  26JAN93   2FEB93     0
       8     3  Fabricate exterior wall panels   26JAN93  28JAN93  29JAN93   2FEB93     3
      19     1  Erect purlins                    26JAN93  26JAN93   1FEB93   1FEB93     4
      21     1  Rod bridging                     27JAN93  27JAN93   2FEB93   2FEB93     4
      23     2  Nail planks to purlins           28JAN93  29JAN93   3FEB93   4FEB93     4
       7     2  Fabricate exterior doors         29JAN93   1FEB93   3FEB93   4FEB93     3
       3     2  Fabricate interior partitions    1FEB93    2FEB93   5FEB93   8FEB93     4
      26     1  Tar paper                        1FEB93    1FEB93   5FEB93   5FEB93     4
      28     2  Corrugated roofing               2FEB93    3FEB93   8FEB93   9FEB93     4
      22     2  Erect exterior wall panels       3FEB93    4FEB93   3FEB93   4FEB93     0
      24     1  Erect doors                      5FEB93    5FEB93   5FEB93   5FEB93     0
      25     1  Waterproof & insulate            5FEB93    5FEB93   5FEB93   5FEB93     0
      27     2  Paint exterior walls             8FEB93    9FEB93   8FEB93   9FEB93     0
      29     1  Erect interior partitions        8FEB93    8FEB93   9FEB93   9FEB93     1
      30     2  Clean up                         10FEB93  11FEB93  10FEB93  11FEB93     0
```

FIGURE 13-33
CPM schedule results after resource leveling.

In working with computer-based scheduling programs, we need to consider their facilities for defining activities, preparing and organizing a network, performing schedule calculations, and preparing reports. Commercial programs vary considerably in their power and ease of use in these respects, and some are strong in both.

Although space did not permit us to go into detail, this chapter briefly described advanced features available on some programs. Some of the most important ones include flexibility in network logic, calendar-based computations, resource and cost analysis, graphical user interface and reporting capabilities, and integration with other software such as CAD, estimating, cost control, and materials management.

The success of critical path methods begins with a clear, unambiguous definition of activities, the involvement of users in preparation of the schedule, and the physical organization of the network itself. Once a good basis for schedule control has been developed, the main criterion for updating the schedule should be how well it represents the actual operations on the project. Where corrective action is warranted, the critical path concept can aid in its most effective application.

Finally, in our two detailed examples we saw the powerful and elegant capabilities of today's modern computer-based scheduling systems, but these examples have given us only a small glimpse of their real power. Programs such as Primavera and MacProject can help project managers deal with schedules, resources, costs, and other constraints posed by today's increasingly complex construction environment. This capability can best be appreciated by applying these programs on real projects.

REVIEW QUESTIONS

1. What key type of information included on a critical path network is missing from a bar chart? In light of this omission, what is the main advantage of bar charts in field construction?
2. Briefly explain why a cumulative progress curve typically takes the shape of an elongated S.
3. If we assume that a project remains on schedule, where will the actual progress curve fall relative to the progress curves corresponding to the early-start and late-start scheduled progress curves? Briefly explain your answer.
4. Which of the following are computed results for an activity in a CPM schedule?

   - Activity duration
   - Late start
   - Resource requirements
   - Free float

5. Briefly define the critical path in a CPM network schedule.
6. Which of the following distinguish arrow networks from the main alternative type?

   - List of immediate following activities
   - I-J node numbering
   - Activities defined at the nodes of the network
   - Dummy activities required to show some logical relationships
   - Cyclical loops not permitted
   - Activities defined as the connecting lines in the network

7. How should the defined scope of schedule activities correspond to the managerial organization on a construction project?
8. List the key steps that take place in preparing a network diagram.
9. Which of the following data-entry modes are allowed by modern scheduling programs such as Primavera and MacProject?

   - Form filling (one form per activity)
   - Tabular data entry (multiple activities in rows)

- Direct graphical entry by "drawing" on the screen with a mouse
- Indirect entry via importing a file prepared in a spreadsheet or database

10. If the project duration computed by the forward pass is 90 days, and the contractual completion date of 100 days is used as the starting point for the backward pass, what is the minimum total float for the activities on the critical path?

11. If an activity can be delayed up to five days without delaying any of its immediate followers, what do we call that scheduling cushion of five days in a CPM network?

12. State at least two ways in which the scheduled times for a network are impacted by introducing a calendar to the computation process.

13. If the start of a following activity must wait at least 12 days until one of its immediate predecessors has started, what type of lag factor is used to define this relationship?

- Start-to-start
- Start-to-finish
- Finish-to-start
- Finish-to-finish

14. Briefly explain two main ways in which the Project Evaluation and Review Technique (PERT) differs from conventional CPM scheduling.

15. Explain the differences between CPM program capabilities for resource aggregation, resource allocation, and resource leveling.

16. What are the pros and cons of interactive versus batch approaches to computer-based resource allocation and leveling?

17. In the case of network-based cost optimization, how can the concept of time/cost trade-offs most efficiently be applied to reduce the overall schedule for a project?

18. Outline two main approaches for combining materials procurement functions with the main project work schedule. Briefly indicate the main disadvantage of showing detailed procurement steps directly on the CPM work schedule.

19. Some construction companies are secretive about revealing the amount of float shown by their network computations to the owner, to subcontractors, and even to their own supervisors. What are some of the reasons that might justify such a policy, and what are some of its advantages and disadvantages?

20. What is the role of CPM scheduling in the documentation of claims, change orders, and disputes in construction? What network-based technique is useful in this process?

APPLICATION PROBLEMS

The first step with any new scheduling program should be to work through its tutorial material and try it with examples that may have been provided by the vendor. For hands-on experience with the program after that step, you might consider the following:

1. Reenter the warehouse example from this chapter and use it as a basis for further exploration. Look well beyond the limited number of processing capabilities and reports that were included in this chapter.

2. Look up any of a number of standard CPM scheduling books—preferably ones with examples that include costs and resources as well as activity durations and logical relationships—and enter selected examples into your computer program.

SUGGESTIONS FOR FURTHER READING AND EXPLORATION

Catapult, Inc. *Microsoft Project for Windows, Step by Step*. Redmond, Wash.: Microsoft Press, 1993. Microsoft Project is a relatively inexpensive and well-configured package available on MS-DOS PC and Macintosh platforms. This book describes the Windows version.

Clough, Richard H., and Glenn A. Sears. *Construction Project Management*, 3d ed. New York: John Wiley & Sons, 1991. This is a straightforward text that focuses on CPM applications in construction.

MacProject Pro User's Guide. Claris Corp., 5201 Patrick Henry Drive, Santa Clara, Calif. 95052-8168, 1992.

Moder, Joseph J., Cecil R. Phillips, and Edward Davis. *Project Management with CPM, PERT, and Precedence Diagramming*, 3d ed. New York: Van Nostrand Reinhold, 1983. This is a classic text for introducing both basic and advanced concepts of network-based planning and control.

Primavera Project Planner—Getting to Know P3. Primavera Systems, Two Bala Plaza, Bala Cynwyd, Pa. 19004, 1991.

Timeline. Cupertino, Calif.: Symantec, 1992. This is another popular microcomputer-based scheduling program.

CHAPTER
14

ACCOUNTING
AND COST
ENGINEERING

Annual income twenty pounds, annual expenditure nineteen nineteen and six, result happiness. Annual income twenty pounds, annual expenditure twenty pounds ought and six, result misery.

Charles Dickens*

Accounting and cost control are two of the most basic management functions on today's construction projects. This chapter will review the key functions and principles of each and show how computer applications have greatly improved their effectiveness.

Construction accounting systems usually deal with maintaining the general ledger, accounts payable to the numerous vendors and subcontractors, accounts receivable from various clients, the payroll, and sometimes capital asset management for major items of equipment. These systems also commonly get into job cost control, but in doing so they delve into the domain of cost engineers, which we will treat separately.

Cost engineering provides the analytical methods and procedures for monitoring, analyzing, forecasting, and controlling the costs on a construction project. Like estimating and schedule and resource control, it is but one of several highly interdependent parts of the project planning and control system.

BASIC CONCEPTS OF ACCOUNTING

There is often a friendly rivalry between the accounting and engineering staffs in construction organizations. The engineers refer to the accountants as "bean counters" who seem to worry more about the money in the company than what it is

---

David Copperfield, 1850, Chapter 12.

actually building, and the accountants have long struggled to obtain good financial information from field people who give little priority to such paperwork. Most books and papers on construction management also tend to gloss over accounting aspects of this business and to emphasize scheduling and control systems instead. But in the business world the financial reports are the primary measures of success or failure, and too often contractors who have paid too little attention to the "bean counting" are sorry afterward. Construction professionals owe it to themselves to acquire some understanding of financial accounting, and today that means that they must learn about how computers have become pervasive in this field.

Almost all business-oriented accounting software uses double-entry bookkeeping and divides company accounts into *assets* (things the business owns, e.g., cash, accounts receivable, buildings, equipment), *liabilities* (things the business owes, e.g., loans, unpaid taxes, invoices payable), and *equity* (the difference between assets and liabilities, e.g., common stock, retained earnings). The fundamental equation of accounting that relates these items in an accounting system is

$$\text{Assets} = \text{Liabilities} + \text{Equity}$$

Accountants further subdivide these categories, often in great detail. Temporary subdivisions of equity accounts are *revenues* (e.g., money paid to us by an owner for work completed by our construction company) and *expenses* (e.g., salaries). A coding system that provides a detailed and systematic classification within the various categories is called a *chart of accounts*. It should not be confused with the detailed cost code for cost engineering that we will introduce later, but there is a connection between them.

We should also briefly note the difference between *cash-based* accounting systems and *accrual* accounting. In cash-based systems, we recognize expenses and revenues when the money actually changes hands, not when the obligations are incurred. For example, we might order and take delivery on some lumber in one month but not pay for it until the next. The expense is not recognized in a cash-based accounting system until the second of those two months. On the other hand, accrual systems recognize expenses and revenues when they are incurred. Thus the lumber purchase would be recorded as a liability (or payable) in the first month, and additional entries would be required when we pay for them in the second month. Most sizable businesses use accrual accounting.

Construction project-oriented accounting also requires distinctions between the *completed-contract method* and the *percentage-of-completion method*. In the completed-contract method, profits and losses are not recognized until a project ends. This is easier to use for small projects of short duration, but for large projects spanning fiscal years it can distort the company's financial picture. Also, because it defers profits into future periods, this method has faced growing restrictions from government tax bodies. In the percentage-of-completion method, income and losses are recognized incrementally based on components of each project's work—work underway as well as complete. It takes more accounting effort but better reflects a company's financial position.

Laypersons tend to get especially confused when accountants talk in terms of *debits* and *credits*. Essentially, amounts added to an asset account are called debits and amounts paid out of an asset account are called credits. On the other hand, the reverse is true in the case of liability and equity accounts. We will leave it to accounting texts listed at the end of this chapter* to unravel these riddles further, and move on to examine the typical components of a computer-based accounting system.

A business accounting system can consist of some or all of the components shown in Fig. 14-1, in which the usual interrelationships and information flows among the components also appear. Accounting software can implement some or all of these components in an integrated or modular form. Integrated systems contain all implemented functions in a single program that is sold as a package. In modular systems you can buy and implement one or more of the modules separately, although those from a given vendor will usually work together in an integrated fashion. In either approach, the breakdown of accounting functions will be similar to that of Fig. 14-1.

The *General Ledger (G/L)* is the central focus of the system. It records financial transactions via direct accounting entries and through information received from other modules. The G/L also produces the basic financial statements upon which a business is judged—the balance sheet and income statement. The *balance sheet* is a snapshot of the assets, liabilities, and equity of a firm at a given point in time. It is basically an organized and detailed breakdown and presentation of the fundamental equation of accounting that reveals the current state of the business. The *income statement*, also called a *profit and loss statement,* records income and expenses over a defined period of time, typically a month, quarter, or year. It better reflects the business' recent performance history. Both statements are necessary to reveal fully the state of and trends for the health of a business organization.

In *Accounts Payable (A/P)* we keep track of money owed to others, including materials suppliers, equipment rental agencies, and so forth. This function tracks invoices received, checks authorizations required for payment, notes due dates and discount possibilities, prints checks, updates the General Ledger, and tallies year-to-date purchases and payments. Its interface to Job Cost can enable cost engineers and accountants to distribute purchase costs to various cost accounts in the Job Cost module.

We can use the *Accounts Receivable (A/R)* module to keep track of money that clients owe to a construction company for work performed and projects completed. Its entries and records could be broken down by project and could also identify deferred payments such as contract retainage. It can also prepare statements and, via the *Invoicing* module, send invoices to clients. As with A/P, it updates Job Cost and the General Ledger and can track cumulative amounts over time.

---

*For example, Peachtree's *Accounting and DOS Primer* provides a clear and concise introduction written by retired Harvard Business School professor and textbook author Robert N. Anthony.

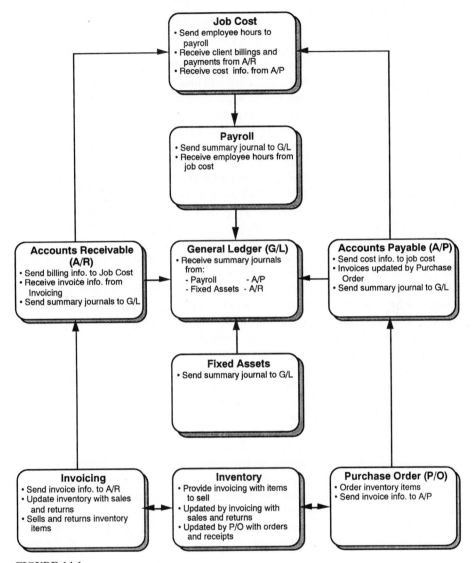

FIGURE 14-1
Components of a business accounting system. (Adapted from Peachtree Software, Inc., *Getting Started,* Norcross, Georgia, 1992, pp. 1–5.)

The *Job Cost* module monitors the direct costs of work items in a project—including labor, equipment, materials, and subcontractors—and compares them to a budget. The budget may have come from an estimating program. The Job Cost module can provide labor hours for the workers to the Payroll module (and also to equipment management and maintenance systems—not shown here because they usually are separate from a typical accounting system). Job Cost can receive information from Accounts Payable and provide information needed to prepare invoices to clients for progress payments. We will further address this function in the cost engineering section and in one of the application software examples later in this chapter.

The *Payroll* module calculates each employee's salary or wages and his or her deductions for taxes, insurance, benefits, and so forth, and supports a secure and confidential process for printing paychecks and quarterly and year-end tax reports. Some payroll systems can also produce the certified payroll reports required by some owners to prove compliance with prevailing wage laws or other regulations. The payroll module should accrue the liabilities to federal, state, and local agencies for withheld income taxes, social security payments, unemployment and disability insurance, and the like and relay that information to the general ledger.

The *Purchase Order* module is an important one in construction, owing to the myriad supplies and materials that go into even a modest-size project. As shown in Fig. 14-1, this module serves as a bridge between Accounts Payable and Inventory modules (or other parts of a construction materials management system).

Inventory modules in typical accounting systems are designed more for retail sales or manufacturing businesses, but they can be adapted to warehousing and related functions in construction. They keep track of stocks on hand, alert management when supplies are running low, and supply information for the Purchase Order module. The connection to Invoicing is shown in Fig. 14-1.

The *Fixed Assets* module mainly deals with real property and equipment. It computes depreciation, keeps track of investment tax credits and taxes, and otherwise helps the company to account for and recover these important costs of doing business. Particularly in heavy and industrial construction, where companies have substantial fleets of equipment and may manage and deploy them from an internal profit center, this module may have interfaces to sophisticated equipment management systems.

This overview of the various modules that go into a typical accounting software package has been fairly brief. Different vendors may package the modules in slightly different ways and leave out certain modules altogether. Not all provide Payroll, which is often delegated to outside services anyway. Many accounting programs also omit the Job Cost module, although that is a major omission in construction unless the company has a separate way (such as integrated scheduling and cost control system) to perform that function.

BASIC CONCEPTS OF COST ENGINEERING*

It is important that *cost engineering* not be confused with accounting. Accountants and cost engineers are in separate professions, and both may be employed on construction projects. The professions are closely interrelated, especially where cost engineering provides information for general ledgers, billings, and for payroll

---

*Parts of this section have been adapted and condensed from Donald S. Barrie and Boyd C. Paulson, Jr., *Professional Construction Management*, 3d ed., McGraw-Hill, New York, 1992, Chapter 13.

purposes, but they are also different. For example, the word "engineering" in cost engineering is not mere window dressing. To monitor and report costs properly and to forecast trends, one must be able to read plans and specifications and have a technical understanding of the work going on in the field.

General objectives for a cost engineering system designed to aid management in the planning and control of construction projects may be stated as follows:

1. To provide an organized and efficient means of measuring, collecting, verifying, and quantifying data reflecting the progress and status of cost items on the project
2. To provide standards against which to measure or compare progress and status
3. To provide an organized, accurate, and efficient means of converting operations data into useful cost information reported in a form and at an appropriate level of detail that can best be interpreted by management
4. To identify and isolate the most important information for a given situation, and to get it to managers in time for consideration and decision making so that corrective action may be taken, if necessary

The topics to be introduced here include the development and application of a *work breakdown structure* (WBS), standard cost codes and project cost codes, conversion of a cost estimate to a control budget, guidelines for obtaining good input cost data, and use of the system for reporting and control. Additional information on these subjects is given in Chapters 10 and 13 of *Professional Construction Management* by Barrie and Paulson.

Work Breakdown Structure

A WBS describes the work elements of a project in a logical hierarchy that can be used for a number of related management and control activities. Integrated management control systems are designed to reflect the interdependence of estimating, cost control, scheduling, materials management, and other systems. The WBS sets forth a common numbering system that can structure both planning and reporting at various levels of detail. With the help of computers, full detail can be made available at lower levels and consistent summary information can be distributed up the management ladder.

Cost codes provide the basic part of the WBS framework upon which a cost-engineering system is built. In a sense, they provide a structural discipline analogous to what critical path networking methods give to schedule and resource planning and control. Like networks, a good cost code can facilitate the cost-engineering process and hence aid project management; a poorly developed code can cause nothing but trouble. Management as well as cost engineers must understand the proper development and application of cost codes.

Cost codes typically fall into two major categories: (1) standard and (2) project. The first provides for uniformity, transfer, and comparison of information among projects. The second serves as the framework for the control budget on a specific project. Both are also often used to interface with the numbering of drawings and specifications, materials procurement documents, activity labels on schedules, quality assurance reports, and so forth.

A *standard cost code* is a systematic classification and categorization of all items of work or cost pertaining to a particular *type of work*. There may be different standard codes for different types of work, such as general building, heavy construction, and process plants.

In some sectors of construction, industrywide standard cost codes are more or less accepted. The best-known example is the Masterformat published by the Construction Specifications Institute,* which was developed through a joint effort of eight industry and professional associations. It is primarily designed for building construction and is commonly incorporated into estimating and cost-engineering software.

In developing a standard cost code, it is appropriate to create a relatively exhaustive checklist of all the items that might be found in the standard code's generic type of construction. No one project would require more than a fraction of the items, but over a period of several years a diversified contractor might encounter most of them. The standard cost code provides a good checklist to help prevent such items from being overlooked in the estimate, budget, and materials procurement schedule.

A *project cost code* is a systematic classification and categorization of all items of work or cost pertaining to a particular *project*. There is normally a different project cost code for each project, but each should be derived from the standard cost code so that similar work packages on different projects can be compared, and especially so that meaningful information can be maintained for estimating purposes.

A project code, however, is adapted to incorporate the particular features and characteristics of the specific project. It thus contains some components not found in the standard code, and it deletes anything not required for the job at hand. In contrast with the exhaustive checklist desirable for the standard code, the project cost code is a day-by-day working document; for practical purposes it must be kept as concise and simple as possible in keeping with the objectives for planning, documentation, and control.

The item code taken from the standard cost code becomes just one part of the number for a typical project code item. This part we will refer to as the *work-type code*. In addition to this, it is common to add a *project number,* an *area-facility code,* and a *distribution code* to form the complete code for a particular item in the project code. An organization will normally have standards specifying how these additional elements are formed, but the actual codes, at least for the project number and area-facilities, will be unique to the project at hand.

The project number identifies the costs collected for this code specifically with the particular project from which they came. The area-facility code recognizes

---

*Masterformat, The Construction Specifications Institute, 601 Madison, Alexandria, Va. 22314-1791.

certain distinct geographic and physical features that logically separate one part of a project from another (e.g., the dam, tunnel, and power plant on a hydroelectric project). Often the management of a project is also structured according to these physical or technological features. These major breakdowns are often referred to as *areas,* and within each area there may be further logical subdivisions into *facilities. Phase code* is another term used for this interim level of descriptive coding. The work-type code is the part based on the standard cost code but with appropriate adjustments for relevance and level of detail for the project at hand. The distribution code breaks out the resource components of the various types of work, such as separating out the labor, materials, equipment, and subcontract costs.

Normally the *account hierarchy* reflects the desired level of detail in a cost code. This concept can apply in a standard cost code and in each of the four or more elements that make up a typical project cost code. Terminology commonly encountered here includes prime account and subaccount. The *prime account* is the highest level of enumeration in a cost code. Subdivisions for greater detail within prime accounts are *subaccounts* and these in turn can have their own subhierarchies. Accounting software should handle such hierarchies.

Direct and Indirect Costs

In general terms, *direct costs* are those that we can immediately associate in the field with work directly contributing to the physical completion of the permanent facility contracted for by the owner. Examples include (1) finishing labor for a concrete floor slab, (2) materials for a structural steel frame, (3) equipment for a foundation excavation, and (4) a subcontractor's charges for installing the air-conditioning system.

Indirect costs are those that necessarily contribute to the support of a project as a whole but cannot be identified directly with specific work items in the permanent facility. Examples commonly include (1) job and office personnel salaries; (2) temporary field buildings and utilities; (3) staff vehicles; and (4) safety and first-aid expenses. However, there is no real uniformity in identifying exactly what is direct and what is indirect, particularly for payroll burdens such as fringe benefits, payroll-based insurance, and social security payments, or for support equipment such as a hoist in a high-rise building. Actual practice varies widely from one company to another, and software should provide for such differences.

Control Budgets

A good, definitive *control budget* is the basic document against which the cost engineer measures and compares actual progress. Reports to management must be accurate, current, and measured against this baseline standard if they are to have much value. The structure for the control budget is the project cost code. It is derived from the cost estimate and includes the quantities associated with each item of work in the code.

Figure 14-2 is part of a simple control budget. There is a trade-off in such reports between the amount of information to supply and the danger of confusing or inundating the user. Simplicity and consistency are the best guidelines.

For tight project control, simply taking figures from the estimate and plugging them into cost codes is not enough. First, for various good reasons (e.g., to match pay items in a unit-price bid), the estimate may not even be organized by the standard cost code. There must then be a redistribution process to get the costs into their appropriate project code categories. Second, management may wish to set control targets more optimistically or realistically than the expectations reflected in the estimate. With good computer software, these conversions and adjustments are not so difficult as they sound.

Sources of Data for Cost Control

No control system, whether computerized or not, and regardless of the skill of its developers, is of any value without accurate, timely input data. Field cost control especially requires good data for materials, equipment, and labor. The last two are particularly important, since these are the resources whose productivity and costs can change most rapidly and are the ones over which a contractor has the most control.

The main sources of data for field cost control are (1) labor and equipment time cards and (2) field surveys of quantities of work in place. Each of these is important, and comparisons between them are essential to evaluate project status satisfactorily.

Labor and equipment time cards are usually filled out daily and submitted either daily or weekly by forepersons, operators, superintendents, or timekeepers, depending on company policy. They normally contain the following information:

Labor time sheets	Equipment time sheets
Employee name(s) and/or number(s)	Machine description(s) and/or number(s)
Date(s) worked	Date(s) worked
Craft or classification(s)	Type of work done
Hours worked (straight time [ST] and overtime [OT])	Hours worked (ST and OT)
Classification by cost code	Classification by cost code
Hourly rates (ST and OT)	Hourly rates
Total hours and dollars, by day and by code	Total hours and dollars, by day and by code
Special conditions (weather, etc.)	Special conditions (breakdowns, etc.)

The time cards are generally preprinted on standard forms and are organized so that they are partially self-checking through crossfooting (adding and extending both horizontally by rows and vertically by columns). On daily cards, it is common to list several employees or machines on the same card, since the two-dimensional

EXAMPLE CONSTRUCTION CO.
Job Cost
JOB COST ESTIMATE REPORT

PAGE 2

JOB NUMBER 93-4	Mountaintown Warehouse	START 08/24/92	P/O 030146/93-4	ORIGINAL	5532494.00	BILLINGS
SUPERVISOR	I. M. Super	DUE 05/31/93		REVISED	5532494.00	PAYMENTS
CUSTOMER # C-22	Easyway Food Company	CLOSE		TOT COST	1960999.40	INVOICES

PHASE/COST CODE	DESCRIPTION	DUE DATE	UNIT MEAS	UNIT COST	ESTIMATED UNITS	ESTIMATED DOLLARS	COST ESTIMATE CALCULATED	COST ESTIMATE DIFFERENCE	BILLING ESTIMATE RATE	BILLING ESTIMATE DOLLARS
04	MASONRY									
04200S	Concrete Block	02/28/93	SF	10.60	4050.00	42930.00	42930.00		12.19	49369.50
05	METALS									
05110S	Structural Steel Framing	01/31/93	TON	1398.00	480.00	671040.00	671040.00		1608.00	771840.00
05300S	Metal Decking	02/15/93	SQFT	1.00	165000.00	165000.00	165000.00		1.15	189750.00
05850S	Misc Metals	04/01/93	EACH	1.85	35670.00	65989.50	65989.50		2.13	75977.10
	PHASE CODE 05 TOTAL				201150.00	902029.50	902029.50			1037567.10
06	WOOD AND PLASTICS									
06110L	Studs - 2x4	03/31/93	HOUR	15.00	162.00	2430.00	2430.00		22.43	3633.66
06110M	Studs - 2x4	03/31/93	MB	820.00	2.70	2214.00	2214.00		943.00	2546.10
06115L	Studs - 2x6	03/31/93	HOUR	15.00	96.00	1440.00	1440.00		22.43	2153.28
06115M	Studs - 2x6	03/31/93	MB	820.00	1.60	1312.00	1312.00		943.00	1508.80
06230L	Plywood Sheathing	03/31/93	HOUR	15.00	60.00	900.00	900.00		22.43	1345.80
06230M	Plywood Sheathing	03/31/93	SQFT	1.20	2000.00	2400.00	2400.00		1.38	2760.00
06310L	Roof Nailers	02/28/93	HOUR	15.00	1200.00	18000.00	18000.00		22.43	26916.00
06310M	Roof Nailers	02/28/93	MB	1160.00	10.00	11600.00	11600.00		1334.00	13340.00
06420L	Sheetrock	04/30/93	HOUR	16.00	80.00	1280.00	1280.00		23.92	1913.60
06420M	Sheetrock	04/30/93	SQFT	0.60	3200.00	1920.00	1920.00		0.69	2208.00
06700S	Cabinets & Shelves	05/15/93	LF	50.00	40.00	2000.00	2000.00		57.50	2300.00
	PHASE CODE 06 TOTAL				6852.30	45496.00	45496.00			60625.24

.... *Phases 01 to 03 and 07 to 16 omitted* ....

LABOR					31736.00	484225.00	484225.00			725840.80
MATERIAL					59035.90	419542.00	419542.00			482578.70
EQUIPMENT					2130.00	47772.50	47772.50			54944.00
SUBCONTRACT					1378834.00	3438299.60	3438299.60	0.00-		3940074.90
OTHER					1.00	10000.00	10000.00			11500.00
JOB 93-4 TOTAL					1471736.90	4399839.10	4399839.10			5214938.40

FIGURE 14-2

An example of a project budget (partial) created using Peachtree Accounting's Job Cost System.

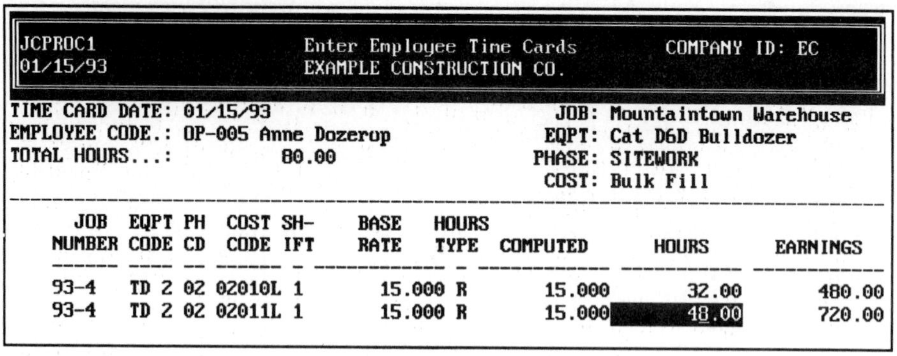

```
┌─────────────────────────────────────────────────────────────────────┐
│ JCPROC1                    Enter Employee Time Cards    COMPANY ID: EC │
│ 01/15/93                   EXAMPLE CONSTRUCTION CO.                    │
├─────────────────────────────────────────────────────────────────────┤
│ TIME CARD DATE: 01/15/93                    JOB: Mountaintown Warehouse│
│ EMPLOYEE CODE.: OP-005 Anne Dozerup         EQPT: Cat D6D Bulldozer    │
│ TOTAL HOURS...:          80.00             PHASE: SITEWORK             │
│                                             COST: Bulk Fill           │
│ ─────────────────────────────────────────────────────────────────────│
│     JOB  EQPT PH  COST SH-    BASE   HOURS                            │
│   NUMBER CODE CD  CODE IFT    RATE   TYPE  COMPUTED      HOURS  EARNINGS│
│   ────── ──── ──  ──── ───    ────   ────  ────────      ─────  ───────│
│    93-4  TD 2 02 02010L 1    15.000  R     15.000        32.00   480.00│
│    93-4  TD 2 02 02011L 1    15.000  R     15.000▓▓▓▓▓▓▓ 48.00   720.00│
└─────────────────────────────────────────────────────────────────────┘
```

FIGURE 14-3

Peachtree Accounting's labor and equipment time card input screen.

matrix allows for both these and the cost codes in which they worked. Weekly cards need one dimension for days of the week, and hence only one employee or machine is listed per card. Figure 14-3 is an example of a time card input screen for labor. It is from a biweekly entry and has the total hours in the heading. The operator's D-6 machine time is also picked up on this form.

Data collection can be greatly facilitated if a computer preprints all information on the cards except for the time entries and special condition notes. Anything that can be done to minimize the time field people need to write will generally improve the accuracy and completeness of reporting. Once input to the job cost system, this data can flow into the payroll system and from there into the general ledger.

In order to associate labor and equipment costs with physical work achieved, it is necessary to estimate or measure the quantities of the elementary work items that have been accomplished during the corresponding period. The same estimates will serve not only for cost reports on labor, materials, and equipment but also for scheduling, procurement, and other parts of the control system.

Normally such quantity reports are a mixture of actual measurements and judgment estimates. In between measurements, it will normally suffice to project from recent production rates, to count truck loads, or to use some similar approximations. These can then be readjusted if necessary when the results of the next physical survey are obtained.

Once the quantity data are combined with the expenditure of labor, equipment, and material resources, one can compute unit costs, make projections of costs at completion, and apply corrective action to operations that are in trouble.

Information Processing and Reporting

Information processing systems take progress and status data, compare them against reference standards such as budgets, and convert the results to information needed by the managers and supervisors on the project. The level of detail, the variety, and the frequency of reports to be produced should be appropriate to the people who will use them. The system should be fast, efficient, and accurate.

Reporting can take many forms, ranging from exception reports on CRT screens through tabular presentations of cost information and graphical presentations. Regardless of the form, in order to be effective for control purposes, a complete report should have five main components:

1. Budget estimates—either total, to-date, or this period, that provide a reference standard against which to compare actual or forecast results
2. Actuals—what has already happened, either this period or to-date
3. Forecasts—based on the best knowledge at hand, what is expected to happen to the project and its elements in the future
4. Variances—how far actual and forecast results differ from those initially planned
5. Reasons—anticipated or unexpected circumstances that account for variances in the actual and forecast behavior of the project and its operations

Selectivity and subreporting are important here. Since time is among their scarcest resources, construction managers and supervisors simply cannot afford to wade through piles of extraneous data to obtain the information they need. By focusing on operations within their responsibility and on those with variances exceeding certain predefined limits, reports can direct management attention to those items most in need of control.

ADVANCED SOFTWARE CONCEPTS AND FEATURES

Both accounting and cost engineering programs are available with a wide range of features and capabilities. They must be matched to the needs of a company and its projects.

Most commercial accounting software is designed for retail or manufacturing firms and may not be well suited to the needs of a larger construction organization. Indeed, just handling the needs of a typical general contractor would be considered "advanced" by the standards of many microcomputer-based accounting packages. But some low-cost mass-market programs are fine for small and medium-size firms, so in this section we will focus on deficiencies to watch out for. Key capabilities that may be missing include the following:

- A Job Cost module that can keep costs of projects separate from one another
- Capacity of a Payroll module to handle the diverse crafts, wages, fringe benefits, and work schedules found in construction, particularly if processing payrolls to meet the varying laws and regulations of multiple states
- The ability to use a computer network or other means to decentralize accounting functions to separate departments (e.g., payroll, accounts payable, etc.) and

even to the job-site level, and also to consolidate information back into the main office and General Ledger for corporate financial analysis and reporting

- The ability to handle separate profit centers (e.g., an equipment department) and even separate companies (e.g., for joint-venture projects)

Also important in this personnel- and money-oriented application are security and password protection, especially if check writing abilities are decentralized to the project level.

Turning to cost engineering, let us consider some advanced capabilities:

- The ability to handle and convert between multiple cost codes simultaneously, sometimes necessary when the contractor needs to report costs in an owner's categories as well as in its own.
- Flexibility in report design to serve the needs of project personnel and also various clients and subcontractors.
- Powerful and flexible work breakdown structures.
- Graphical reporting, such as charts and diagrams to communicate information more clearly that is difficult to understand in tabular reports.
- Cash flow projections.
- The ability to integrate with databases or directly to other software to exchange information or develop analyses that shed new insights on project performance. For example, estimates should feed into control budgets; schedules could combine with cost data to produce cash flows; input documents in common such as time cards should feed both accounting and cost engineering systems; and materials management data should be available to include materials costs in the system.

Other more general modern software features like online help and tutorial files, macro programming for routine procedures, and consistent graphical user interfaces are also desirable in accounting and cost engineering software and are commonly available.

APPLICATIONS AND LIMITATIONS

The dominant issues regarding applications and limitations of accounting and cost engineering software mostly have to do with the scope and capacity of the software versus the needs and abilities found in a given company or project. Smaller firms and projects can be well served by simpler software—as long as it is suited for construction—and indeed they may be overwhelmed by the training and usage demands of software designed for bigger firms. On the other hand, there are relatively few microcomputer-based commercial packages that will adequately cope with the demands of large and complex projects.

Teleware's M.Y.O.B. is a low-cost, integrated, double-entry accounting package that is well regarded and widely used for small-business accounting on both Macintosh and PC platforms. It is a single-user program. Its strengths are ease of use; tight integration; some extra functions including an electronic card file for maintaining information about clients, suppliers, subcontractors, and others; and the ability to produce routine form letters drawing on accounting information. It has some 90 different reports. However, until recently it lacked a payroll module and it is weak on job costing, so at best it might only serve the needs of a small construction company. But it will serve our purposes well to illustrate general financial accounting procedures and reports.

Peachtree Software makes a range of accounting products for both the Macintosh and for PC DOS and Windows systems. They have small integrated packages similar in scope and price to M.Y.O.B., and some that are more advanced. In our second example we will use Peachtree Complete Accounting, which has separate modules, optionally comes in a network version that will support a decentralized accounting organization, and is well-designed and easy to use. Whereas many modular systems cost $500 or more per module and thus several thousand dollars for a system, this whole system, including all of the modules shown in Fig. 14-1, costs less than $300. It is used in construction and can perform most of the functions of microcomputer-based accounting packages that have been especially designed for the construction industry. It may fall short in some specialized union-craft payroll needs; it does not have some of the advanced cost-control features, such as graphical reporting, that are found in a few integrated project control systems; and it is not designed to integrate directly with estimating software. But it is a proven system and is among the most widely used of all computer-based accounting packages. We will mainly use it to illustrate job-cost procedures in our second application example, but all of its modules work well.

Should general business accounting programs like these prove insufficient, vendors like Timberline Software and Software Shop Systems make accounting systems designed for construction contractors. But whereas such systems might add capabilities to integrate with estimating, scheduling, equipment management, and other construction software, their smaller specialized market volume also gets some of them into the higher cost ranges and cumbersome protection schemes discussed for estimating programs in Chapter 12.

Although we are focusing on accounting and the Job Cost control functions available in accounting packages, it is worth reemphasizing that project cost control is also inherent in many of the planning and control packages such as those described in Chapter 13. There are also specialty packages such as Primavera's Expedition that include a wide variety of project administrative and control functions.

FINANCIAL ACCOUNTING WITH M.Y.O.B. ON A MACINTOSH

We will use Teleware's M.Y.O.B. to illustrate some general accounting concepts and functions. It is a fairly simple program to use and thus is suitable for self-teaching. Yet it has enough capabilities to be highly rated for use in general small-business accounting. Its Macintosh and PC-based Windows versions are nearly identical.

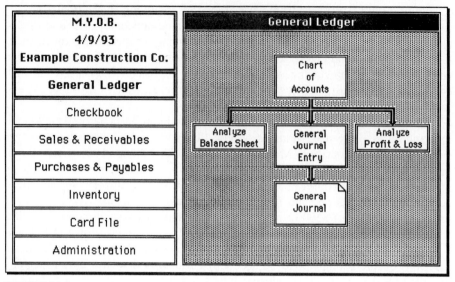

FIGURE 14-4
Command Center for M.Y.O.B.'s General Ledger section.

The main user interface in M.Y.O.B. is a pull-down menu, plus a series of seven **Command Centers** listed at the left side of Fig. 14-4. The pull-down menu options include the usual **File**, **Edit**, and **Window** options; a pull-down version of the seven-item Command Centers menu; an **Inquiry** menu to enable one to search records by account, invoice, purchase order, job, or card; and a **Reports** menu that is context-sensitive in listing the reports available that relate to the specific command center currently in use.

Most of the Command Center options shown in Fig. 14-4 are self-explanatory. **General Ledger**, **Sales & Receivables**, **Purchases & Payables** and **Inventory** are similar to components found in other programs. Apart from some combinations of functions (e.g., purchasing with payables), they correspond fairly well to the Peachtree components that we saw in Fig. 14-1. M.Y.O.B. handles both check writing and cash deposits, reconciliation of bank statements, as well as some related cash flow needs analysis, in a separate module called **Checkbook.** An extension called **Card File** enables us to store information about clients, vendors, employees, and the like, print address labels, and even produce form letters for various accounting-related purposes. We will return to it later. **Administration** is where we go to set up a new M.Y.O.B. system, customize screen and report formats and colors, and enter and edit information about separate jobs. The Administration section also has a **Business Calendar** function similar to that of a personal organizer, and further links events to a **To Do List**. The latter can include items automatically created by other modules, such as the latest dates to take discounts on invoices, reminders to pursue overdue receivables, next due date for recurring items, and even reminders of when to order new stock to maintain warehouse inventories.

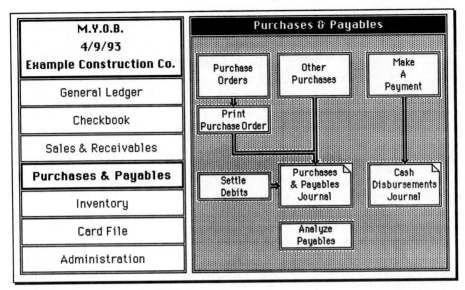

FIGURE 14-5
Command Center for M.Y.O.B.'s Purchasing and Payables section.

Selecting one of the Command Center options with a mouse causes a graphical flow chart to appear in the right-hand panel of the Command Center control screen. For example, in Figure 14-4 is the Command Center flow chart for the General Ledger. The General Ledger item in the menu on the left has a dark border to show that it is the item whose flow chart appears on the right. In Fig. 14-5 is a similar screen for Purchases and Payables. Note how the flow chart gives a general idea of the functions available in this section and shows how information flows from one part to another. Pressing any of the boxes in the flow chart takes you directly to a window for performing the associated function.

For example, clicking the box called **Chart of Accounts** in Fig. 14-4 takes you to a menu and display of the chart of accounts that appears in Fig. 14-6. The headings at the top left of the screen appear like folder tabs and are the menu to the eight main divisions in M.Y.O.B.'s chart of accounts. Clicking on one of the tabs, such as **Liability**, displays the detailed list of accounts for that division in the scrollable window below. Clicking on one of those accounts shows its setup information on the right side of the window. To create a new account, as is happening in Fig. 14-6, one simply types a new number in the field to the right of the "Account #:" prompt. Pressing Return brings us to the line labeled "Name" (where we are entering "Misc. Accounts Payable") and simultaneously enters the new account number in the list to the left (shown in black highlight). At this stage we can also state whether this is a **Detail account** that is postable or a higher-level **Section header** that aggregates totals from subordinate accounts. We can also specify the **Level** of the item (corresponding to the four levels of detail in the chart of accounts) for use in formatting and summarizing reports.

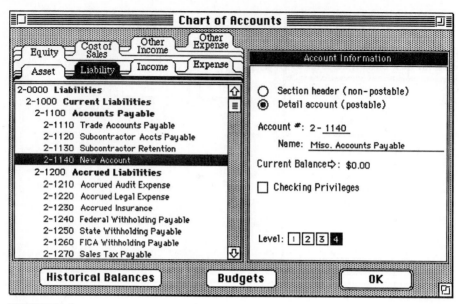

FIGURE 14-6
Chart of accounts.

There are a few other things worth noting in Fig. 14-6. First, the eight tabs limit us to the following primary fixed divisions and numbers for the Chart of Accounts:

1. Asset
2. Liability
3. Equity
4. Income
5. Cost of Sales
6. Expense
8. Other Income
9. Other Expense

We cannot change these numbers or titles, even to use the missing number 7. By default they are level 1 in the organization and report hierarchy. Subordinate section headers are optional and can be at level 2 or level 3; and detail accounts can be at levels 2, 3, or 4.

Pressing the **Budgets** button takes us to a screen where we can enter monthly budgets for each account and see how they compare to last year's and this year's monthly actuals. Similarly, pressing the **Historical Balances** button takes us to a screen where we can review the month-end account balances from the most recently closed month back to the first month of the preceding fiscal year.

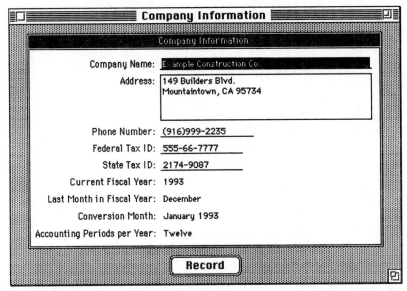

FIGURE 14-7
Company accounting information.

This example has been but one of many screens that are accessible from the various boxes in the seven Command Center panels. Others take us to various journals (e.g., General Journal in Fig. 14-4 and Cash Disbursements Journal in Fig. 14-5), entry forms (e.g., Purchase Orders in Figure 14-5), analytical functions (e.g., Balance Sheet and Profit & Loss in Fig. 14-4 and Payables in Fig. 14-5), and output forms and reports (e.g., Print Purchase Order in Fig. 14-5).

Let us now turn to what happens when we set up a new company in M.Y.O.B. and then enter and process data in this system. The first thing we have to do is use the administrative section to specify some general information about our company and its accounting preferences. Fig. 14-7 is the data-entry form. In addition to the obvious identification information, we specify the current fiscal year, the last month in our fiscal year (we used a calendar year here), the month in which we did our conversion, and whether to use a regular 12-month system or include a thirteenth month (as some accountants do) for making end-of-year adjustments.

We then set up a chart of accounts and enter the initial balances. The numbers and structure of the chart of accounts typically come from an existing system, and the opening balances come from a trial balance. We enter the chart of accounts using the screen form that appears in Fig. 14-6. In Fig. 14-8 we see the resulting Chart of Accounts for the Example Construction Company. In this listing we have used all four levels of detail, as reflected in the levels of indentation in the figure. Note that the levels of detail need not correspond directly to the five digits of the account coding scheme. Apart from the eight main divisions at level 1, the

```
                          Example Construction Co.
                             149 Builders Blvd.
                           Mountaintown, CA 95734

                          Chart of Accounts [Summary]
  7/31/93                                                            Page 1

  Account                                                    Current Balance

  1-0000   Assets                                             $3,650,795.00
           1-1000   Current Assets                            $2,315,295.00
                    1-1100   Cash                               $189,000.00
                             1-1110   Petty Cash                    $250.00
                             1-1120   Cash in Bank               $188,750.00
                    1-1200   Contract Receivables             $1,904,795.00
                             1-1210   Due from Clients - Current  $1,631,295.00
                             1-1220   Due from Clients - Retention  $305,000.00
                             1-1230   Allowance for Bad Debt      ($31,500.00)
                    1-1300   Earnings in Excess of Billings      $99,850.00
                    1-1400   Prepaid Expenses                    $77,900.00
                             1-1410   Prepaid Insurance           $52,500.00
                             1-1420   Other Prepaid & Deferred Items  $25,400.00
                    1-1500   Tools & Supplies -unassigned        $43,750.00
           1-2000   Property & Equipment                     $1,124,500.00
                    1-2100   Land & Buildings                   $433,750.00
                    1-2200   Furniture & Fixtures               $245,250.00
                    1-2300   Vehicles, Machinery & Equip.     $2,377,000.00
                    1-2500   Accumulated Depreciation        ($1,931,500.00)
                             1-2501   Accum Deprec-Buildings    ($344,000.00)
                             1-2502   Accum Deprec-Furn & Fix   ($154,750.00)
                             1-2600   Accum Deprec-Veh,Mach,Equip  ($1,432,750.00)
           1-3000   Other Assets                               $211,000.00
                    1-3100   Deposits                           $211,000.00
  2-0000   Liabilities                                        $1,801,350.00
           2-1000   Current Liabilities                       $1,318,850.00
                    2-1100   Accounts Payable                   $774,250.00
                             2-1110   Trade Accounts Payable     $394,000.00
                             2-1120   Subcontractor Accts Payable  $316,500.00
                             2-1130   Subcontractor Retention     $63,750.00
                    2-1200   Accrued Liabilities                 $95,600.00
                             2-1210   Accrued Audit Expense          $800.00
                             2-1220   Accrued Legal Expense        $1,200.00
                             2-1230   Accrued Insurance            $9,200.00
                             2-1240   Federal Withholding Payable  $52,500.00
                             2-1250   State Withholding Payable    $8,400.00
                             2-1260   FICA Withholding Payable    $22,000.00
                             2-1270   Sales Tax Payable            $1,500.00
                    2-1300   Income Taxes                        $12,000.00
                    2-1500   Notes Payable-Current              $244,500.00
                             2-1510   Working Capital Loan       $140,000.00
                             2-1520   Other Current Debt          $25,000.00
                             2-1530   Accrued Interest Due         $2,000.00
                             2-1540   Equipment Contracts         $77,500.00
                    2-1700   Billings in Excess of Costs        $192,500.00
           2-2000   Other Liabilities                          $482,500.00
                    2-2100   Deferred Income Taxes              $115,000.00
                    2-2200   Notes Payable-Long Term            $367,500.00
```

FIGURE 14-8

Listing of the chart of accounts for Example Construction Co.

```
3-0000    Stockholders' Equity                          $1,849,445.00
          3-1000  Common Stock                            $750,000.00
          3-8000  Retained Earnings                       $298,150.00
          3-9000  Current Year Earnings                   $801,295.00
4-0000                                                  $8,717,681.00
          4-1000  Contract Revenues Earned              $8,717,681.00
                  4-1100  Detail by jobs-1110,1120, etc $7,631,700.00
                  4-5000  Overhead Distributed to Jobs    $184,803.00
                  4-6000  Surplus & Salvage Sales (net)   $102,299.00
                  4-7000  Income from Joint Ventures       $437,579.00
                  4-8000  Equipment Rentals Earned         $361,300.00
5-0000                                                  $7,551,329.00
          5-1000  Cost of Revenues Earned               $7,551,329.00
                  5-1100  Detail by jobs-1110,1120, etc $6,944,422.00
                          5-1111  Labor                 $2,054,174.00
                          5-1112  Material              $1,954,046.00
                          5-1113  Subcontracts          $1,652,952.00
                          5-1114  Equipment Rental      $1,105,070.00
                          5-1115  Overhead                $178,180.00
                  5-3000  Equipment Ownership Costs       $614,192.00
                  5-4000  Discounts Taken (contra)         ($9,385.00)
                  5-5000  Loss on Bad Debts                 $2,100.00
6-0000                                                    $365,112.00
          6-1000  General & Administrative                $365,112.00
                  6-1010  Advertising                       $1,530.00
                  6-1020  Automotive Expenses              $17,967.00
                  6-1030  Bank Charges                        $735.00
                  6-1040  Contributions                     $2,375.00
                  6-1050  Deprec. on Office & Engr Equip    $8,910.00
                  6-1060  Dues & Subscriptions                $700.00
                  6-1070  Employee Benefits                $55,417.00
                  6-1080  Insurance                        $13,202.00
                  6-1090  Office Expense                    $3,970.00
                  6-1110  Postage & Freight                 $1,447.00
                  6-1120  Professional Fees                 $4,717.00
                  6-1130  Rent                             $15,750.00
                  6-1140  Repairs & Maintenance             $3,230.00
                  6-1150  Salary                          $195,158.00
                  6-1160  Taxes & Licenses                 $13,113.00
                  6-1170  Telephone                         $3,236.00
                  6-1180  Travel & Entertainment           $22,275.00
                  6-1190  Utilities                         $1,380.00
8-0000                                                      $1,805.00
          8-1000  Interest Income                           $1,805.00
9-0000                                                      $1,750.00
          9-1100  Interest Expense                            $250.00
          9-2100  Income Taxes-Current                      $1,500.00
          9-2200  Income Taxes-Deferred                         $0.00
```

FIGURE 14-8 *(continued)*

accountant is given considerable flexibility in assigning the numbers in a manner that best suits the needs of the company.

Sections 1, 2, and 3 contain the accounts that appear on the balance sheet (assets, liabilities, and equity). We have used sections 4 and 5 for revenue and cost accounts that will summarize the results reported back from our projects. Section 6 applies to our home office overhead expenses, and sections 8 and 9 apply to revenues and expenses that are not directly associated with our principal business activities.

Once we have our system set up, we can start to use it on a day-to-day basis. Space here does not permit us to explore all of the detailed activities of staff members involved in purchasing, payables, inventory management, and so forth. Suffice it to say that M.Y.O.B. offers them considerable assistance in performing their duties and even in substantially automating recurring functions.

Figure 14-9, a General Journal Entry, is typical of the screens they will use. Here we have posted three debits for material, subcontracts, and equipment rental accounts and assigned them to Job 93-4 (the Mountaintown Warehouse Project). We have also posted credits to three payables accounts and thus kept the accounts in balance. Once we fill in the account number, M.Y.O.B. automatically retrieves the description, so we then just tab over to the debit or credit field to make our entry. Should we have made an error in any of our entries, we would have seen a debit or credit to the right of the "Out of Balance" item at the bottom of the form. Note also that should any of the entries have been for recurring items (e.g., rent), M.Y.O.B. would have also pre-filled the appropriate debit or credit field for us. We can either accept it by tabbing to the next field or override it by typing in the desired entry to replace the default. This type of convenient feature occurs throughout the program. Once we have these account entries in balance, we click

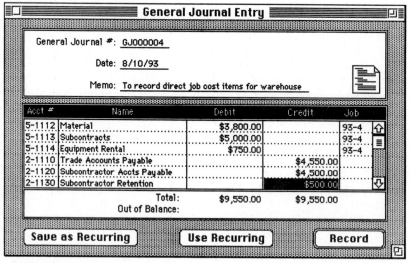

FIGURE 14-9
General Journal Entry form.

Record to post them to the General Ledger. That also clears this entry form and gets it ready for another set of entries. Note that, although we have made these debit entries directly in the General Ledger, they could have come in through the Accounts Payable part of the system.

Once we have our system running, we can select from close to a hundred different reports. Here we will present two of the most essential management reports: an end-of month balance sheet in Fig. 14-10 and a profit and loss (P&L)

Example Construction Co.
149 Builders Blvd.
Mountaintown, CA 95734

Balance Sheet

July 1993

Assets		
Current Assets		
Cash	$189,000.00	
Contract Receivables	$1,904,795.00	
Earnings in Excess of Billings	$99,850.00	
Prepaid Expenses	$77,900.00	
Tools & Supplies -unassigned	$43,750.00	
Total Current Assets		$2,315,295.00
Property & Equipment		
Land & Buildings	$433,750.00	
Furniture & Fixtures	$245,250.00	
Vehicles, Machinery & Equip.	$2,377,000.00	
Accumulated Depreciation	($1,931,500.00)	
Total Property & Equipment		$1,124,500.00
Other Assets		
Deposits	$211,000.00	
Total Assets		$3,650,795.00
Liabilities		
Current Liabilities		
Accounts Payable	$774,250.00	
Accrued Liabilities	$95,600.00	
Income Taxes	$12,000.00	
Notes Payable-Current	$244,500.00	
Billings in Excess of Costs	$192,500.00	
Total Current Liabilities		$1,318,850.00
Other Liabilities		
Deferred Income Taxes	$115,000.00	
Notes Payable-Long Term	$367,500.00	
Total Other Liabilities		$482,500.00
Total Liabilities		$1,801,350.00
Stockholders' Equity		
Common Stock	$750,000.00	
Retained Earnings	$298,150.00	
Current Year Earnings	$801,295.00	
Total Stockholders' Equity		$1,849,445.00
Total Liability & Equity		$3,650,795.00

FIGURE 14-10
One of M.Y.O.B.'s balance sheet reports.

```
                    Example Construction Co.
                       149 Builders Blvd.
                    Mountaintown, CA 95734

                  Profit & Loss Statement

                        July 1993
```

	Selected Period	% of Sales	Year to Date	% of Sales
Contract Revenues Earned				
Detail by jobs-1110,1120, etc	$1,324,200.00	88.1%	$7,631,700.00	87.5%
Overhead Distributed to Jobs	$32,540.00	2.2%	$184,803.00	2.1%
Surplus & Salvage Sales (net)	$10,942.00	0.7%	$102,299.00	1.2%
Income from Joint Ventures	$72,149.00	4.8%	$437,579.00	5.0%
Equipment Rentals Earned	$62,400.00	4.2%	$361,300.00	4.1%
Cost of Revenues Earned				
Detail by jobs-1110,1120, etc	$1,205,909.00	80.3%	$6,944,422.00	79.7%
Equipment Ownership Costs	$98,842.00	6.6%	$614,192.00	7.0%
Discounts Taken (contra)	($1,420.00)	(0.1%)	($9,385.00)	(0.1%
Loss on Bad Debts	$0.00	0.0%	$2,100.00	0.0%
Total Cost of Revenues Earned	$1,303,331.00	86.8%	$7,551,329.00	86.6%
Gross Profit	$198,900.00	13.2%	$1,166,352.00	13.4%
General & Administrative				
Advertising	$290.00	0.0%	$1,530.00	0.0%
Automotive Expenses	$2,842.00	0.2%	$17,967.00	0.2%
Bank Charges	$110.00	0.0%	$735.00	0.0%
Contributions	$500.00	0.0%	$2,375.00	0.0%
Deprec. on Office & Engr Equip	$1,310.00	0.1%	$8,910.00	0.1%
Dues & Subscriptions	$100.00	0.0%	$700.00	0.0%
Employee Benefits	$7,917.00	0.5%	$55,417.00	0.6%
Insurance	$1,952.00	0.1%	$13,202.00	0.2%
Office Expense	$620.00	0.0%	$3,970.00	0.0%
Postage & Freight	$247.00	0.0%	$1,447.00	0.0%
Professional Fees	$1,242.00	0.1%	$4,717.00	0.1%
Rent	$2,250.00	0.1%	$15,750.00	0.2%
Repairs & Maintenance	$530.00	0.0%	$3,230.00	0.0%
Salary	$28,908.00	1.9%	$195,158.00	2.2%
Taxes & Licenses	$2,153.00	0.1%	$13,113.00	0.2%
Telephone	$586.00	0.0%	$3,236.00	0.0%
Travel & Entertainment	$3,300.00	0.2%	$22,275.00	0.3%
Utilities	$180.00	0.0%	$1,380.00	0.0%
Total General & Administrative	$55,037.00	3.7%	$365,112.00	4.2%
Operating Profit	$143,863.00	9.6%	$801,240.00	9.2%
Interest Income	($420.00)	0.0%	$1,805.00	0.0%
Interest Expense	$250.00	0.0%	$250.00	0.0%
Income Taxes-Current	$0.00	0.0%	$1,500.00	0.0%
Income Taxes-Deferred	$0.00	0.0%	$0.00	0.0%
Net Profit/(Loss)	$143,193.00	9.5%	$801,295.00	9.2%

FIGURE 14-11

One of M.Y.O.B.'s profit and loss statements.

statement in Fig. 14-11. The displayed P&L uses one of several P&L report options, this one to show the given month and the year to date concurrently, with all numbers compared as a percent of sales.

Although Fig. 14-11 is shown in more detail (level 3 in this case) than we would normally use for external financial reporting (level 2 would be about right), we can instantly adjust the level of detail and choice of columns displayed and produce similar reports for any of our most likely intended audiences: stockholders, banks, bonding companies, owners, and so forth. Computer-based accounting can do in seconds or minutes what it used to take accountants and clerks hours, days, or weeks to do manually. With low-cost and easy-to-use tools like M.Y.O.B., even small companies can afford to meet the detailed and accurate recording and reporting standards of today's demanding business world.

Before closing, let us look at the Card File, a useful extension included in the M.Y.O.B. system. It can maintain records such as those shown in Fig. 14-12 in categories for **customers, vendors, employees,** and **personal.** With a working card file one can then link these entries to forms (e.g., invoices, statements, and checks) and substantially automate the process of sending transmittal letters, reminders, or even promotional information. Connected to a modem, M.Y.O.B. can even dial the telephone of a client or customer.

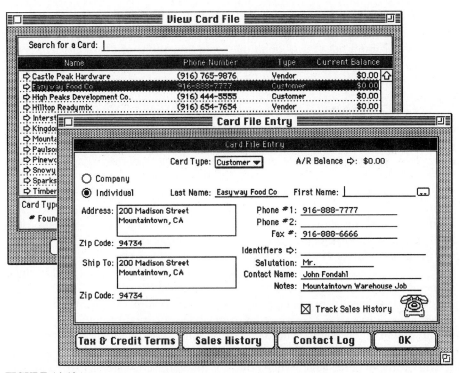

FIGURE 14-12
Card file listing (rear window) with an example card for a customer (front window).

We should reemphasize that M.Y.O.B. is a small-business accounting system. It could work well as a teaching tool, but in construction it would serve the needs of only very small firms like local home builders and specialty contractors. Significant weaknesses from our industry's viewpoint are in job costing and in payroll. (Although the manual for version 3 showed how to run a small payroll by setting up a payroll checking account, expense account, liability accounts for accrued taxes, etc. and a card file for employees, you still had to perform complex payroll calculations for benefits, withholdings, etc. offline. Version 4, released as this book went to press, added a payroll module.)

We will now move on to a more powerful system that does include payroll, job costing, and more. In doing so we will carry on from the basic chart of accounts and balances that we set up in M.Y.O.B. and will concentrate on project-oriented accounting with a fully implemented job cost module. But we will also be moving up somewhat in complexity as well. Again we must emphasize that the choice of a specific application package should be based on a study such as that described in Chapter 5 and on the needs of the specific users and their company.

JOB COST ACCOUNTING WITH
PEACHTREE SOFTWARE ON A PC

The Job Cost module of Peachtree Complete Accounting (PCA) is somewhat typical of those found in general-purpose accounting packages for microcomputers. We use it here not only to show how contractors can apply such programs for project cost control but also to point out some limitations of this approach.

The top half of Fig. 14-13 is a snapshot of a screen that shows part of the **PCA Main Menu**. From **General Ledger** to **Purchase Orders**, it offers all of the key functions of a business accounting system. As with many PC/DOS programs, one can choose among available options in the active menu either by using the arrow keys on the keyboard to move a highlight bar up and down and pressing Return or by typing the letter shown by the option description. Now let us take a closer look at some of the data entry, processing, and reporting functions of the PCA software.

Selecting the **Job Cost** module pops up the four options of the **Job Cost Main Menu** in the center window. The first option from that menu produces the **Maintenance Menu** shown on the right side of the screen. These options mostly take us to data-entry screen forms that are the primary means by which we enter and maintain the overall standard phase codes (P) and cost codes (T) for a given company or division; set up or modify general information about a project (J); enter or modify information about the company's equipment (Q), employees (L), and customers (S); specify percentage or dollar allocations (A) of overhead, waste, or other auxiliary costs to cost codes; and review or modify configuration and default options (O) related to the Job Cost module (e.g., printer assignments for each report, manager and staff passwords, overtime multipliers, shift differentials, payroll overhead percentage, whether or not to interface to payroll, A/P and A/R modules).

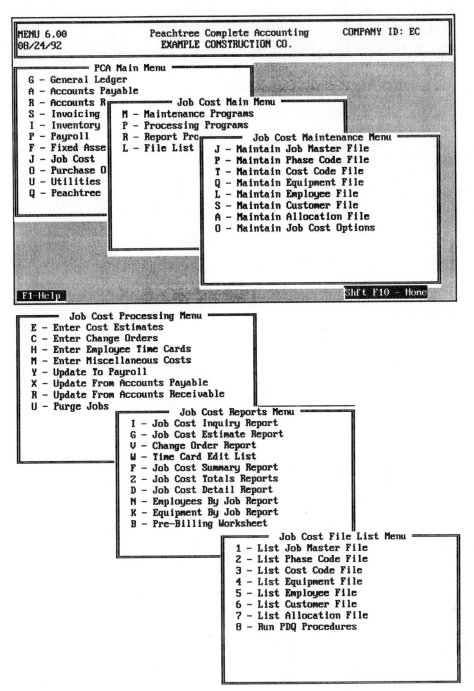

```
MENU 6.00              Peachtree Complete Accounting    COMPANY ID: EC
08/24/92                   EXAMPLE CONSTRUCTION CO.

        ══ PCA Main Menu ══
  G - General Ledger
  A - Accounts Payable
  R - Accounts R ══════════ Job Cost Main Menu ══════════
  S - Invoicing   M - Maintenance Programs
  I - Inventory   P - Processing Programs
  P - Payroll     R - Report Pro ══════ Job Cost Maintenance Menu ══════
  F - Fixed Asse  L - File List    J - Maintain Job Master File
  J - Job Cost                     P - Maintain Phase Code File
  O - Purchase O                   T - Maintain Cost Code File
  U - Utilities                    Q - Maintain Equipment File
  Q - Peachtree                    L - Maintain Employee File
                                   S - Maintain Customer File
                                   A - Maintain Allocation File
                                   O - Maintain Job Cost Options


  F1-Help                                        Shft F10 - Home

        ══ Job Cost Processing Menu ══
  E - Enter Cost Estimates
  C - Enter Change Orders
  H - Enter Employee Time Cards
  M - Enter Miscellaneous Costs
  Y - Update To Payroll
  X - Update From Accounts Payable
  R - Update From Accounts Receivable
  U - Purge Jobs      ══════ Job Cost Reports Menu ══════
                    I - Job Cost Inquiry Report
                    G - Job Cost Estimate Report
                    V - Change Order Report
                    W - Time Card Edit List
                    F - Job Cost Summary Report
                    Z - Job Cost Totals Reports
                    D - Job Cost Detail Report
                    N - Employees By Job Report
                    K - Equipment By Job Report
                    B - Pre-Billing Worksheet
                                ══════ Job Cost File List Menu ══════
                              1 - List Job Master File
                              2 - List Phase Code File
                              3 - List Cost Code File
                              4 - List Equipment File
                              5 - List Employee File
                              6 - List Customer File
                              7 - List Allocation File
                              8 - Run PDQ Procedures
```

FIGURE 14-13
Menu structure for Peachtree Complete Accounting.

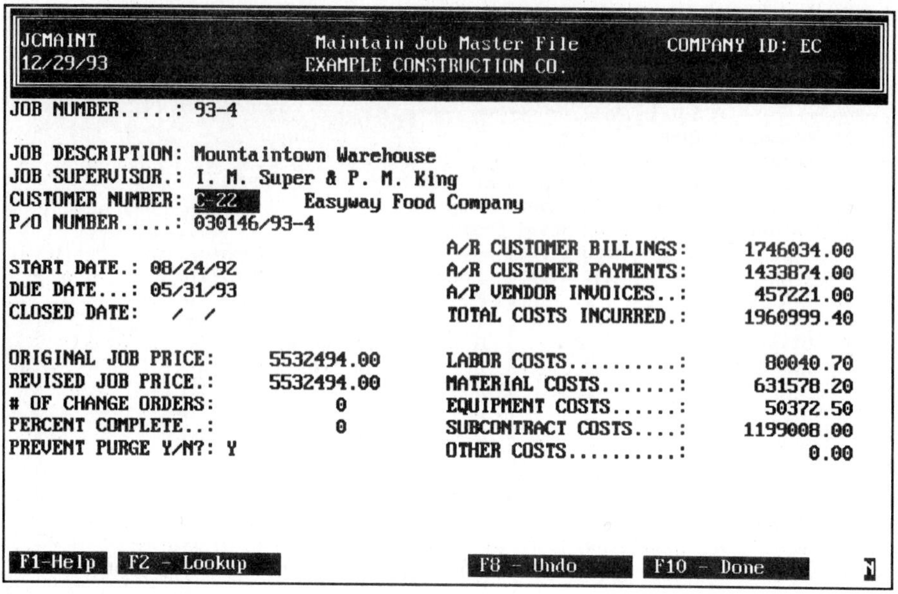

```
┌────────────────────────────────────────────────────────────────────────┐
║ JCMAINT              Maintain Job Master File      COMPANY ID: EC        ║
║ 12/29/93             EXAMPLE CONSTRUCTION CO.                            ║
╟────────────────────────────────────────────────────────────────────────╢
│ JOB NUMBER.....: 93-4                                                    │
│                                                                         │
│ JOB DESCRIPTION: Mountaintown Warehouse                                 │
│ JOB SUPERVISOR.: I. M. Super & P. M. King                               │
│ CUSTOMER NUMBER: C-22        Easyway Food Company                       │
│ P/O NUMBER.....: 030146/93-4                                            │
│                                   A/R CUSTOMER BILLINGS:     1746034.00  │
│ START DATE.: 08/24/92             A/R CUSTOMER PAYMENTS:     1433874.00  │
│ DUE DATE...: 05/31/93             A/P VENDOR INVOICES..:      457221.00  │
│ CLOSED DATE:   /  /               TOTAL COSTS INCURRED.:     1960999.40  │
│                                                                         │
│ ORIGINAL JOB PRICE:   5532494.00  LABOR COSTS.........:        80040.70  │
│ REVISED JOB PRICE.:   5532494.00  MATERIAL COSTS......:       631578.20  │
│ # OF CHANGE ORDERS:           0   EQUIPMENT COSTS.....:        50372.50  │
│ PERCENT COMPLETE..:           0   SUBCONTRACT COSTS...:      1199008.00  │
│ PREVENT PURGE Y/N?: Y             OTHER COSTS.........:            0.00  │
│                                                                         │
│                                                                         │
│ ┌──────────┐ ┌──────────────┐        ┌──────────┐ ┌──────────────┐     │
│ │ F1-Help  │ │ F2 - Lookup  │        │ F8 - Undo│ │ F10 - Done   │   ℕ │
│ └──────────┘ └──────────────┘        └──────────┘ └──────────────┘     │
└────────────────────────────────────────────────────────────────────────┘
```

FIGURE 14-14
Entering project information in the Job Master File.

In Figs. 14-14 and 14-15 are two of the data-entry screens that one encounters when selecting options within the Job Cost Maintenance Menu. General data for one of the projects contained in the **Job Master File** are presented in Fig. 14-14. This is the only one among the Maintenance Menu options that contains project-specific (as opposed to companywide) information. Some of the data are entered on the screen form (description, supervisor, etc.) and others are updated based on activity in other parts of the system (e.g., change orders will update the revised job price; time card and miscellaneous cost entries will update the indicated total costs for labor, material, equipment, etc.; interfaces to other modules can update A/R billings and payments, and A/P). One can use either the Tab or the Return key to move from field to field where data entry is permitted. Note also the Function keys at the bottom of the screen. **F1-Help**, **F8-Undo**, and **F10-Done** are self-explanatory. In data-entry fields that link to information that might exist in other files, pressing **F2-Lookup** pops up a list in which one can just move the cursor to the item desired rather than type it in, thus improving accuracy as well as being convenient. These types of function-key options recur in most data-entry and report specification screens.

Figure 14-15 is the data-entry screen for standard cost codes. Here you can not only enter the code, its description, cost type, and unit of measure but also provide a default unit cost and billing rate. PCA pulls these items in automatically when you select cost codes for the control budget of a specific project, but you can override the cost and billing rates if the project conditions differ from the defaults. This screen and its options to specify default values are also typical

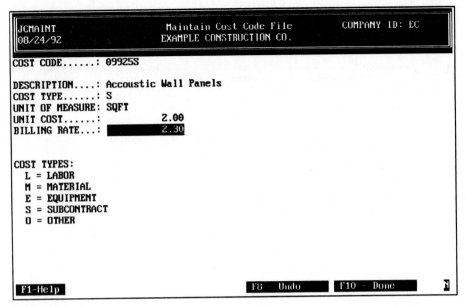

FIGURE 14-15
Entering a job cost code in Peachtree Complete Accounting.

of those encountered when you enter phase codes, equipment, employees, and customers, so they will not be reproduced here.

The lower half of Fig. 14-13 contains submenus for the other three main options in the Job Cost Main Menu. In the **Job Cost Processing Menu** one enters and processes the information for a specific project. The **Job Cost Reports Menu** enables us to select from the 10 main types of reports available in this module. These reports are for a specific project or range of selected projects. The **Job Cost File List Menu** is also a kind of reporting option, but these reports echo the file information entered under the Job Cost Maintenance Menu shown at the top of the figure. Most of the reports under both the Reports menu and the File List menu enable one to direct reports not only to a printer but also to the screen or to a disk file that might be printed later or be included in a word-processing document such as a monthly project status report.

Figure 14-16 is a partial listing of a report that is available from the File List Menu once you have used the data-entry screen from Fig. 14-15 to set up a project cost code. As you can see, each line simply echoes the data that we entered.

Once someone has set up the general information for a company, we can use this information to start performing operations for specific projects. One of the first steps is to transfer information from a cost estimate such as the Mountaintown Warehouse Project in Chapter 12 to a cost budget to be used for cost accounting. If we have an integrated estimating and cost control system, much of this transfer can be performed automatically. However, there is no easy integration path between

```
RUN DATE: 01/29/93          EXAMPLE CONSTRUCTION CO.              PAGE   1
RUN TIME:  5:42 PM                   Job Cost
                             COST CODE FILE LIST

--------------COST--------------  ---COST----  ------UNIT---------  ---BILLING---
  CODE        DESCRIPTION           TYPE       MEAS      COST           RATE
  ------  ----------------------  -----------  ----  -------------  -------------

  01101L  Superintendent          LABOR        HOUR         26.25          40.00
  01140L  Project Manager         LABOR        HOUR         27.50          42.00
  01175L  Office Administrator    LABOR        HOUR          8.00          12.00
  01205M  Temporary Phone         MATERIAL     MO          600.00         690.00
  01210M  Temporary Electrical    MATERIAL     MO         1000.00        1150.00
  01220M  Temporary Heating       MATERIAL     MO          500.00         575.00
  01240M  Temporary Water         MATERIAL     MO          400.00         460.00
  01301M  Temporary Field Office  MATERIAL     MO          500.00         575.00
  013020  Move-in Expense         OTHER        LS        10000.00       11500.00
  01315M  Temporary Toilets       MATERIAL     MO          250.00         288.00
  01345M  First-aid Equip.&Supplies MATERIAL   MO          125.00         144.00
  01601M  Small Tools & Equipment MATERIAL     MO         1000.00        1150.00
  01602E  Vehicles                EQUIPMENT    HOUR          3.75           4.31
                       . . . . . . . . . . . . . . . . .

  01730M  Office Supplies & Equip. MATERIAL    LS         4000.00        4600.00
  01735M  Blue Prints             MATERIAL     LS         1600.00        1840.00
  01740M  Travel & Lodging        MATERIAL     MO          275.00         320.00
  01750M  Job Sign                MATERIAL     LS          300.00         345.00
                       . . . . . . . . . . . . . . . . .

  04200S  Concrete Block          SUBCONTRACT  SF           10.60          12.19
  05110S  Structural Steel Framing SUBCONTRACT TON        1398.00        1608.00
  05300S  Metal Decking           SUBCONTRACT  SQFT          1.00           1.15
  05850S  Misc Metals             SUBCONTRACT  EACH          1.85           2.13
  06110L  Studs - 2x4             LABOR        HOUR         15.00          22.43
  06110M  Studs - 2x4             MATERIAL     MB          820.00         943.00
  06115L  Studs - 2x6             LABOR        HOUR         15.00          22.43
  06115M  Studs - 2x6             MATERIAL     MB          820.00         943.00
  06230L  Plywood Sheathing       LABOR        HOUR         15.00          22.43
  06230M  Plywood Sheathing       MATERIAL     SQFT          1.20           1.38
  06310L  Roof Nailers            LABOR        HOUR         15.00          22.43
  06310M  Roof Nailers            MATERIAL     MB         1160.00        1334.00
  06420L  Sheetrock               LABOR        HOUR         16.00          23.92
  06420M  Sheetrock               MATERIAL     SQFT          0.60           0.69
                       . . . . . . . . . . . . . . . . .

  08510S  Metal Windows           SUBCONTRACT  EACH        128.00         147.00
  08810S  Glass & Glazing         SUBCONTRACT  SQFT          5.00           5.75
  09925S  Accoustic Wall Panels   SUBCONTRACT  SQFT          2.00           2.30
  09940S  Paint-Caulk-VWC         SUBCONTRACT  SQFT          0.79           0.91
                       . . . . . . . . . . . . . . . . .

  15300S  Fire Sprinklers         SUBCONTRACT  SQFT          1.99           2.29
  15400S  Plumbing                SUBCONTRACT  SQFT          0.88           1.01
  15500S  HVAC                    SUBCONTRACT  SQFT          0.80           0.92
  16001S  Electrical              SUBCONTRACT  SQFT          2.35           2.70

        TOTAL COST CODES      92
```

FIGURE 14-16
Partial project cost code file listing for Mountaintown Warehouse.

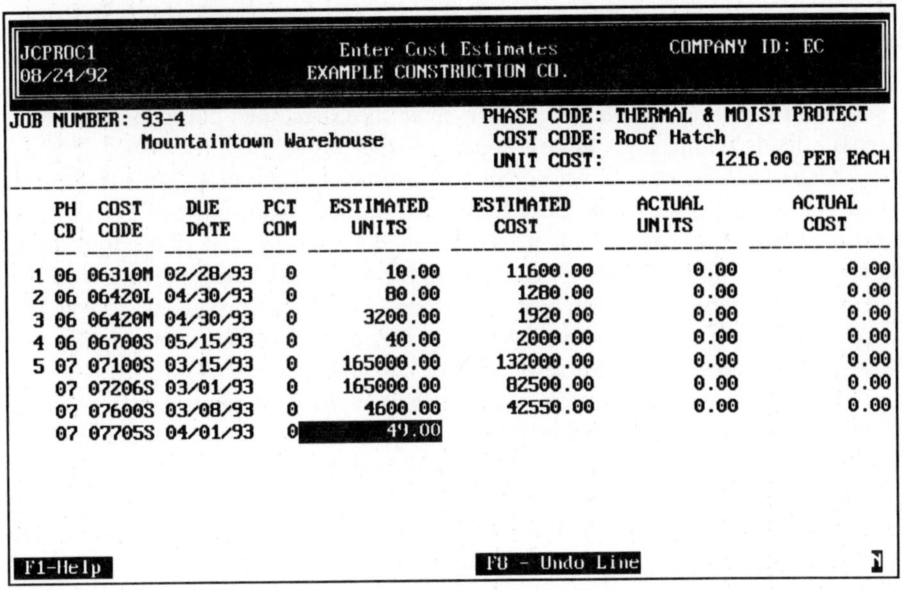

```
╔══════════════════════════════════════════════════════════════════════════╗
║ JCPROC1                    Enter Cost Estimates        COMPANY ID: EC      ║
║ 08/24/92                 EXAMPLE CONSTRUCTION CO.                          ║
╚══════════════════════════════════════════════════════════════════════════╝
```

JOB NUMBER: 93-4   PHASE CODE: THERMAL & MOIST PROTECT
              Mountaintown Warehouse   COST CODE: Roof Hatch
                                          UNIT COST: 1216.00 PER EACH

PH CD	COST CODE	DUE DATE	PCT COM	ESTIMATED UNITS	ESTIMATED COST	ACTUAL UNITS	ACTUAL COST
1 06	06310M	02/28/93	0	10.00	11600.00	0.00	0.00
2 06	06420L	04/30/93	0	80.00	1280.00	0.00	0.00
3 06	06420M	04/30/93	0	3200.00	1920.00	0.00	0.00
4 06	06700S	05/15/93	0	40.00	2000.00	0.00	0.00
5 07	07100S	03/15/93	0	165000.00	132000.00	0.00	0.00
07	07206S	03/01/93	0	165000.00	82500.00	0.00	0.00
07	07600S	03/08/93	0	4600.00	42550.00	0.00	0.00
07	07705S	04/01/93	0	49.00			

```
┌──────────┐                    ┌─────────────────┐                          ┐
│ F1-Help  │                    │ F8 - Undo Line  │                          │
└──────────┘                    └─────────────────┘
```

FIGURE 14-17
Entering a cost estimate for the Mountaintown Warehouse Project.

Timberline's estimating packages and Peachtree Complete Accounting. Thus we have to reenter the estimate. To do so, we use the screen form in Fig. 14-17. We begin by selecting the number of the project we are working on (93-4) and start entering items. We can use F2-Lookup to pick the job number, phase codes, and cost codes from a pop-up list, or type them in directly. We can optionally specify an estimated finish (due date) for each item and also a percentage complete if the project is already underway. We type in the estimated units, and Job Cost then looks up the unit cost and automatically fills in the field for estimated cost. We can override this if we wish. If the project is underway, we get one chance to enter the actual units to date and the actual cost to date, but once entered they can only be modified through transactions in the job cost module (time card or miscellaneous cost entry) or other parts of the accounting system. Once set up, they indeed do reflect actual transactions and are not subject to direct tampering by someone in the field who might like to make the reports look more favorable.

While looking at Fig. 14-17, we should also point out a useful feature and a couple of limitations of this particular job cost system. The feature is the **Phase Code**, which a company might use for various purposes, such as organizing its projects into on-site areas, dividing supervisory responsibility, and so forth. Here we have been relatively unimaginative and simply used it to repeat the first two digits of the CSI Masterformat's standard cost code. But the reason we did this also points out a limitation of this job cost system. Unlike the four levels of summarization we saw in M.Y.O.B.'s chart of accounts and reporting options, Peachtree's Job Cost module is relatively limited in this regard. Thus we used the phase code to force at least one level of summarization according to the major work types within our cost code. Another limitation is evident in the way we duplicated the first five digits before appending a letter to some code numbers, such as 06420L and 06420M. This job-cost system requires a separate cost code for each line item, including the subcomponents of labor, materials, equipment, and the like; it does not let us simply enter these resource component costs as separate fields within a given cost code.

Once we have entered all of this cost estimate information, we can produce what Peachtree calls its **Job Cost Estimate Report**. We used it in Fig. 14-2 to illustrate part of a control budget in our general discussion of cost engineering. Refer to it now and compare its contents to the data-entry form in Fig. 14-17 and the cost code definitions from Figs. 14-15 and 14-16 that produced it. Note that it shows a billing estimate as well as the cost budget, provides subtotals for each phase code, and totals up the resource cost categories.

Figure 14-18 appears after selecting **Enter Miscellaneous Costs** from the Processing section of the Job Cost Main Menu, and we use it to enter the nonlabor and nonequipment costs incurred for accounts that had activity during a given period. In Fig. 14-3 we showed the Time Card form we use for entering labor and equipment costs. We get to that form by selecting **Enter Employee Time Cards** from the Processing section. In both cases we can distribute a given employee's, machine's, subcontractor's, or supplier's costs and quantities to multiple accounts and automatically tap into or else override the existing default unit costs. In the Time Card form we deal with one employee or machine at a time, with one or more cost codes, whereas in Fig. 14-18 we can handle multiple transactions on a single screen before moving on to the next. Note that if we had established links to the Payables and Receivables modules in our accounting system, many of the

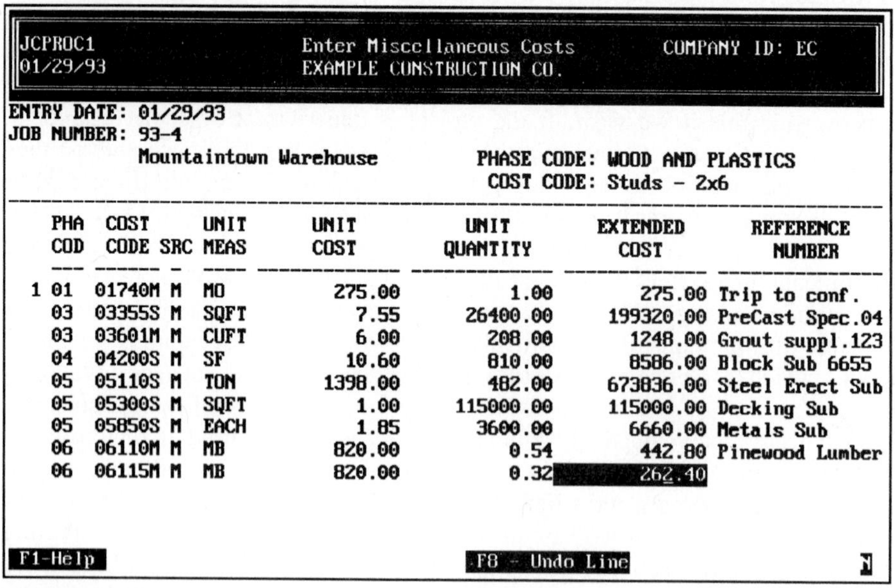

```
┌──────────────────────────────────────────────────────────────────────────┐
│ JCPROC1                      Enter Miscellaneous Costs       COMPANY ID: EC │
│ 01/29/93                     EXAMPLE CONSTRUCTION CO.                       │
├──────────────────────────────────────────────────────────────────────────┤
 ENTRY DATE: 01/29/93
 JOB NUMBER: 93-4
            Mountaintown Warehouse        PHASE CODE: WOOD AND PLASTICS
                                          COST CODE: Studs - 2x6
 ─────────────────────────────────────────────────────────────────────────
   PHA   COST    UNIT    UNIT         UNIT         EXTENDED      REFERENCE
   COD   CODE SRC MEAS   COST        QUANTITY        COST         NUMBER
   ───   ────── ─  ────  ─────────   ─────────     ─────────    ──────────
  1 01   01740M M  MO     275.00          1.00        275.00 Trip to conf.
    03   03355S M  SQFT     7.55      26400.00     199320.00 PreCast Spec.04
    03   03601M M  CUFT     6.00        208.00       1248.00 Grout suppl.123
    04   04200S M  SF      10.60        810.00       8586.00 Block Sub 6655
    05   05110S M  TON    1398.00        482.00     673836.00 Steel Erect Sub
    05   05300S M  SQFT     1.00     115000.00     115000.00 Decking Sub
    05   05850S M  EACH     1.85       3600.00       6660.00 Metals Sub
    06   06110M M  MB      820.00          0.54        442.80 Pinewood Lumber
    06   06115M M  MB      820.00          0.32        262.40

 ┌──────────┐                      ┌──────────────┐
 │ F1-Help  │                      │ F8 - Undo Line│                      ↵
 └──────────┘                      └──────────────┘
└──────────────────────────────────────────────────────────────────────────┘
```

FIGURE 14-18
Entering miscellaneous costs for the Mountaintown Warehouse Project.

transactions shown in Fig 14-18 for subcontractors, suppliers, and so on would come from those modules and their entry would not be permitted in this section of the Job Cost module.

Once we have entered the time card and cost calculations for a given period (day, week, etc.), we next use the Reports menu to double-check the **Time Card Edit List**, and then go back to the Processing menu and select **Update To Payroll** to calculate costs to date for each cost code, phase code, and resource category for that period. If we had set up the Payroll module, the time for each employee would also be transferred there. If we have established interfaces to Payables and Receivables modules in our system, we would also run those updates at this stage.

Finally, we turn to the Job Cost Reports Menu (see Fig. 14-13) to obtain the information needed to see how we are doing on our project and its various cost accounts. The reports that are of particular interest for control purposes are the **Job Cost Totals Reports** and the **Job Cost Detail Report**. Either choice brings us to screens wherein we further narrow our reporting desires. For example, we get the following three overall choices within Totals Reports:

Totals by job number
Totals by phase code
Totals by cost code

We then encounter choices such as those shown in Fig. 14-19 that further narrow the focus of our reports to specific categories of jobs, ranges of time or cost codes, resource costs, and the level of transaction detail. But as we mentioned before, we are limited on the level of detail we can specify within the hierarchy of cost code digits. Nevertheless, with careful use of phase codes and report option choices, some degree of selectivity can be obtained in directing information to the people for whom it is most relevant.

In Fig. 14-20 are selected parts of a cost report specified under the **Totals by Job Number** choice under the Job Cost Totals Report option. This particular

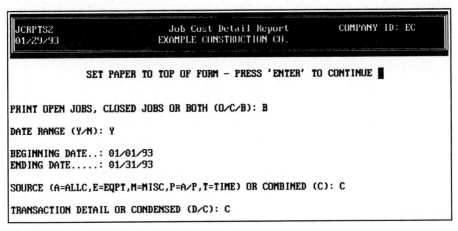

FIGURE 14-19
Specifying the output content for a Job Cost Detail Report.

RUN DATE: 01/29/93
RUN TIME: 5:26 PM

EXAMPLE CONSTRUCTION CO.
Job Cost
TOTALS BY JOB NUMBER REPORT

PAGE 1

						ORIGINAL	5532494.00	BILLINGS
JOB NUMBER 93-4	Mountaintown Warehouse	START 08/24/92	P/O 030146/93-4			REVISED	5532494.00	PAYMENTS
SUPERVISOR I. M. Super	DUE 05/31/93					TOT COST	1960999.40	INVOICES
CUSTOMER # C-22	Easyway Food Company	CLOSE						

PHASE/COST CODE	DESCRIPTION	MEAS	UNITS ESTIMATE	ACTUAL	DIFFERENCE	DIF%	DOLLARS ESTIMATE	ACTUAL	DIFFERENCE	DIF%
01	GENERAL CONDITIONS									
01101L	Superintendent	HOUR	1400.00	872.00	528.00-	38-	36750.00	29757.00	6993.00-	19-
01140L	Project Manager	HOUR	1400.00	872.00	528.00-	38-	38500.00	22339.00	16161.00-	42-
01175L	Office Administrator	HOUR	1400.00	872.00	528.00-	38-	11200.00	7379.00	3820.80-	34-
01205M	Temporary Phone	MO	9.00	5.00	4.00-	44-	5400.00	3000.00	2400.00-	44-
01210M	Temporary Electrical	MO	9.00	5.00	4.00-	44-	9000.00	5132.00	3868.00-	43-
01220M	Temporary Heating	MO	9.00	5.00	4.00-	44-	4500.00	2432.00	2068.00-	46-
01240M	Temporary Water	MO	9.00	5.00	4.00-	44-	3600.00	2176.00	1424.00-	40-
01301M	Temporary Field Office	MO	9.00	5.00	4.00-	44-	4500.00	2610.00	1890.00-	42-
013020	Move-in Expense	LS	1.00	1.00			10000.00	9850.00	150.00-	2-
01315M	Temporary Toilets	MO	9.00	5.00	4.00-	44-	2250.00	1284.00	966.00-	43-
01345M	First-aid Equip.&Supplies	MO	9.00	5.00	4.00-	44-	1125.00	647.00	478.00-	42-
01601M	Small Tools & Equipment	MO	9.00	5.00	4.00-	44-	9000.00	4778.00	4222.00-	47-
01602E	Vehicles	HOUR	1400.00	872.00	528.00-	38-	5250.00	3190.00	2060.00-	39-
01603M	Temporary Lights	LS	1.00	1.00			12000.00	12338.00	338.00	3
01701L	Layout	HOUR	840.00	702.00	138.00-	16-	12600.00	10530.00	2070.00-	16-
01701L	Layout	LS	1.00	1.00			1400.00	1350.00	50.00-	4-
01705L	Current Clean-up	HOUR	335.00	310.00	25.00-	7-	4020.00	4296.00	276.00	7
01710L	Final Clean-up	HOUR	833.00		833.00-	100-	9996.00		9996.00-	100-
01730M	Office Supplies & Equip.	LS	1.00	1.00			4000.00	3867.00	133.00-	3-
01735M	Blue Prints	LS	1.00	1.00			1600.00	1676.00	76.00	5
01740M	Travel & Lodging	MO	9.00	5.00	4.00-	44-	2475.00	1507.00	968.00-	39-
01750M	Job Sign	LS	1.00	1.00			300.00	420.00	120.00	40
	PHASE CODE 01 TOTAL		7695.00	4551.00	3144.00-	41-	189466.00	130558.20	58907.80-	31-

.... *Phases 02 to 03 omitted* ....

04	MASONRY									
04200S	Concrete Block	SF	4050.00	810.00	3240.00-	80-	42930.00	8586.00	34344.00-	80-

FIGURE 14-20

An example of a job cost report (partial) from Peachtree Accounting's Job Cost System.

EXAMPLE CONSTRUCTION CO.
Job Cost
TOTALS BY JOB NUMBER REPORT

```
JOB NUMBER  93-4     Mountaintown Warehouse    START 08/24/92   P/O 030146/93-4      ORIGINAL   5532494.00   BILLINGS
SUPERVISOR           I. M. Super               DUE   05/31/93                        REVISED    5532494.00   PAYMENTS
CUSTOMER # C-22      Easyway Food Company      CLOSE                                 TOT COST   1960999.40   INVOICES
```

CODE	DESCRIPTION	MEAS	ESTIMATE (UNITS)	ACTUAL (UNITS)	DIFFERENCE (UNITS)	DIF%	ESTIMATE (DOLLARS)	ACTUAL (DOLLARS)	DIFFERENCE (DOLLARS)	DIF%
05	METALS									
05110S	Structural Steel Framing	TON	480.00	482.00	2.00	0	671040.00	673836.00	2796.00	0
05300S	Metal Decking	SQFT	165000.00	115000.00	50000.00-	30-	165000.00	115000.00	50000.00-	30-
05850S	Misc Metals	EACH	35670.00	3600.00	32070.00-	90-	65989.50	6660.00	59329.50-	90-
	PHASE CODE 05 TOTAL		201150.00	119082.00	82068.00-	41-	902029.50	795496.00	106533.50-	12-
06	WOOD AND PLASTICS									
06110L	Studs - 2x4	HOUR	162.00	32.00	130.00-	80-	2430.00	624.00	1806.00-	74-
06110M	Studs - 2x4	MB	2.70	0.54	2.16-	80-	2214.00	442.80	1771.20-	80-
06115L	Studs - 2x6	HOUR	96.00	32.00	64.00-	67-	1440.00	624.00	816.00-	57-
06115M	Studs - 2x6	MB	1.60	0.32	1.28-	80-	1312.00	262.40	1049.60-	80-
06230L	Plywood Sheathing	HOUR	60.00	16.00	44.00-	73-	900.00	312.00	588.00-	65-
06230M	Plywood Sheathing	SQFT	2000.00	400.00	1600.00-	80-	2400.00	480.00	1920.00-	80-
06310L	Roof Nailers	HOUR	1200.00		1200.00-	100-	18000.00		18000.00-	100-
06310M	Roof Nailers	MB	10.00		10.00-	100-	11600.00		11600.00-	100-
06420L	Sheetrock	HOUR	80.00		80.00-	100-	1280.00		1280.00-	100-
06420M	Sheetrock	SQFT	3200.00		3200.00-	100-	1920.00		1920.00-	100-
06700S	Cabinets & Shelves	LF	40.00		40.00-	100-	2000.00		2000.00-	100-
	PHASE CODE 06 TOTAL		6852.30	480.86	6371.44-	93-	45496.00	2745.20	42750.80-	94-

.... *Phases 07 to 16 omitted* ....

			ESTIMATE (UNITS)	ACTUAL (UNITS)	DIFFERENCE (UNITS)	DIF%	ESTIMATE (DOLLARS)	ACTUAL (DOLLARS)	DIFFERENCE (DOLLARS)	DIF%
	LABOR		31736.00	27561.00	4175.00-	13-	484225.00	430999.70	53225.30-	11-
	MATERIAL		59035.90	53869.46	5166.44-	9-	419542.00	375453.20	44088.80-	11-
	EQUIPMENT		2130.00	1682.00	448.00-	21-	47772.50	50372.50	2600.00	5
	SUBCONTRACT		1378834.00	148982.00	1229852.00-	89-	3438299.60	1094324.00	2343975.60-	68-
	OTHER		1.00	1.00			10000.00	9850.00	150.00-	2-
	JOB 93-4 TOTAL		1471736.90	232095.46	1239641.44-	84-	4399839.10	1960999.40	2438839.70-	55-

FIGURE 14-20 *(continued)*

483

EXAMPLE CONSTRUCTION CO.
Job Cost
JOB COST DETAIL REPORT

JOB RANGE ALL TO END   CONDENSED   DATE RANGE 01/01/93 TO 01/31/93

JOB NUMBER 93-4	Mountaintown Warehouse	START 08/24/92	P/O 030146/93-4	ORIGINAL 5532494.00	BILLINGS
SUPERVISOR	I. M. Super	DUE 05/31/93		REVISED 5532494.00	PAYMENTS
CUSTOMER # C-22	Easyway Food Company	CLOSE		TOT COST 1960999.40	INVOICES

PHASE/COST CODE	DESCRIPTION	TRANSACTION DATE	SOURCE	REFERENCE	MEAS	UNITS QUANTITY	COST DOLLARS	% TOT	UNIT COSTS STANDARD	ACTUAL
				.... Phase 01 omitted						
02	SITEWORK									
02010E	Bulk Cut	01/15/93	EQPT	CONSOL. TRANS.	HOUR	80.00	4660.00	0.2	58.25	58.25
02010L	Bulk Cut	01/15/93	TIME	OP-005	HOUR	32.00	624.00	0.0	15.00	19.50
02011L	Bulk Fill	01/15/93	TIME	OP-005	HOUR	48.00	936.00	0.0	15.00	19.50
	PHASE CODE 02 TOTAL					160.00	6220.00	0.3		38.88
03	CONCRETE									
03355S	Precast Concrete	01/29/93	MISC	PreCast Spec.04	SQFT	26400.00	199320.00	10.2	7.55	7.55
03601L	Grout			CONDENSED TOTAL	HOUR	160.00	3328.00	0.2	15.00	20.80
03601M	Grout	01/29/93	MISC	Grout suppl.123	CUFT	208.00	1248.00	0.1	6.00	6.00
	PHASE CODE 03 TOTAL					26768.00	203896.00	10.4		7.62
04	MASONRY									
04200S	Concrete Block	01/29/93	MISC	Block Sub 6655	SF	810.00	8586.00	0.4	10.60	10.60
05	METALS									
05110S	Structural Steel Framing	01/29/93	MISC	Steel Erect Sub	TON	482.00	673836.00	34.4	1398.00	1398.00
05300S	Metal Decking	01/29/93	MISC	Decking Sub	SQFT	115000.00	115000.00	5.9	1.00	1.00
05850S	Misc Metals	01/29/93	MISC	Metals Sub	EACH	3600.00	6660.00	0.3	1.85	1.85
	PHASE CODE 05 TOTAL					119082.00	795496.00	40.6		6.68
				.... Phase 06 omitted						
	LABOR						23510.20	1.2		
	MATERIAL						7083.20	0.4		
	EQUIPMENT						5290.00	0.3		
	SUBCONTRACT						1003402.00	51.2		
	OTHER									
	JOB 93-4 TOTAL						1039285.40	53.0		

FIGURE 14-21

An example of a job cost detail report (partial) from Peachtree Accounting's Job Cost System.

484

option does not show the percent complete by each code (other options do), but it does show the estimates, actual costs, and variances for each cost code. Unfortunately, without the due date and percent complete for each item, it is difficult to know whether the actuals are good or bad for items that are now underway, although it does give us an accurate picture of how we did on completed items. The **Totals by Cost Code** does show the percentages and due dates for active items, but unfortunately it leaves off the code descriptions. Given the constraints of the printed page, every cost reporting system makes such trade-offs in content and detail, and these are the types of things that you have to evaluate in selecting a system to meet the needs of a particular company.

Part of a Job Cost Detail Report appears in Fig. 14-21. Its primary purpose is to be a transaction history to show exactly what components went into each account over a user-specified period of time. Here we have a few items for the month of January 1993.

To some, these reporting options may seem fairly limited for the needs of any but small construction projects. It is thus worth pointing out that Peachtree Complete Accounting includes the **Peachtree Data Query** (PDQ) module (option Q on the main menu of Fig. 14-13). This provides one with a macro programming capability to create custom sorting, selection, query, and reporting procedures that can draw on any of the data files contained in the system and produce custom outputs, including simple graphical reports. It can also export data in various spreadsheet and ASCII file formats for use by other programs. Although Peachtree Complete Accounting and its Job Cost Module are already quite useful for construction accounting and cost control, the added flexibility of PDQ can considerably extend their capabilities.

In closing, note that in this example we have only used one of the 10 main modules available in the Peachtree Complete Accounting system. Recall that all of this practical business capability comes in an integrated accounting package that costs only a few hundred dollars and runs on a modest PC! Even so, there are many good alternatives in this competitive accounting software marketplace.

SUMMARY

Accounting and cost control are fundamental to the successful operation of any construction company. Two important and different aspects of this subject include financial accounting for the company as a whole and cost engineering or job-cost accounting for its projects.

Financial accounting deals with the assets, liabilities, and equity of a firm as well as the revenue and expense transactions that determine its profitability. The basic structure for this information is the chart of accounts. Basic approaches to accounting include the cash-based method, which is limited to smaller firms, and the accrual method, which is widely used by almost all of the others. The latter recognizes revenues and expenses when they are incurred, which is not necessarily the same time as when money actually changes hands. In the project-oriented construction industry, we make a further distinction between the completed-contract method and the percentage-of-completion method in determining when to recognize profits and losses from ongoing projects.

Typical computer-based accounting systems consist of several modules that focus on different accounting functions. Central to these is the General Ledger, wherein we maintain the chart of accounts and the information to produce the balance sheets and income statements that are the fundamental tools used by company executives and external entities such as banks and bonding companies to assess the financial health of the firm. Other common modules include Accounts Payable and related purchasing functions to handle our financial transactions with suppliers and subcontractors; Accounts Receivable to invoice for and track money owed to us by our clients; and Payroll, not only to issue employee paychecks but also to take responsibility for the related tax, insurance, and benefit accounting. Many programs further provide capabilities for Fixed Asset accounting, Inventory control, and Job Cost accounting. The latter connects us to the cost engineer's role in construction.

Cost engineering is one of the most vital control functions on any project, and it is important that the reader who will be working in this field seek more in-depth knowledge of the subject. Current nomenclature on major projects sets forth a work breakdown structure that is used to interrelate costs, schedules, responsibilities, and other factors utilizing a common code. Cost codes provide the basic framework upon which a cost-engineering system is built. Two types were introduced. A standard cost code contains and systematically categorizes a relatively complete enumeration of all of the types of work in a generic type of construction, such as general building. The Masterformat is one of the best-known examples. A project cost code, although derived from the standard code, is developed for a specific project and hence contains additional elements for the project number, major areas and facilities within the project, and a means of distributing costs for resources, such as labor, materials, and equipment.

Control budgets derive from estimates of costs and quantities, and they usually provide a structure for recording actual performance, making forecasts, showing variances, and documenting reasons for unexpected problems. There is often a need to recategorize the costs from the estimate to serve the needs of effective cost control or for owner or agency requirements.

Field construction cost control concentrates mainly on labor and equipment. The main sources of data here are labor and equipment time cards, and estimates or surveys of quantities of work in place. Data from other parts of the control system — including scheduling, accounting, procurement, and quality assurance — are also important for consistency, comparisons, and interrelating factors.

In considering advanced concepts and features for financial accounting, we mainly focused on deficiencies to watch out for in terms of construction industry needs. Payroll, job cost accounting and the ability to decentralize accounting functions to the project level and to separate profit centers are important to all but the smaller firms. Provisions for security are also important, especially in decentralized systems.

Advanced features to consider in job-cost accounting or cost-engineering systems include the ability to cross-reference multiple cost codes (e.g., our company's and the client's), flexible work breakdown structures, versatile reporting, including graphical charts for cash flows and the like, and interfaces to common databases and other programs such as estimating, scheduling, and materials-management systems.

A wide variety of commercial software products are available for construction accounting and cost control, and each has its appropriate applications and limitations. There are trade-offs in capabilities and complexity, in software costs and suitability to construction, and in the computer platforms on which different packages will run. Low-cost general-purpose accounting packages like Teleware's M.Y.O.B. and Peachtree's Complete Accounting will adequately serve the needs of some small and medium-size firms, but larger contractors may want to go to construction-focused accounting packages from companies like Timberline Software and Software Shop Systems.

In this chapter, the Macintosh version of M.Y.O.B. was used to illustrate general financial accounting procedures and reports. We chose the Job Cost module from Peachtree Complete Accounting, running on a DOS/PC platform, to show how to set up a control budget and run some basic job-cost functions. For lack of space, we barely scratched the surface in examining the capabilities of these programs. We thus strongly encourage the reader to acquire one of the low-cost packages (several are available for around $100) and experiment directly to obtain a better understanding of this important capability for managing construction companies.

REVIEW QUESTIONS

1. State the fundamental equation of accounting and explain each of its key components.
2. How do the components of a balance sheet relate to the fundamental equation of accounting?
3. What types of information are reflected in the income statement?
4. Outline the major sections of a typical General Ledger chart of accounts, and show which sections track the information contained in (*a*) the balance sheet and (*b*) the income statement.
5. How does the accrual method of accounting differ from the cash-based method? Which is more suitable for larger construction companies?
6. How does the completed-contract method of construction project accounting differ from the percentage-of-completion method? Which is more suitable for larger projects of longer duration? Why?
7. List and briefly state the function of the major modules in an integrated computer-based financial accounting system. Which ones are most important at the corporate level? Which ones are most important at the project level?
8. What are the essential differences between the abilities of a cost engineer and an accountant in monitoring and forecasting project costs? Include strengths and weaknesses for each.
9. Discuss the use of standard and project cost codes in specifying the work breakdown structure for a construction project.
10. How does a standard cost code differ in purpose and content from a project cost code?

11. How do the indirect costs on a project differ from the direct costs? Give an example of each.

12. Which main part(s) of the chart of accounts in the General Ledger for a construction company contain costs that are likely to be distributed to the overhead category in a project cost budget?

13. What is the main purpose in subdividing the control budget for a construction project?

14. Describe the primary mechanisms for capturing and entering the labor and equipment costs incurred on a construction project. How do these costs get allocated to the specific cost accounts on which they were incurred?

15. State the main objectives of the reporting function in a job-cost engineering system.

16. Briefly list some of the most important guidelines for effective cost reporting.

17. Why is a well-implemented payroll module particularly important in a construction accounting system?

18. What is the usefulness of an advanced job-cost system feature that allows the concurrent maintenance and cross-referencing of multiple cost code structures?

19. What are some other project engineering and management computer programs that usefully could be interfaced to a construction accounting and cost control system? Indicate what key types of information might be transferred and the advantage(s) gained in doing so.

20. Briefly describe some of the trade-offs between using popular general business accounting systems in the construction environment and those developed specifically for this industry.

APPLICATION PROBLEMS

The first step with any new accounting program should be to work through its tutorial material and try it with examples that may have been provided by the vendor. For hands-on experience with the program after that step, you might consider the following:

1. Set up a small construction company such as the one we used with the M.Y.O.B. example. Create a chart of accounts, enter the initial balances, and then enter some typical transactions for a month or two. Once you have it running, print out a balance sheet and income statement.

2. Depending on the capabilities of your program, first do the tutorials and then go on to create the data and do some transactions in the accounts receivable module, accounts payable module, and perhaps the payroll module. Also, enter some of your company's major items of equipment in the fixed-asset section and see what methods the program allows for depreciation.

3. If your program has a job cost module, first enter parts of a standard cost code (if it does not already have one), and then set up a project such as the Mountaintown Warehouse project (a more complete set of estimating data is contained in Chapter 4 and Appendix A of *Professional Construction Management,* 3d ed., by D. S. Barrie and B. C. Paulson, Jr.). Enter the project control budget and produce its report. Then run a few months of sample transactions and print out various cost control reports. Examine each carefully and assess which contain the best information for various different needs on the project (forepersons, superintendent, project manager, accountant, home office project operations manager, client, etc.).

SUGGESTIONS FOR FURTHER READING

Apart from the two software manuals, the references below provide general background on accounting, cost control, and cost engineering. The books by Anthony and Horngren are some of the most popular accounting and cost accounting books used in business schools, and are well-proven. Clough provides a good introduction to and overview of construction accounting in two chapters. Coombs and Palmer and Halpin focus on general accounting in the construction industry. Ahuja, Humphrey, Jelen, Kharbanda, and Mueller concentrate mainly on cost accounting at the field project level.

Ahuja, Hira N. *Successful Construction Cost Control*. New York: John Wiley & Sons, 1980.

Anthony, Robert N. *Accounting and DOS Primer*. Norcross, Ga.: Peachtree Software, 1992.

Anthony, Robert N. *Essentials of Accounting*, 5th ed. Reading, Mass.: Addison-Wesley, 1993.

Anthony, Robert N., and James S. Reece. *Accounting Principles*, 6th. ed. Homewood, Ill.: Richard D. Irwin, 1989.

Clough, Richard H. *Construction Contracting*, 5th ed. New York: John Wiley & Sons, 1986.

Coombs, W. E., and W. J. Palmer. *Construction Accounting and Financial Management*. New York: McGraw-Hill, 1977.

Halpin, Daniel W. *Financial and Cost Concepts for Construction Management*. New York: John Wiley & Sons, 1985.

Horngren, Charles T., and George Foster. *Cost Accounting: A Managerial Emphasis*, 7th ed. Englewood Cliffs, N.J.: Prentice-Hall, 1991.

Horngren, Charles T., and Walter T. Harrison, Jr. *Accounting*, 2d ed. Englewood Cliffs, N.J.: Prentice-Hall, 1993.

Humphrey, Kenneth K., ed. *Project and Cost Engineers' Handbook* (sponsored by the American Association of Cost Engineers), 2d ed. New York: Marcel Decker, 1984.

Jelen, F. C., and James H. Black. *Cost and Optimization Engineering*, 2d ed. New York: McGraw-Hill, 1983.

Kharbanda, O. P., E. A. Stallworthy, and L. F. Williams, revised by James T. Stoms. *Project Cost Control in Action*. Englewood Cliffs, N.J.: Prentice-Hall, 1981.

M.Y.O.B. User Guide. Rockaway, N.J.: Teleware, Inc., 1992.

Mueller, F. W. *Integrated Cost and Schedule Control for Construction Projects*. New York: Van Nostrand Reinhold, 1986.

Using Job Cost: Peachtree Accounting. Norcross, Ga.: Peachtree Software, 1992.

CHAPTER
15

OPERATIONS
SIMULATION

It would follow that "significant form" was form behind which we catch a sense of ultimate reality.

Clive Bell*

Production operations involving complex and subtle interactions between machines, workers, and materials often can be evaluated only crudely using manual methods, and experienced professionals must resort to efficiency factors such as "45-minute hours" to allow for probable errors and uncertainties. Computer-based simulation models can be built to analyze such complex systems much more accurately, and they can be refined as additional experience and information are added to them. They also can model quite disparate things, such as the effect of various changes in a company's bidding strategy on its volume and profitability.

Since such capabilities are available, it is surprising that the technique is not more widely used. Until recently, however, the software available for simulation modeling either has been too complex for most working professionals to master without inordinate time and effort, or has been too limited for the type of complex construction problems that most warrant simulation modeling, or has not been readily available at reasonable cost for today's low-cost computers. But all of this has changed. Powerful simulation software is readily and economically available for the most widely used computers, and some modern programs are as easy to learn and run as spreadsheet and database packages.

This chapter introduces the fundamental concepts of simulation modeling and then shows how we can apply today's software to construction problems. As more and more potential users become aware of what simulation software is available and learn how it works, we can hope that the tools will be more widely used in the future.

---

*Art, 1985.

BASIC CONCEPTS OF SIMULATION SOFTWARE

This section first uses a simple example to introduce a few phenomena that are suitable for simulation modeling and then introduces some components common to most simulation programs. It also mentions some commercial package programs that permit the type of simulation modeling described in this chapter.

Introductory Example

Manual methods of computation are limited by time and cost to deterministic analyses of relatively simple problems. In production analysis such problems might include the individual machine output and fleet-balancing calculations for earthmoving, for a crane-and-bucket concrete operation, and so forth. As a very simple example of such calculations, assume that a backhoe can load one truck every 5 minutes, 8 hours per day. How many trucks can it load in a day? How long should it take to make 72 loads? The deterministic answers to these questions, which we can even work out in our heads, are as follows:

$$\text{Number of loads} = \frac{60 \text{ min/h}}{5 \text{ min/load}} \times 8 \text{ h/day} = 96 \text{ loads per day}$$

$$\text{Number of hours for 72 loads} = \frac{72 \text{ loads}}{12 \text{ loads/h}} = 6 \text{ h}$$

The answers are so simple as to seem trivial. Why then, when we go to the field to observe the results of such calculations, do they seem to be worse than we expected? In this case, let us assume that the actual arrivals appear like those in Table 15-1a: We could imagine some plausible reasons for this arrival pattern. Perhaps somebody forgot to call the haulage broker to have a suitable number of trucks ready at 8 A.M.; after a frantic call from the foreman, three arrived by the end of the hour. All were promptly loaded and sent on their way.

TABLE 15-1
Tabular analysis of actual production observations

Hour	8	9	10	11	12	1	2	3	4	
(a) **Arrival data**										
Arrivals	3	15	11	11	8 [lunch]	7	6	8	3	$\Sigma = 72$
(b) **Hourly processing calculations**										
No. to process	3	15	14	13	9	16	10	8	3	
No. processed	3	12	12	12	0	12	10	8	3	$\Sigma = 72$
No. left	0	3	2	1	9	4	0	0	0	

They returned empty about the time others were turning up, so there was a waiting queue at various times in the second hour. Perhaps traffic problems or competing operations reduced truck availability after 2 P.M., reflected in the lower number of arrivals. In any case, the total number of loads carried in the 8 hours was 72 rather than the forecast 96. Now if the estimator had assumed a 45-minute hour, then the forecast would have come out right on target. That is,

$$(45/60) \times 96 = 72$$

But such an accurate result would have been mostly luck given the uncertainties of the job-site environment.

If we were able to manually analyze the production impact of what happened, we might build a table such as that shown in Table 15-1*b*. We have assumed here that the loader worked away at a constant 1 load per 5 minutes whenever it had trucks available. Its operator took an hour off for lunch, at which point one truck was ready to be loaded, and eight others turned up (perhaps after stopping at a fast-food place) by the end of that hour. With the queue full at 1 P.M. and most of the trucks making it back by the end of the lunch hour, a total of 16 loads was possible between 1 P.M. and 2 P.M., but the loader could only handle 12, leaving four trucks in the queue at the end of the hour. After that, fewer trucks were available than could be loaded, so there were no trucks left in the hourly queue for the rest of the afternoon.

In developing this example, we have been doing some of the type of clerical work that a computer simulation program handles very rapidly. The example is still very simple, the main uncertainties implied being truck availability in the first hour and in the afternoon, and variations in their haul and return times. We have not even looked at the materials being processed through the system and the inevitable variations in bucket loads and truck loads. Nor have we considered any variations in the loader cycle times. Breakdowns would add further complications. Needless to say, if we added many of these phenomena, the tabular calculations would become enormously tedious, particularly so if we looked at loader output by the bucket scoop rather than by the hour. Facing such situations, we revert to simple calculations like those given at the beginning of the example and apply efficiency factors to make the answers come out closer to reality.

But with a simulation program, we could easily model all of these complications and more, and do so for much more complex systems. Running the model would take about the same time as doing the first two calculations manually and far less time than developing the table. Although this simple case may not benefit much from such detailed analysis, simulation programs can really be useful and practical when more complex production systems need analysis.

Components of Simulation Analysis

Simulation programs use mathematical and logical models to represent the real world they describe and emulate. Typical components in a simulation model include the following:

- Graphical networks to show the *logical interactions* among the various resources and constraints in a system. These relationships can also be described in mathematical and logical formulas. Unlike CPM networks, where loops in activity chains violate the one-way progression of construction schedules, loops or iterations in logic are often deliberate in simulation to show, for example, repetitive cycles of production that take place in a day's work. In Fig. 15-1 are partial models of networks using the notation of two of the simulation programs to be introduced later in this chapter—Stella and Cyclone.

- *Logical constraints* on models can be of different types. In the truck loading example, there is a combination input constraint (called an *AND* operation in logic terminology). This means that *all* of the input resources (truck and loader in this case) must be present for the subject activity to take place. In Cyclone this constraint is shown as the rectangle with the slash in the upper left corner on **Load Truck**, a notation that it calls a **COMBI** node. In the Stella example, the arc going from the **Truck queue** back to the **Scoops** activity constrains the model to hold the next scoop unless a truck is there to receive it. Alternatively,

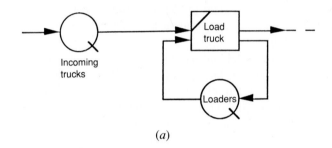

(*a*)

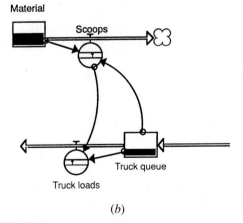

(*b*)

FIGURE 15-1
Partial graphical networks of an earthmoving operation: (*a*) with Cyclone graphical symbols, and (*b*) with Stella graphical symbols.

some activities can be permitted to take place if *any* of the resources are ready. This is a logical *OR* operation. As an example, trucks could enter and exit a haul road, and even proceed concurrently, regardless of which loader (in a multiple loader situation) they came from. The activity **Haul** (not shown) would start any time that any of the inputs were available. A faster truck might even be allowed to pass a slower one, in the model as well as in reality.

- The *progression of time* in the real world is simulated in the model by ticks of a clock; the ticks, in turn, either refer to the host computer's electronic clock or use calculations internal to the simulation program itself. Simulated time can be greatly compressed or expanded relative to real time in order to move quickly through operations—like a month of field production—or to slow something down—such as a delayed blasting sequence. In some programs (e.g., Cyclone) the clock is *event-driven* and advances in variable amounts of time corresponding to each discrete change (event) in the status of the model (e.g., a truck departs). In other cases (e.g., Stella), the underlying pacing is called *clock-driven;* here time advances in uniform increments (e.g., 1 minute) and the state of the model is updated to correspond to what happens within each time increment. Both methods can record the state of the model at each increment in time so that later statistical analyses can process the data to evaluate the behavior of any resource or activity.

- Probability distributions often model the *uncertainty* and *variations* in real-world empirical data, such as changes in bucket volumes and cycle times for a loader, the occasional breakdown of a truck, and the like. Statistical models such as normal, uniform, lognormal, Poisson, and exponential distributions are defined mathematically to represent the range of a measure (time, load, etc.) and then are randomly sampled upon each iteration through a model to produce the specific numbers to be used in a given cycle. The central tendency of models such as the normal distribution (the infamous "bell curve" of student exam life) ensures that more samples will be near the average than the extremes, but high and low numbers are still possible, just as in the real world. In Fig. 15-2 we see a typical normal distribution for the volume of earth in truckloads. The sample taken on iteration i is 21.5 m^3.

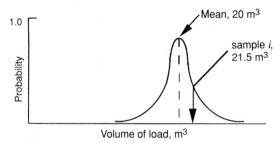

FIGURE 15-2
Sampling a normal distribution on iteration i.

Another important concept in simulation modeling is the distinction between *discrete* and *continuous* phenomena. Systems involving a finite number of identifiable units (e.g., trucks, cranes, workers, pumps, etc.) or activities with clear beginnings and ends to their durations are usually modeled as discrete activities. Fluid flow, the growth of organisms, and chemical reactions are usually modeled as continuous phenomena. However, systems for which large numbers of discrete entities are involved—as is the case for global population studies, monetary systems, and even construction processes such as installing thousands of segmental liners in a long tunnel—can be approximated as continuous for purposes of simulation modeling. Simulation programs tend to be oriented toward either discrete (e.g., Cyclone) or continuous (e.g., Stella) processes. However, there is usually enough flexibility in them that phenomena of the opposite type can be modeled also.

Commercial Software

Simulation programs do not have the large market volumes of the major word-processing, database, and spreadsheet packages, nor are people generally as familiar with them. Most of the commercial simulation packages are produced by smaller firms, some specialized to support only the one package they produce. Several are spin-offs of university research. Nevertheless, a variety of quality software is available, and several firms offer good user support. This section briefly mentions a few of those that, at the time of this writing, offered strong capabilities for the type of modeling most applicable in construction.

Many simulation programs available for today's microcomputers descended from packages originally developed for mainframes and minicomputers or are closely based on such programs. One of the best known is GPSS (General Purpose System Simulator), developed in the 1960s by IBM and now marketed in microcomputer versions. Another package dating from the 1960s, and one with university research roots, is SLAMSystem by the Pritsker Corporation (Indianapolis, Ind.). One developed specifically with construction in mind is Cyclone and its personal-computer variant called MicroCyclone. It is marketed by Learning Systems, Inc. of West Lafayette, Ind. Its flexibility and ease of use have made it popular in college construction programs, and it has had some success in industry. A variant called UM-Cyclone was developed at the University of Michigan and has also been made available to academic users. Another early program that continues to be popular is Simscript II by CACI Products Co. (La Jolla, Calif.). All of the preceding packages are available in versions for IBM and compatible MS-DOS-based personal computers.

A few programs developed for the Macintosh environment are noteworthy for their ease of use and strong graphical interface. Stella II, by High Performance Systems of Hanover, N.H., is particularly good in both of these respects, yet it allows users to model very complex systems. Its heritage goes back to Dynamo, developed by Jay Forester at MIT in the 1960s. Extend, by Imagine That, Inc. of San Jose, Calif., is more difficult to use, but it compensates with a built-in programming language that is somewhat like C. This allows great flexibility and power in model design and implementation.

Caterpillar Inc. developed a program called VEHSIM, a simulator for earth-moving vehicle performance under realistic job conditions. It runs on MS-DOS PCs. Unlike the others mentioned here, it is not a stochastic simulator in that it does not use a random number generator or iterate through repetitive cycles of an operation. Although it is thus only deterministic and its application is quite narrowly focused, it models machines at a level of detail encompassing their engine, tire, and transmission characteristics, and it allows users to input data for very complex, multisegment haul-road designs. One can use it to compare alternative fleet configurations (e.g., pushers and scrapers versus loaders and trucks, each in various sizes) over different possible road designs (e.g., shorter and steeper versus longer and flatter) to determine the most economical choice for a given project. Its precise output can serve as a basis for activity modeling within programs such as SLAM, Stella, or MicroCyclone, and thus is a good example of how two tools can work together for a common objective.

Later in this chapter we illustrate the use of VEHSIM, MicroCyclone, and Stella in typical field construction problems. First, however, we should develop a general approach to simulation modeling using any of the tools.

WORKING WITH SIMULATION PROGRAMS

This section introduces some of the general procedures needed and capabilities provided for working with simulation software. We begin with an overview of the typical steps in simulation modeling:

- Define the problem and establish the components and boundaries of the system to be modeled
- Develop logic and flow diagrams
- Estimate deterministic and/or stochastic values of time durations (e.g., load time), production quantities (e.g., earth in a truckload), and other variables
- Build the model using computer software
- Test, refine, and validate the computer simulation model
- Use the simulation model for planning and decision making

The following sections discuss each of these steps in more detail.

Define the System

This step usually takes place before one uses computer simulation software. It has been said that, in the real world, everything affects everything else. For example, consider an urban highway project for which we are interested in an excavation and hauling operation. Slow production on the operation might delay subsequent activities into the winter months; this might in turn cause the highway completion to be late; this could compound traffic problems in the affected city; more congested traffic might make the city less attractive and cause some businesses to locate elsewhere; this might help "elsewhere" to become more prosperous . . . and so on. But in practical situations, we need to put boundaries on our models. This stage of model definition is somewhat like making a free-body diagram in engineering mechanics: Only the components of interest are modeled, and everything else is simplified to inputs and outputs at the boundaries of the model.

In production systems, it is generally evident that we should examine the physical resources: machines, workers, materials, and such. But one often needs to consider some abstract entities as well, such as inspection checks before continuing work, restrictions on haul times on city streets, space to store materials, and so forth. Some of these can be difficult to visualize but can have significant effects on the accuracy of model behavior.

As an example, consider the earthmoving operation in Fig. 15-3.* In it we see a loader working from a stockpile to load earth into trucks that haul it to a fill operation. We just assume that the stockpiled material will be available (thus ignoring an excavation operation that might be feeding into it), and we also ignore what might be done with the material at the destination (e.g., spreading, compacting and grading). This system is thus defined to focus on three resources: loader, earth, and trucks. As long as there is enough material in the stockpile, and sufficient spreading, compaction, and grading capacity at the other end, one could model just the loading and hauling operation and get good results. If one of the operations outside of the boundary falls short of its needed capacity, then the model will be too optimistic in its results. The boundary of the model should then be enlarged to include the operations that might significantly affect system behavior.

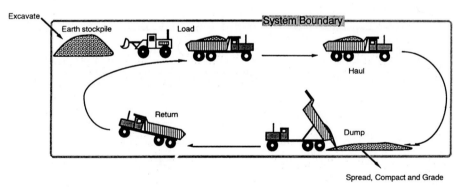

FIGURE 15-3
Examining an earthmoving system.

---

*This example is partly based on one described in *Planning and Analysis of Construction Operations,* by D. W. Halpin and L. S. Riggs, John Wiley & Sons, New York, 1992, Chapter 6, p. 111.

Develop Logic and Flow Diagrams

One normally approaches the development of logic and resource flows by first looking at individual elements of a model and then successively integrating them into an evolving system. For example, picking up from the earthmoving example, we see the flows of the earth, loader, and trucks in Figs. 15-4*a, 4b,* and *4c,* respectively. The earth is the only resource that flows into and out of the system boundaries, and it is passive in the sense that it depends on the actions of the machines to move. The truck and the loader are each given a **wait** state (i.e., a queue) to accommodate the condition that each needs the other before the **load** operation can take place. These partial flows also show an OR branch to indicate that, if the counterpart resource is available, they can skip the wait state and go directly to the load operation. We assume adequate capacity at the boundaries, so no wait state is needed before the **scoop** and **dump** operations. Similarly, we assume no constraints on **haul** and **return.**

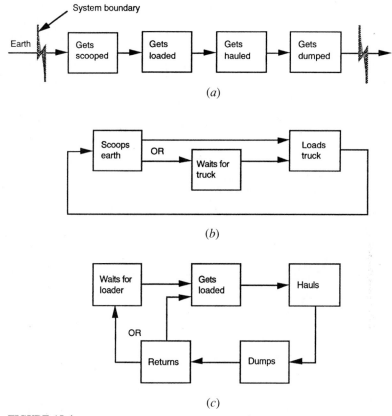

FIGURE 15-4
Individual resource flows: (*a*) earth cycle; (*b*) loader cycle; and (*c*) truck cycle.

In examining the partial cycles in Fig. 15-4, we note several conditions that enable us to integrate the components into a single model:

- Common activities intersect (**Load** is common to all three resources, and several activities involve two of the three).
- There are input constraints (e.g., both a truck and a loader are needed to **Load**).
- Actions proceed in a logical sequence (**Scoop** precedes **Load**, **Load** precedes **Haul**, etc.).
- Redundancies can be eliminated to simplify logic (e.g., the arrows from **Load** to **Haul** and from **Haul** to **Dump** occur twice, as do some others).

We can use these observations to integrate the models into an overall system, such as that shown in Fig. 15-5. Twelve activities from the separate diagrams in Fig. 15-4 consolidate into seven on Fig. 15-5. Patterned lines have been superimposed to identify the three separate resource flows in the diagram. Again, only the earth moves in and out of the system boundaries. The types of logic constraints are not precisely identified in Fig. 15-5 except for using the word OR in two places, but the overall flow cycles are fairly clear. Later we will render such figures in the actual graphical notation of the Cyclone and Stella simulation modeling programs, and they will be more comprehensive and precise. The reader might also consult *Planning and Analysis of Construction Operations* by Daniel W. Halpin and Leland S. Riggs for a more detailed description of how to build up such models from their components. Good model-building tutorials also

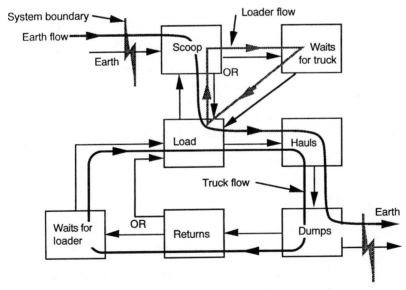

FIGURE 15-5
Integrated resource flows.

appear in simulation software manuals, such as *Stella II Applications* by High Performance Systems.

As mentioned earlier, the type of network in Fig. 15-5 differs from a CPM logic diagram in that there are definite loops in the logic (e.g., the trucks and loader) as well as one-pass components (the earth in this case). Also, in this case there is conditional logic (the two OR branches) to anticipate and account for alternative situations. Once underway, the model will run as long as there is earth entering the system. Indeed, there is nothing in this diagram to keep the machines from loading and hauling air if the earth in stockpile is gone! We will take care of that type of contingency in our simulation examples later in this chapter.

At this stage of model definition, one can also decide where to initiate resources in order to get the model started. In this case one would need to put some earth in the stockpile, a loader in its wait queue, and one or more trucks in their wait queue, and the model would be ready to go. The numbers of resources to use are not important since these can be optimized once the model is running. As an initial case, one could use a simple arithmetic calculation (divide estimated truck cycle time by loader cycle time) to get the number of trucks to put into the system that one loader could manage.

Determining Durations and Production Quantities

Various activities and resources in a model have associated with them quantitative variables. They may represent time durations, money flows and interest rates, production numbers, and so forth. Such variables can either be fixed deterministically at the beginning of a simulation run or be subject to probabilistic variations. The latter are usually represented by statistical distributions, such as uniform, normal, skewed, and the like. Examples used in MicroCyclone appear in Fig. 15-6. Things like truckloads and mix times typically follow normal distributions. The time delay when a traffic light is in a system can be represented by a uniform distribution. Skewed distributions such as beta, lognormal, and the like usually represent quantities where the variation in one direction can be much greater than in the other. For example, a lognormal might represent a haul time. The higher haul durations are likely to be much further from the average than the shorter ones.

Since some of the mathematical representations of distributions are theoretically bounded only by infinity at the extremes, in practical simulation models it is also common to specify upper and lower limits to override values that might fall outside of reasonable ranges. Perhaps one or two standard deviations from the mean might be chosen for a normal distribution representing load volume.

In estimating variables for simulation models, one should look at the likely values of each parameter in its own right, without considering the impacts of events in other parts of the system. This recommendation is a bit different from the efficiency or contingency factors that are commonly included at this stage in conventional estimating. In this case, the interactions when the model runs

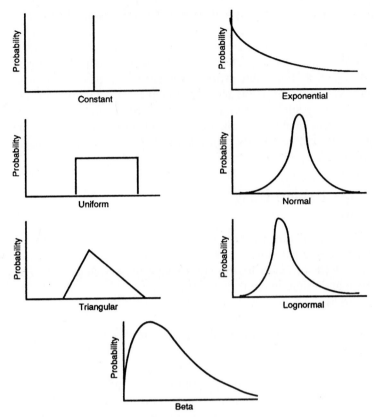

FIGURE 15-6
Statistical distributions used in MicroCyclone.

will take care of the interdependency impacts. It is sometimes sufficient to use a constant for parameters that do not greatly influence system behavior. It is also relatively simple to use a three-point estimate to get an idea of the mean and likely deviation, as is done in PERT scheduling. The three estimates are the most likely, optimistic, and pessimistic numbers that seem appropriate to the parameter at hand. The most likely is usually weighted more heavily, as in the following formula:

$$\text{Weighted mean} = \frac{\text{optimistic} + 4 \times \text{most likely} + \text{pessimistic}}{6}$$

But with most advanced simulation software, one can simply specify the parameters of an appropriate mathematical distribution and let the simulation program use a random number generator to pick off actual numbers to use for each parameter in each iteration through a model's execution. Examples will be given in the Cyclone and Stella cases later in this chapter. Finally, if no standard mathematical formula fits, some programs allow the user to enter graphically whatever shape seems to make sense. This input could be represented as a line graph, a histogram, or otherwise.

The sources of actual numbers to use as parameters in the statistical models are basically the same sources already used by professionals working in the area to which the model applies. For future works, estimators use past experience, books, records, and even intuition. If the model is to represent operations already underway, one can use time studies, automated data recorders, and other means to measure the parameters in the field. In the most advanced cases, real-time data collection systems might be directly connected to feed a model continuously running in a computer.

Data Entry to Build a Model Using Computer Software

Data entry for a given simulation package can be done either by typing a text file using an internal or external editor (both are used in Cyclone), by filling in online screen forms (used in VEHSIM), or by drawing the activities and logic directly on the screen and filling in the relevant descriptions and parameters using interactive forms (this is done in Stella). We illustrate these approaches later in this chapter.

The means for building a model in a particular program are similar to those de-scribed in Chapter 9 for spreadsheets. The options typically include the following:

- General file manipulation (start new file, open old file, save file, close file, etc.)
- Data entry (including the option to duplicate and modify an existing model in order to make a new one—e.g., a loader-truck model can easily be converted to a pusher-scraper model)
- Editing (cut, copy, paste, insert, delete, etc.) of both text and graphical data and components
- Model execution (e.g., specifying run time, clock time increments, termination conditions, etc.)
- Reporting (display or print tables, graphs and diagrams resulting from the ex-ecution of a model, or export results to other programs such as a spreadsheet for further analysis or to a word processor to become a table or figure in a document)

Once a model has been started, and even if it is only partially complete, it can be checked, modified, and expanded until it reaches the desired scope and complexity. Before turning the model over to users, the developer should test and validate it.

Testing, Refinement, and Validation

In testing, the developer manipulates the variable input parameters (resources, time increments, etc.) to gain an understanding of the behavior of the model and check that the behavior conforms to what was intended. Tabular and graphical reports are checked to be sure that they are working and that their results appear reasonable. Graphs are particularly valuable as a means to highlight spikes and gaps in output that may result from errors in modeling assumptions or data input. Once the obvious errors have been corrected, the developer can proceed with testing, running input variables throughout their likely ranges, and exploring the sensitivity of model behavior to different inputs.

Validation is most feasible when one is modeling a system whose real-world counterpart is already in operation. In this case, one can compare model results to the behavior of the real-world counterpart before attempting to alter and optimize the model to explore improved alternatives. For systems that are planned for the future—as is typical when estimating new work—the model results can still be compared with conventional deterministic calculations, and be modified with efficiency or contingency factors as appropriate. Significant differences in model behavior should be accounted for, and modifications and retesting may be necessary.

Once a model has been constructed and tested, it can be turned over to other users for repetitive applications, or it can be adapted to other situations. Over time, experience will accumulate in an organization and simulation applications may become as common and diverse as those of spreadsheets. A valuable library of models may also be accumulated.

Using the Simulation Model for Planning and Decision Making

All that we have discussed thus far leads to the real purpose of simulation models: planning, analysis, and decision making. Once a model has satisfactorily replicated the behavior of an existing or planned system, users can start exercising the model to explore, usually very rapidly and economically, all important facets of system behavior. The actual process of building and exercising such models usually results in a much deeper understanding of the system being modeled, even by experienced people, and the models can serve as an excellent training tool for new users as well.

If we discover inefficiencies in and obstacles to achieving the objectives of the real-world system in the course of simulation, we can copy and modify the model to explore alternative logic, resources, schedules, or other possibilities. Changes in fleet configurations, site layouts, bidding strategies, and others can be explored without risk and without impacting the systems being modeled. Once we have explored, tested, and approved satisfactory alternatives based on model simulation, we can implement the resulting changes in the real-world system. The proof of the pudding comes when the real-world system conforms to the improved manner forecast in the simulation. In doing so, however, it often happens that new constraints that were previously placed outside the boundaries of the model now come into play, or internal interactions that were not previously significant may become influential. For example, improvements in the loader-truck operation from

Fig. 15-4 may overwhelm the spreading and compaction activity and thus cause backups in the system, so this activity should now be added to the model. Changes and new additions to the model should be tested and brought into conformance with observed real-world behavior. It should be evident that using a model as a decision-making tool for improving ongoing systems is an iterative process in which both the model and its real-world counterpart improve over time. Examples of this type have saved millions of dollars in actual construction operations.

ADVANCED SIMULATION MODELING CONCEPTS AND FEATURES

There is such a variety in simulation modeling programs that there is not a standard base configuration to use for comparison. Nonetheless, one can consider differences in some categories that are similar to those we examined for spreadsheets and databases.

Enhancements to Standard Functions

A basic program should allow the construction of models involving logical sequences and cycles of events, and probabilistic variations on key parameters. Some, however, are very limited in the number of activities, resources, and logical connections that can be used in a single model. More advanced programs allow models of considerable complexity to be built and may be limited only by memory in the computer. A basic program is also likely to focus on either discrete or continuous systems. More advanced programs may have features to accommodate both types. Certain programs only model deterministic behavior (e.g., VEHSIM), and some would say this makes them more of a calculator than a simulator. Other programs may limit the available distributions to just a few (e.g., constant and normal). More advanced programs allow a variety of distributions to be used and may even allow the user to define nonstandard parameter behavior via tables, histograms, or line graphs.

Flexibility versus Ease of Use

Simulation programs such as Stella and Cyclone are particularly easy to use and are thus excellent for teaching purposes and for nonexpert users. But they only offer a limited number of resource processing and activity types and logical constructs; these limits can make it awkward and sometimes impossible to express certain kinds of system characteristics. Other programs, including GPSS, SLAM, and Extend, tend to be more powerful in their respective domains, but their flexibility typically results from having model elements at a more rudimentary level and thus they are more difficult to learn and use. Indeed, tapping their full power can be almost like using a procedural language, and thus these programs tend to be more suitable for simulation specialists and experts. For most users, ease of use takes priority over flexibility and power. Fortunately, a wide variety of construction applications can be modeled even with the simpler programs.

Graphics

One's understanding of the behavior of a simulation model can be enhanced greatly with the aid of graphics. Some programs, however, mainly use tabular output, which can require expert interpretation. Simulation graphics is sometimes limited to static representations of model components and logic; these can make it easier to construct models and appreciate their overall structure but they convey little about the dynamic behavior of a model while it is running. Some programs allow the model developer to employ drawing programs to make pictures that can be placed on a diagram (e.g., trucks and loader like those in Fig. 15-3 can be drawn and placed on a logic diagram near the circles and boxes that might otherwise represent them). This feature can further amplify understanding of a model, or at least help in communicating its structure to people not yet familiar with it.

The most helpful form of graphics, however, is dynamic displays that are synchronized to track the changes in various parameters during model execution. The basic Stella program, for example, juggles bar graphs on each queue (resource stack) and moves little dials on each converter and flow regulator that change while the model runs. It also has separate line graphs that plot the values of selected parameters versus time. Stella can also be configured to drive other software such as Apple's HyperCard to provide genuine animation of realistic drawings.

Even more advanced are Bechtel's Walkthru and Construction CAE programs. Walkthru enables one to "move about" in a 3-D CAD solid model of a project, and Construction CAE further allows one to manipulate realistic-looking pieces of equipment in a 3-D model while keeping track of schedule time and thus to simulate and document the construction sequence of a facility. Advanced programs of this type are described in more detail in Chapter 16.

Programmable Macro Commands

In basic simulation programs one just builds the model and then executes it. More advanced programs essentially are programming languages in their own right that allow full automation of all of the model execution or parts of it.

Integrated Software

No simulation program has yet been, or is likely to be, integrated with other common applications in major packages such as those that combine spreadsheets, databases, word processing, and communications. However, most have the capability of exporting their tabular or graphical results to other programs (e.g., to a spreadsheet), and some can be executed from within other programs (e.g., Stella can be run from HyperCard), so some degree of integration among applications is possible. This feature is particularly useful where deeper analysis or more elaborate reporting of tabular results is desired, or when more realistic graphical animation might be helpful. Some programs have external function links so that capabilities that are not available in the standard program can be added by writing routines in a procedural language and tying them into the simulation program's menu of options.

APPLICATIONS AND LIMITATIONS

Simulation programs are most suitable for modeling dynamic systems. Within this broad guideline, the possibilities are limited mainly by one's imagination. Physical systems, whether or not bounded by Newton's laws, are readily modeled. Numerous socioeconomic models have been developed to study the interrelationships of birth rates, food production, climate, health, and economic growth. Models exist for psychological behavior for individuals and groups, and for financial systems ranging from starting a small company to international banking. Biology and chemistry and their applications from agriculture through medicine to zoology are all suitable for simulation models that will improve our understanding of complex systems. The concept of a stock or queue can model physical entities like discrete numbers of people or machines, money, stockpiles of aggregates, water in a reservoir, or even abstract concepts like momentum, greed, commitment, or frustration.

Past and potential applications in construction fall into several categories:

- Production systems of workers, machines, and materials. These are the ones most commonly modeled and will be the focus of the examples in this book.
- Management of tools and materials. Tool rooms, warehouses, and the procurement and distribution of materials that flow into and out of them are good candidates for simulation modeling to develop better management policies.
- Financial models. Simulation models can help us study the impact of alternative marketing plans on company growth and profitability, or focus on project finance, or even help us make investment decisions regarding machinery and materials.
- Organizational models. These can be used to study departmental structures, incentives for efficiency or safety, training programs, and so forth.
- Environmental impacts. Nature can affect a project, and vice versa. For example, simulation has effectively modeled the hydrology of rivers in designing diversion tunnels, cofferdams, and other works. Increasingly, the noise, dust, runoff, and other impacts of the project on its surroundings must be modeled as well.

Simulation models are usually more suitable for cyclical or iterative systems than for situations where a one-pass calculation is sufficient. A sequence of calculations, such as a schedule, an estimate, or a cash flow, would normally be easier with a specialty application program or a general tool such as a spreadsheet. Simulation software is also of limited use in database applications.

Perhaps more to the point is determining when an application is complex enough to justify going to the trouble of building a simulation model. The answer is partly a function of the complexity of the modeling tool; and tools like VEHSIM, Stella, and Cyclone make it worthwhile to build models even of fairly simple systems. Having on hand an adaptable library of models of common operations also makes it attractive to use them. Given this, anything over a day's work involving a couple of workers and a machine probably can benefit from the type of structured thinking and planning that occur in the course of building and using a simulation model.

At the other extreme, detailed modeling of complex real-world systems can become very expensive and even overwhelm the capacity of most computer programs. But is the preferable alternative little analysis at all? It would seem better to break the system down and model what appear to be the more critical components. Then perhaps one can build a summary model that aggregates the results of the component models as a means to look at the overall system.

The following sections illustrate some small but typical applications where simulation programs can help us better plan and manage construction operations.

EQUIPMENT SELECTION EXAMPLE
IMPLEMENTED IN VEHSIM ON A PC

This example employs the Caterpillar VEHSIM simulation program for analyzing alternative machines and haul routes. Although not all capabilities needed for this application are included within VEHSIM, it could be nicely complemented using a spreadsheet.

VEHSIM is a fairly straightforward pull-down menu-based PC application package that has three main components:

- A library of available machines and their specifications. This is basically a set of tables with the type of information contained in manufacturers' handbooks (e.g., *Caterpillar Performance Handbook**). The user can enter additional machines, including those of other manufacturers, into the tables.
- A form for entering characteristics of the haul and return roads, plus related job-site information.
- A form for entering machines selected from the tables, and a form for adding or modifying information (e.g., load and dump times) specific to the operation under study.

For purposes of this illustration, we will select from the machines and use the parameters in the Table 15-2.

Other parameters and assumptions in this VEHSIM model follow:

- Inputs are expressed in metric units.
- A 45-minute efficiency hour is used.
- Return road is the same as the haul road (i.e., reverse grades, same rolling resistance).
- One uses 2 kph as both initial and final velocities in haul and return.

---

*Published by Caterpillar Inc., Peoria, Ill.

TABLE 15-2
Machines selected for VEHSIM example

	769C Truck	773B Truck	621E Scraper	627E Scraper	631E Scraper
Machine Code	C 202	C 205	C 135	C 137	C 133
Cost/hour	$75.00	$95.00	$80.00	$95.00	$110.00
Payload	31,800 kg	45,400 kg	15.3 m^3	15.3 m^3	23.7 m^3
Power	336 kw	485 kw	246 kw	336 kw	336 kw
Fixed Time	3.6 min	4.0 min	1.2 min	1.2 min	1.3 min
Tires	18 × 33	21 × 35	33.25 × 29	33.25 × 29	37.25 × 35

- Material hauled weighs 2,000 kg per bank cubic meter.
- Total volume for excavation = 100,000 bank cubic meters.
- No corrections for speed or altitude need be entered.
- A loader for a 769C costs $90 per hour; it loads in 2.8 min.
- A loader for a 773B costs $120 per hour; it loads in 3.2 min.
- A pusher for a 621E or a 627E costs $80 per hour; it takes 1.0 min. to load and 0.7 min. to backtrack.
- A pusher for 631E costs $100 per hour; it takes 1.0 min. to load and 0.7 min. to backtrack.

This example involves a site that requires some terrain analysis and haul-road design before we can use VEHSIM to evaluate the alternatives. Figure 15-7 is a contour map that shows the origin (borrow area) and destination (fill area) for an earthworks operation. Our first task is to plot two alternative haul roads that meet the following criteria:

- Neither haul road can involve significant amounts of cut or fill. Basically, they must follow the contour of the terrain.
- In both of the alternative roads, assume that the rolling resistance will be 70 kg/tonne.
- In both cases, do not go into the environmentally protected area defined by the swamp and the pond.
- The first haul-road option should minimize the distance, but in doing so its up and down grades should not exceed 20 percent over any length.
- The second haul-road option should involve a minimum of grade change from the borrow area to the fill area. To the extent possible, it should have a steady descending grade from the borrow area to the fill.
- Keep the speeds of the vehicles under 20 kph when within 200 m of the borrow and fill sites.
- Assume the center of the borrow and fill area for computing distances.

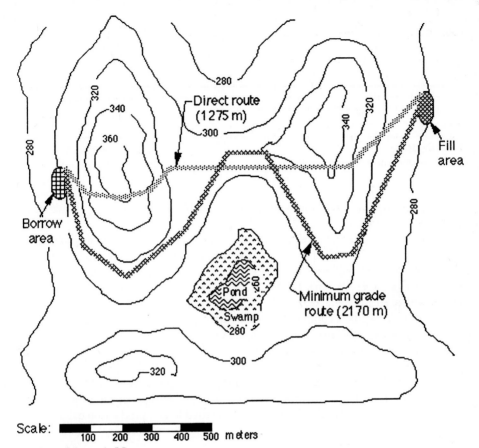

Scale: ▮▮▮▮▮▮▮▮▮
100 200 300 400 500 meters
Contour interval: 20 m

FIGURE 15-7
Contour map showing alternative routes from Borrow area to Fill area.

In Fig. 15-7 two possible routes that fit these criteria are superimposed. In Fig. 15-8 we see their profiles. The shorter route is 1275 m but involves steep grades. The flatter route stretches out to 2170 m (70 percent longer). The lengths and grades of the individual segments for each road are listed in Table 15-3.

In using VEHSIM for this problem, we first select the machines specified in Table 15-2, enter the information for each road, and determine the following:

- The machine with the highest production
- The machine with the lowest cost per unit of production
- The machine with the shortest overall cycle time

Figure 15-9 is a screen that shows the main pull-down menu of VEHSIM along the top, and the haul-road data-entry submenu at the left. Note also that the

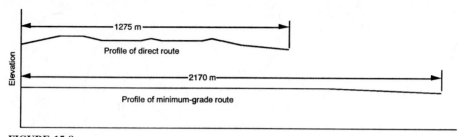

FIGURE 15-8
Profiles of alternative routes from Borrow to Fill.

function-key options and other commands available at this time are shown at the bottom of the screen.

Figure 15-10 is the screen where the selection of scrapers is in progress.

The input screen used to specify the materials to be hauled and to define the haul route precisely appears in Fig. 15-11. These data apply to the short, steep route. In this case, we have taken the default option to make the return route a mirror image of the haul road, with only the grades reversed.

Figure 15-12 is a comparative summary report provided after each machine has been run over the two haul roads under the indicated conditions. It clearly shows that, in this particular case, the Cat 773B truck on the short steep road has both the highest production rate and lowest production unit cost, whereas the powerful Cat 627E scraper on the short road has the lowest total cycle time. Note that this report also shows the type of loader or pusher selected for each haul vehicle, even though VEHSIM does not do detailed loader production calculations.

TABLE 15-3
Segment lengths and grades for alternative haul routes

Shorter route		Flatter route	
Segment length, m	Grade, %	Segment length, m	Grade, %
200	20.0	200	0.0
120	0.0	1400	0.0
100	−20.0	370	−3.5
150	0.0	200	−3.5
100	8.0		
100	−8.0		
145	0.0		
60	10.0		
60	−10.0		
240	−8.3		

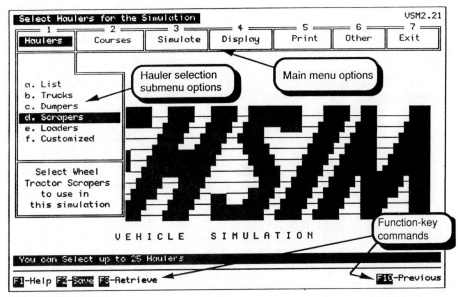

FIGURE 15-9
VEHSIM menu and submenu for machine selection.

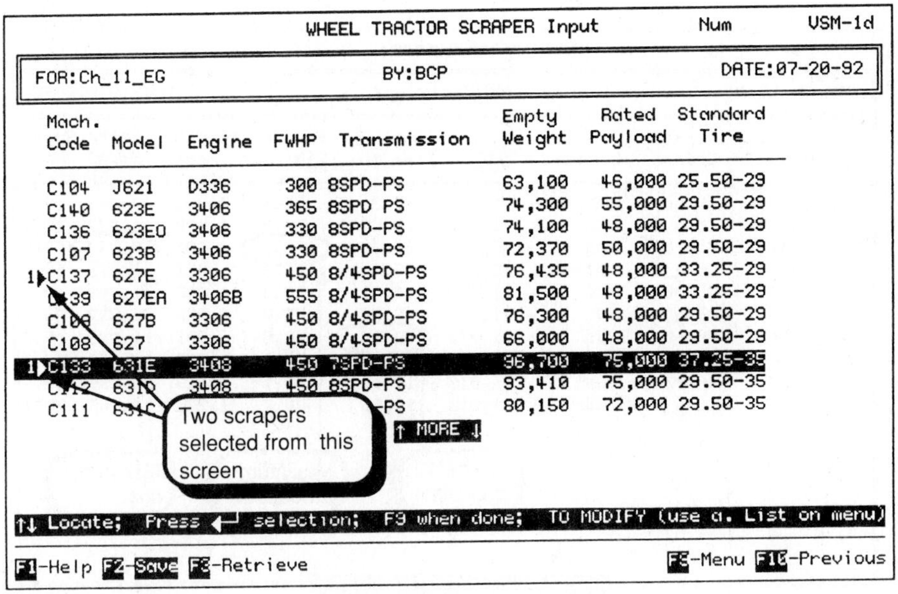

FIGURE 15-10
VEHSIM machine selection.

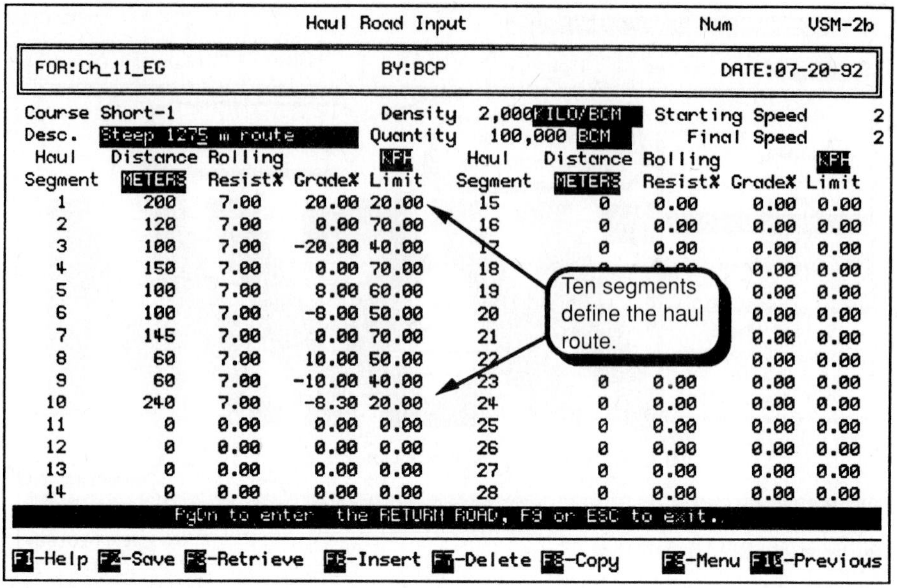

Haul Road Input   Num   VSM-2b

FOR:Ch_11_EG	BY:BCP	DATE:07-20-92

Course Short-1   Density 2,000 KILO/BCM   Starting Speed   2
Desc.   Steep 1275 m route   Quantity 100,000 BCM   Final Speed   2

Haul Segment	Distance METERS	Rolling Resist%	Grade%	KPH Limit	Haul Segment	Distance METERS	Rolling Resist%	Grade%	KPH Limit
1	200	7.00	20.00	20.00	15	0	0.00	0.00	0.00
2	120	7.00	0.00	70.00	16	0	0.00	0.00	0.00
3	100	7.00	-20.00	40.00	17	0	0.00	0.00	0.00
4	150	7.00	0.00	70.00	18	0	0.00	0.00	0.00
5	100	7.00	8.00	60.00	19		0.00	0.00	0.00
6	100	7.00	-8.00	50.00	20		0.00	0.00	0.00
7	145	7.00	0.00	70.00	21		0.00	0.00	0.00
8	60	7.00	10.00	50.00	22		0.00	0.00	0.00
9	60	7.00	-10.00	40.00	23	0	0.00	0.00	0.00
10	240	7.00	-8.30	20.00	24	0	0.00	0.00	0.00
11	0	0.00	0.00	0.00	25	0	0.00	0.00	0.00
12	0	0.00	0.00	0.00	26	0	0.00	0.00	0.00
13	0	0.00	0.00	0.00	27	0	0.00	0.00	0.00
14	0	0.00	0.00	0.00	28	0	0.00	0.00	0.00

Ten segments define the haul route.

PgDn to enter the RETURN ROAD, F9 or ESC to exit.

F1-Help F2-Save F3-Retrieve F4-Insert F5-Delete F6-Copy   F8-Menu F10-Previous

FIGURE 15-11
VEHSIM haul-road input screen.

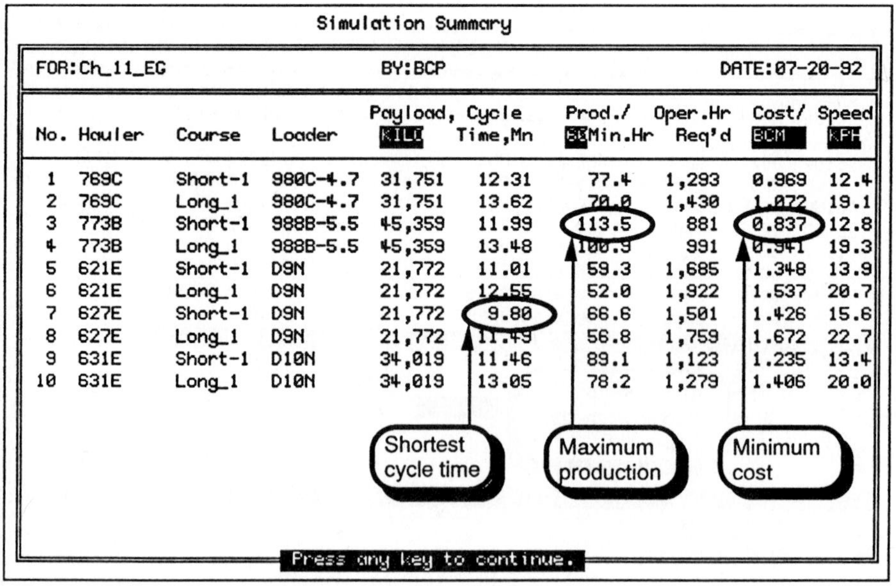

Simulation Summary

FOR:Ch_11_EG	BY:BCP	DATE:07-20-92

No.	Hauler	Course	Loader	Payload, KILO	Cycle Time,Mn	Prod./ BCMin.Hr	Oper.Hr Req'd	Cost/ BCM	Speed KPH
1	769C	Short-1	980C-4.7	31,751	12.31	77.4	1,293	0.969	12.4
2	769C	Long_1	980C-4.7	31,751	13.62	70.0	1,430	1.072	19.1
3	773B	Short-1	988B-5.5	45,359	11.99	113.5	881	0.837	12.8
4	773B	Long_1	988B-5.5	45,359	13.48	100.9	991	0.941	19.3
5	621E	Short-1	D9N	21,772	11.01	59.3	1,685	1.348	13.9
6	621E	Long_1	D9N	21,772	12.55	52.0	1,922	1.537	20.7
7	627E	Short-1	D9N	21,772	9.80	66.6	1,501	1.426	15.6
8	627E	Long_1	D9N	21,772	11.49	56.8	1,759	1.672	22.7
9	631E	Short-1	D10N	34,019	11.46	89.1	1,123	1.235	13.4
10	631E	Long_1	D10N	34,019	13.05	78.2	1,279	1.406	20.0

Shortest cycle time   Maximum production   Minimum cost

Press any key to continue.

FIGURE 15-12
VEHSIM production and cost summary report.

In Fig. 15-13 we see the detailed analysis of the production and cost for the Cat 773B truck hauling 100,000 bank cubic meters (bcm) on the short road. In Fig. 15-14 are details of the performance of the truck over every segment of the haul road. A similar report is available for the return road. Note the amount of detail that it provides (including the theoretical maximum steady state velocity and the top velocity obtained on each segment), and consider how long it would take using a calculator to produce this information for all five machines on both roads. The computer does it in seconds.

One could use this information plus some assumptions about the loader and the pusher production and costs together with a spreadsheet to compute additional items such as the following:

- The overall fleet costs and production for various alternatives
- Allowances for the fixed costs of building each alternative haul road
- The impact of doing some earthworks to flatten the grades on the shorter route

These are but a few of many possibilities that can readily be analyzed with the aid of two powerful tools like a spreadsheet and a simulation program. This type of information can also feed into specialized equipment cost estimating packages such as Caterpillar's Fleet Production and Cost.

```
                    Travel Time and Production
 ┌───────────────────────────────────────────────────────────────────┐
 │ FOR:Ch_11_EG              BY:BCP                  DATE:07-20-92      │
 ├───────────────────────────────────────────────────────────────────┤
 │ Mach.              Empty   Payload ---Tire--- Corrections   O&O Cost│
 │ Code Model Ident.  Weight  KILO    Size  Type Speed Alt.    $/Hr    │
 │ C205 773B  TKD-042 39,295  45,359 21.00-35 E3  1.00  1.00   95.00   │
 │       Engine  FWHP  Transmission                                    │
 │       3412DI  650 7 SPD AUTO                                        │
 ├───────────────────────────────────────────────────────────────────┤
 │ Course Short-1              Density    2,000 KILO/BCM               │
 │ Desc.  Steep 1275 m route   Quantity   100,000                      │
 │                                                                     │
 │           LOAD TIME =    3.20        LOADER = 988B-5.5              │
 │           HAUL TIME =    5.07        PAYLOAD =    22.7 BCM          │
 │           DUMP TIME =    0.70                                        │
 │           MANU. TIME =   0.10    PROD./60 MIN.HR. =   113.5 BCM     │
 │           RETURN TIME =  2.92                                        │
 │           CYCLE TIME =  11.99     OPER. MIN/HOUR =    60 MIN        │
 │           TRIPS/HOUR =   5.00    PROD./60 MIN.HR. =   113.5 BCM     │
 │           CYCLE DIST. = 2,550                                        │
 │           AVG.SPEED =   12.8     OPER. HR. REQ'D. =    881          │
 │ AVG.SPEED 60 MIN =      12.8     COST PER BCM    =    0.837         │
 ├───────────────────────────────────────────────────────────────────┤
 │          Press any key - Next      PgUp - Previous                  │
 └───────────────────────────────────────────────────────────────────┘
```

FIGURE 15-13
VEHSIM production and cost detail report for the Cat 773B truck on the short road.

```
┌─────────────────────────────────────────────────────────────────┐
│              Travel Time and Production - Haul Road               │
├─────────────────────────────────────────────────────────────────┤
│  FOR:Ch_11_EG              BY:BCP              DATE:07-20-92       │
├─────────────────────────────────────────────────────────────────┤
│  Course Short-1                Density  2,000 █KILO/BCM█           │
│  Desc.  Steep 1275 m route     Quantity   100,000                 │
├─────────────────────────────────────────────────────────────────┤
│  RUN NO.   MODEL              █HAUL ROAD█                          │
│     3      773B            INITIAL VEHICLE SPEED         2 █KPH█   │
│  HAUL                                                             │
│  SEG    DIST   ROLL   GRADE    VEL     MAX SS    TOP    LAST  ACCUM│
│  NO.   █METERS█ RES     %     LIMIT     VEL      VEL    VEL   TIME │
│                               █KPH█    █KPH█    █KPH█  █KPH█  MIN  │
│   1     200   7.00   20.00    20.00     5.37     5.37   5.37  2.24 │
│   2     120   7.00    0.00    70.00    22.87    22.80  22.80  2.64 │
│   3     100   7.00  -20.00    40.00    61.88    40.00  40.00  2.80 │
│   4     150   7.00    0.00    70.00    22.87    40.00  28.84  3.07 │
│   5     100   7.00    8.00    60.00    11.68    28.84  11.68  3.48 │
│   6     100   7.00   -8.00    50.00    61.88    40.60  40.60  3.69 │
│   7     145   7.00    0.00    70.00    22.87    40.60  29.31  3.94 │
│   8      60   7.00   10.00    50.00     9.14    29.31  10.29  4.18 │
│   9      60   7.00  -10.00    40.00    61.88    32.72  20.00  4.33 │
│  10     240   7.00   -8.30    20.00    61.88    20.00   2.00  5.07 │
├─────────────────────────────────────────────────────────────────┤
│          █↓↑-Scroll   Any key-Next   PgUp-Previous█               │
└─────────────────────────────────────────────────────────────────┘
```

FIGURE 15-14
VEHSIM production and travel time report for the Cat 773B truck over each segment of the short haul road.

CONCRETE PLACING EXAMPLE USING MICROCYCLONE ON A PC

This example involves a simulation model of a concrete column-placing operation using the Cyclone methodology. The operation consists of concrete delivery by 8-m^3 ready-mix trucks, hoisting using a crane and 1-m^3 buckets, and placing and vibration by a labor crew, for a total of 24 columns on the second floor of a building. A sketch of the operation appears in Fig. 15-15. Each column requires 2 m^3 of concrete, for a total of 48 m^3, or 6 truckloads. Cycle time components include 1.0 minute to load and hoist a concrete bucket, 2.0 minutes to place and vibrate a bucket load, 0.5 minutes to return the bucket to be loaded, and, after every two bucket loads to fill a column, 3.0 minutes to move the placing setup to the next column.

We will use the simulation model to identify the physical input resource that most constrains production in this case, to identify the physical input resource that least constrains production, and to determine how long it takes to pour the 24 columns. The model's network logic also allows the next truckload of concrete to move up and start filling the bucket while the crew is moving to the next column following that column which used the last cubic meter from the previous truck and computes the production (in columns per hour) we would expect to get from the system.

FIGURE 15-15
Placing concrete in column forms.

Model Definition

In Fig. 15-16 is presented the model structure using Cyclone's standard notation. Activities 4, 6, and 10 are COMBI nodes (i.e., using **AND** input logic), activity 8 is a NORMAL activity, QUEUES are at nodes 2, 3, 5, 7, 9, and 11, CONSOL-IDATE FUNCTIONS are at nodes 12 and 13, and a COUNTER FUNCTION is at node 14. Note that both a crane (at 3) and a concrete truck (at 2) are required as inputs before activity 4 can take place; a bucket of concrete (at 5), a crew (at 7), and an available column (at 11) are required as inputs before activity 6 can begin; and a crew (at 7) must be available and a column must be poured (at 9) before starting activity 10. Also observe that the crew at Queue 7 is shared between two activities (6 and 10). GENERATE 8 at Queue 2 indicates that the arrival of one truck makes available 8 cubic meters of concrete. It is balanced by a CONSOL-IDATE 8 at function node 12, meaning that 8 cubic meters must be placed from the current truck before another truck moves up to unload. Similarly, GENERATE 2 at Queue 11 means that the availability of a new column form makes room for two buckets of concrete to be placed. It is balanced by a CONSOLIDATE 2 at function node 13, which indicates that two buckets must be placed in a column to make it complete. Function node 14 counts completed columns.

Running MicroCyclone

One runs MicroCyclone for the most part using a hierarchy of function-key menus such as the processing menu and the report selection menu shown in Fig. 15-17. F1, F2, and so forth refer to the function keys on a typical IBM PC keyboard. As

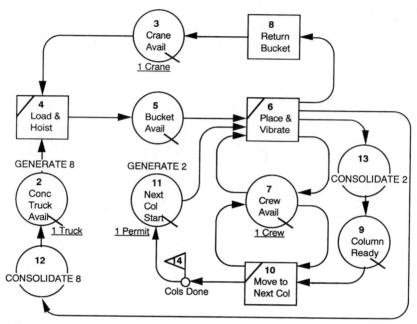

FIGURE 15-16
Cyclone COLPOUR model.

with most application programs, one can create new data files, edit existing ones, choose from a number of processing options, and produce a variety of reports. Both text and graphical options are available for running a model and presenting reports.

We can create or edit a new simulation data input file internally using a very simplified text editor embedded in MicroCyclone, or externally using a more powerful editor such as the standard EDIT utility that comes with DOS. The resulting file for this particular concrete-column-placing example appears in Fig. 15-18. Note that it opens with a **NAME** record that puts the model name in single quotes, and it specifies the upper bound on the run length as the lesser of the maximum time in minutes (here 480) or the number of cycles measured at the Counter Function (24 in this case, counted at node 14).

The **NETWORK INPUT** section defines each node in the model, describes the logical relationships, and specifies associated parameters such as the duration data sets that go with Combi and Normal activities, and Generate and Consolidate amounts (at Queues 2 and 11, and Functions 12 and 13, respectively).

The **RESOURCE INPUT** section specifies the initial configuration of resources and their variable costs (fixed costs can also be specified in MicroCyclone). For example, the first line says to initialize one truck at Queue node 2 with a variable cost of $60 per hour.

In the **DURATION INPUT** section, one specifies parameters for the duration data sets that are associated with Combi and Normal activities. For clarity,

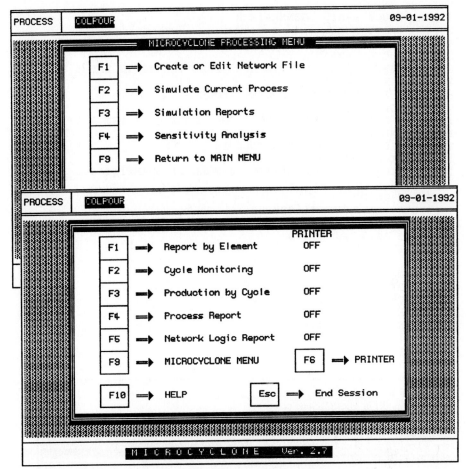

FIGURE 15-17
An example of MicroCyclone menus.

the set numbers were chosen here to be the same as their corresponding activity numbers (4, 6, 8, and 10), though this is not required. Indeed, one data set can be used with multiple activities. MicroCyclone offers several alternative statistical distributions, as was shown in Fig. 15-6. Only constant (in sets 4 and 8, where DET = deterministic) and normal (in sets 6 and 10) are used in the COLPOUR model. In set 10, the parameters show that Combi 10 has a mean duration of 3.0 minutes and a variance of 1.6.

The data file ends with the key word **ENDDATA**.

One can run MicroCyclone using either a text or graphics mode. Text mode produces an event-by-event record of what happened at each stage of model execution, with the option to print the complete history. The graphics mode produces a one-screen summary diagram while the model runs. An example taken midway through the execution of COLPOUR appears in Fig. 15-19.

```
NAME 'COLUMN_POUR' LENGTH 480 CYCLE 24

NETWORK INPUT
2 QUE 'TRUCK AVAIL' GENERATE 8
3 QUE 'CRANE AVAIL'
4 COMBI SET 4 'LOAD & HOIST' FOLL 5 PRE 2 3
5 QUE 'BUCKET AVAIL'
6 COMBI SET 6 'PLACE & VIBRATE' FOLL 7 8 12 13 PRE 5 7 11
7 QUE 'CREW AVAIL'
8 NORMAL SET 8 'RETURN BUCKET' FOLL 3
9 QUE 'COLUMN READY'
10 COMBI SET 10 'MOVE TO NEXT COL' FOLL 7 14 PRE 7 9
11 QUE 'NEXT COL START' GENERATE 2
12 FUNCTION CONSOLIDATE 8 FOLL 13
13 FUNCTION CONSOLIDATE 2 FOLL 9
14 FUNCTION COUNTER QUANTITY 10 FOLL 11 QUANTITY 1.0

RESOURCE INPUT
1 'TRUCK' AT 2 VAR 60.00
1 'CRANE' AT 3 VAR 100.00
1 'CREW' AT 7 VAR 90.00
1 'PERMIT' AT 11

DURATION INPUT
SET 4 DET 1.0
SET 6 NORMAL 2.0 1.0
SET 8 DET 0.5
SET 10 NORMAL 3.0 1.6

ENDDATA
```

FIGURE 15-18
COLPOUR model in MicroCyclone notation.

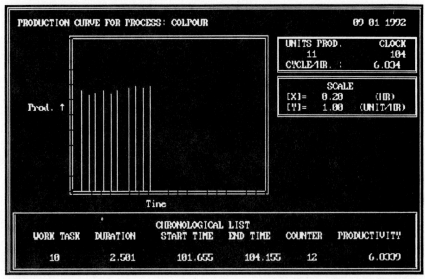

FIGURE 15-19
Graphical MicroCyclone execution display.

518

Executing the model produces vast amounts of data that can be summarized in a variety of summary and detail reports in both graphical and text forms. In Fig. 15-20 are three reports on the behavior of the activities, queues, and functions, respectively. These reports not only summarize what happened in each activity but also give an idea of how well balanced the model is and help identify logic and resource bottlenecks.

From the Queue report it appears that the truck least constrains production, whereas the crane is the main constraint. Though not shown here, the production report said that it would take about $3\frac{1}{2}$ hours to place concrete in the 24 columns.

TYPE	LABEL	DESCRIPTION	STATISTICS				
			COUNT	MEAN DUR	AR.TIME	AVG.NUM	% BUSY
COMBI	4	LOAD & HOIST	49	1.00	4.14	0.24	23.9
COMBI	6	PLACE & VIBRATE	48	2.07	4.20	0.48	48.3
NORMAL	8	RETURN BUCKET	48	0.50	4.21	0.12	11.7
COMBI	10	MOVE TO NEXT COL	24	2.87	8.55	0.34	33.6

TYPE	LABEL	DESCRIPTION	STATISTICS			
			AVG.WAIT	AVG.UNIT	UNITS END	% OCCUPIED
QUE-GEN	2	TRUCK AVAIL	13.81	3.8	7	91.5
QUE	3	CRANE AVAIL	0.00	0.0	0	0.0
QUE	5	BUCKET AVAIL	0.67	0.2	1	16.1
QUE	7	CREW AVAIL	0.51	0.2	1	18.1
QUE	9	COLUMN READY	0.00	0.0	0	0.0
QUE-GEN	11	NEXT COL START	1.73	0.4	2	41.7

TYPE	LABEL	DESCRIPTION	STATISTICS		
			COUNT	BETWEEN	FIRST
FUN-CON	12		48	4.20	2.55
FUN-CON	13		48	4.20	2.55
FUN	14		24	8.55	9.01

(End of Report !!)

Command	Turn_pg	Back_pg	Element_MENU	Report_MENU

FIGURE 15-20
MicroCyclone activity reports.

Other reports, not shown here, explain the logical relationships between model components, detail each step of model execution, and summarize productivity and costs at the model level. The first two can be especially useful in developing, debugging, and tuning model logic, and the productivity and cost summaries are particularly helpful in trying to optimize the results of the model itself.

HIGHWAY PAVING EXAMPLE IMPLEMENTED IN STELLA ON A MACINTOSH

This example model describes a paving project for a 4-km highway section. The layout of the project and the present location of the equipment are shown in Fig. 15-21. This section first provides a description of the problem and then shows how to model and analyze the system using Stella on a Macintosh computer.

Description

We can describe this operation in terms of both its activities and its constraints. The key activities are as follows (mean durations in parentheses):

- Batch plant batches and mixes concrete (8 m^3, 2.0 minutes).
- Concrete is loaded into dump truck (one batch per truck every 2.0 minutes).
- Truck hauls concrete to paver/spreader (time variable with distance; average speed 48 kph). Up to 5 trucks are available.
- Truck dumps (1 minute).

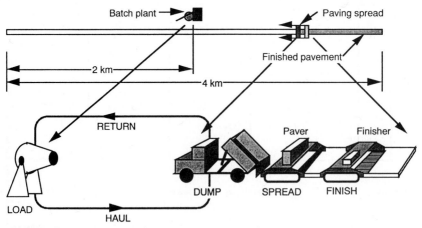

FIGURE 15-21
Highway paving operation.

- Paver/spreader receives and spreads load in layer 20-cm thick by 8 m wide (advances 5 m for an 8-m^3 truckload).
- As soon as it dumps, truck returns to batch plant (time variable with distance; average speed 60 kph).
- Finisher follows paver/spreader, finishes concrete surface, and sprays on curing compound (not implemented in initial model).

The key constraints are as follows:

- Batch and mix concrete in plant located at the halfway point along the road. The plant only has one mixer, and the operator is instructed not to dump a mixed batch unless a truck is under the plant ready to load.
- The paving operation starts at a point 2000 m east of the plant and paves westward toward the plant.

Building the Model in Stella

This section briefly indicates how Stella helped us to build the paver model. The main menu of Stella and the model's diagram under construction are shown in Fig. 15-22. The upper pull-down menus are similar to those of other Macintosh applications such as Excel and FileMaker. Significant differences are the selection among four alternative display types under the **Windows** menu (diagram, graph, table, and equations), and the **Display** and **Run** menus, which have options specific to simulation and Stella. The **File** menu has a context-sensitive **Save Diagram** instruction (or **Equation, Graph,** or **Table**) to export these items in forms intelligible to other Macintosh programs.

Most of the model building is done graphically. The four "building block" icons on the upper part of the left bar are objects that one can position anywhere in the working area of the screen. The circle is a **Converter** that can hold various types of formulas containing variables or constants that influence the behavior of the model. It can also hold a graph of one parameter (input) versus another (output). The **Connector** is the arrow-shaped symbol that logically connects different objects in the model together. The half-shaded rectangle is one of four types of holding places (Stella calls them **Stocks**; two of the most important ones are discrete [Queue] and continuous [Reservoir], and two others are stocks that behave like a Conveyor and Oven, respectively). The double-line arrow with a T (valve) and circle (control mechanism) attached is a **Flow Regulator Assembly**, which is usually just called a **Flow**. Stella further uses a **Cloud** icon at the beginning or end of a flow to represent inputs or outputs at the boundary of a model (sometimes called **sources** and **sinks** in other simulation programs). The Cloud appears when a Flow is placed but not connected at one end. An example appears in the lower right corner of the diagram.

The four icons midway down on the left bar are tools we use to manipulate the objects on a screen. The **hand** is a pointer that can move the aforementioned

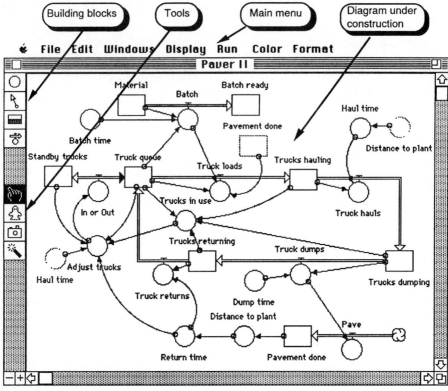

FIGURE 15-22
Entering the model on the Stella diagram screen.

graphical objects about on the screen or indicate an object to be selected so that additional information can be entered via an associated dialog box. The **ghost** is used to duplicate and half-fade one or more objects in a model so that they can appear again in a remote location on the screen without cluttering up the diagram with connector lines drawn clear across existing objects. (Note the use of duplicate ghost icons for **Distance to plant** where it inputs to **Haul time**, **Haul time** where it inputs to **Adjust trucks**, and **Pavement done** where it inputs to **Truck loads**. These avoid the need to draw the equivalent connector arrows across the diagram from the original copies of these objects.) The **camera** icon can position graphical drawings imported from a drawing program. The **dynamite stick** (some Macintosh developers get carried away in the cute use of icons) points to and deletes elements no longer needed in a model.

As objects are placed on the screen, one can select them (either in the **Edit** menu or by double-clicking) to enter additional information. For example, double-clicking the Flow icon labeled **Truck_returns** produces the data entry and editing window shown in Fig. 15-23. Here we can enter a constant or formula that regulates the behavior of the Flow. Note the availability of a small calculator

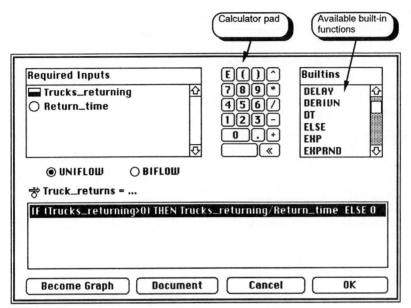

FIGURE 15-23
Entering formula for the **Trucks_returning** flow.

and a scrollable box that provides access to dozens of built-in functions (similar to those found in a spreadsheet program). One can also specify whether it has unidirectional (e.g., this one) or bidirectional (e.g., Flow labeled "In or Out") flow. If other objects have inputs to this icon, one must also show how those inputs affect its behavior. In this case, **Trucks returning** and **Return time** influence **Truck returns**. Objects that have not yet been named have the placeholder "Noname *i*" attached, and those for which needed equations or initial values have not yet been defined have a "?" in the symbol.

Clicking on a Converter or Stock icon produces a similar data input screen. Within the Graph window one can also specify range(s) for the scale(s) not only of the graph itself but also of the dynamic graphical display associated with each object icon that appears in the Diagram window when the model runs.

Description of the Application

Fig. 15-24 is a diagram of the completed paver model partway through a typical simulation run. The main load-haul-dump-return cycle is in the center of the diagram. There are a few extra components in the load and dump queue areas to accommodate some limitations of the Stella program. The material batching section is at the top, and the paving operation is at the bottom of the diagram.

The snapshot of this drawing was taken at the time when the haul and return distance was short enough that three trucks had pulled off into the **Standby trucks** stock. The **Trucks in use** converter maintains the total of all the trucks in the

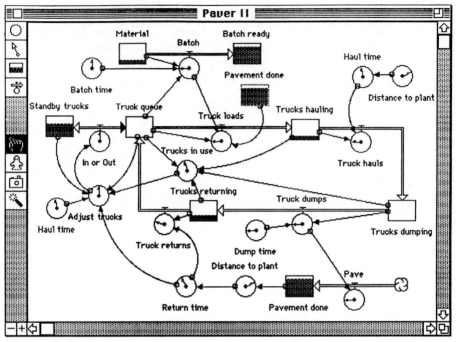

FIGURE 15-24
Diagram of the Paver model during execution.

active stocks (not in the Standby) and this total in turn is used for a calculation in **Adjust trucks** that in turn controls the two-way **In or Out** flow between the **Truck queue** and the **Standby trucks** stock. There is also an input from this adjustment from the current **Haul time** and **Return time**, and also from the **Truck queue** itself to prevent it from being pulled negative. The model adjusts **Haul time** and **Return time** as a function of the **Distance to plant**, which in turn changes as **Pavement done** progresses. Details will be given in the formulas.

When this Stella model is running, the various shaded bars on the stocks move up and down in proportion to the number of resources they contain at any given moment. The needles on the little gauges for the Flow controllers and Converters also move around the upper half of their circles in proportion to the scale of whatever parameter they represent. Any of these displays can be turned off to speed up execution, but when they are running they do help one to understand better what is going on in the model.

The equations for the paver model are listed in Fig. 15-25. It would be tedious to enter, check, and correct these equations manually in a conventional text-oriented simulator, but Stella generates these equations automatically as a by-product of building the graphical model in the diagram window (as was shown in Fig. 15-22) and entering equations for objects via their dialog boxes (as was shown in Fig. 15-23). Rather than making it difficult to get the equations right, it is hard to get them wrong.

```
Batch_ready(t) = Batch_ready(t - dt) + (Batch) * dt
        INIT Batch_ready = 0
Batch = IF ((Truck_queue≥0.8) AND (Material>0)) THEN
        PULSE(8.0,0,Batch_time)  ELSE 0
Material(t) = Material(t - dt) - (Batch) * dt
        INIT Material = 6400
Batch = IF ((Truck_queue≥0.8) AND (Material>0)) THEN
        PULSE(8.0,0,Batch_time)  ELSE 0
Pavement_done(t) = Pavement_done(t - dt) + (Pave) * dt
        INIT Pavement_done = 0
Pave = Truck_dumps*5
Standby_trucks(t) = Standby_trucks(t - dt) + (In_or_Out) * dt
        INIT Standby_trucks = 0
In_or_Out = Adjust_trucks
Trucks_dumping(t) = Trucks_dumping(t - dt) + (Truck_hauls - Truck_dumps) * dt
        INIT Trucks_dumping = 0
Truck_hauls = IF (Trucks_hauling>0) THEN
        Trucks_hauling/Haul_time ELSE 0
Truck_dumps = IF (Trucks_dumping>0) THEN MIN (Trucks_dumping/Dump_time,1)  ELSE 0
Trucks_hauling(t) = Trucks_hauling(t - dt) + (Truck_loads - Truck_hauls) * dt
        INIT Trucks_hauling = 0
Truck_loads = IF ((Truck_queue≥0.8)  AND (Pavement_done ≤ 4000))
        THEN (Batch/8)  ELSE 0
Trucks_returning(t) = Trucks_returning(t - dt) + (Truck_dumps - Truck_returns) * dt
        INIT Trucks_returning = 0
Truck_returns = IF (Trucks_returning>0) THEN Trucks_returning/Return_time ELSE 0
Truck_queue(t) = Truck_queue(t - dt) + (Truck_returns - Truck_loads - In_or_Out) * dt
        INIT Truck_queue = 5
Truck_returns = IF (Trucks_returning>0) THEN Trucks_returning/Return_time ELSE 0
Adjust_trucks = IF (Trucks_in_use > ROUND(0.4+(3+Haul_time+Return_time)/2))
        AND (Truck_queue ≥ 1)  THEN PULSE(1,0,100)  ELSE
                IF (Trucks_in_use < ROUND(-0.7+(3+Haul_time+Return_time)/2))
                AND (Standby_trucks ≥ 1) THEN PULSE(-1,0,100)  ELSE  0
Batch_time = MAX(1.4,NORMAL(2,0.25))
Distance_to_plant = -2000+Pavement_done
Dump_time = MAX(0.6,NORMAL(1,0.25))
Haul_time = MAX (ABS(Distance_to_plant)/800,0.2)
Return_time = MAX (ABS(Distance_to_plant)/1000,0.15)
Trucks_in_use = Truck_queue+Trucks_hauling+Trucks_dumping+Trucks_returning
```

FIGURE 15-25
Internal equations automatically generated for the Paver model.

Some equations of particular interest include the following:

- **Truck_loads:** two controls necessary to ensure that (1) at least one truck is in the queue and (2) the paving is not yet complete (i.e., at 4000 meters)
- **Haul** and **Return** times (The absolute value of distance is used because it starts at -2000 m from plant and goes to $+2000$ m. The assumed speeds for haul and return in meters per minute are 800 and 1000, respectively.)
- **Adjust_trucks** (Note the conditional bounding on the current desired number of trucks, where the number of trucks is based on their total cycle time divided by their load time, plus a constant [0.4 and -0.7] to keep the integer number of trucks in use consistent with the distance yet without oscillating with successive random duration times near the transition points.)

The equations dealing with *dt* and −*dt* are those where stocks are adjusted with each increment of the simulation clock to correspond to the user-specified time interval, *dt*. The INIT values are the initial quantity of resources placed in stocks before simulation begins. PULSE, NORMAL, MAX, MIN, ABS, and ROUND are standard Stella functions that behave similarly to those in spreadsheets and procedural languages. Several equations use IF...THEN...ELSE logic, which is also similar to other development programs.

The reader will likely not fully understand all of these equations, but examining them for a few moments will improve your understanding of the model's behavior. Again, note that we did not type these equations in directly; they were a by-product of building the model graphically and providing responses to inputs requested in easy-to-understand dialog boxes when needed.

Application of the Model

This section shows a few results of using the model. Fig. 15-26 is a graph generated when the **Distance to plant, Haul time, Return time,** and **Trucks in use** variables have been selected. Note especially how the number of trucks varies in integer jumps with the distance of the paving operation from the batch plant, first decreasing and then increasing. Haul and return times also vary, but continuously. The flat spots at the bottoms of the haul and return time curves correspond

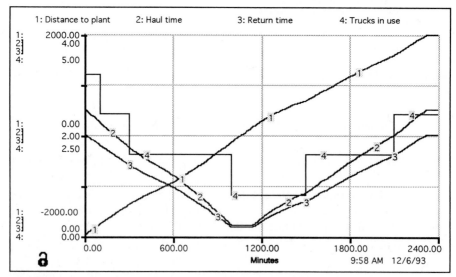

FIGURE 15-26
Graphical plot of the Paver model.

to an arbitrary constant (0.2 minutes for haul and 0.15 minutes for return) to recognize that there is some start and stop overhead even for a short haul or return. Also note how multiple scales are placed on the vertical axis. This graph grows on the screen as it is formed, left to right, while the model runs its allotted time. The interplay of the various parameters can become especially clear in this manner. The model virtually entices the user into an explorative "What if?" type of analysis, and one's understanding of system behavior grows rapidly in this way. With a large enough screen, the development of this graph can be observed in one window while the dynamic model diagram runs in another, and tabular results grow in still another. Stella allows several of these graph pages to be produced and displayed simultaneously, so all of the variables need not be squeezed up and cluttered on one page.

Table 15-4 is a small section of the tabular output for the Paver model. The user can specify which variables are to be displayed and their column order. There is a line entry for every increment of the system clock. Although it looks a bit tedious, the tabular results can be invaluable for examining the details of the model, step-by-step, and identifying exactly where events changed course and what was happening at the time they did. Stella can also export the tabular results to spreadsheet, statistics, or mathematical modeling programs for virtually unlimited post-run analyses of the data.

Stella II is an excellent example of a continuous system simulator. As shown in the Paver model, it can also adapt to some discrete system modeling conditions. More recent versions of Stella offer even more flexibility in this respect. Compared with VEHSIM and MicroCyclone, it is particularly notable for its dynamic graphics displays, both in the Diagram and the Graph windows. This real-time animation is very helpful to a person trying to build and analyze a model.

SUMMARY

Computer-based simulation packages have numerous potential applications in construction, yet they have lagged spreadsheets, databases, scheduling packages, and other types of software in acceptance and implementation in construction. In part this lag may have occurred because such programs were too complex or time-consuming for most construction professionals, but recent developments in commercial simulation packages have made them as accessible as the more widely used programs. Several well-developed packages are available on both the IBM PC-compatible and Apple Macintosh platforms.

Components of simulation modeling are similar to those of CPM scheduling. Most use graphical network diagrams to define activities and their logical interdependencies, and one can also add resources and costs to the models. However, simulation networks usually have cyclical logic, where the activities execute repeatedly for a certain time period or number of cycles. Probability distributions often represent uncertainty and variations in time estimates, logic alternatives, and other parameters in a simulation model.

TABLE 15-4
A small section of tabular output for the Paver model

Minutes	Material (cubic meters)	Truck queue	Distance to plant (meters)	Trucks in use	Pavement done (meters)
0	6400.00	5.00	−2000.00	5.00	0.00
1	6392.00	3.00	−2000.00	4.00	0.00
2	6392.00	3.00	−2000.00	4.00	0.00
3	6384.00	2.00	−1998.17	4.00	1.83
4	6384.00	2.18	−1997.08	4.00	2.92
5	6376.00	1.38	−1994.40	4.00	5.60
6	6376.00	1.75	−1991.98	4.00	8.02
7	6376.00	2.18	−1987.77	4.00	12.23
8	6376.00	2.82	−1987.61	4.00	12.39
9	6368.00	2.15	−1986.68	4.00	13.32
10	6368.00	2.41	−1986.05	4.00	13.95
⋮	⋮	⋮	⋮	⋮	⋮
997	3456.00	2.24	−162.16	2.00	1837.84
998	3456.00	2.24	−162.16	2.00	1837.84
999	3456.00	2.24	−157.16	2.00	1842.84
1000	3456.00	4.30	−152.16	2.00	1847.84
1001	3456.00	3.30	−148.88	1.00	1851.12
1002	3448.00	2.30	−148.88	1.00	1851.12
1003	3448.00	2.30	−148.88	1.00	1851.12
1004	3448.00	2.30	−148.88	1.00	1851.12
⋮	⋮	⋮	⋮	⋮	⋮
2316	24.00	1.38	1980.36	3.00	3980.36
2317	16.00	0.73	1983.10	3.00	3983.10
2318	16.00	1.18	1983.59	3.00	3983.59
2319	8.00	0.45	1986.24	3.00	3986.24
2320	8.00	0.85	1987.46	3.00	3987.46
2321	0.00	0.17	1990.98	3.00	3990.98
2322	0.00	0.69	1992.45	3.00	3992.45
⋮	⋮	⋮	⋮	⋮	⋮
2334	0.00	2.99	1999.98	3.00	3999.98
2335	0.00	2.99	1999.99	3.00	3999.99
2336	0.00	2.99	1999.99	3.00	3999.99
2337	0.00	3.00	2000.00	3.00	4000.00
2338	0.00	3.00	2000.00	3.00	4000.00
2339	0.00	3.00	2000.00	3.00	4000.00
2340	0.00	3.00	2000.00	3.00	4000.00
⋮	⋮	⋮	⋮	⋮	⋮
2399	0.00	3.00	2000.00	3.00	4000.00
Final	0.00	3.00	2000.00	3.00	4000.00

Simulation models and their associated programs can be clock-driven, where time steps are uniform during execution, or event-driven, where the clock is updated at every significant change in the state of the model. Models can also be continuous or discrete in nature.

The usual steps in simulation modeling are (1) to define the problem, its main components, and its boundaries; (2) to develop logic and flow diagrams; (3) to estimate activity durations, initial resource needs, and resource productivity; (4) to build the model using computer software; (5) to test, refine, and validate the model; (6) to use the model for planning and decision making.

Advanced features available in some packages include (1) enhancements to the standard functions; (2) improved flexibility and ease of use through better user interfaces and other means; (3) graphical techniques for building models, monitoring their execution, and displaying their results; (4) programmed automation via macros; and (5) integration with other software such as spreadsheets.

Applications in construction include, among many possibilities, (1) production systems of workers, machines, and materials; (2) management of tools and materials; (3) financial modeling at the company and project levels; (4) models of organizations and their behavior; and (5) environmental impact studies. Packages have become easy and economical enough to use that any potential operation requiring more than a few days of work or a few thousand dollars of resources might benefit from the time invested in building a model. Very complex problems can also be approached by breaking them down into a hierarchy of models.

This chapter illustrated three different applications that fall into the production systems application area. The first was an earthmoving selection case that employed Caterpillar's VEHSIM program to show how an estimator could quickly cut through the formidable calculations that would otherwise be required to check five alternative machines over two different multisegment haul routes. The second employed MicroCyclone to model a concrete column-placing operation and check for logic and resource bottlenecks. The third used Stella II to illustrate the case of a paving spread with variable haul distances and times from a central batch plant. None of these models was complex relative to the capabilities of the simulation software it used, but each provided a much more detailed analysis than one would want to do with conventional methods.

Looking to the future, we see that far more powerful modeling environments are coming into the construction industry, not the least of which is the type of *virtual reality* system that will move from design visualization into construction planning and management. Chapter 16 looks at some of these more advanced simulation modeling concepts. Meanwhile, any project can benefit from relatively easy-to-use simulation programs such as MicroCyclone and Stella, and earthmoving operations can certainly benefit from VEHSIM. These kinds of tools should join spreadsheets, databases, CPM scheduling packages, and word processors as basic tools on the construction professional's computer workbench.

REVIEW QUESTIONS

1. Compared with conventional production calculations, how do simulation models make it much easier to deal with estimates of activity durations that are uncertain and variable?

2. In defining a simulation model, should one also apply a contingency factor by shortening the number of minutes available per operating hour, for example to 40 or 45 minutes instead of 60 minutes, to allow for the uncertainties that will take place in the real-life situation? Briefly explain your answer.

3. What is a simulation modeling convention where time in the simulated process is advanced in uniform increments as the state of activities in the model is changed called?

   - Clock-driven
   - Time-based
   - Event-driven
   - Real-time

4. Which is the dominant modeling technique used in the MicroCyclone simulation system?

   - Continuous operations simulation
   - Queuing theory
   - Discrete operations simulation

5. Saying that a simulation model is *discrete* (i.e., not continuous) means what?

6. Do simulators of continuous systems provide a better approach than discrete simulation systems to solving production system problems where resource behavior in the model can be described as fluid-like flows? Briefly justify your answer.

7. In the *problem definition* stage of simulation modeling, what is the most important first step?

   - Count the number of resources observed
   - Define system boundaries
   - Determine the logical relationships between activities
   - Figure out which resources are static and which are not

8. Resources that might need to be included in a simulation model can be either physical or abstract. Give two examples of *abstract* resources and briefly indicate how they might be included in a simulation model.

9. Which of the following types of statistical models would best describe the travel times for trucks in city traffic?

   - Constant
   - Normal (i.e., bell curve)
   - Uniform
   - Skewed (e.g., lognormal)

10. Which *three* of the following types of continuous statistical distributions best model a variable activity duration where some durations tend to be much longer than the mean time?

> Normal
> Uniform
> Beta
> Constant
> Lognormal
> Histogram
> Exponential

11. When inputting the haul course for a VEHSIM simulation, you have the option to make the return road a mirror image of the haul road. If you choose this option, which one of the input parameters for haul road data is handled differently for the computations related to the return time of the machine?

12. The VEHSIM program can handle up to five different resources simultaneously. How can this help you to match a loader or pusher to the correct number of trucks or scrapers so that you can recommend the fleet that will give the most economical production? What is its main limitation in doing this type of analysis?

13. What is the most important piece of information that should be added to the VEHSIM program to make it more useful in doing economic comparisons of alternative machines for a given application?

14. In the MicroCyclone simulation modeling program, several types of graphical symbols are used to define network logic. What is the type of node called that represents an activity that consumes time but requires all of its input resources to be available before it can commence?

    What is the symbol that defines precedence relationships between activities?

    What is a place to hold idle resources called?

15. True or false: In the MicroCyclone simulation program, the COUNTER node can be used as a multiplier to convert production to another unit of measure.

16. True or false: One of MicroCyclone's output reports in Figure 15-20 focuses on NORMAL and COMBI activities. In this report, a number near 100 percent in the "% BUSY" time column will usually indicate that the activity could use more input resources to make it more productive.

17. In the Stella simulation modeling program, several types of graphical symbols are used to define network logic. What is the type of node called that represents an activity that consumes time and input resources to produce an output? What symbol defines precedence relationships between activities? What is a place to hold idle resources called?

18. In the Stella simulation program, briefly describe the technique used to convert production in one unit of measure (e.g., truckloads) to another unit of measure (e.g., cubic meters).

19. State the main advantages of Stella's dynamic graphical diagram and line graphs relative to VEHSIM and MicroCyclone.

DEVELOPMENT PROBLEMS

For development experience with a simulation program to which you have access, you might first consider the following:

1. Reimplement this chapter's concrete column-placing example in a system other than MicroCyclone.
2. Reimplement the paving example in a system other than Stella.

You could then set up simulation models for similar production systems, such as the following:

3. **Truck loading, haul, dump, and return.** Assume that the loader has a 2-cubic-yard (c.y.) bucket, cycles at 0.4 minutes per scoop, and thus loads a 10 c.y. truck in about 2 min. The trucks have a total cycle time of about 10 minutes, thus making 5 trucks (10/2) seem like a good balance in a deterministic calculation. However, extensive use is made of probabilistic sampling to throw uncertainty into material and time computations. Check out the effect on production and cost. (See the Pascal example in Chapter 8 for background information on this application.)

4. **Scraper production.** This is similar to the preceding model, except that the volume corresponding to the load time for the scraper is picked from a load-growth curve. You might try to figure out how to modify the model to feed back the *optimum* load time. (See the Lotus 1-2-3 example in Chapter 9 for background information on this application.)

5. **Concrete pumping.** In this problem you will define a small model of a concrete-placing operation. Trucks arrive and discharge into a pump, which in turn feeds concrete to workers placing concrete in mat foundation for a new building. Some of the relevant production information is given below:

   (a) The trucks arrive at approximately 15-minute intervals, normally distributed with a standard deviation of 2 minutes. Each truck holds 10 cubic yards.
   (b) The pump can handle up to 80 cubic yards per 60-minute hour.
   (c) There is a total of 600 cubic yards in the pour.

6. **Pile placing.** Make a neat sketch of a logic diagram, using the appropriate Stella or MicroCyclone symbols, for the pile-placing example shown in Fig. 15-27. Your model should be sufficiently detailed to account for the main elements in the description that follows. Also show approximate formulas for any elements that significantly affect the behavior of the model. Do not embellish the model beyond what is requested (e.g., with breakdowns, random variables, etc.).

   The operation consists of three main steps. First, an auger bores a hole into the soil approximately 6 meters deep by 0.6 meters in diameter (about 1.7 m^3 of earth is removed). This takes about 16 minutes per hole, of which 12 minutes are actual boring and 4 minutes are repositioning and setup. Second, a mobile crane fetches a 6-m pile and drops it into the hole. This takes about 4 minutes, of which 1 minute is placing the pile in the hole. After about 5 holes and piles are ready, an 8-m^3 transit-mix truck arrives and places lean concrete into the holes. This takes about 2 minutes per hole,

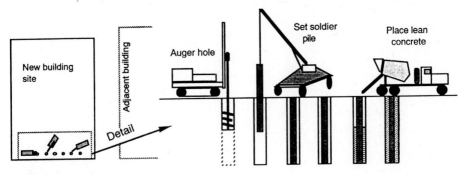

FIGURE 15-27
Sketch of pile-placing operation.

plus 1 minute to position or reposition to each hole, for a total of 15 minutes to unload the truck. The truck (assume there is only one) then takes 45 minutes to go back to the batch plant and return with another load of concrete.

SUGGESTIONS FOR FURTHER READING AND EXPLORATION

Halpin, Daniel W., and Leland S. Riggs. *Planning and Analysis of Construction Operations*. New York: John Wiley & Sons, 1992. This book is well-written and provides an excellent introduction to building simulation models for construction applications. It also provides extensive information about using the MicroCyclone program.

Halpin, Daniel W. *MicroCyclone User's Manual*. West Lafayette, Ind.: Learning Systems, Inc., 1990.

Ioannou, Photios G. *UM-Cyclone Discrete Event Simulation System Reference Manual, User's Guide*. Department of Civil Engineering, University of Michigan, Ann Arbor, Mich. 48109, September 1989.

Stella II Tutorial and Technical Documentation. Hanover, N.H.: High Performance Systems, 1992.

VEHSIM User's Manual. Peoria, Ill.: Caterpillar Inc., 1987.

PART
VI

FUTURE
TRENDS

Any sufficiently advanced technology is indistinguishable from magic.

Arthur C. Clarke*

W hat can be more interesting than to think about the future? What can be more fraught with peril and prone to error?

Part VI will take a quick look ahead with only a single chapter. Even its brief glimpse should stimulate one's imagination as to what might be possible. Students of construction engineering and management who are reading this book today will see much of that future unfold, and indeed many will be among those who create and shape it. This small look ahead will most certainly pale when compared retrospectively to what they will actually see and do.

---

*The Lost Worlds of 2001, 1972.

CHAPTER

16

BUILDING FOR TOMORROW ON TODAY'S TECHNOLOGY AND APPLICATIONS

I never think about the future. It comes soon enough.

Albert Einstein*

Computer applications began evolving in larger construction companies in the late 1950s, little more than a decade after the first electronic computer was built, and almost as soon as commercially supported procedural-level programming languages became available. Breakthrough computer-based project management techniques, such as the critical path method (CPM) for scheduling, were developed at about the same time and also moved quickly into the construction industry. It seemed like a promising start.

The next two decades saw steady progress, particularly in home office applications such as accounting and finance, though the impact did not match the revolutions that took place in banking, airlines, and telecommunications. Project-oriented applications such as CPM scheduling, estimating, and simulation continued to evolve, with enhancements such as graphical plotter output, but well into the 1980s their overall success remained far below early expectations. Initial

This chapter is based partially upon a lecture that I presented at the 25th Anniversary Symposium of the Institut für Maschinenwesen im Baubetrieb at the Universität Karlsruhe, Germany, in June 1992, and the Kudroff Memorial Lecture that I gave at The Pennsylvania State University on November 2, 1992.
*Interview, 1930. (Quoted in *The Macmillan Dictionary of Quotations*.)

problems were related to costs of and access to computer hardware, complexity of the software, and difficulties in interpreting output, as well as gaps in training for site-level people who were almost totally unfamiliar with computers.

Now that we are in construction's fourth decade of computer applications, it is clear that things have been changing more rapidly. Why? In the 1980s microcomputers became widespread and their cost rapidly decreased; powerful but easy-to-use application-development software, such as spreadsheets and databases, was introduced and greatly increased the number of people who could create applications and do so economically; and thus the overall level of user experience and sophistication with computers grew enormously. We now see microcomputers at work even on smaller job sites, and the people using them effectively know how to apply software tools for estimating, scheduling, and cost control. Even CAD is becoming more common in site construction offices, both for interpreting and modifying construction drawings and for designing temporary facilities. Some applications, such as simulation, have yet to become widely adopted, but their acceptance is probably just a matter of time. Apparently computers are at last fulfilling their promises to the construction industry, and much of what was confidently predicted years ago is finally coming to pass, albeit a decade or more behind the early optimistic forecasts.

What can be more interesting now than to think about the future? But given the imperfect record of making projections about computer applications in this industry, what can be more difficult to foresee? This chapter will attempt some predictions, first for the near term and then for the more distant future.

TRENDS IN THE NEAR FUTURE

Some trends are fairly easy to predict because their prototypes are here now and have already overcome the main technological barriers to implementation. Although it is more difficult to predict their diffusion in the face of human and organizational barriers, the high interest levels that some prototypes have generated should help them on that front as well. The next three sections examine three cases: the integration of existing applications, 3D electronic modeling and simulation, and multimedia-based systems.

INTEGRATED APPLICATIONS

There were efforts in the 1970s and 1980s to develop mainframe and minicomputer databases to integrate construction applications, but—perhaps because they remained in arcane mainframe software environments—they met with limited success. For the most part, construction applications have been implemented in relatively self-contained packages, and there are some major compatibility barriers to exchanging even file-level data among many of them. But these barriers are breaking down for many reasons. First, developers of system and application-development software are making it easier for designers of application software to interface to other applications. Capabilities such as Apple's *publish* and *subscribe* in the more recent versions of its Macintosh system are a good example, and they have already been exploited by spreadsheets, word processors, and graphics programs. Second, commercial developers of related packages, such as estimating, scheduling, cost control, and CAD, have been working together to provide better interfaces for moving data between their packages, and users have already begun exploiting such capabilities. Thus even now it is possible to take a set of standard package applications and link them together in a manner that begins to resemble an integrated construction information system.

Figure 16-1 has been adapted from a diagram prepared by Kelar Corporation, a marketer and integrator specializing in computer applications for architecture, engineering, and construction *(AEC)*. Names of commercial software packages marketed by Kelar can be attached to almost all of the boxes in Fig. 16-1, though the interface success ranges from smooth to awkward. Nevertheless, each arrow on the diagram shows that some degree of interfacing has been accomplished, and several of these packages make the sum of the components in the diagram even more useful than the pieces taken separately.

At the top of the diagram, the functions of planning, scheduling, and performance measurement can be handled by some of today's more powerful network-based project management packages. The interface to CAD developed by Kelar serves mainly to use a CAD package to plot a network, although others have experimented with enabling project planning software to interrogate the CAD database to extract objects used in automated planning (Froese and Paulson, 1994; Ito, Ueno, Levitt, and Darwiche, 1989). The interface from CAD to estimating facilitates the calculation of lengths, areas, and volumes (e.g., earthwork and concrete) and transfers a bill of materials for items that can be priced directly (e.g., plumbing

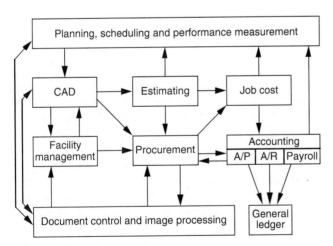

FIGURE 16-1
Integration of application packages. (Adapted from a diagram prepared by Kelar Corporation, 5456 McConnell Avenue, Los Angeles, Calif. 90066.)

and light fixtures) and entered into the procurement system. CAD, particularly if it evolves into an accurate as-built model during construction, is increasingly being used by owners and operators for facility management. The estimate has several useful outputs to other systems. Its crew production estimates send resource requirements and time durations to the scheduling system; its categorized estimate of costs forms the basis of a budget for the job-cost system; and its quantities and materials feed into the procurement system. Procurement information flows into both the job-cost system and the accounts payable component of the accounting system; both of these are well-established links that are available in most of the high-quality accounting packages. Job cost provides input data to the accounting system, particularly for the payroll and accounts receivable billing modules. All of the accounting modules, in turn, feed into the general ledger. The diagram also shows interfaces to document-control and image-processing packages. Not shown, but readily added, could be construction equipment management.

In just a few years, we will probably look back on patchwork integration such as that shown in Fig. 16-1 as being fairly inefficient. The most important things now are recognition of the importance of such integration, its acceptance in industry, and the demands that acceptance will create for increasing sophistication. The cooperation that developers have established to build the one-to-one links shown in the figure will probably lead to tighter integration in the future, particularly if the present trends continue toward mergers and buyouts in the software industry. More advanced software technologies, such as object-oriented databases and multitasking operating systems for microcomputers, will also facilitate better integration. In any case, the trend toward the better integration of construction applications should lead to the use of some excellent overall application environments in construction companies by the end of this decade.

3D ELECTRONIC MODELING AND SIMULATION

Computer-aided design software is finding its way onto construction project sites, but already designers in the home office are making regular use of three-dimensional (3D) walk-through simulations. These graphical simulations provide realistic views into designed space long before it is constructed. Problems ranging from component interferences to subtle problems with lighting can be caught well before their real-world costs and resources are committed.

Some recent products have made these 3D simulations useful for the construction site. Examples are Bechtel's Construction CAE* and Jacobus Technology's Simulation Session Manager.† By adding a time base to the simulation, engineers and managers can see how the components are installed over time,

---

*Bechtel Group, Inc., 50 Beale Street, San Francisco, Calif. 94105.
†Jacobus Technology, Inc., 7901 Beechcraft Ave., Gaithersburg, Md. 20879.

experiment with different sequences, and play the result like a video animation to communicate the plan clearly to field supervisors and crew members. Some systems also add scaled 3D images of major pieces of construction equipment, such as cranes and trucks, to help users explore with the computer actual materials-handling methods. These systems also interface to CAD software such as Intergraph's IGDS and McDonnell Automation's GDS; project management software such as Primavera and Panorama; to relational database software such as Informix and Oracle; and to expert systems such as KES-II. By-products of these simulations can be short-interval schedules, resource requirements, and materials logistics details that can be sent to scheduling and database software.

Figure 16-2 is a sketch based on a screen image from Bechtel's Construction CAE taken at the time an installation sequence is being developed. The actual screen contained a complex 3D solid model shown in colors. The animated picture appears in one or more views in the main part of the screen; schedule information appears at the top; and control menus and supporting details about objects in other windows are displayed on the right side. Note that this animation is far more than a sequenced cartoon. The three-dimensional space is fully modeled to

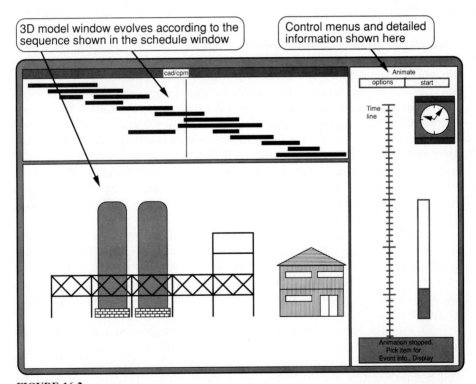

FIGURE 16-2
Typical screen configuration of Construction CAE. (Adapted from Bechtel Corp., 50 Beale Street, San Francisco, Calif. 94105.)

high precision, each object has an identity and can be linked to attributes in a database (materials specifications, costs, etc.), and the system behaves according to scalable real-time durations.

Basic elements in this system include (1) a preprocessor and spooler to transfer information from a 3D CAD program to Construction CAE; (2) a library of graphical models of various kinds of materials-handling equipment; (3) the construction simulator; (4) CPM scheduling calculations and interfaces; and (5) a scope-control database. The system has the following capabilities: (1) to model construction openings, (2) to group elements into modules, (3) to group related work items into bundles, (4) to include temporary facilities (e.g., scaffolding) in the model for a defined period of time, (5) to display and manipulate materials-handling equipment in the same 3D graphical environment and time scale as the construction facility on which it is working, (6) to perform online interference checking between constructed objects or between processes and objects, (7) to animate work operations, (8) to use color to represent the status (e.g., complete or underway) of objects in the model, (9) to alter the visibility (hide, fade, high-light) of subsystems (e.g., piping) in the model, and (10) to provide control over windows displayed on the screen (e.g., simultaneously show window plan, eleva-tion, section, or perspective views). The outputs provided by Construction CAE include (1) graphical CPM schedules at various levels of detail, (2) annotated graphical views (e.g., to explain a change request to a fabricator), (3) tabular schedule and resource reports, and (4) animated videos (e.g., to show a field crew today's work objectives). Although this is a somewhat complex system, it still can be used directly by engineering professionals and provides them with a powerful tool for design, planning, analysis, and management.

To illustrate an application, Fig. 16-3 is a representation of the software con-figuration that was used in Jacobus' Simulation Session Manager (which evolved from Bechtel's Construction CAE) implementation on Boston's massive and highly complex Central Artery/Tunnel Project, which will be under construction for sev-eral more years. It provides an integration between Bechtel's Walkthru, Primavera's scheduling package, Oracle's database, and the simulation system to help planners develop the best methods to design and sequence their construction operations in a busy urban environment.

Construction CAE and Simulation Session Manager are but a taste of things to come in this area. As workstations become faster and virtual reality devices become practical, more complex, continuous, and realistic images will look and behave almost like the construction process itself. Virtual reality will have impor-tant applications in operations planning, worker training, and facility management for future construction projects.

In a recent lecture, Steven Fenves (1991) extrapolated technologies and concepts like those presented in this section to describe the scenario of a Virtual Master Builder. This seems an ironic but fitting juxtaposition of the apex of medieval architecture and the technological threshold we are about to cross as we near the end of this millennium.

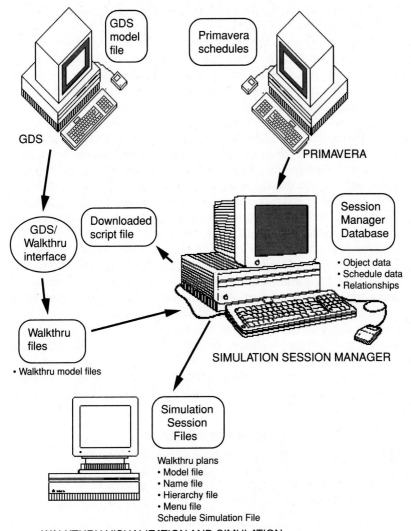

WALKTHRU VISUALIZATION AND SIMULATION

FIGURE 16-3
Architecture and interfaces in Simulation Session Manager. (Adapted from figure by
Jacobus Technology, Inc., 7901 Beechcraft Avenue, Gaithersburg, Md. 20879.)

MULTIMEDIA-BASED SYSTEMS

Multimedia computing systems are moving rapidly into training functions in
many industries. Their economics and teaching effectiveness are already considered superior to traditional training methods, even though they are still in a fairly
early stage of development and their costs will continue to fall. As the technology

becomes a standard part of future workstations, it will become common in construction (El-Bibany and Paulson, 1991).

A multimedia computer system such as that sketched in Fig. 16-4 includes a powerful microcomputer, a large color graphics monitor, a video source such as an optical disc player or CD-ROM, and a video interface that will display not only computer text and graphics but also a window containing the TV pictures coming from the video source. A speaker, mouse, and keyboard are also standard.

There are three windows in the screen in Fig. 16-4. On the upper right is the main control program window. It provides control menus and graphical icons that enable the user to navigate through and interact with the application software, display text, and graphics that contain the content of a course module, and also display controls for the videodisc and provide access to other software. The second window (lower right) displays video stills and motion segments from material stored on the videodisc. When segments are running, concurrent audio may come from the speaker. This supplements the application content with actual pictures of objects, processes, or other images pertinent to the course module. The third window is the display of another application called from within the control module. For example, the control module might start a simulation program to run a dynamic network model of a complex construction process, or a spreadsheet model to enable an engineer to analyze an engineering problem.

Attractive as this technology may seem, until recently it has required expensive hardware and software additions to conventional microcomputers, and application development has been complicated by the demands of orchestrating so many new input sources into an integrated training module. Several packages are now available that can help training developers work more directly and more

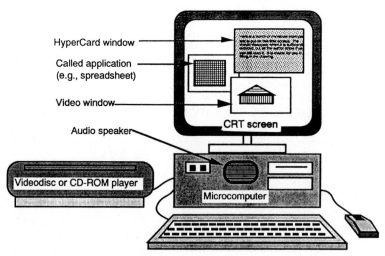

FIGURE 16-4
Multimedia computer configuration.

easily on their application, including Macromedia's Authorware and AimTech's IconAuthor.

Some of the most productive applications of multimedia technology could go well beyond training. In project-oriented industries like construction, where the production process and products are large in scale and require unique solutions to field problems, passing on the methods and techniques developed on one project to people who could use them on other projects is difficult; the industry is thus criticized for its lack of *corporate memory*. Multimedia technology could record innovative and productive methods to build up a corporate memory that is accessible to the planners and managers of future projects. There are major challenges, however, to integrating such a powerful source of information into other project software systems such as CAD and project databases.

ON THE HORIZON

The past decade saw several new computer applications move onto the construction scene, but it is not yet clear where they are headed, or how soon they will begin having a major influence on this industry. Examples to be considered in the following three sections include advanced artificial intelligence applications, automation and robotics, and fully integrated construction project planning, design, management, and production control systems.

ADVANCED ARTIFICIAL INTELLIGENCE APPLICATIONS IN CONSTRUCTION

The main reason for trying to apply AI and, specifically, expert systems to construction is to deal with the qualitative and judgment-based types of problems that are prevalent in this industry. Perhaps the most valuable career asset for a construction professional is not mathematical or scientific skill of the type taught in engineering schools but rather is experience and the good judgment to use that experience to solve new problems. One major objective of construction artificial intelligence researchers is to capture that type of experience in computer programs so that other construction engineers and managers can access it and apply it, perhaps even after the experts who provided the knowledge are no longer available. Such programs also provide a means to integrate and validate the knowledge and experience of many experts and thus provide a means for accumulating and improving a body of knowledge over time.

Implicit in this type of computer application is the need to deal with uncertainty and ambiguity in the information needed to design, build, and operate a constructed facility. For example, design starts with only general conceptual knowledge of what a project will look like when it is completed. Yet as early design decisions evolve into commitments for configurations, materials, and systems, they can adversely affect construction costs and schedules, and compromise the efficiency and effectiveness of facility operation. AI techniques can capture knowledge of construction methods that an operation needs and make this available at the design stage. For example, if a designer has a choice of configurations for a concrete structure, an expert system could provide advice as to which would be most economical to build. Perhaps more than any other area of computer technology, AI-based applications have real potential for becoming powerful mind extendors for human professionals, particularly to enhance human judgment, intuition, pattern recognition, and integration abilities in a manner similar to the way earlier generations of computer applications amplified limited human abilities for calculations and for the organization, storage, and retrieval of information.

A promising area for future applications of AI technology is in planning, monitoring, and controlling the construction process itself. Already there have been some good attempts to build construction planners of various types; and other applications have analyzed construction contracts, prepared construction cost estimates, and selected construction methods. Other important applications will be in helping people to coordinate the vast amount of documentation that is generated in a large construction project and to negotiate the long and complex permitting procedures that are now required for most projects. Perhaps the most interesting applications will occur when AI techniques supplement or replace the procedural programming that is now used for automated machines and robots. We return to this topic at the end of the next section.

CONSTRUCTION AUTOMATION AND ROBOTICS*

Although construction technology evolved slowly for many decades until the 1980s, the last few years mark the beginning of a renaissance in this field, and there is reason for optimism about the future. Just as microcomputers are now widely accepted for engineering and administrative applications in construction, in the future they and other electronic technologies will be used more and more to automate real-time data collection from and process control of the physical construction operations themselves.

This section briefly examines the unique characteristics that make construction especially challenging for computer-based automation and robotics in the field. Next, it describes the general nature of different applications and reviews the progress and obstacles faced by the first generation of construction automation and *robotics*. It also mentions the greatly increased potential for automation that is evident in automated building systems that are now being implemented. Finally, looking to the future, we focus especially on the knowledge and software characteristics that will give robots the higher levels of intelligence and capability they will need to become more widely useful in construction.

---

*This section is based on Paulson (1985) and Paulson, Babar, Chua, and Froese (1989).

Challenges for Construction Robotics

The challenges of developing robots for construction job sites are much greater than those for most factories. For instance, the products of construction are much more complex and vastly larger. Furthermore, in contrast to the repetitive products that flow down production lines, the design of the construction product and the process to build it are usually uniquely adapted in each case. Whereas the manufacturing process is largely steady-state once production has started, that in construction is ever-changing. The physical environment of construction is often much more hostile to machines as well as people, so machine design must account for extremes of weather, dust, and unexpected forces.

In contemplating the future of automation and robotics in construction, we must define and classify needs for and barriers to implementation of automated data acquisition, process control, and robotics in several areas. Categories could include large versus small projects, labor-intensive versus capital-intensive operations, industry sectors (buildings, civil works, process plants, housing), phases and technologies within projects (site work, foundations, structural, piping, electrical, etc.), and types of firms (design-construction, general contractor, specialty contractor, etc.). Where will industry's progress be slowed if new automation and robotics technologies are not implemented? Where will costs and schedules be unnecessarily high and quality low?

It will also be important to consider potential industry barriers. In a field like automated process control and robotics, there are certainly some very real social and economic problems as well as technical obstacles that must be identified and overcome or accommodated if research efforts are to succeed eventually in development and implementation. In brief, the challenges to technological advances are many in construction and relate as much to institutional problems—like craft, company, and process fragmentation; risk and liability; codes and standards—as they do to purely technological or economic concerns.

In spite of the obstacles, there is reason to believe that computer automation and robotics technologies will meet the challenges. First, robots by nature are programmable and thus should be adaptable to changing environments, rather than always be restricted to repeatable, very structured tasks by the present state of the art. Second, most construction jobs do indeed require more judgment than those in factories, but is this not then an area where new developments in artificial intelligence and expert systems might best emerge? Finally, the environment is certainly a factor. Construction's robots must handle not only a few days of gentle rain but also a winter on Alaska's North Slope, platform operations in the North Sea, work under 1000 feet of water off the California coast, 50°C and sandstorms in the Middle East, or framing of steel on a cold winter day 100 stories up in Chicago. In many of these situations, robots could provide welcome relief from the safety and health problems construction workers face every day. Construction machinery is manufactured to cope with such environments, so surely construction robots could be also.

Construction Automation and Robotics in Practice

Computer-based automation and robotics technologies have already made some small but noteworthy inroads on construction job sites. Although only a few have been immediate economic successes, in almost all cases the developers and experimenters have been accumulating valuable experience and knowledge that leave them well-poised to capitalize on the inevitable new advances in the future. It is thus worth taking a look at what has been accomplished to date.

The following paragraphs review eight areas where automated data acquisition, process control, and robotics have been applied: computer-based data acquisition systems for field engineering operations, automated monitoring of construction quality control, monitoring of field parameters related to safety, automated monitoring of production rates and quantities for field operations, on-site automated process control for fixed plants, partial or full automation of mobile equipment, fixed-base or dimensionally constrained *manipulators,* and mobile robots and androids.

COMPUTER-BASED DATA ACQUISITION FOR FIELD OPERATIONS. Real-time computer systems can assist in monitoring field engineering parameters during construction (e.g., soil movement toward a foundation or tunnel excavation; alignment and dimensional tolerances for structural members and subgrade thicknesses; and construction live and dead loads on partially completed structures). In this area, structural and geotechnical engineers have made considerable progress, but few have integrated their analyses and design-based instrumentation into ongoing construction decision making. Some contractors, Japanese in particular, have put considerable emphasis on integration, and they routinely instrument and monitor deep foundation excavations, offshore works, tunnels, and similar structures.

AUTOMATED MONITORING OF CONSTRUCTION QUALITY CONTROL. Many quality control procedures produce after-the-fact rejection, with consequent delays, interruptions, and expensive rework. Computer-based monitoring systems could provide immediate feedback to enable remedial action to be taken while production is underway and thus minimize the consequences of defects.

An early integration of data acquisition and quality control was concrete batch plants. A more recent example is a continuous weld-quality monitor developed by the Army Construction Engineering Research Laboratory. Offshore platform projects, among others, have used various remote sensing devices to position piles and other parts of their structures. Lasers and other automated guidance technologies serve not only to improve excavation and grading quality but also to improve significantly the placement accuracy of materials and components. Other possibilities include bolt tightness, structural member alignment and deflection, and soil compaction. In general, however, this area is in its infancy in construction, and future research will further explore its potential.

AUTOMATED MONITORING OF CONSTRUCTION SAFETY. There are several construction situations in which computer-based monitoring contributes to construction safety. The aforementioned instrumentation of tunnels and deep excavations certainly has such implications. A particularly good illustration is that of construction cranes, where excess tipping moments and structural failures can result in catastrophes. It has thus become routine in Europe and also to some extent in the United States to use load cells on cables, inclinometers (angle sensors), and length indicators on booms, and pressure cells or strain gauges on outriggers to monitor the key moment and load parameters to keep a crane within safe operating limits, as shown in Fig. 16-5. A computer interface in the cab keeps the operator informed, and automatic routines sound alarms or can even limit controls to help prevent the operator from working unsafely.

AUTOMATED MONITORING OF PRODUCTION RATES AND QUANTITIES. Good field management requires monitoring of production rates and quantities, with measures of associated resource consumption and productivity. Often, such data are collected through manual measurement (surveying) techniques and through administrative procedures like labor and equipment time cards. Although the electronic revolution has brought rapid advances in surveying equipment for measuring distances, angles, and volumes, their use can still be too expensive to justify collection of detailed quantities, and the administrative procedures are notoriously vulnerable to delays and inaccurate recording. Automated monitoring can help provide good information quickly and economically.

Existing examples include mining operations that use sensors to count vehicles such as trucks to estimate production volumes and even optionally to reroute trucks to idle loaders. Some manufacturers attach automatic recorders to tunneling machines to record excavation volumes, advance rates, jack pressures, and

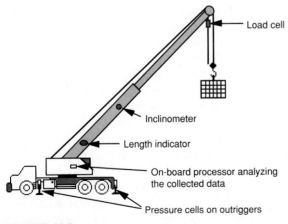

FIGURE 16-5
Sensors employed to improve crane safety.

other parameters to guide the operation. Computers monitoring the concrete batch plants mentioned before also automatically record production and keep track of their stockpiles of cement and aggregates. Factory-like production operations, such as precast concrete plants, are also seeing automated production monitoring more frequently. The potential for automation of this type is great, and with improved information should come more efficient planning and management of the field operations.

ON-SITE AUTOMATED PLANT CONTROL. The most progress to date in automation for construction has been made in control of temporary on-site shops and plants for batching concrete, carpentry, cutting and bending reinforcing steel, fabricating pipe spools, aggregate crushing and screening, and making precast concrete elements. These types of applications resemble factory automation, but with provisions for rapid and economical set-up and take down, severe environmental conditions, and a frequently changing product mix. For example, in batch plants for higher-volume, high-quality concrete production, a computer controls the selection, transport, weighing, charging, and mixing of cement, sand, aggregates, water, and admixtures for a batch that meets specified design criteria for a specific application and simultaneously handles administrative reporting for delivery, quality, and cost control. To the extent that more construction processes and components can be redesigned for prefabrication in plant-type facilities, whether on-site or off-site, more construction processes can benefit from automated process control of this type.

AUTOMATION OF MOBILE EQUIPMENT. Manufacturers and contractors are already beginning to exploit the potential for partial or full automation of mobile construction equipment, including trucks ranging from light utility vehicles to large off-highway haulers; excavating and grading equipment such as scrapers, loaders, shovels, compactors, and graders; materials hoisting equipment such as cranes and forklifts; and a wide range of specialty equipment such as paving machines, tunnel boring machines, construction railways, cableways, and pipelayers. Manufacturers are experimenting with and producing machines that include onboard microprocessors for monitoring performance, maximizing engine power and fuel economy, optimizing gear shifts, positioning loader buckets, keeping crane loads within safe tolerances, and so forth. These applications are beneficial but fairly limited relative to the real potential for partial or fully automated control of machines as whole units in the overall production process.

More dramatic is the application of automated excavation grade control using laser surveying equipment, combined with electrohydraulic feedback control systems mounted on bulldozers, motor graders, scrapers, and the like. In the 1970s, some government agencies and construction contractors helped pioneer these applications of partial automation. In applications such as grading highways, constructing large parking lots, and digging canals, these techniques have reduced costs, in some cases by over 80 percent, and have improved quality (e.g., achieving subgrade thickness tolerances of 2 percent versus 10 percent to 20 percent otherwise achieved on normal work). They also permit the substitution of lower-cost machines (e.g., small bulldozers for motor graders) and lower-skilled operators while giving quality improvements. Figure 16-6 is a schematic diagram showing the laser transmitter and the laser receiver mounted on a post connected to the blade of a motor grader.

The next step, moving to more fully programmed automation of all degrees of motion, is much more difficult. At this stage the machines, in effect, would become robots with some limited programmed intelligence for self-guidance and control over complex routes and surfaces, making decisions on what to do about obstacles and responding (if only by stopping) to unanticipated changes. Automation of independent machines needs much more research before it becomes safe and practical, but the potential is there.

SPATIALLY CONSTRAINED MANIPULATORS. Much of the publicity about robotics in manufacturing industries currently centers around multijointed robot arm and hand mechanisms that are attached either to a fixed base or to a platform such as a gantry that covers a clearly defined and limited area. With such a well-prescribed, three-dimensional frame of reference, they can be programmed for operations requiring high precision.

Although the construction environment is often loosely constrained and frequently reconfigured, it still has considerable potential for this type of automation. The processes can be redesigned to fit the tools, and the tools can evolve to handle more flexibly a wider variety of processes. Japan's Shimizu Construction Company has mounted such an arm on a mobile platform and used it to apply sprayed insulation in building construction. Other possible applications of such robots are in tunnels where operations (e.g., liner erection) are highly repetitive and fit into well-defined geometric constraints, in concrete masonry, and in painting.

One significant difference between the design of much construction machinery and that used in factories is worth mentioning here. Factory machines tend

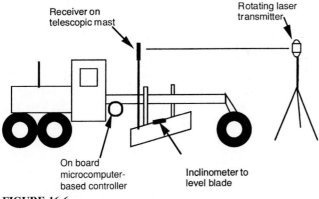

FIGURE 16-6
Laser-controlled motor grader.

to be large and massive relative to the products they work on and depend on their rigidity to maintain positional tolerances precisely. Construction machines—such as cranes, hydraulic excavators, and others that look somewhat like factory robots in their motions—have relatively high payload-to-weight ratios but as a result are fairly limber. They depend on the human operator to maintain positional compliance by making small adjustments in position as the load nears its goal. Robotics people sometimes call this end-point feedback. We should keep these rigid-versus-compliant design differences in mind when we think of moving factory technologies into field construction.

MOBILE ROBOTS AND ANDROIDS. Mobile robots and *androids* (robots that approximate the physical form and some functions of humans) include those with wheel, track-type, and walking transporters. Although such robots often contain one or more manipulator arms like their fixed-base counterparts, they are distinguished by a high degree of mobility, normally unconstrained by tracks, guide wires, or other fixed references; they have a wider variety of sensors (sonic, light, touch, temperature, etc.) to cope with changing and less predictable environments; they are often battery-powered to move without external power sources; and they typically have more onboard computer power to allow more independent programmed analyses of and responses to their environment and tasks. The term *android* is included here so that *robot* can apply to a wider range of mobile devices without regard to their physical shape or functionality.

Apart from Shimizu's platform-mounted robot arm mentioned previously, there have been few attempts at developing and applying mobile, walking-type robots and androids even in manufacturing, let alone construction. However, owing to the nature of project sites, such devices might have even greater potential in construction than in plant-based manufacturing. Prototypes include a remotely controlled, track-mounted robot that assisted in the cleanup at the Three Mile Island nuclear plant; others for assembly of space platforms and work on other planets; and Odetics, Inc.'s ODEX-I, a promising six-legged robot with a high power-to-weight ratio. The ODEX-I weighs 400 pounds, can walk at a fast pace, and can lift 1000 pounds at its center point. However, there is no real prototype as yet for a general-purpose robot that might be flexible enough to be a general utility tool on field projects. This is an attractive area for research.

In summary, field project-oriented industries like construction have taken only a few, loosely related steps toward automated data acquisition, process control, and robotics for field operations. Initially there will be a lot to learn from aerospace research and plant-based manufacturing industries, but in the long run both the challenges and the potential rewards are even greater in large-scale field operations.

Moving beyond Task-Specific Construction Robots

Given the difficult and complex environment of construction, it is remarkable that robots are already performing routine tasks on some job sites at the level just described. The first construction robots were derived by adding sensors and computer-based controls to existing construction equipment (e.g., to control the cutting edges or screeds on various types of earthmoving and paving equipment), by adapting the comparatively rigid factory-type robots to construction (e.g., for spraying fireproofing material or painting), or by developing hybrids of the two (e.g., robot arms mounted on tunnel machines). Although the sophistication of their mechanisms and sensors has often been quite high, these robots have had only rudimentary forms of programmed "intelligence," and some machines that have been called robots are really just teleoperated devices without any programmed automation at all.

Most of the construction robots developed to date are standalone devices designed to perform narrowly defined tasks without the need to communicate or cooperate with other machines. The concept of a construction *crew* does not really apply yet to construction robots. However, coordinated teams of robots quite commonly perform sequential operations on factory assembly lines, and there are some formal communication mechanisms linking them together; it was only a matter of time before similar technology also moved to construction.

A major step toward an integrated system of robots is now being undertaken by some of the largest contractors in Japan. The large Japanese general contractors, like Obayashi, Shimizu, and others, have moved well into development of systems that will substantially automate the construction of high-rise buildings; and early versions of these systems are already deployed on building projects. Basically, they consist of a jack-up frame on which to mount a variety of robots for materials handling (e.g., cranes, hoists), fabrication (welding, cutting, finishing), and inspection. The frames have all-weather enclosures to enable work to continue around the clock, at any season of the year. This framework is initially positioned at the first of a series of repetitive floors to be built. Next, the whole frame jacks itself up to the next level, and the robots build another floor. The idea is somewhat like a slip-form for constructing a concrete structure, except that a whole building, not just a concrete structure, is "extruded" from the system. This process continues until the building is done, then the automated components are removed, leaving the frame in place to become the structure for the top floor of the building. Figure 16-7 is a conceptual diagram of this system.

The use of these systems is in part motivated by an expected shortage of skilled labor in Japan, but over time they will have economic and quality advantages similar to those of an automated factory. About 90 percent of present labor requirements will be replaced by automation. Those workers who remain will probably be highly skilled technicians who can program and maintain the robots. The systems provide for substantial integration of structural, mechanical, electrical, and finishing operations that are used in the construction of a building. There are also obvious interfaces to and interactions with design. In this way they are analogous to the computer-integrated-manufacturing (CIM) systems that have revolutionized other industries.

Impressive as such automated building systems will be, there remain many challenges facing the advancement of construction automation and the development of more capable construction robots. Perhaps the most difficult is that of

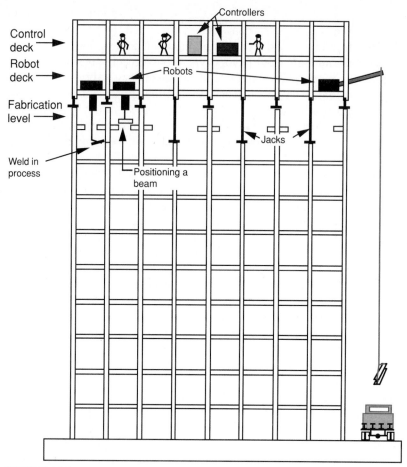

FIGURE 16-7
Automated building construction system. (Adapted from Obayashi Corp. brochure.)

developing the intelligent software to integrate future machines into the complex environment where they will work. Let us take a closer look at the environment of construction from the intellectual perspective of a future construction robot.

Toward More Intelligent Robots

Before considering what should go into the core of construction robot software, we must think about some bounds on this software. For the most part we are looking at the intelligence to support successful execution of construction tasks, and this probably will not require much in the way of, for example, aesthetic appreciation, ability to dance, or innumerable other facets that characterize human existence. Relative to the intelligence and human dimensions of a typical construction worker, the core of a construction robot's software is still a most rudimentary kind of "intelligence."

In general, what is needed is some way of modeling *within* robot agents some "understanding" of their environment, such as key characteristics of objects and other agents, in ways useful for reasoning. Among other things, researchers should seek to reduce the knowledge that needs to be encoded in machine systems a priori by enabling them to tap the vast knowledge sources in their environment when needed. This capability is termed extensibility, which some might call a simple form of learning. Automatons should be able to assemble knowledge, enlist other agents needed to perform a task, and respond dynamically to change. Robot reasoning and control software should deal with unexpected obstacles, road conditions, failure of a machine-positioning system, damaged material, improper tools, or imprecise instructions.

Examples of some basic types of *cultural knowledge* that would be common to and not task-specific for such robots could include the following:

- Knowledge and abilities to deal with space and time:
    Interpretation of and reasoning about geometric 3-D space
    Motion planning
    Newtonian mechanics (velocity, forces, time, etc.)
    The ability to monitor the location and status of other objects
- General abilities to receive, analyze, and respond to input from sensors
- Communications abilities to access knowledge and data in the environment:
    Access to design data and project information data bases
    Knowledge of organizational structure and access to other agents
    Ability to communicate and cooperate with other agents
- Task planning abilities:
    Understand and use design specifications for task planning
    Seek and interpret project administrative information for schedules and the like
    Be able to select and locate the appropriate tools and materials
    Do the low-level planning for the task based on high-level inputs
- Extensibility and learning, that is, the ability to assemble information from sources beyond the robot and to retain the information for the future
- Self-knowledge, that is, to know the robot's own limits and capabilities and relate these to the demands of its tasks and to requests from other agents in the environment

Although this list represents a fairly ambitious set of research and development objectives, good progress is already being made in several areas. It will take decades, however, to achieve robotic intelligence at anything approaching the scope described here.

ADVANCED INTEGRATION

In future construction field environments, intelligent machines, like their human counterparts, will need to harness considerable knowledge to plan and control autonomous tasks in spite of the fact that they will be limited in their own knowledge and abilities. Not only the robots but also most of the intelligent agents will need a unifying core of intelligent software and a framework for defining and communicating knowledge about designs and field operations in a way that can effectively be utilized for their production tasks.

A broad conceptual view of the construction knowledge environment is presented in Fig. 16-8, along with some ways in which the core software of an intelligent robot might interact with the environment. The organizational context in which the robots might be working, the interfaces to computer-aided design (CAD) databases and reasoning, interactions with other field agents—both human and machine—and interfaces to knowledge sources in the world beyond the field are illustrated.

In an environment of this type, all of the agents—both human and machine—could be working in the context of an integrated model, possibly one that may evolve from today's research on distributed databases and knowledge bases, object-oriented systems, constraint-based systems, neural networks, and other advanced topics in computer science and engineering. In this context, from project conception through design and construction and on to facility management over the life cycle of the project, the virtual model would evolve and change to reflect the history, present state, and future plans for the facility accurately.

The scope of research needed to build theories and core software to support integrated design, construction, and facility management is vast. Each step in this research should lead toward a general architecture that organizes, manipulates, and communicates the knowledge that agents need to function productively in a

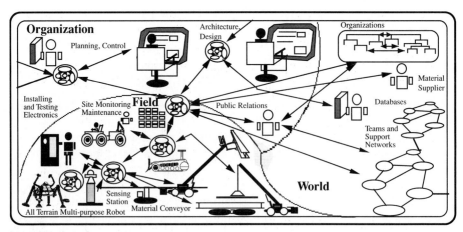

FIGURE 16-8
Integrated environment for construction automation. (Sketch by Lai Heng Chua.)

knowledge environment. The resulting software could then be extended by developers of applications-oriented agents to handle particular areas of expertise, whether in carrying out design, managing other machines, doing specific physical tasks, or monitoring and controlling a facility's operation. The ultimate objective should be to design and develop the general theory and software core for machine agents—the rudimentary "brains" of the beasts—which can then be embedded in agents specialized for particular tasks. Thus parallel virtual and physical models of our built environment may evolve, as well as intelligent agents, both artificial and biological, who could survive and work cooperatively to sustain and enhance the world in which they coexist.

SUMMARY AND CONCLUSION

Computer applications began in construction in the late 1950s and have evolved ever since. Clearly computer technologies will continue to advance at breathtaking speed, and they will continue to reshape the world in which we live. In this chapter we reviewed the present state of three rapidly evolving areas—integration of application software, 3D modeling and simulation, and multimedia systems—and projects where these areas might be headed in the near future. We then discussed three other areas whose full impact in construction is likely to be felt later, including artificial intelligence, robotics, and computer-based integration of design, construction, and facility management. Progress in all of these areas is technically and economically feasible, but how quickly and effectively the construction industry will adapt to and exploit this technology for its own advancement is less certain.

Numerous obstacles inhibit technological advances in construction. Examples include concerns about liability, building codes that are not under the control of the innovators, organizational and contractual fragmentation between and within designers and constructors, and many others. These factors tend to make consulting engineers and some construction contractors overly conservative when faced with an opportunity to adopt advanced technology, even when the direct economic benefits seem obvious. The result is what some perceive as a somewhat backward industry.

Fenves (1991) noted that on average a 17-year period elapses from the introduction of a new building technology until it becomes generally accepted by the construction industry and its regulatory agencies. Even then, most new building technologies have a relatively modest effect on construction costs or schedules. He contrasted this to the well-known fact that the performance/cost ratio of computer technologies doubles every two years. The relative gap could be 2^8 or more by the time adoption takes place—a factor measured in the hundreds.

On the one hand, more rapidly advancing technologies will continue to command a larger share of national economies than construction, whose share is declining in most industrialized countries. On the other hand, if the advanced technologies could more rapidly be harnessed to improve the state of technology in construction, then construction could more effectively help these same countries to restore and enhance their infrastructure the better to meet the needs for housing, public works, and industrial improvement. More than any other present or foreseeable technology, the improved application of computers could help the construction industry better the world in which we live.

REVIEW QUESTIONS

1. After decades of slow progress, what stimulated the rapid proliferation of computers in construction in recent years?
2. Briefly describe some useful steps that can be taken to integrate applications such as those that were presented in Part V, Chapters 12 to 15.
3. How does simulation of the type described for Bechtel's Construction CAE differ from the Cyclone and Stella simulation systems in Chapter 15?
4. Briefly outline an application whereby a multimedia computer-based system could be used to help record and make accessible some innovative construction methods as they occur on construction projects.
5. What characteristics of construction problem solving make it particularly well-suited to advanced applications of artificial intelligence?
6. What are the primary challenges and obstacles to the introduction of automated process control and robotics to construction?
7. In Fig. 16-5 we saw how automated instrumentation can improve the safe operation of a crane. Indicate how similar instrumentation could be applied to (a) a forklift and (b) a front-end loader.
8. How does the level of integration in the automated building system differ from that of earlier generations of construction robots?
9. Where might technologies from artificial intelligence and expert systems best be applied in improving the "intelligence" of an automated building system?
10. Outline some of the characteristics of a knowledge environment that would support substantially higher levels of integration among the various parties and systems involved in the design and construction of major projects.

REFERENCES AND SUGGESTIONS FOR FURTHER READING

El-Bibany, Hossam, and Boyd C. Paulson, Jr. "Collaborative Knowledge Integration Systems: A Tool for AEC Design, Management and Coordination." *Microcomputers in Civil Engineering*, Vol. 9, No. 1, February 1994, pp. 29–40 (in press).

El-Bibany, Hossam, and Boyd C. Paulson, Jr. "Microcomputer/ Videodisc System for Construction Education." *Microcomputers in Civil Engineering*, Vol. 6, No. 2, 1991, pp. 149–160.

Fenves, Steven J. "The Potential of Computer-Based Technologies in Civil Engineering." The Twenty-Ninth Henry M. Shaw Lecture in Civil Engineering, North Carolina State University, Raleigh, N. C., December 1991.

Froese, Thomas M., and Boyd C. Paulson, Jr. "An Object-Model-Based Project Information System." *Microcomputers in Civil Engineering*, Vol. 9, No. 1, February 1994, pp. 13–28 (in press).

Howard, H. C. "Project-Specific Knowledge Bases in AEC Industry." *Journal of Computing in Civil Engineering*, Vol. 3, No. 1, January 1989, pp. 25–41.

Howard, H. C., R. E. Levitt, B. C. Paulson, Jr., J. G. Pohl, and C. B. Tatum. "Computer Integration: Reducing Fragmentation in AEC Industry." *Journal of Computing in Civil Engineering*, Vol. 3, No. 1, January 1989, pp. 18–32.

Ito, Kenji, Yasumasa Ueno, Raymond E. Levitt, and Adnan Darwiche. *Linking Knowledge-Based Systems to CAD Design Data with an Object-Oriented Building Product Model.* Technical Report No. 017, Center for Integrated Facility Engineering, Stanford University, Stanford, Calif., August 1989.

Paulson, Boyd C., Jr. "Automated Control and Robotics for Construction." *Journal of Construction Engineering and Management*, ASCE, Vol. 111, No. 3, September 1985, pp. 190–207.

Paulson, Boyd C., Jr., Nadeem Babar, Lai-Heng Chua, and Thomas Froese. "Simulating Construction Robot Agents and Their Knowledge Environment," *Journal of Computing in Civil Engineering*, ASCE, Vol. 3, No. 4, October 1989, pp. 303–319.

Sanvido, Victor E., and Deborah J. Medeiros. "Applying Computer-Integrated Manufacturing Concepts to Construction." *Journal of Construction Engineering and Management*, ASCE, Vol. 116, No. 2, June 1990, pp. 365–379.

Watson, G. W., R. L. Tucker, and J. K. Walters, eds. *Automation and Robotics in Construction X.* Proceedings of the Tenth International Symposium on Automation and Robotics in Construction (ISARC).Amsterdam, The Netherlands: Elsevier Science Publishers, B.V., 1993.

APPENDIX

A

ANSWERS TO SELECTED QUESTIONS

Chapter 1

1. (*a*) This answer could show how gains in productivity from using CAD will either displace people who fail to adapt to new technologies or enable those who use it to design to much higher standards of quality and detail as well as work more productively. However, the person using CAD requires time and ability to acquire new and advanced skills, but it can also make design jobs more interesting and rewarding.

    (*b*) Computers can help estimators be much more productive and accurate in their work. These improvements could enable a company to expand its estimating volume and reduce the risk of errors in bidding too low or too high. Some firms also use the added analytical ability to assist designers and owners in assessing the cost trade-offs of design alternatives before getting to the construction stage.

    (*c*) Computer applications in cost engineering can reduce much of the clerical drudgery that used to accompany the jobs of cost engineers and enable them to spend more time on analyzing the reasons behind cost deviations. They also make it possible to keep more detailed records more accurately and lead to clear and consistent standards of input and reporting.

(*d*) Direct use of personal computers by construction engineers and managers has encouraged the broader acceptance of advanced network-based scheduling software. This in turn helps them to manage schedules better and thus reduce the costs of their projects.

3. Better schedule, cost and quality planning and control tools can help contractors to provide facilities more quickly and at less cost and to meet higher quality standards than they might otherwise. These gains are particularly important in the computer and electronics markets, where companies expand rapidly when they have successful products and can have the facilities to move new products to market ahead of their competitors.

Chapter 2

1. *Corporate office uses*

- Accounting and payroll. (Generally these are a home office application, with some subordinate work done in the field.)
- Company and project finance. (Corporate finance is done in the home office, as is the project feasibility analysis and monitoring.)
- Equipment management. (Most firms maintain centralized control over their equipment fleets, though detailed maintenance management may be delegated to the field.)
- Human resources management. (This is almost always coordinated centrally, with detailed hiring and firing delegated to the field.)
- Office administration. (Usually this function is done in the home office, with selected functions delegated to the field.)
- Education and training. (Usually these are conducted by the home office, with some implementation aspects delegated to the field.)

Project management

- Cost engineering. (Detailed monitoring is usually a project management responsibility, with summary reports and accounting inputs going back to the home office.)
- Project planning and scheduling. (Clearly these are a field project management application, usually with only technical support from the home office.)
- Materials management. (Identification of needs and requisitions usually happens at the project level, but firms vary as to whether actual purchasing and expediting are done at the project or home office level.)

Field operations people

- Generally the forepersons and superintendents do not work directly with computers, but they do provide inputs to payroll and costs systems, materials management systems, schedules, and so forth, and thus need some understanding of how they work.

3. In the near term, construction professionals need to be aware of new developments in automation and the basic technologies that support them. Journal articles, specialty conferences, and trade shows are useful sources of information. To influence technology

developments, firms can send technically astute and articulate delegates to conferences where standards may be discussed, participate in university research activities, and so on. Although these activities may not offer a short-term payoff, they can build corporate and individual expertise so that the firm can be better prepared to evaluate and capitalize on the new technologies as they become available.

Chapter 3

1.

Device	Characteristic
_E_ ROM	A. High-speed, primary read/write memory
_D_ WORM optical laser disk	B. Removable direct access storage media
_C_ magnetic tape	C. Lowest cost/unit of storage for read-write media
_F_ Winchester drive	D. Removable high-density, read-only storage media
_B_ floppy disk	E. Instant availability of stored programs
_A_ RAM	F. Large capacity, direct access secondary storage

3. The answer is (*b*). The *instruction set* of a particular type of microcomputer is the electronic hardware logic implementation that limits and defines the rudimentary set of operations that the microcomputer can perform.

5.
 (*a*) Additional physical memory modules (reduces need for swapping I/O)
 (*b*) Floating point processor (speeds up this type of application)
 (*c*) ROM-based system software (retains memory when power fails)

7. True (There is not really a dominant node to be the center of a star network.)

Chapter 3 Problems and Essay Questions

1. (*a*)

Design parameter	Ratio
Word size (on data path)	≤ 2.0
System clock (CPU cycles/sec)	2.0
CPU instruction set	Low
Memory size	None
Cache memory for disk I/O	None
Floating-point coprocessor	Low
Memory chip technology	None
Hard disk capacity	None
Hard disk I/O rate	None

(b) System A: (1,000,000 − 600,000 − 200,000) / 200 = 1,000 activities
System B: (4,000,000 − 600,000 − 200,000) / 200 = 16,000 activities
(c) Hard disk I/O rate
500 activities × 200 bytes = 100,000 bytes = 100 KB
100 KB / 1,200 KB/sec = 0.08 seconds (not counting disk access and so on)

3. Refer to Fig. 3-17 for the numbered boxes 1 to 5. The basic computer hardware (box 1) can be similar, but the software (box 2) is designed for real-time I/O applications. The interfaces (box 3) have some characteristics in common with the interfaces for printers, disks, and the like, but in particular for A/D and D/A applications the interfaces will be specific to physical monitoring and process control. The amplification and deamplification (box 4) needs for interfacing to industrial processes tend to be much wider in range than for peripherals designed for use with computers, and the instrumentation and controls for the processes themselves (box 5) are much more varied and unique to such applications.

Chapter 4

1. The answer is (d). A programmer does not have to be concerned about the maximum amount of physical RAM memory available to run a program.

3. The answer is (c), the computer program that interfaces a computer processor and a peripheral.

5. False. (Procedural language statements usually translate into multiple—not single—machine code instructions.)

7. False. See definitions in chapter.

9.

- A multiprogramming OS partitions memory into separate areas for two or more programs to be loaded at once and then supervises their execution.
- Time sharing is somewhat more sophisticated in that it usually works with multiple online users and allocates CPU, memory, and peripheral resources in a way that gives each user the impression of having the individual attention of the computer.

11.

- Multitask batch does not have online users, but a typical real-time, multi-user multitask system does.
- The real-time system can precisely control time-critical applications such as industrial process control.
- The multi-user system would most likely support interactive applications, whereas batch programs proceed automatically to completion once they start.

13. See Chapter 3, Figs. 3-7 and 3-8 for descriptions of tape and disk hardware. Either will work for sequential files, but only disk is suitable for direct access files owing to the long seek times that would often be necessary if such files were kept on tape.

15. A database structure (1) can provide close links between multiple tables of information, (2) can enable users to extract information in more flexible and custom-tailored ways, spanning multiple tables, (3) can provide links between diverse applications in various departments in a company, and (4) can provide more complex relationships between data items.

Chapter 5

1. (*a*)

	Costs	Benefits
Quantitative	Hardware and software acquisition Service and maintenance Training courses Computer support staff Supplies	Estimated 0.5% greater profits 1.0% less in subcontractor costs Tripled productivity for quantity take-off Save staff (4)
Qualitative	Conversion costs Training on job Human feelings Added load on system Vulnerable to failure of technology at key times	Growth potential from more bidding volume Consider more alternatives Productive employees Accuracy and reliability Standardization Quicker access to more information Ease of last-minute changes Improved morale Improved company image

(*b*)

Application	Annual cost	Annual savings	B/C ratio
(1) Quantity take-off	$24,000	$4 \times \$30,000 = \$120,000$	5:1
	or $\rightarrow$	$8 \times \$30,000 = \$240,000$	or 10:1
(2) Productivity and cost	$48,000	$0.5\% \times \$50m = \$250,000$	
			5.2:1
(3) Bid summarization and preparation: with item (2)	$12 \times (\$2000 + \$4000)$ $= \$72,000$	$0.5\% \times 60\% \times \$50m =$ $\$300,000$ $+250,000$ $\$550,000$	7.64:1
without item (2)	$12 \times \$4000 = \$48,000$	$\$300,000$	6.25:1

(*c*) Various arguments could be made here, depending on the funds available and the organizational aspects of the decision. In general, avoid version 1 of application (1) since it has the lowest B/C ratio and involves morale-hurting layoffs. On the other hand, version 2 of application 1 implies that we can get the benefit of eight extra quantity take-off staff with tripled capacity of the existing four, but the B:C ratio of 10:1 would only come about if the rest of the estimating system expands to take advantage of the increased capacity, and if there are enough jobs available to bid. If the company has the capacity to take on two applications at once, the next best scenario is applications 2 and 3, with a B:C ratio of 7.64 : 1. Application 3 is also a fairly conventional extension to typical estimating practice, so this might be a fairly low-risk way to start. Or application 3 could be taken by itself with a B:C of 6.25:1.

Chapter 6

1. *Response time*

There are three factors: computation time, output requested, and load on system by others. It should be fast (2 to 15 seconds) to maintain train of thought, but other psychological factors also apply:

- Assure user that computer is working on problem (e.g., graphic feedback)
- Activity clumping and psychological closure where user can take a break
- Short-term memory limits require quick feedback

Accessibility

- Amount of time one gets a busy signal—discouraging if excessive
- Log-on and start-up procedure—keep simple, possibly allow macros
- Physical location of workstations—make it convenient

Recovery from failures

- Should have regular backups and easy recovery procedures
- Provide instructions to help user correct errors
- "Fail-soft" procedures to give time to save and shut down

Charge structure

- High computer resource charges (CPU and so forth) and low online charges lead to willingness to stay on line and do work as needed
- High online charges cause the inconvenience of frequent log-ons and log-offs, discourage routine use, and encourage more thinking and analysis before logging on

3. (*a*) The answer to this question would basically follow the pattern given in the example developed in this chapter. Figure 6-5 shows a hierarchical chart. For additional information pertinent to this question, refer to the database examples in Figures 4-6 and 4-7, and look ahead to the example developed using the Paradox program in Chapter 10, including Tables 10-2 and 10-3 and Figures 10-6 to 10-8 and 10-14 to 10-21.

(*b*) Numerous details are contained at the lower levels of Figure 6-5 and in Table 10-2. One would consider how such information is generated, processed, and used at the day-to-day level by staff members in the equipment department.

5. (*a*)
- Define the scope of "maintenance" (e.g., parts inventory? scheduling? history? alert system?)
- Scale: how many projects? how many machines? and so forth
- Who are the users? home office administrators? shop personnel?

- Frequency of transactions, periodic or on line, volume, and the like
- Centralized versus decentralized?
- Input sources and output needs?
- Interface to other applications (records, accounting, etc.)

(*b*) Allocate adequate time to plan and design the application before developing — perhaps 20 percent each to feasibility study and analysis, 30 percent to design, 20 percent to programming or acquisition; 10 percent to implementation.

(*c*) This answer could use a hierarchical block diagram similar to that in Fig. 6-8, with the main breakdown of input, processing, and output. As well as the program structure, one must design the input forms, files, and output reports, but not in detail for this problem.

Chapter 7

1. (*a*)

- Start with the feasibility study, employing a group drawn from both user areas and the computer staff in the company. Be sure to obtain management support for the study.
- Prepare a manual of specifications (1) for internal review and approval and (2) for use by vendors in proposals.
- Solicit proposals.
- Evaluate proposals, possibly using a weighted ranking system such as that shown in Table 5-3.
- Hold a follow-up evaluation and audit once the system is installed and in operation.

(*b*)

- Lease: lower initial costs, but may be more cumulatively. No residual value for resale when no longer needed. More flexibility for changes and upgrades. May have better maintenance and service support.
- Buy: High initial capital outlay, but may have tax depreciation advantages. Lower annual cost if equipment has long service life. More risk of obsolescence, less flexible for changes and upgrades, etc.

3. (*a*)

	Advantage	**Disadvantage**
Total conversion	Avoids costly duplicate operations	Risk of disruption if conversion has major problems
Parallel/gradual	Security if new system has problems, cross-check results	Stressful load on staff to run two systems at once; users have alternative to resist change
Phased modules	Can convert and learn in a manageable way over time	Drags out conversion; must be modular type of application
Pilot	Safe way to prototype, with confined impact in case of problems	Not applicable to companywide systems

(b) Financial accounting: total conversion
   Scheduling: pilot

5. (a) Terminal at 4 cps

$$(\text{PR: } 100 \times 300 + \text{Cost: } 200 \times 120)/(4/\text{s} \times 60 \text{ s/min}) = 225 \text{ min/wk}$$

$$225 \text{ min/wk} \times 52 \text{ wk/yr} = 11{,}700 \text{ min/yr}$$

Micro at 480 cps

$$(\text{PR: } 100 \times 300 + \text{Cost: } 200 \times 120)/(480/\text{s} \times 60 \text{ s/min}) = 1.875 \text{ min/wk}$$

$$1.875 \text{ min/wk} \times 52 \text{ wk/yr} = 97.5 \text{ min/yr}$$

Note: Assume 10 bits/character, allowing control signals and parity bits on top of the 8-bit ASCII code.

	Total annual costs	
	Terminal	**Microcomputer**
Buy	$1000	$3000
Maintain	12 × $50 = $600	12 × $150 = $1800
Modem	$300	$300
Communication charges	11,700 × $0.75 = $8775	97.5 × $0.75 = $73
Total	$10,675	$5173

(b) The communications costs are much lower with the microcomputer because it can transmit records collected offline at the full speed allowed by the modem.

   The people costs (not included in the analysis from part (a)) would be about the same for either alternative, because the same keystroke rate of 4 cps would apply whether working locally on the micro or remotely via the terminal.

Chapter 8

1. Integer: 1946; Real: 3.1416; String: "This is a string constant." ; Binary: 10110001; Date: 3/1/94

3. Languages such as FORTRAN and BASIC reserve certain characters to predefine the type of a variable without the requirement for a type declaration statement unless explicitly overridden in a type declaration statement. For example, in FORTRAN variable names starting with the letters I, J, K, L, M, and N refer to integer numbers; in BASIC, variable names with the "$" character at the end refer to strings.

5. The symbols +, −, ÷, and * are arithmetic operators, whereas the rest are Boolean operators. Of the arithmetic operators, ÷ (divide sign) is seldom used because it does not appear on standard keyboards. The slash (/) is usually used instead.

7. A library of procedures or functions contains tested code for commonly used programming tasks that can in turn be used in other programs without having to recreate them from scratch.

9. Pascal enables a programmer to divide a program into logically self-contained modules with local variables and precise calling mechanisms that can be tested and then combined with other modules to make a complete program. In BASIC, at least in its earlier standard version, all of the code for a program was in one big module, with relatively weak structure, which made testing and debugging more difficult for programs of equivalent scope.

11. Output could send information to a disk or tape file for storage, or to a CRT or printer to put it in a form that a user can see.

13. COBOL—COmmon Business Oriented Language. Its statements read somewhat like verbal sentences.

15. ALGOL

17. The C language is rapidly becoming the preferred standard for both scientific and commercial programming applications, particularly for professional programmers.

19. The cylinder test program used a data file for input.

Chapter 9

1. Both spreadsheets and microcomputers are well suited for individual, interactive application by a user. Their low cost makes it feasible to disseminate such applications widely. Both are relatively easy to use compared with earlier computer technologies.

　　Spreadsheet software is considered to be the first new and practical software that actually exploited the potential of microcomputers and broadened their appeal to the business world after their early adoption by computer technophiles and hobbyists. Business market volume in turn provided tremendous support for the rapid evolution of microcomputer technologies, a situation that continues to the present day.

3. Computer spreadsheets use the CRT screen as a "window" that can be moved ("scrolled") around a much larger conceptual space.

5. In spreadsheet programs, the result of the formula usually appears in the cell, while the formula appears in a separate edit bar near the top of the screen. The format of the result is usually determined by menu commands and by manipulating graphical icons in the application software.

7. The big advantage is that, once the framework for an application has been set up, calculations based on input data can be substantially automated. Thus the impact of a single change early in the logic of a large application—such as a labor rate in a cost estimate—would take hours to compute with a calculator, but can be computed almost instantly with a spreadsheet. The user is encouraged to see what happens if certain input parameters vary and thus can find the most critical inputs for analysis in a problem.

9. Context-sensitive help takes the user directly to the section of the online reference material that is pertinent to what the user is trying to do at the moment help is called. The user does not have to start from an index or table of contents to get to that information.

11. Spreadsheet graphics usually focus on presentation charts for business and scientific users: bar charts, line graphs, pie charts, and the like. They are not intended as general drawing tools for artists or technical people. CAD and drawing software are much more capable in being able to render almost any type of graphical image that can be output to a screen or paper.

13. Spreadsheet database manipulations can provide a variety of functions, including sorting, searching, and selection; counts, sums, and averages; and often more complex statistical analyses using specialized statistical functions.

15. Macros enable the developer to automate sequences of spreadsheet commands as well as just the calculations that take place within cells. They can create custom menu items, create graphical "buttons" to perform certain actions, and indeed customize a spreadsheet application in such a way that a user needs to know little about the underlying spreadsheet concepts. They can also automate the start-up and shutdown sequences of applications that would otherwise be difficult for a novice to work with.

17. A custom template still requires the user to add the information pertinent to a given application, whereas an implemented spreadsheet application usually contains all of the constants and general application data to make it ready for day-to-day application.

19. The scraper problem was rather simple by spreadsheet standards, but interesting features included the table look-ups, custom menu and guided data entry implemented via macros, and the graphical plot.

Chapter 10

1. File and database software evolved from mainframe and minicomputer programs of this type, whereas spreadsheets were originally developed for microcomputers. They are similar in that both have evolved to highly interactive, user-driven implementations, with increasing use of graphical user interfaces.

3. Databases offer more complex information structures, whereas file systems are mostly limited to two-dimensional tabular structures. See Chapter 4 for details.

5. The design of a record layout includes the data elements, data types, field sizes, sequence of items within the record, and whether a given field is keyed for sorting and searching or is just another attribute in the record.

7. Typical sections in designing a report layout include the main body of the form, with its record layout and number of records per page, plus report and page headers and footers, totals and subtotals in various sections, and text and graphics to help explain the report.

9. Business graphics extensions can convert tabular information to various types of charts that rapidly and visually convey the relative magnitudes of data contained within a file or database. Business graphics are basically an alternative reporting technique.

11. Text, number, currency, date, graphic, plus computed fields and reference fields that draw upon other fields.

13. An automatic script code generator records a user's actions (keystrokes and mouse clicks for menu selections, parameter entry, and so on) for a finite series of steps in using a program. The generator turns these recorded actions into the corresponding macro or script code that can then be invoked with a single mouse click or keystroke combination when the user wishes to repeat that series of actions in the future. The developer of a custom application can also use this capability to generate code that might otherwise have to be written directly in the script or macro language. The developer usually can modify or extend the automatically generated code.

15. Hierarchical passwords selectively can control access to different tables, records, and fields in a database and can also control the type of access (e.g., user to read only, allow data entry and editing, allow deletion, and so on). Thus various types of users in different departments can work with a coordinated information system within the policies and constraints deemed appropriate by their organization's management.

17. A construction engineering file or database application that could benefit from advanced mathematical and financial functions is estimating, particularly in sections where it draws upon historical cost and productivity information and combines it with quantity and production calculations for various bid items. A complex financial application sometimes implemented with database software is a payroll system, with all of its tax and wage complications.

19. Networked data access software must be designed so that simultaneous or conflicting actions do not cause errors or inconsistencies in the database. This possibility can be handled with temporary locks on files, tables, records, or fields while one user is making changes. Consideration must also be given to the efficiency of software execution when dealing with multiple users.

21. The tool management system illustrated the use of a graphical user interface in a custom application, a graphical data field, and methods for manipulating data within the file to locate specified information.

Chapter 11

1. Many aspects of construction engineering are experience- and judgment-based, which in turn contribute to the expertise that is valued in the industry's top engineers and managers. These characteristics are difficult to capture in traditional procedural programming methods, but some areas of expertise in this field do lend themselves to expert systems.

3. All but the second one (the volume calculation) could be called heuristics, but the first and the fourth come closest to the form of heuristic rules.

5. The knowledge representation methods are certainty factor, logical rule, frame, and multiple inheritance. The explanation facility and the inference engine are components or characteristics of the expert system software that supports the knowledge representation methods.

7. One would start with that milestone and work backward in time to determine what has to be done to stay on schedule (e.g., add resources, make more activities concurrent, work multiple shifts, and so on).

9. The domain expert can provide sample cases, offer answers and advice for test cases and real-life cases that can be compared to the results given by the expert system, and can examine the interim reasoning and the final results given by the expert system to assess their quality relative to human experts.

11. Object-oriented languages encapsulate both procedures and data into self-contained modules called objects, and the objects can communicate with each other via a mechanism called messages.

13. Advanced features that would be most useful are these:

- Automated control of an excavator: real-time monitoring and control
- Uncertainty regarding bid markup: fuzzy set theory

- Front end for an estimating cost database: capability for external linkages to other software
- Constructibility input to structural designer: integration with 3-D CAD systems

Chapter 12

1.

- CAD to estimating: bill of materials, dimensions, and quantities
- Estimating to job-cost accounting: budget for labor, materials, equipment, subcontractors, overhead
- Estimating to scheduling: work breakdown structure, activity durations

3. The contractor can be more flexible in sort and search criteria and can get information as soon as it enters the database. Disadvantages could occur if there are technical problems with the computer linkage. There might also be excessive costs if the database is overused.

5. The estimator could use spreadsheets for calculations, an estimating program such as HCSS to make up crews and check their costs, and a simulation program to simulate a production system.

7. Productivity data in terms of worker hours and equipment hours per unit of production would be more useful than costs, though there will still be variations by region and by project type. Current unit resource costs can then be applied to the productivity rates to make cost estimates for current work.

9. In a unit-price bid, the indirect costs and markup have to be distributed to the direct-cost unit price bid items, whereas in a lump-sum bid they are simply added to the total.

11. A nonlinear distribution can be used to unbalance bids either by front-loading revenues to earlier operations in a project, or by placing more indirects on items that are likely to overrun and taking them from items that are likely to underrun.

    User-definable report forms enable a company to make the reports from a packaged estimating system conform to existing company standards, meet client requirements, or satisfy special requirements of specific projects.

    Integration with published cost databases, such as those provided by R. S. Means and Richardson Engineering, provides ready access to productivity and cost data that can be used to supplement company-generated information, serve as a check on subcontractor bids, and so on.

    Range estimating provides a better feel for the risk, uncertainty, and sensitivity of an estimate and its components.

13. Estimating software is relatively specialized, so the developers' costs have to be recovered over a much smaller volume of sales, hence the higher per unit costs to buy estimating software.

15. Timberline's Light system draws upon a database of work items and unit costs for labor, materials, and a catchall "other." The user selects items from the database that are pertinent to the project being bid, enters dimensions and, with the aid of Timberline, computes quantities. Timberline applies unit costs to the quantities to get totals in various categories. The user can also specify indirect costs in various categories, add a profit, and compute the total costs for a bid. Timberline's software can produce a variety of reports that are useful in analyzing the bid and preparing it for submittal to the owner.

The entry-level Light version has several limitations, including the small number of line items and columns in its worksheets, lack of provision for a work breakdown structure, inflexible handling of indirects and markups, lack of interfaces to other software (e.g., CAD and cost control) and hardware (e.g., a digitizer), inconvenient handling of resource costs, and so on. Most of these limitations are overcome with more advanced versions in Timberline's estimating product line.

Chapter 13

1. Missing: explicit definition of the logical dependencies among activities
   Advantage: simple and widely understood as a means of communicating schedules to clients, field construction people, and others
3. Actual progress will normally fall between the early- and late-start curves because some activities will be delayed within their available float times to even out resource usage or for other reasons, but such activity delays can be caused without delaying the overall project.
5. The critical path is the chain of activities through the CPM network that has the minimum total float (zero if the late finish is set equal to the early finish for the project as a whole). In other words, activities on the critical path are those that cannot be delayed without delaying the project's computed early finish time.
7. Where possible, an activity's scope should place it entirely within one supervisor's responsibility.
9. All four methods of data entry are usually allowed.
11. Free float
13. Start-to-start
15.

   - Resource aggregation simply adds up the utilization across the various activities scheduled in each time interval; resource profiles can be plotted.
   - Resource allocation starts with the aggregated resource profiles and then attempts to reschedule activities in such a way that the project stays within limits on available resources. Such algorithms may use available activity float times to minimize the extension of overall project duration but will delay the project if necessary to stay within resource limits.
   - Resource leveling starts with the aggregated resource profiles and then attempts to reschedule activities within the given overall project duration (i.e., within the activity float times) to smooth out or reduce the variations in resource usage from one time interval to the next.

17. Time/cost trade-off methods put the focus on activities that (a) are on the critical path and (b) can be expedited at the lowest cost per unit time in order to reduce the overall project duration at minimum cost.
19. Float provides the construction managers with flexibility to meet project requirements yet still achieve an efficient utilization of resources, space, and so on. If the float in a computed schedule does not reflect resource constraints, that schedule can also become unrealistic if another party takes over the allocation of the float. The

main disadvantage of the general contractor keeping the float is that this denies similar flexibility to subcontractors and to the owner, so an open and cooperative process can often be best for the overall good of the project.

Chapter 14

1. Assets = Liabilities + Equity
   Assets are things a business owns; liabilities are what it owes; equity, also called net worth, is the stock or other measure of the value of the owners' stake in the company.

3. The income statement reports income and expenses over a defined period of time (year, quarter, etc.).

5. Accrual methods recognize income and expenses when they are incurred, whereas cash-based methods recognize income and expenses when the money actually changes hands. Accrual methods are not only more suitable for larger companies, but also are usually required.

7.

- General ledger—records financial transactions and produces the financial statements (income statement and balance sheet)
- Accounts payable—tracks money owed to others and makes payments
- Accounts receivable—tracks money owed to the company by others for work the company has done; prepares statements and invoices to clients
- Job cost—monitors the costs of work taking place on projects (jobs)
- Payroll—calculates wages and salaries due employees, calculates taxes and other payroll-based expenses, and issues payments to employees and others

   The preceding modules are the important ones found in most accounting systems. Others sometimes included are for purchasing, inventory management, and fixed asset accounting.

   All but job cost are important at the corporate level, whereas job cost is of primary interest on projects. Some companies also decentralize some aspects of payroll, accounts payable, and accounts receivable to the project level, but usually only for large projects.

9. Both types of cost codes should be organized according to the ways that work is typically broken down in their applicable types of construction, such as by craft and subcontractor specialties. The project cost code also ought to correlate to the way the plans and specifications are organized, and to the breakdown of the project schedule.

11. Direct costs are those associated with the work that directly produces the facility that is being built, whereas indirect costs are those associated with support functions, such as the site management, engineering and administrative staff, access roads, temporary storage facilities, and so on.

13. Control budgets can be subdivided to match pay items on unit-price jobs, to fit with subcontractor and supervisory responsibilities, and even to correlate with a breakdown requested by the owner.

15. The primary objectives of job-cost reporting are to get pertinent information to the supervisors responsible for the work generating the costs and to do so in time for them to take corrective action if needed. Secondary objectives include reports to the home office and saving cost information as historical records to update the company's estimating database.

17. Payrolls are complex in construction, owing to the diversity of crafts and dispersed work sites, to complicated wage and fringe benefit calculations, to the frequent hiring and layoffs of workers in this field, and to the linkages to job-cost and other parts of the accounting system. It takes a fairly sophisticated payroll accounting program to handle the needs of this industry.

19. Interfaces to other systems could include estimating (to convert estimates to budgets at the outset and to update the cost database when done), scheduling (to use a common work breakdown structure, keep track of costs at the activity level, and generate cash flows using schedule information), procurement and materials management (to track materials costs), payroll and accounts payable (to track labor, materials, and subcontractor costs), and equipment (to track equipment costs).

Chapter 15

1. Simulation programs allow one to define durations as probability distributions with ranges and likelihood of possible durations for a given activity.

3. Clock-driven

5. A discrete model consists of finite entities (e.g., trucks, workers) that perform tasks that can be subdivided with respect to time into distinct steps (e.g., load, haul, dump, return). Continuous models deal with phenomena that can be defined or approximated as flows that move continuously from one part of the model to another (e.g., a model of an oil refinery, a hydrologic model, a population growth model).

7. Define system boundaries

9. Skewed (to show the greater possibility of durations being much longer than average rather than shorter than average).

11. Grades on the return road have the opposite sign from those of the same segments on the haul road.

13. Adding loader costs and some additional calculation capabilities would enable VEHSIM to perform optimum fleet size and cost computations.

15. False. Resources can be multiplied via a GENERATE attached to a QUEUE node, and multiple resources can be combined back into a single one at a CONSOLIDATE function node, but neither of these coversions can take place at a COUNTER function node.

17.
   - The type of activity that consumes time while processing inputs and outputs is a Flow Regulator, or Flow for short.
   - The symbol that defined prcedence relationships is the double-line arrow that feeds into and out of Flow and Stocks. (Note that the single-line arrow defines functional relationships, not activity sequences.)
   - A place to hold idle resources is a Stock.

19. The main advantage is that the model is animated while the simulation is running, thus providing more intuitive insight into a model's behavior. MicroCyclone does have a concurrent graphing capability (see Fig. 15-19), but it is more limited.

Chapter 16

1. Rapid proliferation of microcomputers in construction in recent years has been stimulated by many factors, but among the most important are steep drops in the costs of computer hardware and software; the development of easily used, versatile, and

practical end-user application software that can solve day-to-day problems; and the increasing sophistication and experience of construction professionals when it comes to using computers.

3. Construction CAE is focused specifically on construction simulations; has much more powerful and realistic graphics; works with an accurate 3-D model of a project; integrates CAD with scheduling, cost, and materials management features; and generally provides the most realistic simulations of construction processes.

5. Construction is an industry where many decisions and processes are based at least in part on the experience and judgment of knowledgeable professionals in the field. AI techniques such as expert systems are more suited than other types of software for capturing these types of decision-making and problem-solving processes so that others can benefit from the expertise of successful professionals.

7. Like cranes, forklifts and front-end loaders have limits on tipping moments, but these limits are compounded by more rapid acceleration and stopping as these vehicles move about. Load cells and inclinometers could apply here, as could accelerometers and other types of transducers. They could be attached to the lifting arms, hydraulics, and other mechanisms on the machines.

9. AI technologies could assist in monitoring instruments that measure the status of production processes to be sure that quality standards are being met and to watch for potential breakdowns before they happen. AI could get involved in coordinating materials procurement and distribution. AI could also help in decision making and problem solving when unplanned events take place (e.g., a breakdown of a key component). These are just a few of many possibilities.

APPENDIX

B

GLOSSARY

This glossary includes some of the most important and frequently used computer terms and a few other technical terms that appear in this book, but space prohibits making it a substitute for a computer dictionary. Some words not included here are defined in context where they first appear in this book and can be found via the Index. For a more complete reference on computer terminology, the reader is encouraged to acquire a computer dictionary. Some of those current at the time of this writing are listed below; most are inexpensive.

Downing, Douglas, and Michael Covington. *Dictionary of Computer Terms,* 3d ed. Hauppauge, N.Y.: Barrons, 1992. Short but mostly adequate definitions; inexpensive.
Microsoft Press Computer Dictionary. Redmond, Wash.: Microsoft Press, 1991. Has more complete definitions; some drawings.
Nader, Jonan. *Prentice Hall's Illustrated Dictionary of Computing.* Englewood Cliffs, N.J.: Prentice Hall, 1992. Has more complete definitions; numerous photos and drawings.
Pfaffenberger, Bryan. *Computer User's Dictionary,* 3d ed. Carmel, Ind.: Que Corp., 1992. Midway between the briefer and more complete dictionaries.
Webster's New World Dictionary of Computer Terms, 4th ed. Englewood Cliffs, N.J.: Prentice Hall, 1992. Short but mostly adequate definitions; inexpensive.
Wyatt, Allen L. *Computer Professional's Dictionary.* New York: McGraw-Hill, 1990. Short definitions; some drawings; moderately priced.

AEC An acronym for Architecture/Engineering/Construction. Used when referring collectively to these functions in the design and construction process.

AI *See* artificial intelligence.

algorithm A tested, step-by-step logical procedure followed by a computer program to carry out a particular task or reach a specified outcome, such as solving a mathematical problem, sorting a set of records, and so forth.

577

alpha testing   Once a computer program is complete, this is the initial phase of testing before turning it over to its intended users. *See also* beta testing.

analog to digital (A/D)   For computers, this refers to the process of converting some measure of a continuous (i.e., analog) phenomenon (e.g., temperature) to a voltage that in turn can be approximated by discrete steps (i.e., digital increments) that can be represented by binary computer numbers. *Compare* digital to analog.

android   A robot designed to represent a human closely in function or appearance.

architecture   The physical design and structural characteristics of a computer system, including system software as well as hardware. *See also* CISC, RISC.

argument   A parameter or expression used with an operator, or passed from one part of a program to another, typically in a parenthesized list following the name of a subroutine, procedure, or function.

artificial intelligence (AI)   The branch of computer science that attempts to make computers perform some of the functions that humans recognize to be "intelligent" behavior, including logical reasoning, symbolic inference and knowledge representation, pattern recognition for vision or speech processing, and so forth.

ASCII   American Standard Code for Information Interchange; a 128-item set of 7-bit binary codes that represent the upper- and lowercase English alphabet, numeric digits, punctuation marks, and some control codes for processing and exchanging information within and between computer systems. Some manufacturers use an eighth bit to define up to 128 more special character codes, which is sometimes referred to as extended ASCII.

assembly language   A set of mnemonic codes and a syntax for representing the machine language of a computer in a form that is intelligible to trained programmers.

attribute   One of the data fields in a file or database; can often be thought of as the name of a column heading in a file or database table.

back up   As a verb, to copy information from one computer storage area (e.g., a hard disk drive) to another (e.g., a tape drive or floppy disk), usually to provide added security in case the first copy is lost or damaged. **Backup,** as a single-word noun, refers to the copy resulting from a backup procedure.

backward chaining   A method of problem solving that first specifies the goal and then searches back from there to match rules with information from a knowledge base that will support or refute that goal. *Compare* forward chaining.

bandwidth   Although similar to the frequency bandwidth in radio communications, in computers this term is sometimes broadened to refer to the throughput rate that data can move through a particular channel (e.g., a local area network).

batch program   A program that requires all input data and processing options to be prepared or specified in advance of execution, and then proceeds automatically for processing and output. *Compare* interactive program.

baud   Historically related to an old French telegraphy scheme called the Baudot code, baud refers to the maximum rate at which a transmission device can change the rate of a signal. Today it is loosely but inaccurately used to mean the transmission rate of a modem measured in binary units (bits) of data per second (e.g., 9600-baud means 9600 binary data bits per second).

beta testing   Once a computer program is nearly complete, this phase of testing usually involves turning it over to a representative sample of its intended users who agree to test and evaluate it and keep detailed notes on program performance and problems. *See also* alpha testing.

bit   One unit in the binary number system (i.e., a logical processing or storage element that is either 0 or 1, true or false, charged or zero, etc.); the smallest unit of information recognized by a computer system.

bit-mapped   Usually refers to a display screen whose output area can be manipulated one element (dot) at a time, rather than as clusters of dots representing predefined symbols, to enable flexible display of characters and graphics.

boot   Short for bootstrap, refers to the process of initially loading core pieces of a computer's operating system into memory once the computer has been turned on or restarted (rebooted).

buffer   In computer hardware, this is usually an area of high-speed memory to or from which data from slower memory or peripheral devices can efficiently be loaded in large blocks so that programs can access that data more efficiently or so the CPU can do other things while the buffer passes data to the slower device; alternatively, an area of memory containing information currently shown on a display screen. *See also* cache.

bug   An error in a computer program or in computer hardware that causes incorrect processing and/or results. *See also* debugging.

bus   The electronic circuit paths (wires or traces) in a computer that carry information from one part of the computer system to another.

button   In graphical user interfaces, this usually refers to a graphical object on a display screen that causes some programmed result to happen when the user clicks on that area with a cursor controlled by a mouse or similar device. *See also* cursor, graphical user interface, mouse.

byte   The combination of eight binary units of data that, together, can represent up to 256 (in base 10) different numbers or binary codes. Typically one byte is equivalent to one ASCII character of information. *See also* ASCII, bit.

cache   A high-speed memory buffer that duplicates large blocks of frequently used data held in lower-speed memory or peripheral devices. The cache memory more closely matches the speed of a high-speed central processing unit (CPU), enabling it to work more efficiently.

CAD (and **CADD**) Computer-aided design and/or drafting. Computer hardware and software systems that enable designers to prepare two-dimensional drawings and three-dimensional models directly on a computer screen.

CAE Computer-aided engineering. Complements CAD, and generally refers to the software that enables engineers to perform the analytical calculations to support a design.

CAM Computer-aided manufacturing. Broadly refers to the process of converting CAD and CAE data into a form that can drive automated manufacturing processes, and to the programs that control the processing equipment.

CASE Computer-aided software engineering. Describes software tools and procedures that aid programmers in designing, developing, testing, and documenting computer programs.

cathode-ray tube CRT for short. Refers to the type of glass-enclosed display device wherein one or more electronic beams are emitted from a cathode in the back of the tube and are electromagnetically guided precisely around a display surface consisting of phosphors that are activated to display white or color dots and patterns when struck by the electron beam. Common examples include television sets and the monitors used by most desktop computers.

CD-ROM *See* compact disc read-only memory.

cell A spreadsheet term that refers to the area where a column and a row of a table intersect. Data or a formula can be placed in a cell.

central processing unit Also called the CPU. The main electronic processing unit that fetches, decodes, and executes instructions to control a computer. *See also* microprocessor.

chip *See* integrated circuit.

CISC *See* complex instruction set computer.

click To use a mouse or similar device to position a cursor over an object on a screen, and then press a button on the device to invoke an action related to that object. *See also* mouse, cursor.

commands Usually refers to preprogrammed actions in a computer program that the user can type as key words or select as menu choices using cursor keys, a mouse, or single keystrokes. Examples in a typical program include SAVE, COPY, QUIT.

compact disc read-only memory (CD-ROM) Optically based storage devices similar to audio CDs that, once data have been placed on them in a manufacturing process, can be read by laser optics contained in CD-ROM drives. User-writable versions are also becoming available. *See also* optical disc.

compiler A computer program that translates the human-readable source code representation of a computer program written in a high-level programming language into machine-readable object code. All code is translated before the program can execute. *Compare* interpreter.

complex instruction set computer (CISC) Usually refers to a CPU design where over a hundred, and often several hundred, distinctly different processing instructions are included. Modern CISC microprocessors can have a million or more discrete electronic devices contained on a silicon chip less than one square centimeter in area. *Compare* RISC.

computer   A complex electronic machine that accepts structured input, processes it according to precisely defined rules coded as programs, and outputs the results. See Chapter 3 for a more detailed description of computer hardware.

context   The part of an expert system that holds the specific information about the current problem being addressed by the expert system and its user.

context-sensitive help   This form of online reference information for a running computer program goes directly to the section of the reference information that most pertains to the command, action, or section of the program with which the user is presently working. *See also* help.

controller (for peripheral)   The electronic hardware interface between a computer and a peripheral device such as a disk, tape drive, or a display screen.

conversion   The process of taking information from an older version of an application and getting it into a form that a new application can use. Can also include the training and other preparations for the people and processes involved. *See also* total conversion.

CPM   *See* critical path method.

CPU   *See* central processing unit.

critical path method (CPM)   A method of project scheduling that is based on a logical network to represent discrete activities and the dependency relationships among them. The purpose of CPM is to produce the earliest and latest starts and finishes for each activity that will achieve the minimum project duration.

CRT   *See* cathode-ray tube.

cursor   A small graphical symbol such as a cross, box, or an arrowhead that shows the user's current location for inputting text or graphics, or that points to an object, in an application program displayed on a computer screen.

database   Generally refers to an information storage and retrieval system wherein the content of and relationships among the items of information stored are more complex than can be represented as one or more independent tabular data files. Sometimes used more loosely to refer to the aggregation of information stored in a computer system, whether file-like or not. *See also* file.

debugging   The process of locating and correcting errors in computer programs or in computer hardware. *See also* bug.

declarative programming   A form of programming where relationships among program and data elements are specified, but without regard to the sequential logical procedure for processing to reach program objectives. Declares knowledge (e.g., $F = MA$) rather than specifies procedures.

development environment   The computer hardware and the software (editors, compilers, debuggers, etc.) that are used together to create computer programs; or a particular combination of tools sold as a commercial package that provides a programmer with all tools needed to create certain types of software.

device driver   The low-level software that performs the logical processing to enable a computer peripheral (e.g., a disk drive or printer) to work with a computer and its operating system software.

digital   In computers, refers to machines whose fundamental logic is binary (on or off, one or zero, true or false, etc.) and thus only allows discrete rather than continuous (or analog) states at any given time.

digital to analog (D/A)   For computers, refers to the process of converting internal data stored in discrete steps (i.e., digital increments) that can be represented by binary computer numbers into a voltage that can in turn be output to represent or control a continuous (i.e., analog) phenomenon outside the computer (e.g., partially open a valve, change a heat setting, vary audio sound output, etc.). *Compare* analog to digital.

digitizer   A device used to convert printed information such as an engineering drawing to an internal electronic form in a computer. *See also* Fig. 12-6 for an example. A device that works without human manipulation is more commonly called a scanner.

direct access   Refers to actions related to a peripheral device such as a disk drive that can position the read or write heads directly to the location where information is to be transferred, rather than having to move first through a range of storage media, as would be the case in a tape drive.

directory   An electronic catalog stored in a computer that contains information about files stored on a disk.

disc   *See* disk.

disk   A platter-shaped storage medium that can hold computer-readable information. Storage method can be optical, electromagnetic, or a combination of these. For optical storage media, common practice is to use the spelling **disc.** *See also* disk drive, floppy disk, hard disk.

disk drive   The enclosure and electromechanical hardware mechanism that reads and writes information contained on a disk. *See also* Fig. 3-8.

disk operating system (DOS)   An operating system that loads in part or in whole from disk into memory when a computer starts. In the PC arena, the acronym DOS often refers to a specific type of operating system developed by Microsoft and IBM for use with microcomputers that use the Intel 80x86 family of CPUs. *See also* MS-DOS, operating system, PC, PC-DOS.

display   A general term for the visual output device connected to a computer, most commonly based on CRT or LCD technology. *See also* cathode-ray tube, liquid crystal display.

distributed system   Refers to a system of two or more interconnected computers that might be placed in diverse locations and yet work together in some aspects of the applications the system is designed to perform.

document   A piece of work created with an application program and stored as a named computer file. *See also* file.

documentation   The manuals and other references that tell how to work and/or maintain computer hardware and software. Also, a stage in application development.

domain   In AI and expert system software, the topical area of expertise that is represented or coded in an application, or the type of application for which the program is intended to be used.

DOS   *See* disk operating system, MS-DOS, PC-DOS.

dot-matrix printer   A printer that uses mechanical force to drive a patterned matrix of small steel pins into an inked ribbon to place images on paper.

double-click   To click a mouse button rapidly twice in a row to invoke an action in software that is different from what happens when clicking only once. *See also* click.

dynamic data exchange (DDE)   Refers to the ability of two or more programs to be able to exchange information and even commands directly while they run concurrently in a computer. The exchange can take place directly in memory locations rather than in the older method of data exchange via files stored on a disk. Microsoft has adopted this term for its form of interprocess communication.

editor   A program that enables the user to enter, change, store, and retrieve text stored in files. Editors usually do not have the powerful formatting capabilities of advanced word processing software, but are often more finely tuned than word processors to the needs of computer programmers writing programs.

electronic mail (e-mail)   Software designed to enable users to compose, transmit, and receive messages over a communications network to and from other users at local or remote sites.

expert system   A computer program that captures the knowledge and thought processes of a human expert, including experience-based heuristic knowledge as well as objective factual knowledge, in a form that can be made available to a user in performing a task or solving a problem drawing upon the encoded expertise. *See also* shell.

expression   A combination of two or more constants, variable identifiers, and operators, usually within a formula or programming language statement, that yields a result upon evaluation.

facility management   Contracting with a third-party firm to manage all or part of a company's in-house computer operations.

field   A component of a data record that holds a single item of data of a given type. Usually used in the context of file or database systems.

file (of computer data)   A collection of related records of similar format; a collection of computer-readable code or data stored as a single named entity on a peripheral device such as a tape or disk drive (could be a program as well as a data or document file).

floating-point coprocessor (FPU)   An auxiliary processing unit specifically designed to add floating-point numerical processing instructions (e.g., log, sine, etc.) to complement the capabilities of a central processing unit and increase processing speed on applicable instructions.

floppy disk (or **diskette**)   A type of disk-based computer data storage medium that uses an electromagnetic metal oxide on a flexible plastic base to hold the data. Typically 3.5 or 5.25 inches in diameter, enclosed in a rectangular plastic cover for protection. One of the most common backup and offline storage devices for microcomputers. *See also* disk, disk drive.

flowchart   A logic diagram using various shapes of polygons, ellipses, and other symbols to represent processing, decision, and storage steps in a computer program, connected by lines or arrows to show the logical flow of data processing. *See also* Fig. 6-8 for an example.

forward chaining   A method of expert system problem solving that starts by examining the rules and facts in a knowledge base in order to proceed toward a solution that is consistent with the facts and rules. *Compare* backward chaining. *See also* expert system.

FPU   *See* floating-point coprocessor.

frame   A method for representing knowledge—consisting of a collection of logically related data elements in a knowledge-based expert system, somewhat analogous to a record in a data file—that is usually suitable for a generalized and hierarchical representation of information and its specific instances.

freeware   Programs that are developed and distributed for free, though the developer often retains a copyright to control the terms of subsequent distribution by others. *See also* shareware.

function   A computer subprogram that is called from other parts of a program and returns a result in the variable represented by the name of the function.

graphical user interface (GUI)   A method by which a user can communicate with a computer program using pictorial objects (such as rectangular windows, scroll bars, buttons, dialog boxes, etc.) that can be manipulated by pointing, clicking with a mouse, typing keys, and so on. Most Apple Macintosh programs and Microsoft Windows programs employ a graphical user interface (GUI). Pre-GUI software could be considered text-based, with a keyboard and printer or character-based display screen as the principal user interface. *See also* interface, user interface.

gray-scale   Printing and screen display output that, although fundamentally black and white, can also vary intensity to produce various shades of gray in between.

GUI   *See* graphical user interface.

hard disk   A type of computer data storage medium that uses an electromagnetic metal oxide on a rigid metal disk as a base to hold the data. Can be fixed either in a sealed enclosure or in a removable cartridge. In fixed hard disk systems, it is common to have multiple disks stacked one above the other on the same drive shaft. *See also* disk, disk drive, Winchester disk.

hardware (computer)   The enclosures, power supplies, mechanical apparatus, and electronic circuitry that make up computers and their peripheral equipment.

help   Disk-based reference information built into or directly accessible by a computer program to enable a user to seek and receive assistance about how to work a program or its specific features while it is running. Also called **online help.** *See also* context-sensitive help.

heuristic   An empirically based rule of thumb, based on long experience, that has been shown to produce good results, but that may not be subject to rigorous scientific proof.

icon   A small graphical object on a display screen whose design suggests a function that can be invoked by manipulating that object with a mouse or other device, usually in a graphical user interface. For example, the "trash can" in the Macintosh interface is used for deleting data files.

impact printer   A printer that uses a mechanical device to drive a formed character or a matrix of small steel pins into an inked ribbon to place images on paper.

index   In reference to files and databases, an array of information that points to the records in a file. Often the index will be sorted in a different order from the main body of data, thus facilitating alternative means of data access.

inference engine   In an expert system, controls the reasoning process that examines the data from the current problem context and consults the knowledge base to produce a solution to the problem and provide a supporting explanation.

information service (or **information utility**)   An organization that collects and archives information on one or more subjects (e.g., engineering journal articles) and makes the information available, often for a fee, to users who are remotely accessing the organization's computers via a network or telephone modem (e.g., Dodge DataLine in Chapter 12).

inheritance   In an object-oriented system, the process by which entities in a subclass of a given class of objects can take on properties from their counterparts higher up the hierarchy.

ink-jet printer   A printer employing a precisely controlled ink-jet spray mechanism to place images directly on paper.

input/output (I/O)   The parts of a program and the related computer hardware that move data to and from the computer's central processor, its peripheral devices, and the computer system users.

integrated circuit (IC)   A complex electronic circuit implemented on a single piece of semiconductor material such as silicon, also called a **chip.**

integrated software   Multiple applications implemented in a single software package, or separate software application packages designed to work together closely, particularly for data exchange, and with a consistent user interface. *See also* user interface.

integration   In architecture, engineering, and construction (AEC), integration generally refers to the linking of applications such as computer-aided design, estimating, scheduling, and cost control so that data can move directly from application to application rather than have to be reentered at each stage.

interactive program   A program that works with a user in an incremental, back and forth manner with respect to input, processing, and output. *Compare* batch program.

interface   In computers, can refer either to the hardware and software to connect together various system components or to the way a computer program has been designed to work with a human user. *See also* graphical user interface, user interface.

interpreter   A program that provides a software development and execution environment wherein lines of source code for an application program are translated and executed only when encountered in the application's flow of logic; they are not precompiled but are translated again whenever they are encountered. *Compare* compiler.

joystick   An electronic pointing device used to control computer programs, especially for video games.

keyboard   An electromechanical, typewriter-like input device used for entering character-based information into a computer, and for controlling program execution via commands and function keys.

key field   A data attribute (or column in a table) that is indexed for more efficiently sorting, searching, and selecting records in a file or database.

knowledge base   The part of an expert system that stores facts, usually as declarative knowledge, and heuristics about the problem domain, commonly as rules. *See also* expert system.

knowledge-based system   *See* expert system.

knowledge engineer   A person who understands an expert system tool and can interface between the knowledge source and the computer to acquire and express the relevant expertise suitably to build the expert system application.

LAN   *See* local area network.

laser printer   A printer employing a focused laser beam and an electrophotographic process to place images onto a light-sensitive drum to create an electrostatic charge that in turn attracts and holds toner in the image pattern. This drum then transfers the full-page image to electrostatically charged paper pressed between the drum and a roller. The paper moves next to a heat source that fuses the image onto the page.

LCD   *See* liquid crystal display.

library (of software)   Organized and accessible sets of named, prewritten program modules that have been successfully tested and stored in a file so they can be used again in other programs. A library of modules performing common functions is often included as part of a development environment for a computer language.

linker A utility program that combines various units of machine-executable object code into a single machine-executable program.

liquid crystal display (LCD) A display screen technology that uses electrically induced phase changes in liquid crystals to polarize light and thus create dot patterns (and even colors). The technology facilitated the development of thin, low-power, flat-panel displays that have been especially popular in small portable computers.

local area network (LAN) The cables and communication signal control devices that permit multiple computers and peripheral devices to be connected together within a finite location, such as an office building or a university campus; the system consisting of interconnected devices. *Compare* wide area network.

machine language Also called **machine code.** The electronic, binary representation of the instructions that control a computer processor of a particular design. Usually translated from assembly language or a compiler-level language written by humans into this form, which is intelligible to the computer. Also refers to printed versions of the instructions displayed as numbers in base 2, or aggregated into base 8 (octal) or base 16 (hexadecimal) notation. *See also* assembly language, compiler, linker.

macro Originally used to define a named procedure made up of more rudimentary steps that can then be called upon by name and automatically be expanded by a macro assembler into its full form to become part of an assembly language or compiler-level program. Also commonly refers to programmed procedures written within application programs such as spreadsheet and database packages that automatically replicate a series of user actions (keystrokes, menu choices, mouse clicks, etc.) when the procedure is subsequently invoked with a single user action or called by another macro.

mainframe The general name given to physically large and powerful computers of the type that do the central processing in large business, government, or scientific organizations. Usually shared by many concurrent users and programs, providing access to widely used databases and applications. *See also* supercomputer.

manipulator The part of a robot that performs a task such as welding, bolting, and the like. Analogous to a human hand.

memory (computer) Sets of semiconductor chips, each typically capable of storing a million or more bits, and organized such that groups of chips store a series of bytes or words in a manner logically accessible to a computer processor. *See also* random-access memory (RAM), read-only memory (ROM).

menu A list of choices displayed by a computer program showing the user's options at a given stage of the program's execution.

microcomputer A computer whose processor normally is implemented on a single small silicon chip and for which the whole machine is usually configured in an enclosure suitable for use on a desktop or as a lightweight portable. *See also* workstation.

microprocessor   The hardware implementation of electronic circuitry for a computer's instruction set, control logic, and sometimes some cache memory, all combined together on a single semiconductor chip.

minicomputer   A computer design midway between mainframes and microcomputers. Until recently their processors consisted of several logic chips on one or more circuit boards, but now they differ from microcomputers mainly in the speed and capacity of their components. They are commonly designed to be shared by multiple users working at CRT terminals. *See also* workstation.

modem   Short for *modulator-demodulator.* A device that converts a computer's digital electronic signal into the type of analog voice-grade signal that can be transmitted over a telephone line, and that receives and decodes such analog signals back into digital data for input to a computer.

module   Generally refers to a self-contained piece of a computer program that can be separately developed and tested to perform a particular task and then be combined with other modules to make a more complex program.

monitor   A video display device, based on a CRT tube, that can receive output signals directly from a computer.

mouse   A hand-held pointing device that generates electronic signals when it is moved along a flat surface; these signals encode motion that computer software can interpret to move a cursor on a display screen. One or more buttons on the mouse can send other signals to the computer. *See also* Fig. 3-10.

MS-DOS   A single-tasking, single-user disk operating system developed by Microsoft, Inc. for microcomputers based on Intel 80x86 CPUs. *See also* disk operating system (DOS), operating system.

multimedia   Systems of computer software and hardware that combine TV-like video and animated graphic displays plus sound and other media to provide a rich, interesting, and highly interactive interface to a human user.

multiplex   A technique that makes it possible for several signals to share the same physical communication channel simultaneously.

network (for scheduling)   *See* critical path method (CPM).

network (of computers)   Refers to a system of two or more interconnected computers, printers, and other devices that might be placed in diverse locations and yet communicate to work together in applications that the system is designed to perform. *See also* local area network, wide area network.

network data access   Technology that enables a user running a program on one computer to have almost transparent and automatic access to information anywhere else in a network of computers without worrying about where the data are actually stored physically.

object   In an object-oriented program, refers to a self-contained module that encapsulates both procedures and data pertinent to its function.

object code   The translated, machine-readable form of a computer program, produced by a compiler or assembler.

operating system   The software that supervises the utilization of the computer's processing and memory resources, loads programs to be executed from storage devices into memory, and in general handles the traffic among peripherals, memory, and the processor. *See also* disk operating system (DOS).

operator   A computer professional whose responsibility is to run medium- and larger-size computers so that others can work with their application programs.

optical disc (or **disk**)   A type of computer data storage medium that uses variations in the reflectiveness of a material just beneath a transparent disc surface to encode computer-readable data. Data are typically retrieved by bouncing a laser beam off the rotating surface into a photocell that detects variations in intensity in the reflected light. Optical discs come in both read-only and read-write (magneto-optical) versions. *See also* compact disc read-only memory (CD-ROM).

package software   Prewritten programs packaged and sold in a standardized form in the retail market (in contrast to custom-programmed software).

parallel data transmission   Data divided and transmitted simultaneously along parallel data channels, such as multiple wires. Can be internal to the computer, as for a data bus, or external, to a peripheral such as a printer or disk drive. *Compare* serial data transmission. *See also* bus, small computer system interface (SCSI).

parity   An error-checking method whereby one or more extra check bits are added to data character codes for checking that each character is transmitted correctly to or from memory or over a communication line. For example, in single-bit parity checking, the added bit is set to 0 or 1 to make the total number of 1s in each fixed-size group of transmitted bits an even number (for even parity) or an odd number (for odd parity).

password   A user- or system-specified character sequence used to control access to a computer system, a program or file, or parts of a program or file.

PC   Generically refers to a personal computer, but commonly used specifically to refer to personal computers designed by IBM or manufacturers of IBM compatibles that are based on Intel 80x86 CPU chips. *See also* microcomputer.

PC-DOS   IBM's version of MS-DOS, with its own modifications and enhancements. *See also* disk operating system (DOS), MS-DOS.

peripheral device   Input, output, storage, and communication devices that connect to a computer system to enhance its functionality. Examples include disks, plotters, printers, modems, CRT monitors, and so forth.

personal computer   A standalone microcomputer system intended for use by one person at a time. *See also* microcomputer.

PERT   *See* Program Evaluation and Review Technique.

pixel (pel)   Short for picture element. A single dot among a rectangular grid of thousands of such elements that is the smallest addressable unit that can be manipulated to form part of character or graphic images on a display screen or printer.

plotter   An electromechanical peripheral device used for producing graphical outputs such as architectural and engineering drawings. *See also* Figs. 3-14 and 3-15.

printer   A computer output device designed to transfer electronically encoded text and graphics images to paper. *See also* dot-matrix printer, ink-jet printer, laser printer.

procedural language   A computer programming language wherein the programmer precisely specifies the sequential control logic of the program's execution. Examples include BASIC, C, FORTRAN, and Pascal.

procedure   A section of a computer program that either specifies a precisely defined process or implements an algorithm to achieve a desired objective. The procedure might be given a name and be callable from other parts of a program. *See also* function, subroutine.

program   A complete set of computer instructions written in a programming language that, when translated to a computer's native instruction set and executed, performs one or more tasks to help achieve the program user's objectives. *See also* compiler, interpreter, software.

Program Evaluation and Review Technique (PERT)   A form of network-based project scheduling that emphasizes key events or milestones in a schedule, and may use multiple time estimates for activities to allow probability analysis for the likelihood of achieving a calculated project duration.

programmable macro   *See* macro script.

programming   The process of developing computer programs, particularly the process of expressing the program in a computer language.

programming environment   *See* development environment.

query by example (QBE)   A fill-in-the-blank type of query language that prompts the user via forms for the inputs needed (fields, criteria, linkages) to perform file or relational database functions to retrieve data for output in screen or printer reports.

query language   A set of keywords and a syntax or grammar that enable a user to manipulate, retrieve, and display or report data from a file or database.

RAM   *See* random-access memory.

random-access memory (RAM)   Memory devices wherein data can be both stored and retrieved by a computer processor. Contents of RAM are usually lost when power is turned off. *Compare* read-only memory. (*Note*: The word "random" is misleading, because memory operations are precisely and explicitly controlled.)

range   A rectangular area in a spreadsheet containing one or more cells. For example, the area from cell B4 on the upper left to F9 on the lower right could be defined as the range (B4..F9).

read-only memory (ROM)   Memory devices wherein data are permanently encoded in semiconductor chip hardware in a manufacturing process and thus can be retrieved by a computer processor but cannot be deleted, written to, or replaced. *Compare* random-access memory.

real-time   For computers, refers to programmed actions that take place according to actual specified times or events in the external environment in which the computer operates. Particularly important in online data collection and process control applications.

record   A set of related data fields, usually with a consistent sequence and format, that make up one of a set of instances for the information kept in a file or database. For example, the fields that define a specific schedule activity or an employee's weekly time card input could be held in records.

reduced instruction set computer (RISC)   Usually refers to a CPU design where fewer than a hundred different processing instructions are included. However, these are highly optimized, thus permitting fewer discrete electronic devices in their silicon chip for lower costs, higher speeds, and easier programming. *Compare* complex instruction set computer (CISC).

relation   A two-dimensional matrix consisting, usually, of columns of data attributes and rows of records (sometimes called tuples) that together form one of the information tables in a relational database.

relational database   One of the most commonly used data organization methods. Organizes and stores information in one or more tables that can be related to each other via attributes held in common by at least two tables for any given pair, with all tables thus related either directly or indirectly.

reserved word   A word that has special significance (e.g., as a command or arithmetic operator) in a computer program or programming language and is thus not allowed to be used by a programmer or user for other purposes in the program.

response time   The time that elapses from the moment that a user or device makes a request of a computer program or device until the moment that the program or device returns the requested information (or gives a message about its inability to comply). Usually is a function of hardware characteristics, program efficiency, and the number of concurrent tasks in the system.

RGB monitor (or **RGB display**)   A CRT display that has separate electron beams to activate screen phosphors that correspond to the Red (R), Green (G), and Blue (B) additive primary colors whose mixture and intensity can produce all other colors.

RISC   *See* reduced instruction set computer.

robotics (in construction)   A relatively new but rapidly evolving field that applies automated or partially automated computer-based control systems to machines that can sense, react to, and manipulate objects in their environment. They range from partial automation of traditional earthmoving equipment to complete new automated high-rise building systems.

ROM *See* read-only memory.

rule-based system A type of expert system that captures the experience-based surface knowledge of one or more experts by means such as heuristic rules. *See also* expert system.

scanner A peripheral device that uses light-sensing technology to input printed text or drawings in paper or slide form and to convert their images to an electronic form that can be stored and then manipulated by computer programs. *See also* digitizer.

script Another term for a macro. A set of user-programmed instructions for a generic application software package such as a database system, typically written using the application's command words and syntax that could otherwise be invoked manually by the user. *See also* macro.

SCSI *See* small computer system interface.

semantic network An AI software technology that represents objects or pieces of information as nodes and expresses relationships between them as links to provide a powerful and flexible way of representing knowledge.

sensor A device used to detect or measure energy-related phenomena (mechanical forces, motions and displacements, hydraulic and pneumatic pressures, light, temperature, etc.) and produce corresponding electrical currents or voltages that can be input to and processed by a computer. Examples include strain gauges, load cells, inclinometers, pressure gauges, flow meters, photocells, and thermocouples.

sequential processing Processing program instructions one step at a time, in the sequence they are encountered, or organizing and processing data records one after another.

serial data transmission Reformatted binary data that can be sent down a single communication line as a linear stream of bits. *Compare* parallel data transmission.

service bureau A firm that offers its computers, application software, and related services for use by others on a fee basis.

shareware A variation of freeware (software distributed for free), but distributed with the intent and request that recipients who actually find the programs useful send a monetary contribution to the developers to help defray costs of distribution, registration, and documentation. *See also* freeware.

shell An expert system development environment without a knowledge base but with a common user interface, explanation facility, knowledge acquisition facility, and inference engine. This combination of components is an expert system shell since it can hold different knowledge bases. The shell user adds the domain-specific knowledge base to create an expert system.

simulation program   Software that enables a user to build analytical models of a real-world process, usually with allowance for uncertainties and variations in the process and its environment, to run and test the models, and to obtain output information the better to understand or improve the process that is modeled.

slot   In an object definition, a place in a frame in which to encode attributes and values of an object's properties, or point to other frames, rules or procedures that can instantiate the values for the slots. *See also* frame, object.

small computer system interface (SCSI)   A high-speed mechanical, electrical, and functional standard for interfacing peripheral devices such as disk drives, scanners, and so forth to a computer processor. Permits devices to be connected in sequence along a parallel electronic communications path.

software (computer)   The encoded instructions that, when executed by a computer, perform useful functions. *See also* program.

sort   To organize a set of data items or records into a specified order determined by one or more of the attributes used as keys. For example, to sort a set of numbers in descending order, or sort employee names by last name, first name.

source code   A computer program in a text-based form that is readable by a human but that needs to be translated into machine language before it can be executed by a computer. *Compare* machine language, object code.

spreadsheet program   A software package that provides a generic problem-solving and application development environment that lends itself to tabular information. This type of program especially employs numeric calculations that involve mathematical or logical relationships between one or more cells in the spreadsheet. A **spreadsheet** is a particular instance of an application built using a spreadsheet program. *See also* cell, range.

SQL   *See* structured query language.

statement   The smallest executable component in a programming language, typically written in one or a few lines. Composed according to a rigorous syntax, it performs a specific step, such as declaring a variable, assigning the result of an evaluated expression to a variable, defining a format for an output, performing a logical test, or transferring control to another location in the program.

string   A set of one or more text characters, possibly delimited by quotes or other special characters, that can be treated as a single entity by a computer program, or be manipulated into other forms. For example, "This is a string of text with 44 characters."

structured programming   A well-organized and clear approach to programming where sections of code are divided into self-contained modules that can be organized in a hierarchical manner into the total program.

structured query language (SQL)   A database language that works with relational databases derived from a standard developed by IBM in the 1970s.

subroutine   A named and callable procedure that specifies a precisely defined process, or implements an algorithm, to achieve a desired objective. Callable from other parts of a program with zero or more arguments passed. *See also* function, procedure.

supercomputer   A very powerful type of mainframe computer that is particularly well optimized for large-scale scientific computing applications. *See also* mainframe.

support   The services provided by computer hardware and software vendors, by third-party firms, or in-house for a computer operation. Services can include repair and maintenance, consulting advice on using hardware and software, upgrades of hardware or software, training, and so forth.

syntax   The grammar and format that must be adhered to in writing program statements, issuing commands to a computer program, providing data input to a program, and the like.

system software   The underlying software that runs the computer itself and supervises the loading and execution of other programs. Close extensions to the system software include peripheral device drivers, file-handling utilities, command-file processors, and some utility programs such as linkers. *See also* operating system.

table   In the database context, *see* relation.

tape drive   The enclosure and hardware mechanism that reads and writes information contained on magnetic tape.

tape, magnetic   An electromagnetic storage medium consisting of a long strip of plastic film coated with a metallic oxide that can contain one or more data tracks on which computer data can be stored.

template   Used in the context of generic development software such as spreadsheets and databases, refers to a previously developed framework for a particular type of document (e.g., a tax form) into which a user can insert his or her own specific data. Also, a flat plastic frame that fits around certain function keys on a keyboard and contains words and symbols to explain the function of those keys in the context of a given application program.

time-sharing   A computer facility that allows multiple users and/or programs to use the same computer simultaneously. Time-sharing operating systems divide up, interleave, and frequently reallocate the available CPU time, memory, and other resources in a way that gives each user or program the sense that they have the computer's undivided attention (as long as the computer is powerful enough to handle the combined load).

total conversion   Sometimes called "going cold turkey." Converting from an old system to a new one with little or no overlap. *See also* conversion.

transducer   A device that converts energy from one form into another. In relation to computers, input transducers are sensors that convert an external form of energy into electrical signals that can be detected by computer circuitry. Output transducers convert the computer's electronic signals into a form that affects the external environment (e.g., exert a force, produce a sound, turn on a light, etc.). *See also* analog to digital, digital to analog, sensor.

user interface   The commands, menus, graphical screen images and icons, hardware input devices, and other features that define how a user interacts with a given type of computer system or a particular computer program. *See also* graphical user interface, interface.

utility program   A program that performs routine functions needed in the normal operation of a computer, or provides useful new functionality and convenience in computer operations. Examples include disk file management, processing of preprogrammed batch commands, program linkers, system error diagnostics, screen savers, communications, and so on.

value-added reseller (VAR)   A vendor that acquires computers and system software at a discount from manufacturers, adds application software or hardware from one or more other vendors, and possibly develops their own software and/or specialized hardware (e.g., CAD extensions or new instrumentation) on top of those products, then integrates it all into a working system that provides added functionality for a customer, usually with ongoing support.

variables   Storage locations, usually in memory, where a series of changeable data values of a predefined type can reside temporarily while a program executes, for example, data read in from a disk file, interim results of computations or character manipulations, and so forth. They are usually named with a set of characters that bear some resemblance to the purpose of the variable.

virtual memory   Refers to the ability of a computer and operating system design to use and manage part of a disk storage unit as if it were an extension of the main RAM memory of the computer, and divide up running programs into "pages" so that only parts of them need be in memory at one time. A virtual memory operating system maps memory into pages and swaps pages of code and data from disk to memory as needed during program execution and usually can do so for multiple programs running concurrently.

virus   A program usually written with the malicious intent of surreptitiously "infecting" a computer to which it is transmitted (usually attached to normal-looking files on a removable disk or sent over a network), and subsequently having some adverse effect on the computer or its operator. Effects range from simply displaying an annoying joke to completely wiping out the contents of the computer's disk drive and disabling the system.

wide area network   A computer data communications network that spans national and even international boundaries, such as Internet, which connects universities and numerous research centers in industry and government. *Compare* local area network.

Winchester disk   A specific type of early IBM hard disk that enclosed the disk and drive mechanism in a hermetically sealed and contaminant-free environment. It had a then-unprecedented 30-millisecond access time to 30 megabytes of data, hence the allusion to the classic 30-30 Winchester rifle of the Old West. The term is still widely used to describe other types of fixed hard disks. *See also* hard disk.

window   In a graphical user interface, a rectangular area on a display screen that contains all or part of the currently displayed information pertaining to a particular program or document. Multiple windows can be displayed simultaneously in window-oriented software environments such as the Apple Macintosh Finder, IBM OS/2, and Microsoft Windows programs.

word   A unit of storage, usually consisting of one or more bytes, that corresponds to the amount of data normally transferred or operated upon in a single computer CPU instruction; typically 2 to 8 bytes (16 to 64 bits) in today's microcomputer hardware designs.

word processor   A software application that enables a user to enter text, and even graphics, into a document, edit the entered information, apply a variety of text styles and formats to the information contained in the document, maintain address files and process form letters, and send the results to a printer. Can include a dictionary, thesaurus, spelling checker, and other accouterments of traditional desktop writing. *See also* editor.

workstation   A powerful microcomputer system that is especially suitable for advanced software development, demanding engineering analysis and design applications, high-speed graphical animation, and other CPU-intensive work. *See also* microcomputer, minicomputer.

INDEX